Advances in Microelectronics: Reviews

Volume 1

Sergey Y. Yurish

Editor

Advances in Microelectronics: Reviews

Book Series, Volume 1

International Frequency Sensor Association Publishing

Sergey Y. Yurish
Editor

Advances in Microelectronics: Reviews, Volume 1

Published by International Frequency Sensor Association (IFSA) Publishing, S. L., 2017
E-mail (for print book orders and customer service enquires): ifsa.books@sensorsportal.com

Visit our Home Page on http://www.sensorsportal.com

Neither the authors nor International Frequency Sensor Association Publishing accept any responsibility or liability for loss or damage occasioned to any person or property through using the material, instructions, methods or ideas contained herein, or acting or refraining from acting as a result of such use.

ISBN: 978-84-697-8633-8
e-ISBN: 978-84-697-8634-5
BN-20171221-XX
BIC: TJFD

Acknowledgments

As Editor I would like to express my undying gratitude to all authors, editorial staff, reviewers and others who actively participated in this book. We want also to express our gratitude to all their families, friends and colleagues for their help and understanding.

Contents

Contributors

Ratshih S. Abd El-Azeem
Microelectronics Department, Electronic Research Institute, Cairo, Egypt

Nahla T. Abou- El-Kheir
University of Ottawa, Ottawa, Canada

Ateyyah Al-Baradi
Physics department, Taif University, Taif, Saudi Arabia

Lorena Anghel
Univ. Grenoble Alpes, CNRS, TIMA Laboratory, F-38031 Grenoble, France

Piotr Augustyniak
AGH-University of Science and Technology, 30, Mickiewicz Ave. 30-059 Kraków, Poland

Mounir Benabdenbi
Univ. Grenoble Alpes, CNRS, TIMA Laboratory, F-38031 Grenoble, France

Mattia Borgarino
Eng. Department "Enzo Ferrari", University of Modena and Reggio Emilia, Via Vivarelli, 10, 41125 Modena, Italy

I. R. Bose
Fraunhofer Research Institution for Microsystems and Solid State Technologies EMFT, Germany
Universität der Bundeswehr München, Germany

Enrique Cantó-Navarro
Automatic and Electronic Engineering, Universitat Rovira i Virgili, Avda. Països Catalans, Tarragona, Spain

Mansun Chan
Peking University Shenzhen SoC Key Laboratory, PKU-HKUST Shenzhen-Hong Kong Institution and Peking University Shenzhen Institution, Shenzhen 518055, P.R.China

Tso-Fu Mark Chang
Institute of Innovative Research, Tokyo Institute of Technology, Yokohama, 226-8503, Japan
CREST, Japan Science and Technology Agency, Yokohama, 226-8503, Japan

Chun-Yi Chen
Institute of Innovative Research, Tokyo Institute of Technology, Yokohama, 226-8503, Japan
CREST, Japan Science and Technology Agency, Yokohama, 226-8503, Japan

Jay Chen
Rudolph Technologies, 550 Clark Drive, Budd Lake, NJ 07828, USA

Johnny Dai
Rudolph Technologies, 550 Clark Drive, Budd Lake, NJ 07828, USA

Sreedharan Baskara Dass
Integrated Circuit Engineering, Malaysia

Erping Deng
North China Electric Power University, Beijing, China

Michael Dimopoulos
Univ. Grenoble Alpes, CNRS, TIMA Laboratory, F-38031 Grenoble, France

Moataz S. El-Kharashi
Department of Electron Devices, University College Southeast Norway, Norway

Magdy A. El-Moursy
Microelectronics Department, Electronic Research Institute, Cairo, Egypt
Mentor Graphics Corporation, Cairo, Egypt

Wayne Fitzgerald
Rudolph Technologies, 550 Clark Drive, Budd Lake, NJ 07828, USA

Yi Gang
Univ. Grenoble Alpes, CNRS, TIMA Laboratory, F-38031 Grenoble, France

Ahmed Gharib
Institute for Electronics Engineering, Friedrich-Alexander-University of Erlangen-Nuremberg, 91058 Erlangen, Germany
EESY-IC GmbH, 90449 Nürnberg, Germany

Woo Young Han
Rudolph Technologies, 550 Clark Drive, Budd Lake, NJ 07828, USA

Jin He
Peking University Shenzhen SoC Key Laboratory, PKU-HKUST Shenzhen-Hong Kong Institution and Peking University Shenzhen Institution, Shenzhen 518055, P.R.China

Xiaomeng He
Peking University Shenzhen SoC Key Laboratory, PKU-HKUST Shenzhen-Hong Kong Institution and Peking University Shenzhen Institution, Shenzhen 518055, P. R.China

Yongzhang Huang
North China Electric Power University, Beijing, China

Michiko Inoue
Nara Institute of Science and Technology, Japan

Anwar Jarndal

Department of Electrical and Computer Engineering, University of Sharjah, 27272 Sharjah, UAE

Jian Ding

Rudolph Technologies, 550 Clark Drive, Budd Lake, NJ 07828, USA

Eliasz Kańtoch

AGH-University of Science and Technology, 30, Mickiewicz Ave. 30-059 Kraków, Poland

Cheolkyu Kim

Rudolph Technologies, 550 Clark Drive, Budd Lake, NJ 07828, USA

Timothy Kryman

Rudolph Technologies, 550 Clark Drive, Budd Lake, NJ 07828, USA

C. Kutter

Fraunhofer Research Institution for Microsystems and Solid State Technologies EMFT, Germany
Universität der Bundeswehr München, Germany

C. Landesberger

Fraunhofer Research Institution for Microsystems and Solid State Technologies EMFT, Germany

Wei Li

California State University, Bakersfield, CA 93311, USA

Mariano López-García

Electronic Engineering, Universidad Politècnica de Cataluña, Avda. Victor Balaguer, 08800, Vilanova i la Geltrú, Spain

Haijun Lou

Peking University Shenzhen SoC Key Laboratory, PKU-HKUST Shenzhen-Hong Kong Institution and Peking University Shenzhen Institution, Shenzhen 518055, P.R.China

Rubén Lumbiarres-López

Electronic Engineering, Universidad Politècnica de Cataluña, Avda. Victor Balaguer, 08800, Vilanova i la Geltrú, Spain

Weichun Luo

Key Laboratory of Microelectronics Devices & Integrated Technology, Institute of Microelectronics, Chinese Academy of Sciences, Beijing 100029, China
University of Chinese Academy of Sciences, Bejing 100049, China

F. Ma

State Key Laboratory for Mechanical Behavior of Materials, Xi'an Jiaotong University, Xi'an 710049, Shaanxi, China

Katsuyuki Machida
Institute of Innovative Research, Tokyo Institute of Technology, Yokohama, 226-8503, Japan
CREST, Japan Science and Technology Agency, Yokohama, 226-8503, Japan

Robin Mair
Rudolph Technologies, 550 Clark Drive, Budd Lake, NJ 07828, USA

Abhijit Mallik
Kansai University, Japan

Mike Marshall
Rudolph Technologies, 550 Clark Drive, Budd Lake, NJ 07828, USA

Kazuya Masu
Institute of Innovative Research, Tokyo Institute of Technology, Yokohama, 226-8503, Japan
CREST, Japan Science and Technology Agency, Yokohama, 226-8503, Japan

Manjusha Mehendale
Rudolph Technologies, 550 Clark Drive, Budd Lake, NJ 07828, USA

Ben Meihack
Rudolph Technologies, 550 Clark Drive, Budd Lake, NJ 07828, USA

Y. Mengb
Shannxi Key Laboratory of Surface Engineering and Remanufacturing, Xi'an University, Xi'an 710065, Shaanxi, China

Amar Merazga
Physics department, Taif University, Taif, Saudi Arabia

Priya Mukundhan
Rudolph Technologies, 550 Clark Drive, Budd Lake, NJ 07828, USA

Mahshid Mojtabavi Naeini
Malaysia-Japan International Institute of Technology, Universiti Teknologi Malaysia

Takashi Nagoshi
National Institute of Advanced Industrial Science and Technology, Tsukuba Ibaraki, 305-8564, Japan

Amin M. Nassar
Electronics Department, Cairo University, Cairo, Egypt

Linwei Niu
Department of Math and Computer Science, West Virginia State University, Institute, WV 25526, USA

Yasuhisa Omura
Kansai University, Japan

Chia Yee Ooi
Malaysia-Japan International Institute of Technology, Universiti Teknologi Malaysia

N. Palavesam

Fraunhofer Research Institution for Microsystems and Solid State Technologies EMFT, Germany

Ahmad Rahati Belabad

Department of Electrical Engineering, Amirkabir University of Technology, Tehran, Iran

Jaspal S. Sagoo

QinetiQ, St Andrews Road, Malvern, Worcestershire WR14 3PS, United Kingdom

Marco Salvaterra

Eng. Department "Enzo Ferrari", University of Modena and Reggio Emilia, Via Vivarelli, 10, 41125 Modena, Italy

Fei Shen

Rudolph Technologies, 550 Clark Drive, Budd Lake, NJ 07828, USA

Gurvinder Singh

Rudolph Technologies, 550 Clark Drive, Budd Lake, NJ 07828, USA

Masato Sone

Institute of Innovative Research, Tokyo Institute of Technology, Yokohama, 226-8503, Japan

CREST, Japan Science and Technology Agency, Yokohama, 226-8503, Japan

Hao-Chun Tang

Institute of Innovative Research, Tokyo Institute of Technology, Yokohama, 226-8503, Japan

CREST, Japan Science and Technology Agency, Yokohama, 226-8503, Japan

Wenwu Wang

Key Laboratory of Microelectronics Devices & Integrated Technology, Institute of Microelectronics, Chinese Academy of Sciences, Beijing 100029, China

University of Chinese Academy of Sciences, Bejing 100049, China

K. W. Xub

Shannxi Key Laboratory of Surface Engineering and Remanufacturing, Xi'an University, Xi'an 710065, Shaanxi, China

Daisuke Yamane

Institute of Innovative Research, Tokyo Institute of Technology, Yokohama, 226-8503, Japan

CREST, Japan Science and Technology Agency, Yokohama, 226-8503, Japan

Hong Yang

Key Laboratory of Microelectronics Devices & Integrated Technology, Institute of Microelectronics, Chinese Academy of Sciences, Beijing 100029, China

University of Chinese Academy of Sciences, Bejing 100049, China

Tomokazu Yoneda

Nara Institute of Science and Technology, Japan

Masaharu Yoshiba
Institute of Innovative Research, Tokyo Institute of Technology, Yokohama, 226-8503, Japan
CREST, Japan Science and Technology Agency, Yokohama, 226-8503, Japan

Zhibin Zhao
North China Electric Power University, Beijing, China

Xingye Zhou
Peking University Shenzhen SoC Key Laboratory, PKU-HKUST Shenzhen-Hong Kong Institution and Peking University Shenzhen Institution, Shenzhen 518055, P.R.China

Preface

Every research and development must be started from a state-of-the-art review, and microelectronics is not an exception. The review is one of the most labor- and time-consuming parts of research. It is strongly necessary to take into account and reflect in the review the current stage of development, including existing technologies, CAD tools and devices. Many PhD students and researchers working in the same area are doing the same type of work. A researcher must find appropriate references, to read it and make a critical analysis to determine what was done well before and what was not solved till now, and determine and formulate his future scientific aim and objectives. The professionally made state-of-the art must take into account not only open access books, articles and conference papers available online for free, but also traditional monographs and journal articles, which are available only off-line and in print (paper) format.

After very successful publication of four volumes of our popular *'Advances in Sensors: Reviews'* Book Series (2012-2016), it was decided to start publication of new Book Series on *'Advances in Microelectronics: Reviews'* along with the other Book Series with advances and reviews, which are coming soon.

It is my great pleasure to present the 1st volume of Book Series titled 'Advances in Microelectronics: Reviews' started by the IFSA Publishing in 2017. The volume is published as an Open Access Book in order to significantly increase the reach and impact of this Book Series, which also published in two formats: electronic (pdf) print (paperback).

The 1st volume of new Book Series contains nineteen chapters written by 72 authors from academia and industry from 16 countries: Canada, China, Egypt, France, Germany, Iran, Italia, Japan, Malaysia, Norway, Poland, Saudi Arabia, Spain, United Arab Emirates, UK, and USA.

This book ensures that our readers will stay at the cutting edge of the field and get the right and effective start point and road map for the further researches and developments. By this way, they will be able to save more time for productive research activity and eliminate routine work.

Dr. Sergey Y. Yurish

Editor
IFSA Publishing   *Barcelona, Spain*

Chapter 1
Efficient Digital Interpolation Filter

**Ratshih S. Abd El-Azeem, Magdy A. El-Moursy,
Amin M. Nassar, Ahmed Gharib, Nahla T. Abou-El-Kheir
and Moataz S. El-Kharashi**

1.1. Introduction

Many applications in digital signal processing (DSP) nowadays need high-performance and low-power VLSI systems. Digital interpolation filters are widely used in DSP applications such as Delta-Sigma based digital to analog converters (Δ-$\sum$ DAC) [1-6]. Conventional interpolation filters use up-sampling and digital low pass filter (LPF) to obtain high precision [7-13]. The interpolation process is used to increase the number of samples of the required signal by an interpolation factor (L). This process is performed by inserting (L-1) uniformly spaced zero values between the sampled data followed by low pass filter [14]. Digital low pass filter requires large number of components such as multipliers, adders, and delay units [15]. Discrete Fourier Transform (DFT) in addition to Fast Fourier Transform (FFT) are used for cost reduction needed to implement the filter [16-19]. This reduction leads to high performance and low power design. Low pass filter algorithm use either Finite Impulse Response (FIR) [20] or Infinite Impulse Response (IIR) [21]. For interpolation, the FIR filter type is appropriate algorithm due to its linearity. For low power design [22-30] a transposed direct form (TDF) FIR filter realization is proposed in [31]. Digital data could be interpolated using different interpolation techniques. Nonlinear interpolation needs storing input samples according to the polynomial order [32-33]. That requires large memory and hardware to achieve high accuracy. On the other hand, linear interpolation requires only two samples that reduce the hardware [34]. The chapter is organized as follows. In Section 1.2, FIR Filter using sharing multiplication technique is presented. In Section 1.3, Computational Filter technique is described. In Section 1.4, simulation results and comparison between the two techniques are provided. Conclusion is summarized in Section 1.5.

Ratshih S. Abd El-Azeem
Microelectronics Department, Electronic Research Institute, Cairo, Egypt

1.1.2. Interpolation

Interpolation can be defined as the estimation of the unknown, signal's samples using a weighted average of a number of known samples at the vicinity points. Interpolators are employed in several forms in most signal processing and systems of decision making. Interpolators' applications comprise conversion of a discrete time signal to a continuous time signal, conversion of the sampling rate in multi-rate communication systems, low bit rate speech coding, Delta Sigma based digital to analog converters (Δ-$\sum$ DAC) [35-37]. This subsection is organized as follows. In Subsection 1.1.2.1 interpolation of a sampled signal is described. In Subsection 1.1.2.2 digital interpolation by a factor of L is described.

1.1.2.1. Interpolation of a Sampled Signal

A general use of interpolation is the rebuilding of a continuous time signal $x(t)$ from a discrete time signal $x(m)$. The restriction for the regaining of a continuous time signal from its samples is defined by the Nyquist sampling theorem [38-44]. The Nyquist theorem states that a band limited signal, with the largest frequency content of F_c (Hz), might be reconstructed from its samples if the sampling frequency is larger than $2F_c$ samples per second. Fig. 1.1 considers a band limited continuous time signal $x(t)$ sampled at rate of F_s samples per second. The discrete time signal $x(m)$ can be depicted as follows:

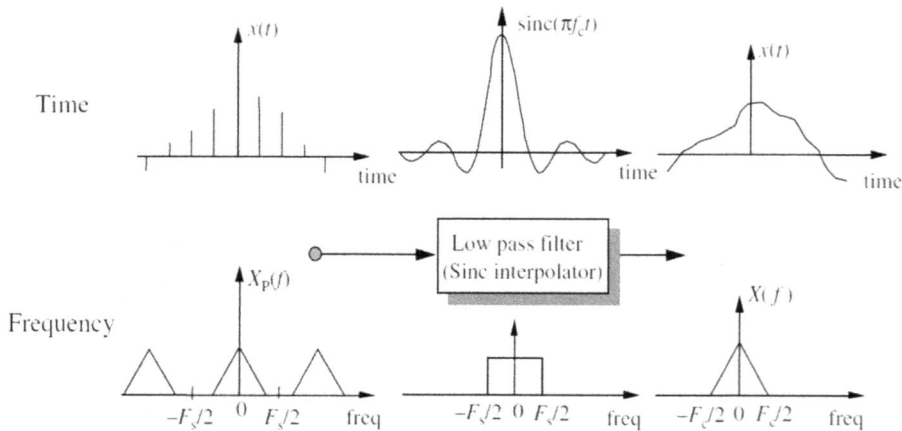

Fig. 1.1. Reconstruction of a continuous time signal from its samples [38].

$$x(m)=x(t)\,v(t)=\sum_{m=-\infty}^{\infty} x(t)\,\delta\,(t-mT_s), \qquad (1.1)$$

where $v(t)=\sum \delta\,(t-mT_s)$ is the sampling function and $T_s=1/F_s$ is the sampling period. Obtaining the Fourier transform [45-46] of Equation (1.1), it can be made known that the spectrum of the sampled signal is defined by

$$X_s(f) = X(f) * V(f) = \sum_{k=-\infty}^{\infty} X(f + k\,F_s), \qquad (1.2)$$

where $X(f)$ and $V(f)$ are the spectra of the signal $x(t)$ and the sampling function $v(t)$ respectively, and * stand for the convolution operation. Equation (1.2), illustrated in Fig. 1.1, explains that the spectrum of a sampled signal consists of the original base band spectrum $X(f)$ and the replications or images of $X(f)$ spaced uniformly at frequency intervals of $F_s=1/T_s$. When the sampling frequency is higher than the Nyquist rate, the baseband spectrum $X(f)$ is not aliased by its images $X(f \pm kF_s)$, and the original signal can be reconstructed by a low pass filter as shown in Fig. 1.1. Hence, the ideal interpolator of a band limited discrete time signal is an ideal low pass filter with a sinc impulse response. The reconstruction of a continuous time signal throughout sinc interpolation can be as follows

$$x(t)= \sum_{m=-\infty}^{\infty} \; x(m) \; T_s \, f_c \, sinc \, [\pi \, f_c (\, t - m \, T_s \,)]. \tag{1.3}$$

In practice, the sampling frequency F_s should be adequately higher than $2 F_c$, to be suitable for the transition bandwidth of the interpolating low pass filter.

1.1.2.2. Digital Interpolation by a Factor of L

Digital interpolators' applications include sampling rate conversion in multi-rate communication systems and up-sampling for enhanced graphical representation. Considering a band limited discrete time signal $x(m)$ with a base band spectrum $X(f)$ as shown in Fig. 1.2. The sampling rate can be go higher by an interpolation factor of L through interpolation of $(L-1)$ samples between each two samples of $x(m)$ [47-50]. Digital interpolation by a factor of L can be realized through a two-stage process of (a) $(L-1)$ zeros are inserted in between each two samples and (b) low pass filtering of the zero inserted signal by a filter with a cutoff frequency of $F_s /2L$, where F_s is the sampling frequency. Consider the zero inserted signal $x_{zi}(m)$ achieved by inserting $(L-1)$ zeros between each two samples of $x(m)$ and given as follows

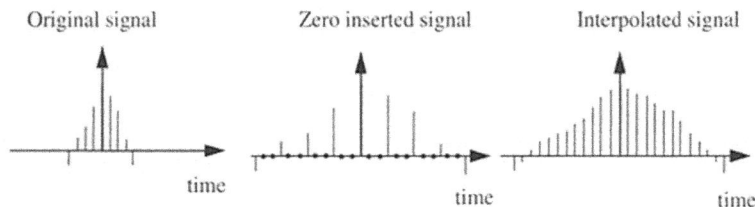

Fig. 1.2. Interpolation using up-sampling by a factor of 3 using zero insertion and digital low pass filtering [38].

$$x_{zi}(m) = \begin{cases} x\left(\dfrac{m}{L}\right), & m = 0, \pm L, \pm 2L, \dots \\ 0, & otherwise \end{cases}. \tag{1.4}$$

The zero inserted signal's spectrum is related to the original discrete time signal's spectrum as follows

25

$$X_{zi}(f) = \sum_{m=-\infty}^{\infty} x_{zi}(m) \, e^{-j2\pi fm},$$

$$= \sum_{m=-\infty}^{\infty} x(m) \, e^{-j2\pi fmL}, \qquad (1.5)$$

$$= X(L.f).$$

Equation (1.5) illustrates that the spectrum of the zero inserted signal $X_{zi}(f)$ is a frequency scaled version of the spectrum of the original signal $X(f)$. Equation (1.5) illustrates that the base band spectrum of the zero inserted signal composes of L replications of the based band spectrum of the original signal. The interpolation of the zero inserted signal is therefore equivalent to filtering out the replications of $X(f)$ in the base band of $X_{zi}(f)$. To keep the real time interval of the signal, the sampling rate of the interpolated signal $x_{zi}(m)$ needs to become higher by a factor of L [51-55].

The interpolation traditional method is used to form a polynomial Interpolator function that move across the known samples. Nonlinear interpolation such as Lagrange use a polynomial of order (the number of involved samples - 1) [56-60]. On the other hand, linear interpolation involves only two samples of the original sequence $x(m)$. The values interpolated between $x(m)$ and $x(m+1)$ lies on the straight line connecting the original two samples, so linear interpolation use the first order polynomial. That makes linear interpolation less in hardware cost, so linear interpolation is used in the proposed design. In the coming subsection different digital filter types are discussed.

1.1.3. Digital Filter

Digital filters and signal processing are being employed in the information systems, and are realized in various areas and applications. Some examples of those applications are video processing, radar, noise reduction, channel equalization, audio processing, and biomedical signal operations [61]. Digital filter algorithms use either Infinite Impulse Response (IIR) or Finite Impulse Response (FIR).

Filters may be classified into the following classes, depending on the form of the filter equation and the structure of implementation:

(a) Linear filters and nonlinear filters.

(b) Recursive and non-recursive filters.

This section is organized as follows. In Subsection 1.1.3.1, methods to describe filters are discussed. In Subsection 1.1.3.2, linear time invariant digital filter is presented. In Subsection 1.1.3.3, Finite Impulse Response filter is described. In Subsection 1.1.3.4, the description of infinite Impulse Response filter is provided. In Subsection 1.1.3.5, filtering, convolution and correlation are discussed. In Subsection 1.1.3.6, other filter structures such as cascaded and parallel are demonstrated.

1.1.3.1. Methods to Describe Filters

Filters can be expressed using the next frequency or time domain methods which can be expressed by the following four methods:

(a) Time domain input-output relationship: A difference equation is employed to illustrate the output of a discrete time filter in terms of a weighted group of the input and previous output samples [62]. For example a first order filter has the following difference equation

$$y(m) = a\, y(m-1) + x(m), \tag{1.6}$$

where $x(m)$ is the filter input, a is the filter coefficient and $y(m)$ is the filter output.

(b) Impulse Response: A filter can be expressed according to its response to an impulse input as follows

$$y(m) = a^m, m=0, 1, 2, \tag{1.7}$$

$y(m) = 1, a, a^2, a^3, a^4, \ldots$ for $m=0,1,2,3, 4 \ldots$ with assumption $y(-1)=0$. Impulse response is helpful because: (I) any signal can be expressed as the sum of a number of shifted and scaled impulses, therefore the response of a linear filter to the input signal is the sum of the responses to all the impulses that constitute the signal, (ii) an impulse input consists of all frequencies with equal energy, and hence it stimulates a filter at all frequencies and (iii) impulse response and frequency response are Fourier Transform pairs.

(c) Transfer Function, Poles and Zeros: The transfer function of a digital filter H(z) is the ratio of the z-transforms of the filter output and input as defined by

$$H(z) = Y(z) / X(z). \tag{1.8}$$

For illustration the transfer function of the filter of Equation (1.6) is defined as

$$H(z) = 1/(1 - a\, z^{-1}). \tag{1.9}$$

A helpful way of obtaining insight into the behavior of a filter is the pole zero description of the filter. Poles and zeros are the roots of the denominator and numerator of the transfer function, respectively.

(d) Frequency Response: The frequency response of a filter illustrates the methodology of the filter modifies the magnitude and phase of the input signal frequencies [63-64]. The response of the frequency of a filter can be acquired by getting the Fourier transform of the impulse response of the filter, or by easy substitution of the frequency variable $e^{j\omega}$ for the z variable $z = e^{j\omega}$ in the z-transfer function as

$$H(z = e^{j\omega}) = Y(e^{j\omega}) / X(e^{j\omega}). \tag{1.10}$$

The frequency response of a filter is a complex variable and can be illustrated in terms of the filter magnitude response and the phase response of the filter. In the next section linear time invariant digital filters are described.

1.1.3.2. Linear Time-Invariant Digital Filter

Linear time-invariant (LTI) filters are class of filters whose output is a linear group of the input signal samples and whose coefficients do not change with time. Assume that input *x(m)* produces the output *y(m)*. If an input *Kx(m)* produces an output *Ky(m)*, the system meets with the restriction of homogeneity, where *K* is any arbitrary constant. If the input is *K1x1 (m)* + *K2x2(m)*, the output is *K1y1 (m)* + *K2y2(m)*, where *K1* and *K2* are arbitrary constants, then the system meets with the superposition property. A system that meets with both homogeneity and superposition is known as a linear system. The term time invariant implies that the filter coefficients and hence its frequency response are fixed and do not change with time [65]. In the time domain the input-output relationship of a discrete-time linear filter is given by:

$$y(m) = \sum_{k=1}^{N} a_k\, y(m\text{-}k) + \sum_{k=0}^{M} b_k\, x(m\text{-}k), \qquad (1.11)$$

where $\{a_k, b_k\}$ are the filter coefficients, and the output *y(m)* is a linear group of the previous *N* output samples [*y(m−1),…, y(m−N)*], the present input sample *x(m)* and the previous *M* input samples [*x(m−1),…, x(m−M)*]. The characteristic of a filter is completely identified by its coefficients $\{a_k, b_k\}$.

For a time-invariant filter the coefficients $\{a_k, b_k\}$ are constants computed to achieve a specified frequency response. The filter transfer function, fetched by getting the *z*-transform of the difference equation (1.11), is defined by:

$$H(z) = \sum_{k=0}^{M} b_k\, z^{-k} / (1 - \sum_{k=1}^{N} a_k\, z^{-k}). \qquad (1.12)$$

The frequency response of this filter can be achieved from Eq. (1.12) by substituting the frequency variable $e^{j\omega}$ for the *z* variable, $z = e^{j\omega}$, as

$$H(e^{j\omega}) = \sum_{k=0}^{M} b_k\, e^{-j\omega k} / (1 - \sum_{k=1}^{N} a_k\, e^{-j\omega k}). \qquad (1.13)$$

Filter Order: The order of a discrete-time filter is the greatest discrete-time delay used in the input-output equation of the filter. For example, in equations (1.11 or 1.12) the filter order is the bigger of the values of *N* or *M*. For continuous time filters the filter order is the order of the greatest differential term employed in the input-output equation of the filter. In the next section finite impulse response filter is described.

1.1.3.3. Finite Impulse Response (FIR) Filters

Finite impulse response filters are the Linear and non-recursive filters which have no feedback and their input output relation is defined by [66-67]:

$$y(m) = \sum_{k=0}^{M} b_k\, x(m\text{-}k). \qquad (1.14)$$

As illustrated in Fig. 1.3, the output *y(m)* of a non-recursive filter is a function only of the input signal *x(m)*. The response of such a filter to an impulse composes of a finite sequence of *M+1* samples, where *M* is the filter order. Hence, the filter is well-known as a Finite-

Duration Impulse Response (*FIR*) filter. Other names for a non-recursive filter include all zero filter, feed forward filter or moving average (MA) filter.

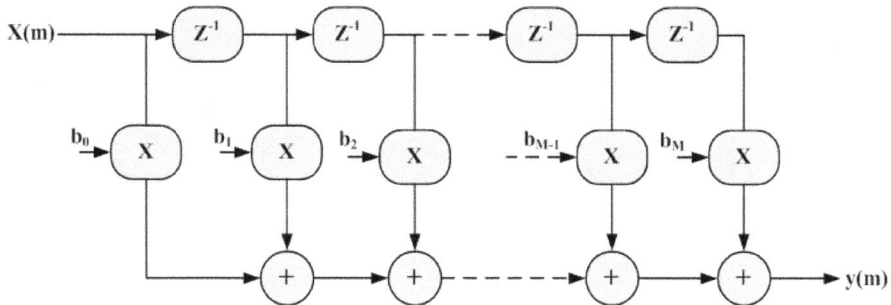

Fig. 1.3. Direct form FIR filter.

There are different structures for realization of the FIR filter. The Direct form realization of the FIR filter is a direct implementation of Equation (1.14). The transposed direct form is illustrated in Fig. 1.4. The transposed direct form is faster than the direct form so, the transposed direct form is used. In the next section infinite impulse response filter is discussed.

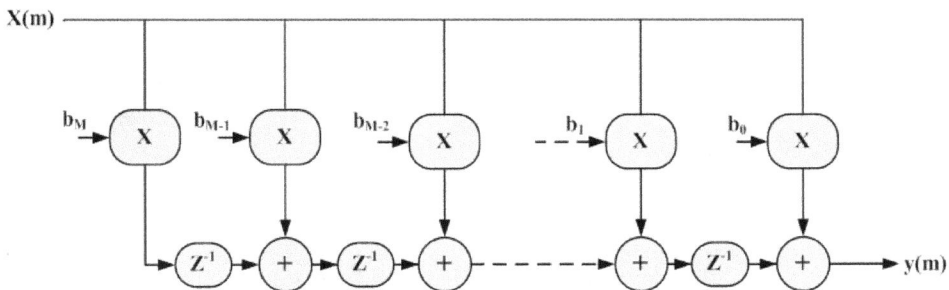

Fig. 1.4. The transposed direct form of FIR filter.

1.1.3.4. Infinite Impulse Response (IIR) Filters

Infinite Impulse Response (IIR) Filters are the recursive filters that have feedback from output to input, in general its output is a function of the previous output samples from a side, and the present with the past input samples from other side as defined by [66-67]:

$$y(m) = \sum_{k=1}^{N} a_k \, y(m\text{-}k) + \sum_{k=0}^{M} b_k \, x(m\text{-}k). \tag{1.15}$$

Fig. 1.5 reflects a direct form implementation of Equation (1.15). Theoretically, when a recursive filter is stimulates by an impulse, the output continues forever. So a recursive filter is also defined as an Infinite Duration Impulse Response (IIR) filter. Other naming for an IIR filter comprise feedback filters or pole-zero filters. Direct form II structure is obtained by sharing the delay terms assuming that *M=N*, as shown in Fig. 1.6.

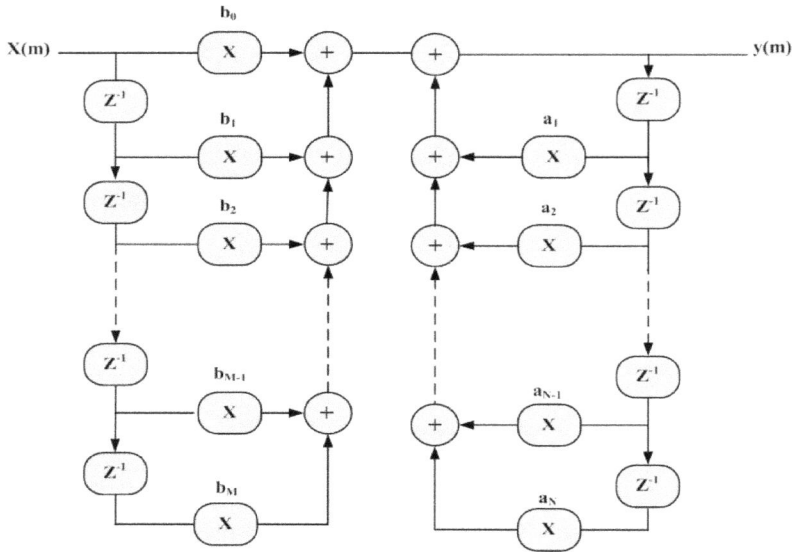

Fig. 1.5. Direct form I structure of IIR filter.

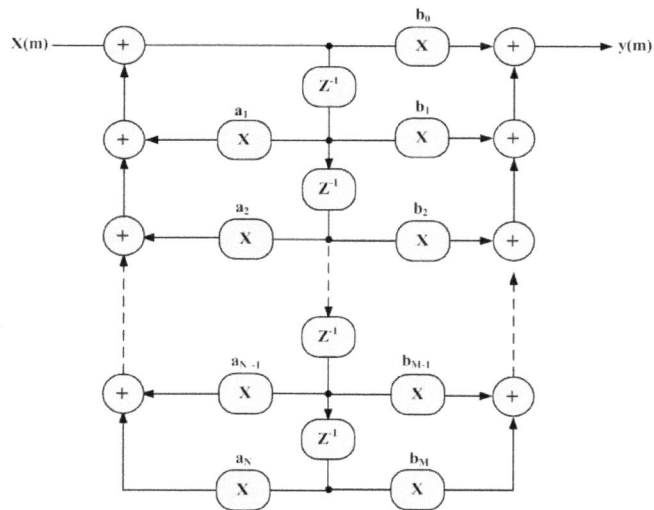

Fig. 1.6. Direct form II structure of IIR filter.

1.1.3.5. Filtering, Convolution and Correlation

The filtering operation, as illustrated in Equation (1.11), implicates summation of the products of the filter coefficient vectors with the input and output signal vectors [62]. The filter output is found as the sum of two vector products: a weighted group of the input samples, $[b_0, ..., b_M][x(m), ..., x(M-1)]^T$, plus a weighted group of the output feedback $[a_1, ..., a_N][y(m-1), ..., y(m-N)]^T$. All signals can be described as a combination of shifted and scaled impulses. Thus, it follows that the process of filtering of a signal $x(m)$ can be mathematically defined as the convolution of the input signal and the impulse response of the filter $h(m)$ as

$$y(m) = \sum_{k=1}^{N} h(k)\ x(m-k). \tag{1.16}$$

The filtering, or convolution process, of Equation (1.16) illustrated in Fig. 1.7, is consisted of the following four sub-operations:

(1) Fold the signal $x(k)$ to get $x(-k)$, this is performed because the samples with the earliest-time index (i.e. most distance past) go into filter first;

(2) Shift the folded input signal $x(-k)$ to yield $x(m-k)$;

(3) Multiply $x(m-k)$ by the impulse response of the filter $h(k)$;

(4) Sum the results of the vector product $h(k)x(m-k)$ to yield the filter output $y(m)$.

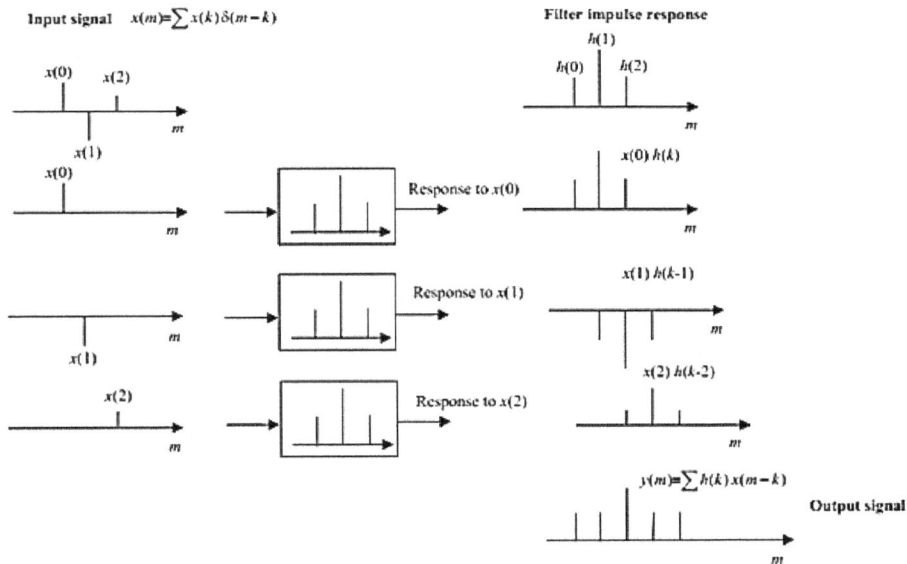

Fig. 1.7. The convolution of input and the impulse response [62].

The frequency domain the convolution operation converts into a multiplication operation, thus Equation (1.16) becomes

$$Y(f) = H(f)\,X(f), \tag{1.17}$$

31

where $X(f)$, $H(f)$ and $Y(f)$ are the input, the filter response at frequency f and the output, respectively.

The relationship between correlation, convolution, and filtering, considers the correlation of the two sequences $x(k)$ and $h(k)$ which is described as

$$r(m) = \sum_{k=1}^{N} h(k) \; x(k-m). \tag{1.18}$$

From Equations (1.16) and (1.18), the difference between convolution and correlation is that in convolution of two signals, one of the two signals, say $x(k)$, is folded in time to become $x(-k)$, as shown in Fig. 1.8. Therefore, the relation between convolution and correlation can be defined as

$$Conv \; (h(k)x(k)) = Corr \; (h(k)x(-k)). \tag{1.19}$$

The Fourier transform of Equation (1.16), of convolution of two sequences $h(k)$ and $x(k)$, tends to equation (1.17) in the frequency domain, repeated here

$$Y(f) = H(f) X (f), \tag{1.20}$$

while the Fourier transform of Equation (1.18), of correlation of two sequences $h(k)$ and $x(k)$, tends to

$$R(f) = H(f) * X (f), \tag{1.21}$$

where the asterisk sign indicates complex conjugate. Note that the folding of a signal in time domain is equivalent to a complex conjugate operation in the frequency domain.

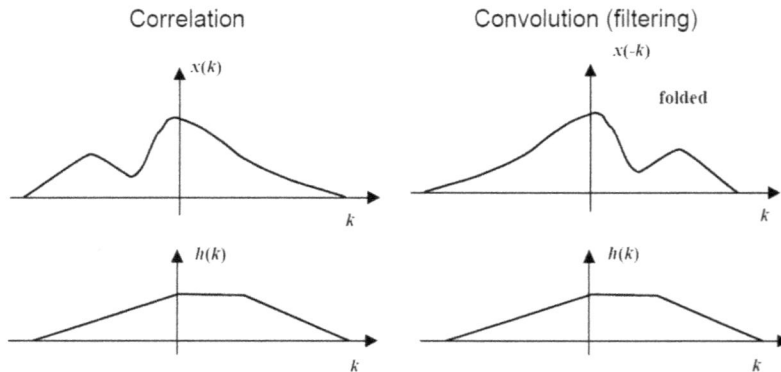

Fig. 1.8. A comparative illustration of convolution and correlation operations [62].

1.1.3.6. Other Filter Structures such as Cascade and Parallel Forms

There are different structures to realize a digital filter. These structures present several tradeoffs between cost of implementation, complexity, computational efficiency, and stability. This subsection is organized as follows. In Subsection 1.1.3.6.1 cascaded filter structure is discussed. In Subsection 1.1.3.6.2 parallel filter structure is described.

1.1.3.6.1. Cascaded Filter Structure

The cascade implementation [68] of a filter is realized by defining the filter transfer function $H(z)$ in a factorized form as

$$H(z)=\frac{\sum_{K=0}^{M} b_k \, z^{-k}}{1-\sum_{K=1}^{N} a_k \, z^{-k}}=G\frac{(1-z_1z^{-1})(1-z_1^*z^{-1})(1-z_2z^{-1})(1-z_2^*z^{-1})......(1-z_{M/2}z^{-1})(1-z_{M/2}^*z^{-1})}{(1-p_1z^{-1})(1-p_1^*z^{-1})(1-p_2z^{-1})(1-p_2^*z^{-1})......(1-p_{N/2}z^{-1})(1-z_{N/2}^*z^{-1})}, \qquad (1.22)$$

where G is the filter gain and the zeros (z_ks) and poles (pks) are either real-valued or complex conjugate pairs. The factorized terms in Equation (1.22) can be gathered in terms of the complex conjugate pairs and defined as cascades of second order terms as

$$H(z)=G[\frac{(1-z_1z^{-1})(1-z_1^*z^{-1})}{(1-p_1z^{-1})(1-p_1^*z^{-1})}]\times[\frac{(1-z_2z^{-1})(1-z_2^*z^{-1})}{(1-p_2z^{-1})(1-p_2^*z^{-1})}]\times..\times[\frac{(1-z_{M/2}z^{-1})(1-z_{M/2}^*z^{-1})}{(1-p_{N/2}z^{-1})(1-p_{N/2}^*z^{-1})}]. \qquad (1.23)$$

Each bracketed term in Equation (1.23) is the z-transfer function of a second order IIR filter as illustrated in Fig. 1.9 and Fig. 1.10. Eq. (1.23) can be defined in a compact notation as

$$H(z)=G \prod_{k=1}^{K} H_k(z), \qquad (1.24)$$

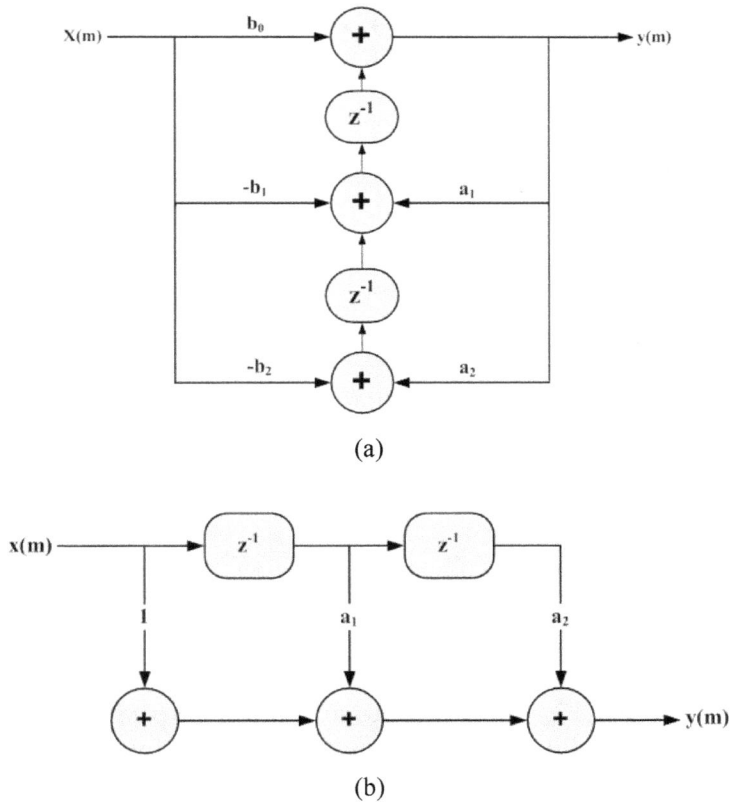

(a)

(b)

Fig. 1.9. Realization of a second order section of: (a) IIR Filter, (b) FIR filter.

where for an IIR filter, assuming that $N > M$, the variable K is the integer part of $(N+1)/2$ and for an FIR filter K is the integer part of $(M+1)/2$. Mention that a cascade filter may also have one or several first order sections with real-valued zeros and/or real-valued poles. When N or M are odd numbers there are at least one first order zero or first order pole in the cascade expression.

For an IIR filter, as illustrated in Fig. 1.9 (a), each second order cascade section has the form

$$H_k(z) = (1+b_{k1}\ z^{-1}+b_{k2}\ z^{-2})\ /\ (1+a_{k1}\ z^{-1}+a_{k2}\ z^{-2}). \tag{1.25}$$

For a FIR filter, as illustrated in Fig. 1.9 (b), each second order cascade section has the form

$$H_k(z) = 1+b_{k1}\ z^{-1}+b_{k2}\ z^{-2}. \tag{1.26}$$

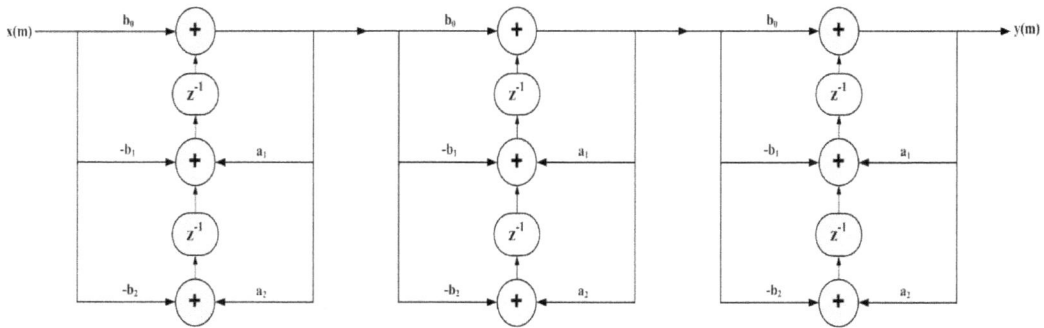

Fig. 1.10. Realization of an IIR cascade structure from second order sections.

1.1.3.6.2. Parallel Filter Structure

An alternative to the cascade implementation is explained in the previous section. The filter transfer function $H(z)$, could be expressed using mathematical approach called partial fraction method. In a parallel form [68] as parallel sum of a number of first order and second order expressed as

$$H(z) = K + \sum_{k=1}^{N_1} \frac{e_{0k}+e_{1k}z^{-1}}{(1-a_{1k}z^{-1}-a_{2k}z^{-2})} + \sum_{k=1}^{N_2} \frac{e_k}{1-a_k z^{-1}}, \tag{1.27}$$

$$H(z) = K + \sum_{k=1}^{N_1+N_2} H_k(z). \tag{1.28}$$

Generally the filter is assumed to have N_1 complex conjugated poles, N_1 real zeros, N_2 real poles, and K is a constant. Fig. 1.11 illustrates structure of the parallel filter. An advanced implementation of an *FIR* filter (*CFIR*) is presented in the next section.

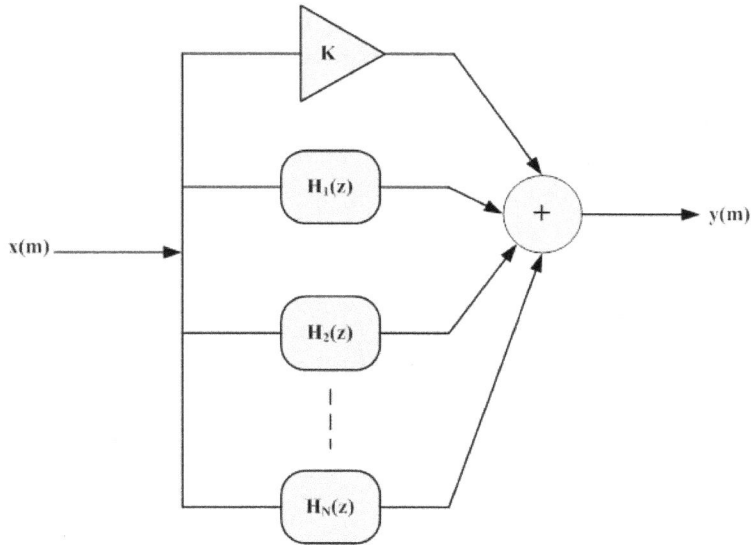

Fig. 1.11. A parallel-form filter structure.

1.2. FIR Filter Using Sharing Multiplication Technique

FIR filters are well known for their linearity, so they are widely used in applications such as interpolation and decimation filters. With the increase in sampling rate, performance and so the device complexity of these applications, power becomes an important issue particularly for portable devices [69-72]. Decreasing the power raises the device reliability and raises the time needed to change or recharge device battery, which is an important issue for many applications [73-76]. Due to high-speed needs and raises complexity of DSP systems, filtering operations become computationally intensive and power expensive, so low power techniques are used to reduce the power consumed by multipliers such as dividing the multiplication process itself [77]. Also high performance adders are essential for high performance design [78-85]. For these reasons, low power techniques have been applied to the conventional structure of FIR filter in order to enhance the power such as the sharing multiplication technique and pipelining. The section is organized as follows. In Subsection 1.2.1, FIR filter using sharing multiplication technique is discussed. In Subsection 1.2.2, the precomputer is described. In Subsection 1.2.3, the shift and add unit is presented. In Subsection 1.2.4, circuit implementation is provided.

1.2.1. FIR Filter Using Sharing Multiplication Technique

An enhanced FIR implementation using computation sharing multiplication technique *CSHM* (*CFIR*) is discussed. The operation of the *CFIR* [86-87] is described. The transposed direct form (*TDF*) of the *CFIR* filter is shown in Fig. 1.12. The output of FIR is the convolution sum of the coefficients and the shifted input samples,

$$y(m) = \sum_{k=0}^{N} C(k) X(m-k) \tag{1.29}$$

The filter is implemented as a product of the coefficient vector $C = [C_0, C_1, \ldots, C_{n-1}]$ with the scalar input samples $X(m)$. The *CFIR* filter reduces the redundancy in FIR filtering operation [88-89]. The multiplication process is performed by the precomputer. The precomputer outputs are shared through shift and add units. This leads to low power and high performance in sharing multiplication technique [86-87]. This technique is based on precomputer and $(N+1)$ number of shift and add units *(S&A)* where N is the Filter order.

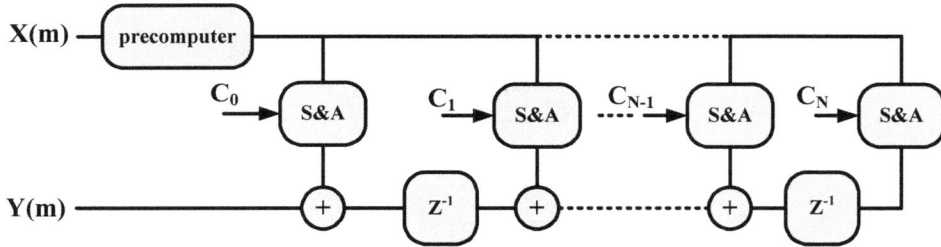

Fig. 1.12. FIR using *CSHM*.

For example, assume two coefficients of 8-bits $C_0 = 01011000$, $C_1 = 00111011$, needs to be multiplied by input sample X each coefficient has to be divided into two segments each of 4 bits. Each segment is multiplied by the input sample regarding to its weight as follows,

$$C_0 * X = (01011000) X = 2^4 (0101) * X + (1000) * X, \quad (1.30)$$

$$C_1 * X = (00111011) X = 2^4 (0011) * X + (0101) * X. \quad (1.31)$$

It is noted that the segment (0101) is common between both multiplication operations that is calculated one time and shared to any tap. The segment (0011) matches one of the alphabets. The segment (1000) has been shifted three times to the right. The segment (0101) in C_0 is compensated by multiplying by 2^4 to meet its weight. The detailed structure of *CFIR* filter is shown in Fig. 1.13. The following blocks are implemented to realize *CFIR*. In the next subsection the precomputer is described.

1.2.2. Precomputer

Precomputer consists of 8 paths X, 3X, 5X, 7X, 9X, 11X, 13X and 15X, that are called alphabets. These alphabets can divide any multiplication operation into groups of multiplications of those alphabets. Paths that have two operands need 1 carry select adder. For instance, 5X =4X+1X, 4X is the input sample shifted left two times, 1X is the input without any shift, its implementation is shown in Fig. 1.14.

Paths that have three or four operands need carry save adder. Carry save adder consist of one row of full adders and half adders then carry select adder for three operand case. Two rows of full adders and half adders are needed for 4 operand case. For instance, the 13X=8X+4X+1X, 4X is the input sample shifted left two times, and 8X is the input sample shifted left three times. 8X+4X+1X added together through one row of full adders and half adders. Carry select adder is used to get the final output, its implementation is shown in Fig. 1.15. The detailed description of the adders is in the following subsection.

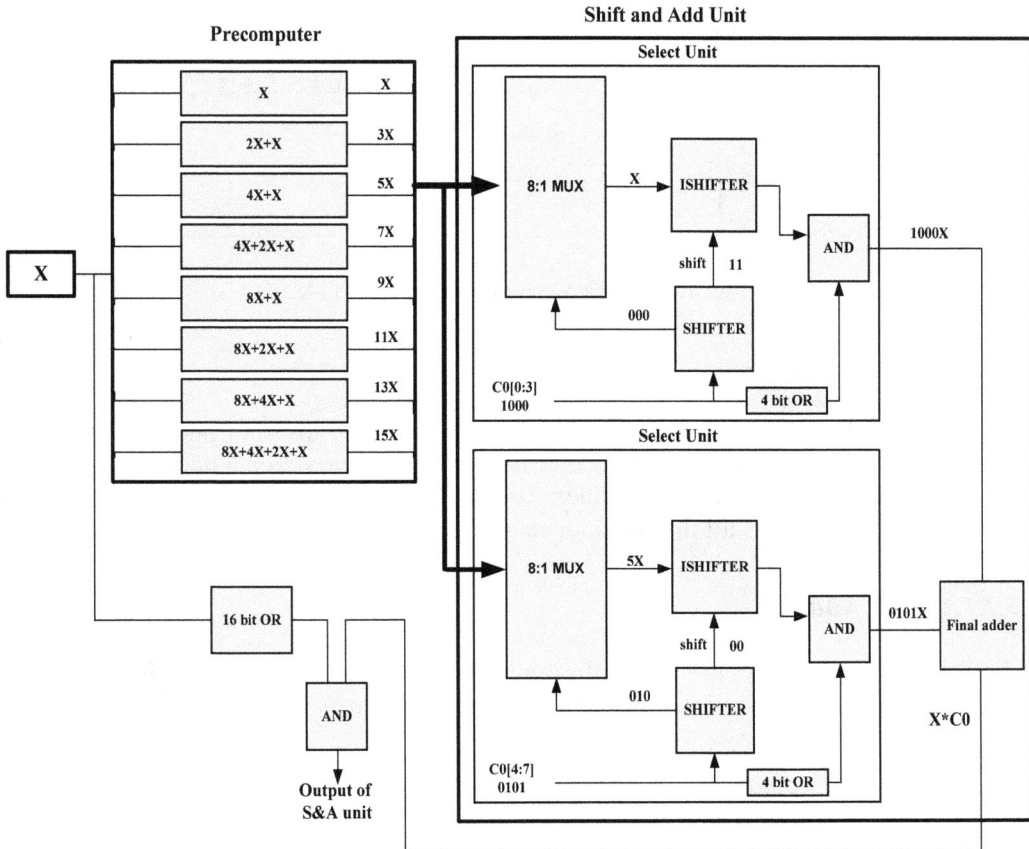

Fig. 1.13. Block diagram of *CFIR* filter.

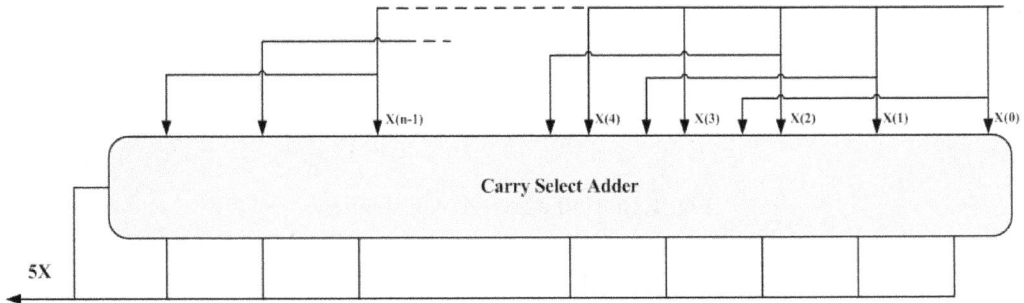

Fig. 1.14. Implementation of the path 5X.

37

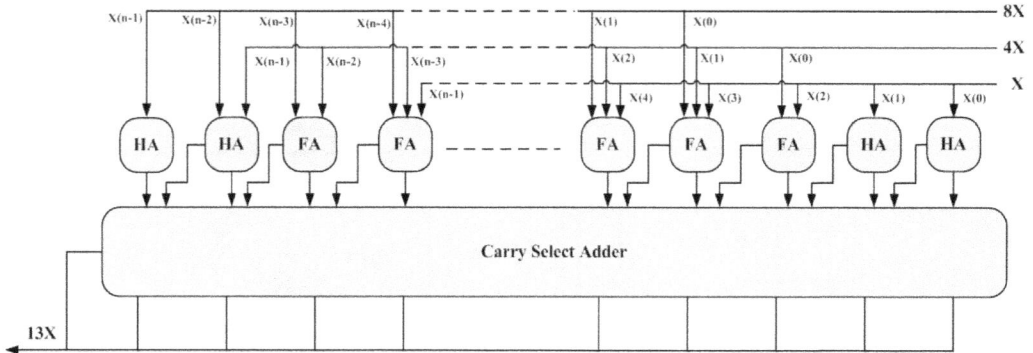

Fig. 1.15. Implementation of the path 13X.

1.2.2.1. Adders Types

Adders are the basic element used in addition, subtraction, division and multiplication operations implementation. The half adder, full adder and carry ripple adder are the basic units used to implement the bigger adders such as carry select and carry save adders. In the following subsections all the previous mentioned adders are described.

1.2.2.1.1. Half Adder

A half adder (*HA*) is a combinational circuit employed for adding two bits, a_i and b_i, without a carry in. Its gate level implementation is shown in Fig. 1.16. The sum s_i and carry output c_i are given by [90]:

$$s_i = a_i \oplus b_i , \tag{1.32}$$

$$c_i = a_i \, b_i . \tag{1.33}$$

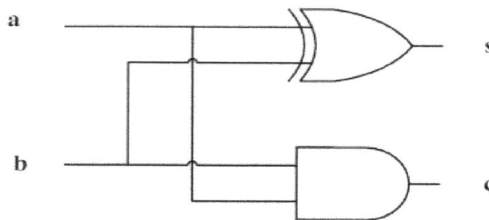

Fig. 1.16. Half adder implementation.

1.2.2.1.2. Full Adder

A full adder (*FA*) employed for adding three bits. Its gate level implementation is shown in Fig 1.17. The sum s_i and carry output c_{i+1} are given by [90]:

$$s_i = a_i \oplus b_i \oplus c_i, \tag{1.34}$$

$$c_{i+1} = (a_i \oplus b_i) c_i + a_i b_i . \tag{1.35}$$

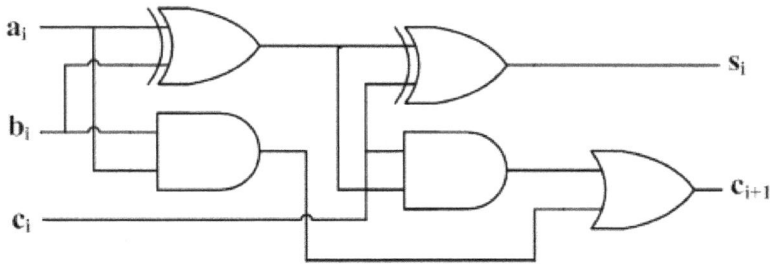

Fig. 1.17. Full adder implementation.

1.2.2.1.3. Ripple Carry Adder

The ripple carry adder is the simplest type of adders. Its implementation is shown in Fig. 1.18. Each stage has to wait for the input carry coming from the previous stage. The delay of the adder is dependent on the number of stages. It can be used effectively when adding small number of bits or in hybrid adders in which it can be used as a part of a big adder [90].

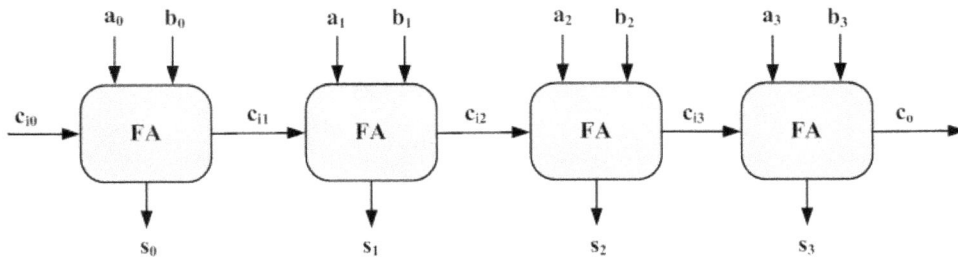

Fig. 1.18. 4 bits ripple carry adder.

1.2.2.1.4. Carry Select Adder

The basic idea of carry select adder (*CSA*) is to place two adders at each stage. One set of adders calculates the sum by assuming a carry in 1, and the other carry in is 0. The actual sum and carry are selected using a 2 to 1 multiplexer based on the carry from the previous group. Fig. 1.19 illustrates a 16 bit carry select adder. The adder is divided into four groups of 4 bits each. As each block is of equal width, their outputs be ready simultaneously. In an unequal width *CSA* the block size at any stage in the adder is set larger than the block size at its less significant stage. This helps in reducing delay further as the carry in less significant stages is ready to select the sum and carry of their respective next stages [91].

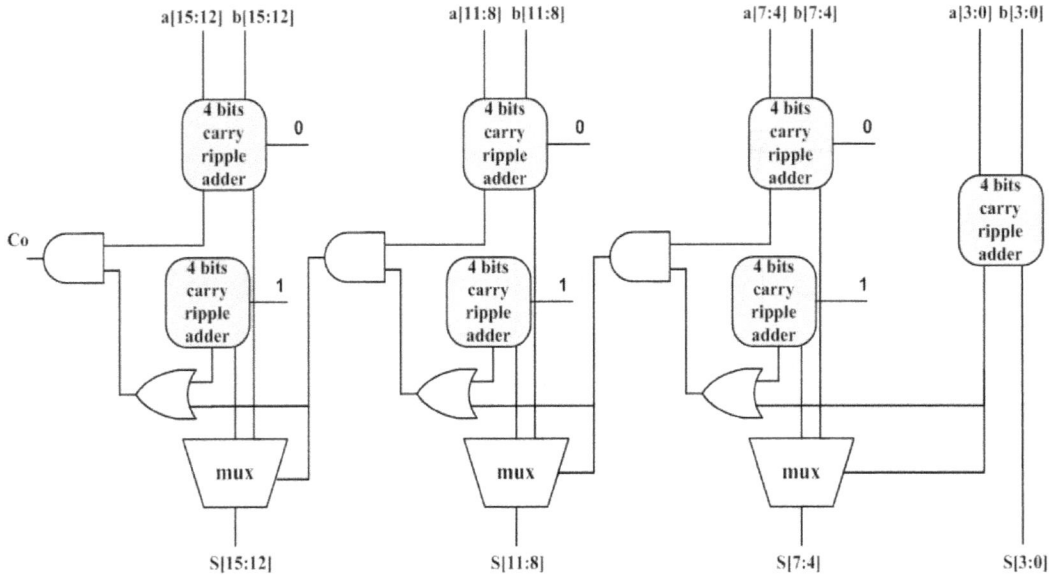

Fig. 1.19. 16 bit carry select adder.

1.2.2.1.5. Carry Look Ahead Adder

Carry look ahead is fast method to add numbers. This method doesn't require the carry signal to propagate stage by stage. Instead it uses additional logic to accelerate the propagation and generation of carry information [92].

$$g_i = A_i \cdot B_i, \tag{1.36}$$

$$p_i = A_i + B_i, \tag{1.37}$$

$$C_{i+1} = g_i + p_i C_i, \tag{1.38}$$

$$Si = pi \oplus ci, \tag{1.39}$$

The expression for the carry-out can be extended by substituting the expression for the carry-out of the previous stage as shown in Fig. 1.20. The carry look ahead adder is shown in Fig. 1.21:

$$C_1 = g_0 + C_0 \, p_0, \tag{1.40}$$

$$C_2 = g_1 + g_0 \, p_1 + C_0 \, p_0 \, p_1, \tag{1.41}$$

$$C_3 = g_2 + g_1 \, p_2 + g_0 \, p_1 \, p_2 + C_0 \, p_0 \, p_1 \, p_2, \tag{1.42}$$

$$C_4 = g_3 + g_2 \, p_3 + g_1 \, p_2 \, p_3 + g_0 \, p_1 \, p_2 \, p_3 + C_0 \, p_0 \, p_1 \, p_2 \, p_3. \tag{1.43}$$

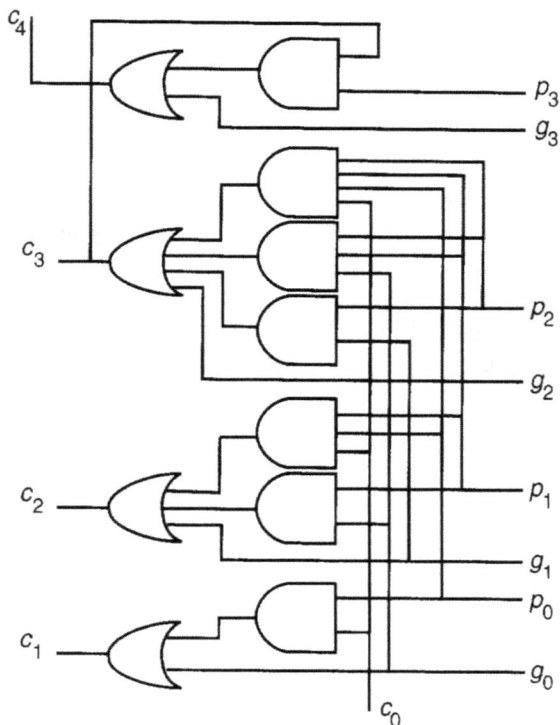

Fig. 1.20. 4 bit carry look ahead logic [92].

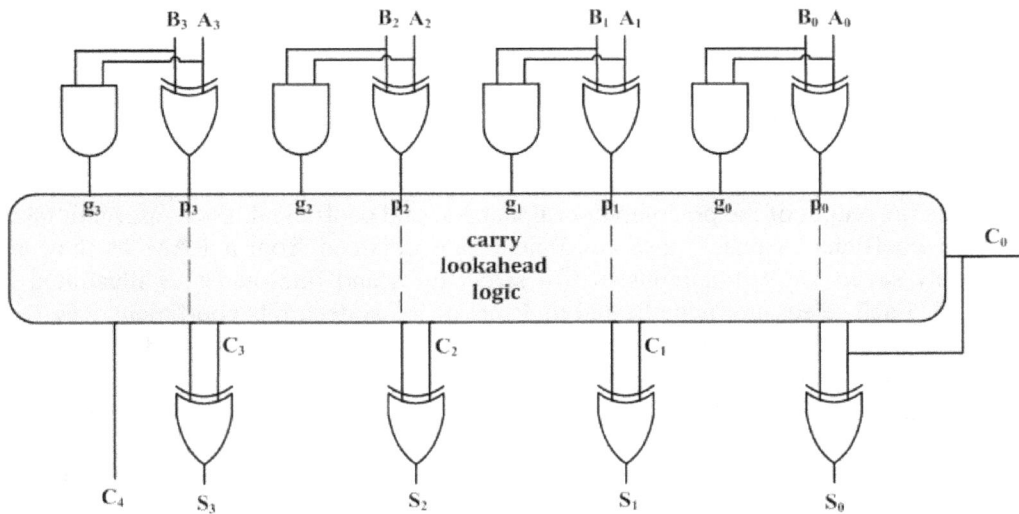

Fig. 1.21. 4 bit carry look ahead adder.

1.2.2.1.6. Carry Save Adder

Carry save adder is commonly used in multiplication operations [92], as it allows the addition of more than one operand. In carry save adder technique [93], a full adder sums three inputs and produces a sum output of unit weight and a carry output of double weight. If N full adders are used in parallel, they can accept three N-bit input for example words $X_{N...1}$, $Y_{N...1}$, and $Z_{N...1}$, and produce two N-bit output words $S_{N...1}$ and $C_{N...1}$, satisfying $X + Y + Z = S + 2C$, as shown in Fig. 1.22. The results correspond to the sums and carries-out of each adder. This is called carry-save redundant format because the carry outputs are preserved rather than propagated along the adder. The carry word C is shifted left by one position (because it has double weight) and is added to the sum word S with carry select or carry look ahead adders.

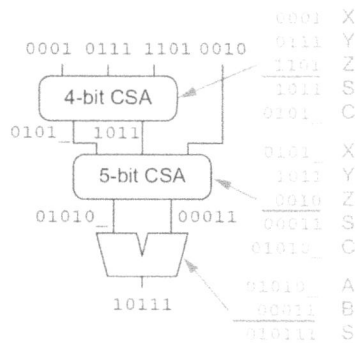

Fig. 1.22. Carry save adder [93].

1.2.3. Shift and Add Unit (*S&A*)

FIR filters have number of taps according to their order. The number of filter taps equals the filter order plus one. Each tap of the filter has one shift and add unit. *S&A* units perform suitable select, shift and add operation to get the correct multiplication output [94]. *S&A* units take the output of the precomputer (alphabets) and coefficient. Each *S&A* unit takes different coefficient value. Those coefficients are obtained from a RAM as they are previously saved. *S&A* units consist of 4 select units and final adder as illustrated in Fig. 1.23. Each select unit is dedicated to 4 bits of the sixteen bits coefficient. The first select unit takes the least significant four bits and the next ones take shifted segments of the coefficient according there weights. The 17th bit of the coefficient is reserved for the sign bit that is XORed with the 17th bit of the input sample to generate the final output sign [95-96]. The final adder adds the output of the 4 select unit according to their weights to get the final product $X*C$. In the following subsections select unit and final adder are discussed.

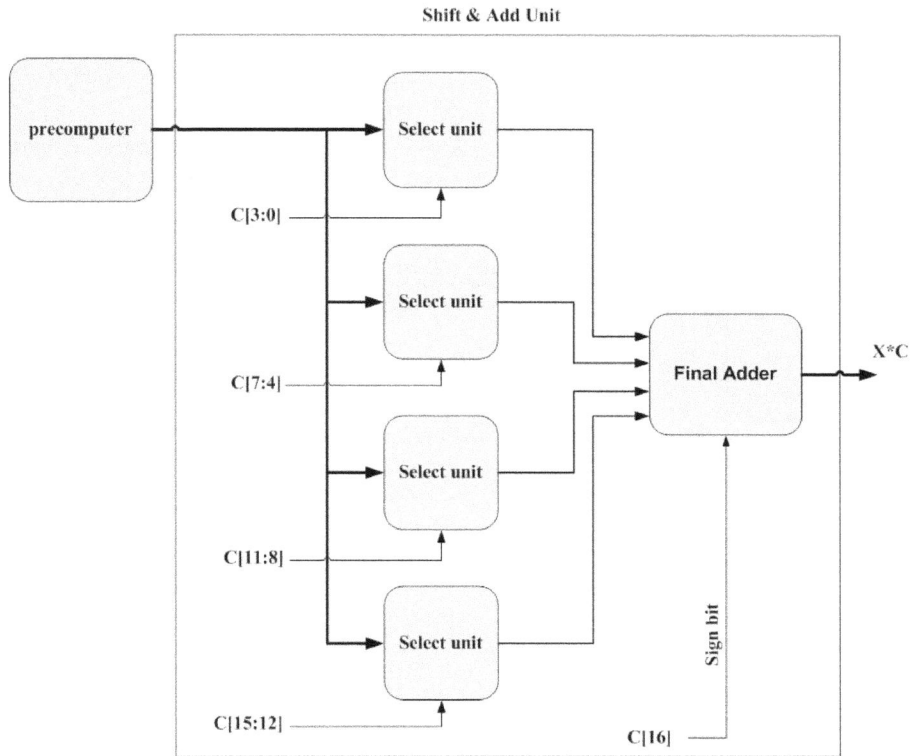

Fig. 1.23. Shift and add unit architecture.

1.2.3.1. Select Unit

Select unit has 8-to-1 MUX and SHIFTER control and ISHIFTER and gate as shown in Fig. 1.24. The inputs are alphabets from precomputer that transfer to 8-to-1 MUX and 4 bits segment that transfers to the SHIFTER. Select unit is used to select the appropriate alphabet to do the multiplication process. Also select unit adjusts the shift to get the final multiplication output. AND gate avoids the select unit calculations for zero input coefficients. In the following subsections the SHIFTER control and ISHIFTER are discussed.

1.2.3.1.1. SHIFTER Control

SHIFTER control performs right shift operation to detect the number of leading zeros. Then sends the suitable select signal through 3 bits select line to 8-to-1 MUX to select the appropriate multiplication path. SHIFTER control sends the shift values through 2 bits line to ISHIFTER to get the final multiplication output. In Fig. 1.24, there is an example for the coefficient segment (1000). The SHIFTER control detects the number of leading zeros which is 3 in this case and sends this number to ISHIFTER. After detecting the number of the needed shifts, the shifted number has to be one of the alphabets. This number is then compared to the alphabets. SHIFTER control sends the suitable select signal to 8-to-1 MUX to select the appropriate alphabet. The shifted number here is (0001) that matches the first alphabet. Fig. 1.25 shows the architecture of SHIFTER control. It contains two OR gates, two inverters, two 2-to-1 MUXs and shifter MUX. The shifter MUX is a logarithmic shifter [87] that is responsible for creating the suitable select signal to 8-to-1 MUX. The shifter MUXs architecture is shown in Fig. 1.26.

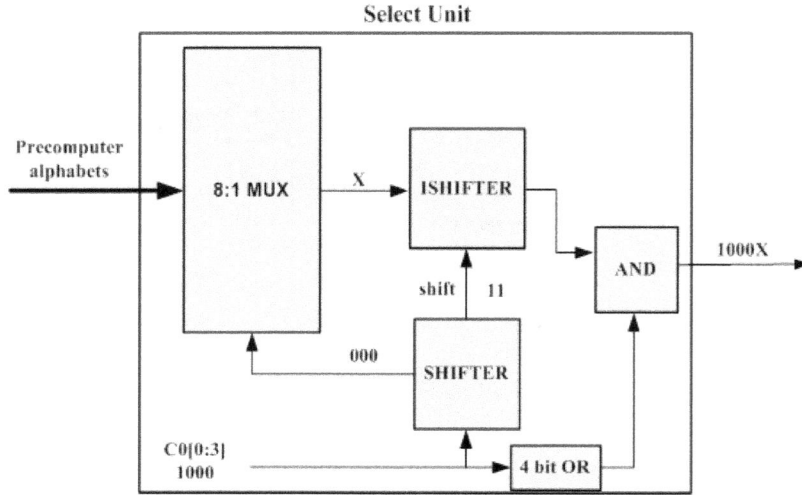

Fig. 1.24. Select unit architecture.

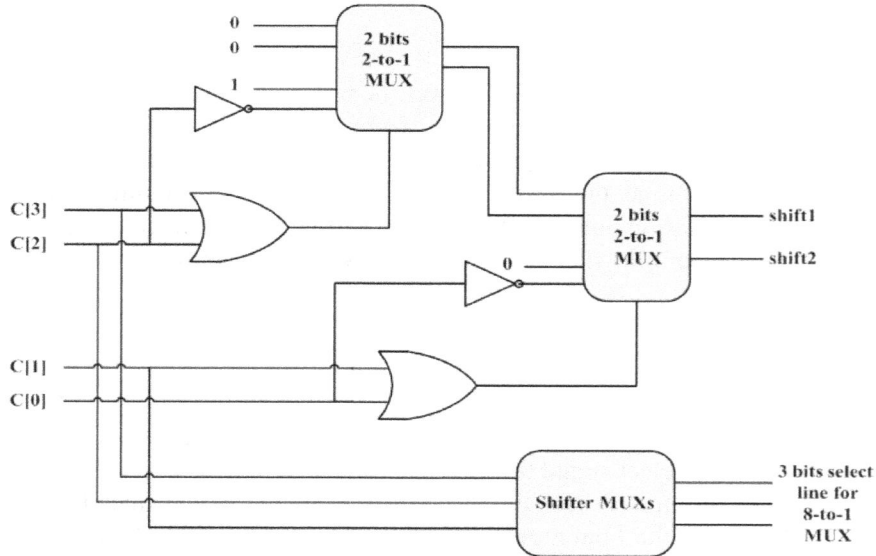

Fig. 1.25. SHIFTER control architecture.

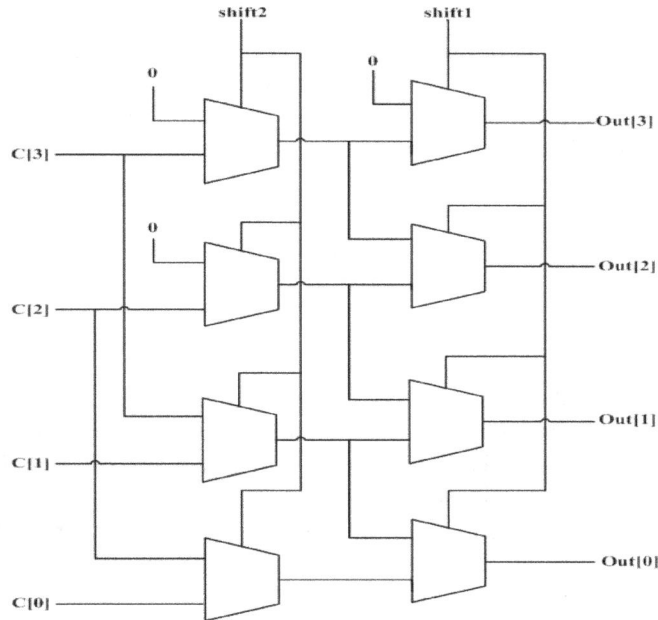

Fig. 1.26. Shifter MUX architecture.

1.2.3.1.2. ISHIFTER

ISHIFTER inverses the operation performed by the SHIFTER control. ISHIFTER uses the shift value to make left shift operation for the value of the selected multiplication path. The output result of the ISHIFTER is the exact equivalent of the input sample multiplied by the coefficient. Then the result is ANDed with the output of the 4 bits OR that indicates zero case coefficients. Finally the output of the select unit is yielded.

1.2.3.2. Final Adder

The final adder is a carry save adder that sums the outputs of the 4 select units. It is composed of two rows of full adder and half adder and carry select adder and XOR gate array. Its output is the final output of the S&A unit.

1.2.4. Circuit Implementation

A 14th order *CFIR* filter is implemented. The number of taps (*G*) and the coefficient values depend on interpolation factor *(L)* [97]. The number of taps is determined by

$$G = 2 L - 1 \tag{1.44}$$

The design has one precomputer, fifteen *S&A* units, fourteen 33 bit carry select adder, and fourteen delay units. In interpolation process, *(L-1)* zero input samples are added so 16 bit OR gate and gate are added to detect zero input samples. Detecting zero input samples eliminates the calculations in *S&A* unit and precomputer. That leads to power consumption reduction and high-performance. Both input and coefficients are represented in fixed point representation and consist of 17 bit where the most significant bit represents the sign. The input samples pass through precomputer. The precomputer outputs are shared by all shift and add units. The output of shift and add units are accumulated through 33 bit carry select adder at the end of each tap. The outputs of carry select adders that propagate through delay units, then yield the final outputs are 33 bits in two's complement format. Computational Filter technique is presented in the next section.

1.3. Computational Filter

The conventional interpolation filter consists of up-sampling followed by low pass filter. A new interpolation filter using direct computation called Computational Filter *CF* is proposed [98-99]. The design is multiplier free. A direct sample calculation is used to achieve significant hardware implementation reduction. The section is organized as follows. In Subsection 1.3.1, computational filter operation is proposed. In Subsection 1.3.2, circuit to realize M_{maj} and M_{min} is presented. In Subsection 1.3.3, calculate the order $O(i)$ block is presented. In Subsection 1.3.4, step selection block is provided. In Subsection 1.3.5, filter implementation is discussed.

1.3.1. Computational Filter Operation

The *CF* calculates output samples *Y(n)* to avoid using digital low pass filter to reduce hardware implementation. Computational filter perform linear interpolation that needs two successive samples of the input data sequence *X(m)* and *X(m+1)*. The main blocks of the *CF* are illustrated in Fig. 1.27.

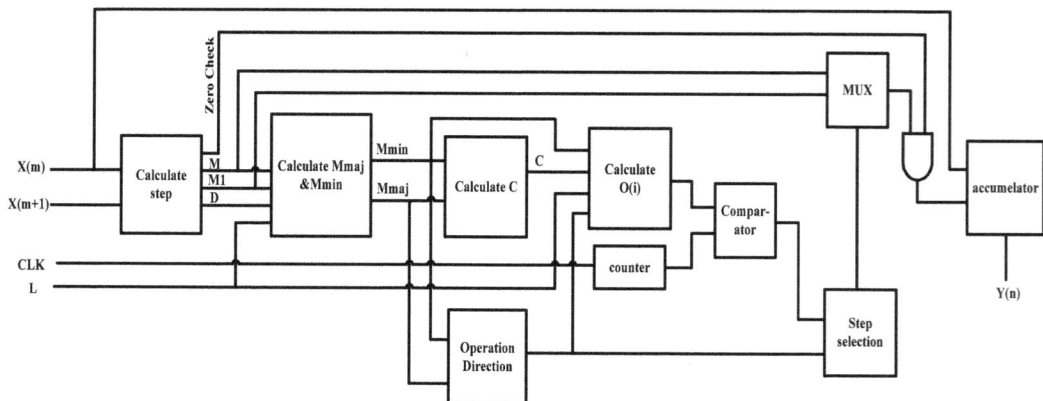

Fig. 1.27. The architecture of Computational Filter.

The difference between two successive integer input sample *D* is determined in step calculation block shown in Fig. 1.28,

$$D = X(m+1) - X(m). \tag{1.45}$$

Assuming $X(m)$ is integer, $Y(n)$ is expected to be an integer possessing the same precision of $X(m)$. The proposed technique is directly extended to fixed point representation. For linear interpolation, D and the interpolation factor L are used to get the step that is given by,

$$Step = (\frac{D}{L}). \tag{1.46}$$

Assuming fixed point representation and normalizing all bits to the available integer numbers of bits, the output samples are calculated. The step could be fraction number, but the output is integer number. In the ideal case the output sequence should be incremented by the step, so the ceiling M_l or the floor M of the step are used [34],

$$M = round \ (\frac{D}{L}), \tag{1.47}$$

$$M_l = M+1. \tag{1.48}$$

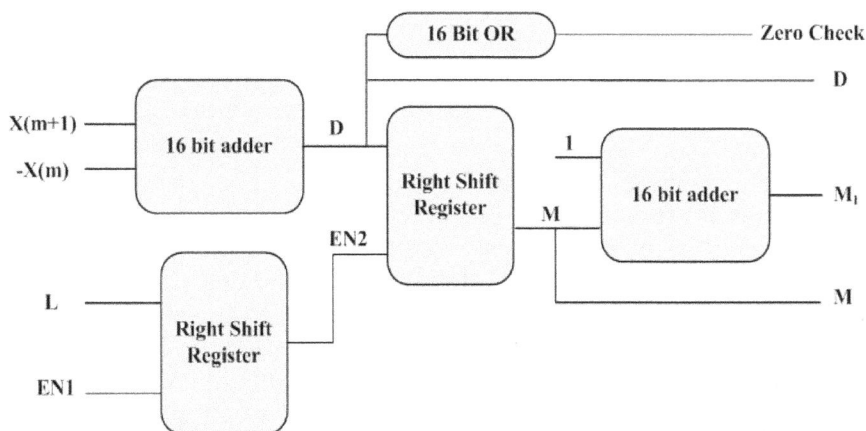

Fig. 1.28. Step Calculation block.

The number of output samples which are incremented by floor of the step M is r. The number of output samples which are incremented by ceiling of the step M_l is s, where,

$$r + s = L, \tag{1.49}$$

$$M * r + M_l * s = D. \tag{1.50}$$

Given D and L, r and s are determined by solving (1.49) and (1.50),

$$r = M_l * L - D, \tag{1.51}$$

$$s = D - M * L. \tag{1.52}$$

Equations (1.47), (1.48), (1.51), and (1.52) are the equations required to achieve the Computational Filter.

The output of the step calculation block is M, M_l, D and Zero Check. To guarantee the uniform distribution of the calculated samples, the output samples are distributed based on two shapes *Shape1* and *Shape2* as shown in Fig. 1.29. The distribution depends on the value of r and s. The majority output samples are the maximum of r and s. The minority output samples are the minimum of r and s,

Sample index	0	1	2			L
Output samples	y_{maj}	y_{min}	y_{maj}			y_{min}

(a)

Sample index	0	1	2			L
Output samples	y_{maj}	y_{maj}	y_{maj}			y_{maj}

(b)

Fig. 1.29. The distribution of the minority and majority in the output samples
(a) *Shape1*, (b) *Shape2*.

$$M_{maj} = Max\ (r,\ s),\ (1.53)$$

$$M_{min} = Min\ (r,\ s).\ (1.54)$$

The outputs of "Calculate Step" block are used to determine M_{maj} and M_{min} through "Calculate M_{maj} and M_{min}" block, M_{maj} and M_{min} are used to determine C and operation direction through "Calculate C" block and "Operation Direction" block, respectively. The "Operation Direction" block is responsible for choosing which shape is used to distribute the output samples. A "Counter" is used to determine the order of output samples. The "Calculate $O(i)$" block is responsible for detecting the order of the minority samples that is replaced with the majority samples and the other way around. The output of the "Calculate $O(i)$" block and the "Counter" are compared through a "Comparator". The output of the "Comparator" and the "Operation Direction" blocks determine the step selection through a MUX. The selected step is added to the input sample through accumulator. The step is ANDed with Zero Check that indicates the special case of equal successive input samples ($D=0$). In this case, the output samples get the same value of the input samples.

Shape1 is used when $M_{min} \geq M_{maj}/2$, this shape also includes $M_{min} = M_{maj}$. The number of minority samples that are swapped with the majority samples (C) in *Shape1* is defined by,

$$C = \frac{M_{maj} - M_{min}}{2}.\ (1.55)$$

The order of the swapped samples is determined by,

$$O_{y\ maj}(i) = 2* [i*L/2\ (C+1)] - 1,\ i = 1, 2, 3, \ldots, C.\ \ \ \ \ \ \ \ \ \ \ \ (1.56)$$

Shape2 is used when $M_{min} < M_{maj}/2$. The order of the minority samples that are replaced with the minority samples is defined by

$$O_{y\,min}(i)= 2* [i* L / 2 (M_{min} +1)] - 1, \; i= 1,2,3, \dots , M_{min} . \qquad (1.57)$$

The entire methodology could be explained by an example as shown in Fig. 1.30 where x-axis indicates the samples order and y-axis indicates the output samples $Y(n)$. Assuming interpolation factor of 8, $X(m)=2$, $X(m+1)=14$ for Fig. 1.30 (a), $D=12$, $M=1$, $M_l=2$. $r=4$, $s=4$. $M_{min} = M_{maj}$ this case follow *Shape1*. The output samples are $Y(n)=$ [2 3 5 6 8 9 11 12 14]. $X(m)=99$, $X(m+1)=5$ for Fig. 1.1.30 (b), $D=-94$, $M=-12$, $M_l=-11$. $r=6$, $s=2$. $O_{y\,min}$ $=2$, 4 so the samples of order 2 and 4 use step M instead of M_l where $M_{min} < M_{maj}/2$ following *Shape2*. The output samples are $Y(n)=$ [99 88 76 65 53 41 29 17 5]. In the next subsection circuit to realize M_{maj} and M_{min} is presented.

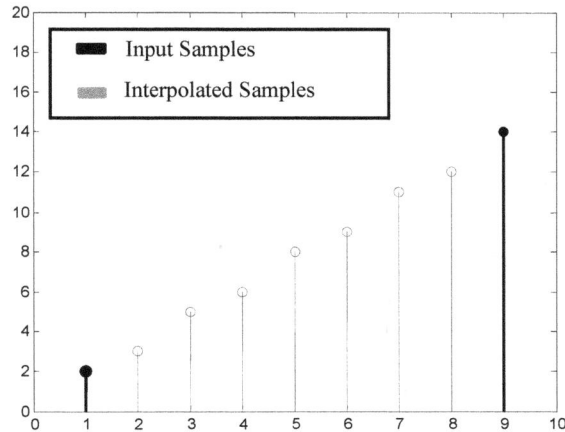

(a)

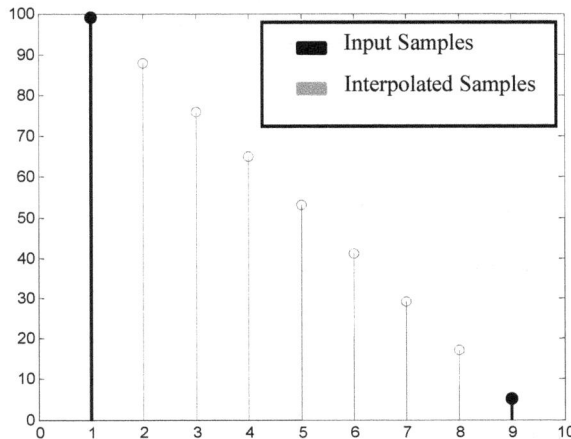

(b)

Fig. 1.30. Matlab simulation results for the interpolation factor of 8: (a) *Shape1* is used; and (b) *Shape2* is used.

49

1.3.2. Circuit to Realize M_{maj} and M_{min}

The Calculate M_{maj} & M_{min} block is shown in Fig. 1.31. It is responsible for determining M_{maj} and M_{min}. This block consists of three shift registers, two adders, comparator, and two multiplexers. The right shift register is followed by an OR gate and is used to enable the two left shift registers. The right shift register is used to divide its input L by two until its output becomes zero. That make the two left shift registers are enabled for L times. The left shift register is used to multiply its input by 2, so the two left shift registers outputs are M multiplied by L and M_1 multiplied by L. The two 16 bits adders are used to get (r and s) in (1.51) and (1.52). A comparator is used to determine the maximum and minimum of both r and s through the selector of the two multiplexers. In the next subsection "Calculate the order $O(i)$" block is presented.

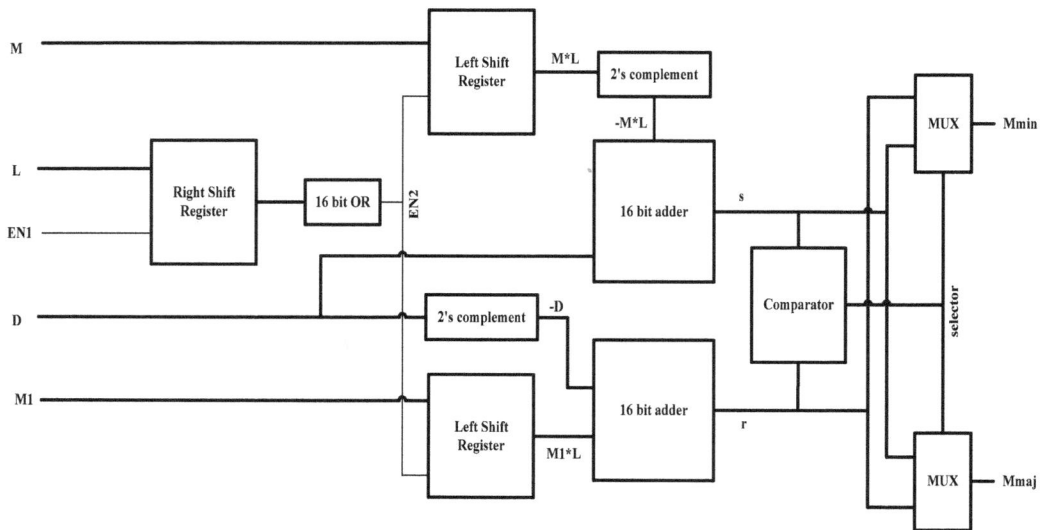

Fig. 1.31. The block diagram of the hardware to calculate M_{maj} and M_{min}.

1.3.3. Calculate the Order $O(i)$ Block

The "Calculate $O(i)$ Block" is responsible for detecting the order of the minority samples that are replaced with the majority samples and the other way around. For *Shape1*, the "Calculate $O(i)$ Block" detects the order of the minority samples that are replaced with the majority samples in the output samples. For *Shape2*, the "Calculate $O(i)$ Block" detects the order of the majority samples that are replaced with the minority samples in the output samples. There are two operations inside this block. The first one is the multiplication operation which is done by addition. There is counter that counts the number of required addition. The output of this counter is compared with the output of multiplexer (C or M_{min} according to the operation direction). When they are equal the addition process is stopped and the multiplication process is completed. The division is done by subtraction. The quotient is the number of subtraction processes that is counted by counter. The quotient is multiplied by 2 to get the final result of $O(i)$, this is done by one left shift. In the next subsection step selection block is presented.

1.3.4. Step Selection Block

Step Selection block is shown in Fig. 1.32. It is responsible for selecting the required step according to the output of the Operation Direction block and the output of the $O(i)$ block. The Operation Direction determines the shape through the *opd* signal. If *opd* = 0, *Shape2* is used. If *opd* = 1, *Shape1* is used. The signal *rev* is the output from the M_{maj} and M_{min} block, this signal is the selector of multiplexers that responsible for determining M_{maj} and M_{min}. If *rev*=0, M_{maj} equals *r* and M_{min} equals *s*. If *rev*=1, M_{maj} equals *s* and M_{min} equals *r*. The signal *T* determines when the step toggle between *M* and M_1, when the order of the sample which is required to toggle its step (the output of $O(i)$ block) equal to the output of the Counter. If *T* = 0, the step follows the normal case of the selected shape. If *T*=1, the step is toggled. This process is illustrated in Table 1.1 and Table 1.2 as an example.

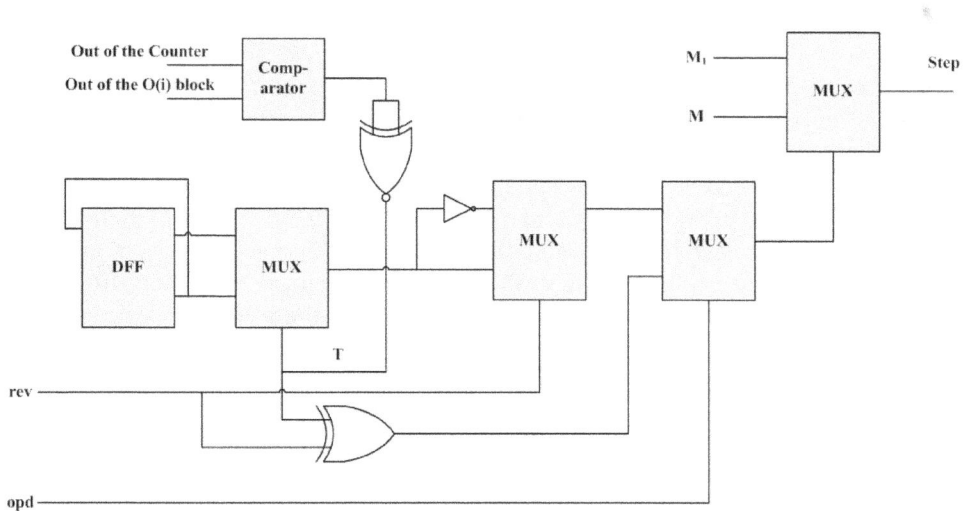

Fig. 1.32. The block diagram of Step Selection block.

Table 1.1. The selection step process.

opd	*rev*	*T*	shape	step
0	0	0		M_1
0	0	1	Shape2	M
0	1	0		M
0	1	1		M_1
1	0	0		NormalStep1
1	0	1	Shape1	Toggle
1	1	0		NormalStep2
1	1	1		Toggle

Table 1.2. Example to illustrate step selection process.

Sample order From the Counter	1	2	3	4	5	6	7	8	9
Sample order From $O(i)$	0	0	3	0	0	6	0	0	0
T	0	0	1	0	0	1	0	0	0
NormalStep1	M_l	M	M_l	M	M_l	M	M_l	M	M_l
Output step1	M_l	M	M	M	M_l	M_l	M_l	M	M_l
NormalStep2	M	M_l	M	M_l	M	M_l	M	M_l	M
Output step2	M	M_l	M_l	M_l	M	M	M	M_l	M

1.3.5. Filter Implementation

The *CF* is multiplier free. *CF* uses three clocks, *Clk1*, *Clk2*, and *Clk3*. *Clk1* is used to sample the input data. *Clk2* is used to perform output samples calculations. *Clk3* is used to trigger the output. *Clk1=(L*Clk3) + 5 Clk3*. Five clock cycles of *Clk3* are needed to determine the order *O(i)*. The division by *L* in (1.47) is done by Z-bit right shift, where $Z = Log_2 L$. The multiplication in (1.51) and (1.52) is performed by Z-bits shift left. Single right shift is required to do division in (1.55). The division in (1.56) and (1.57) are performed by addition and subtraction operations. No multipliers or divisors are required to implement the filter. This reduces the area significantly and accordingly power consumption. The detailed structure of the *CF* is shown in the following figures. The first part of the main circuit of the *CF* that contains "Calculate Step" block, "Calculate M_{maj} and M_{min}" block, "Calculate C" block, "Calculate the order *O(i)*" block and "Operation Direction" block is shown in Fig. 1.33. The second part of the main circuit of the *CF* that contains "select step" block and accumulator as shown in Fig. 1.34. The circuit details of "Calculate M_{maj} and M_{min}" block is shown in Fig. 1.35. The main circuit of "Calculate the order *O(i)*" block, that contain multiplication and division blocks is shown in Fig. 1.36. The sub-circuit which is responsible for multiplication in the *O(i)* block is shown in Fig. 1.37. The sub-circuit which is responsible for division in the *O(i)* block is shown in Fig. 1.38.

Whenever the value of the interpolation factor increases the frequency that sample the input data decrease. To improve the frequency limitations pipelining is done and shown in Fig. 1.39 and Fig. 1.40. The pipelining is expressed as a small shaded square in these figures. The next subsection presents simulation results and comparison.

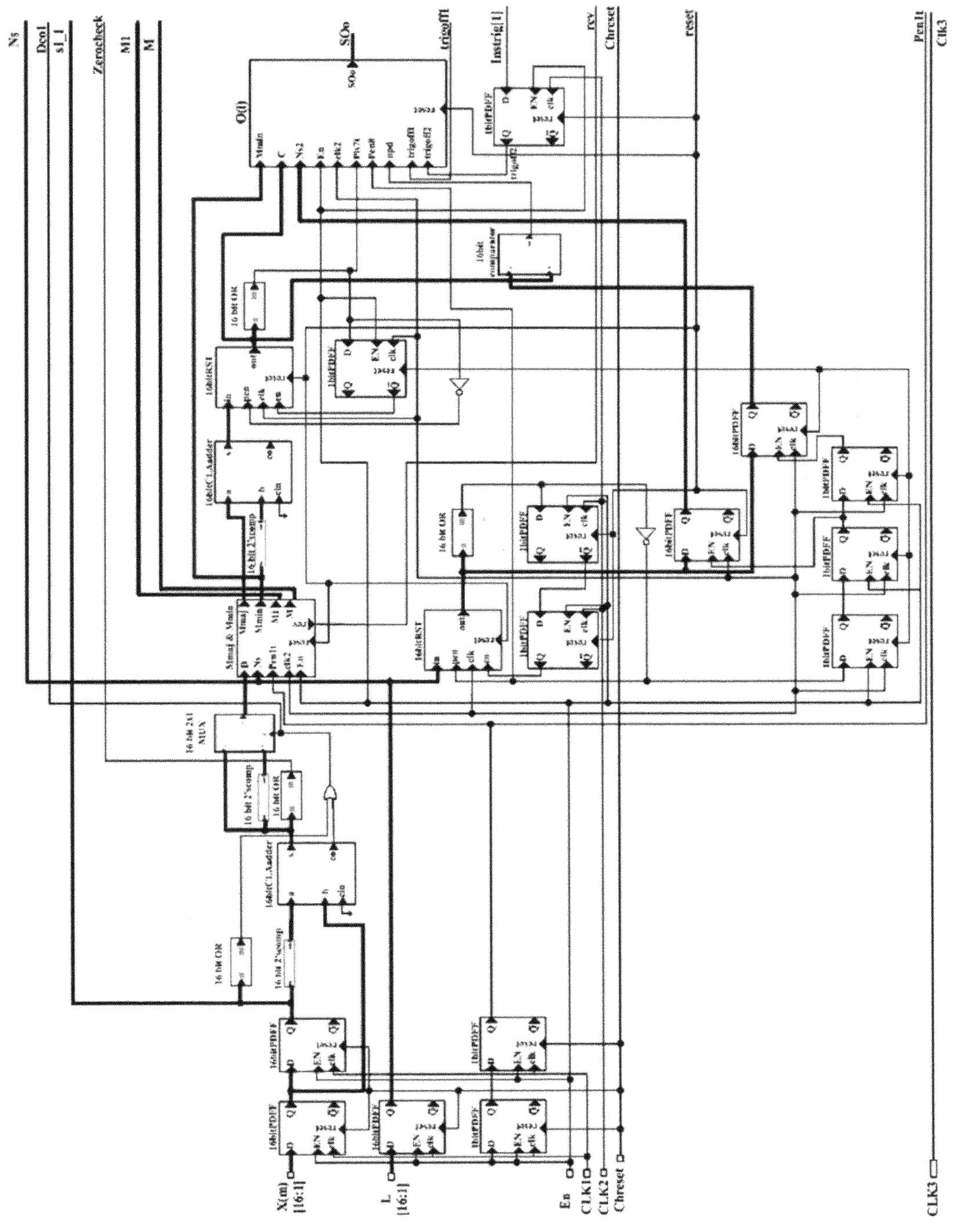

Fig. 1.33. The first part of the main circuit of the *CF*.

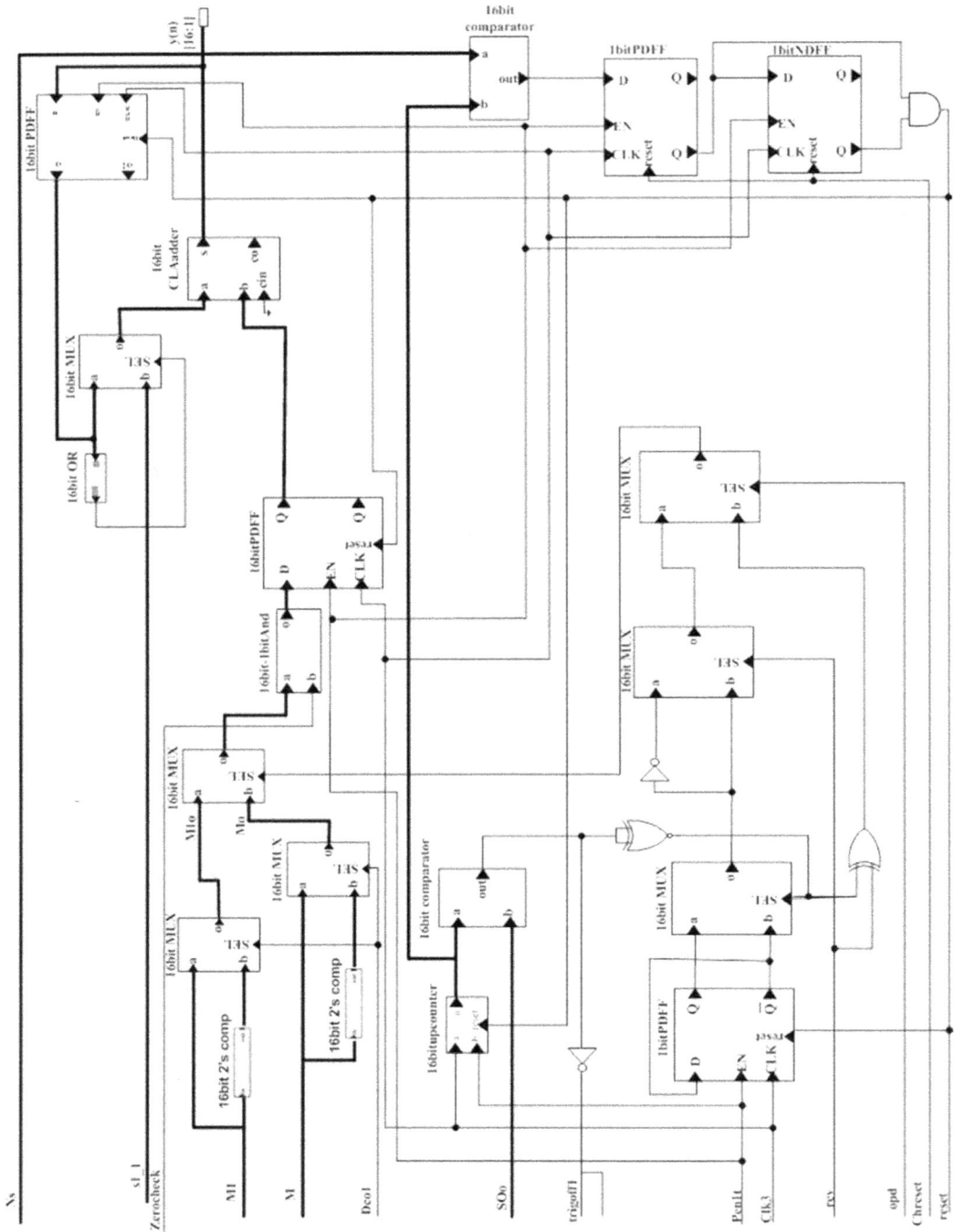

Fig. 1.34. The second part of the main circuit of the *CF*.

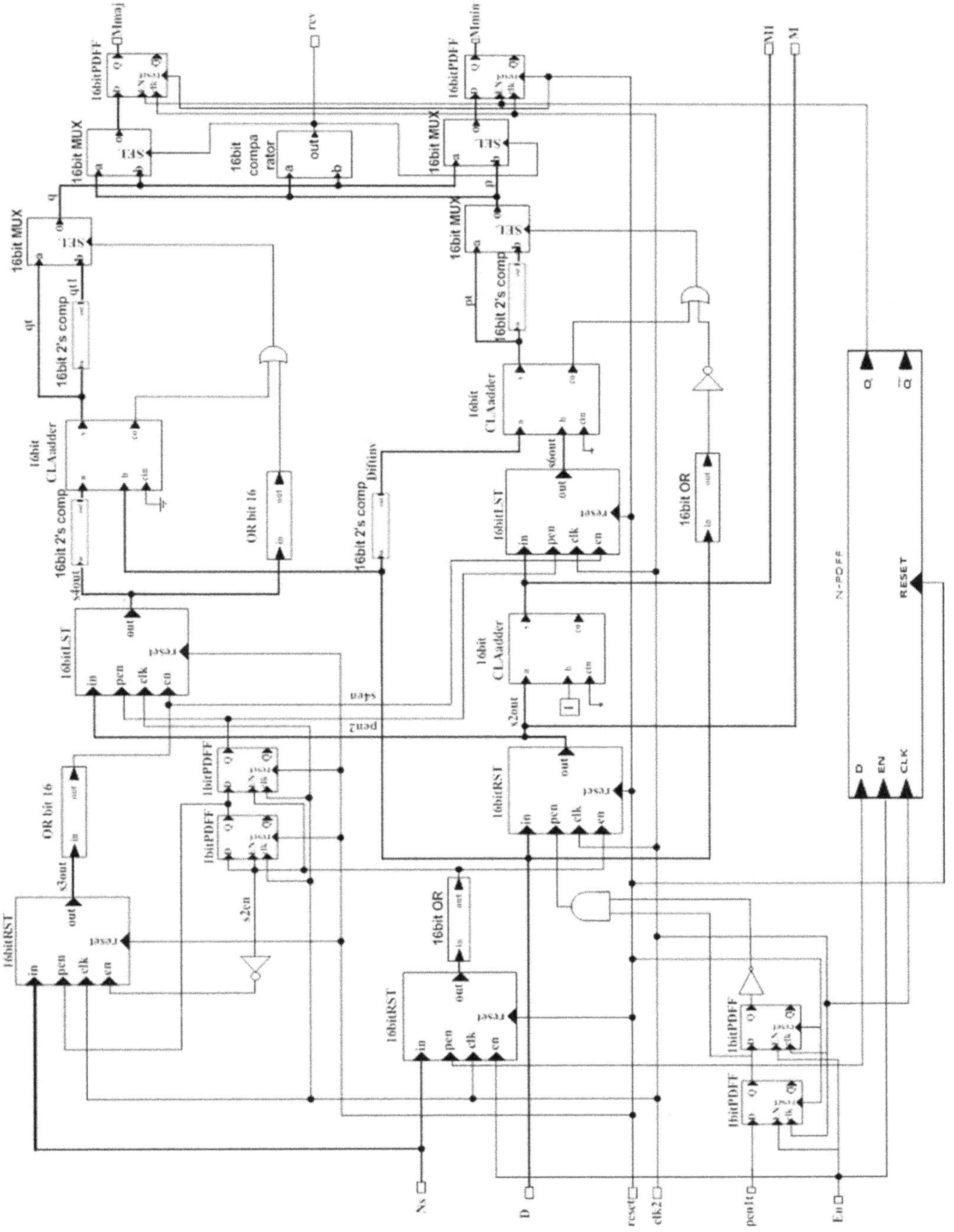

Fig. 1.35. The circuit of "Calculate M_{maj} and M_{min}" block.

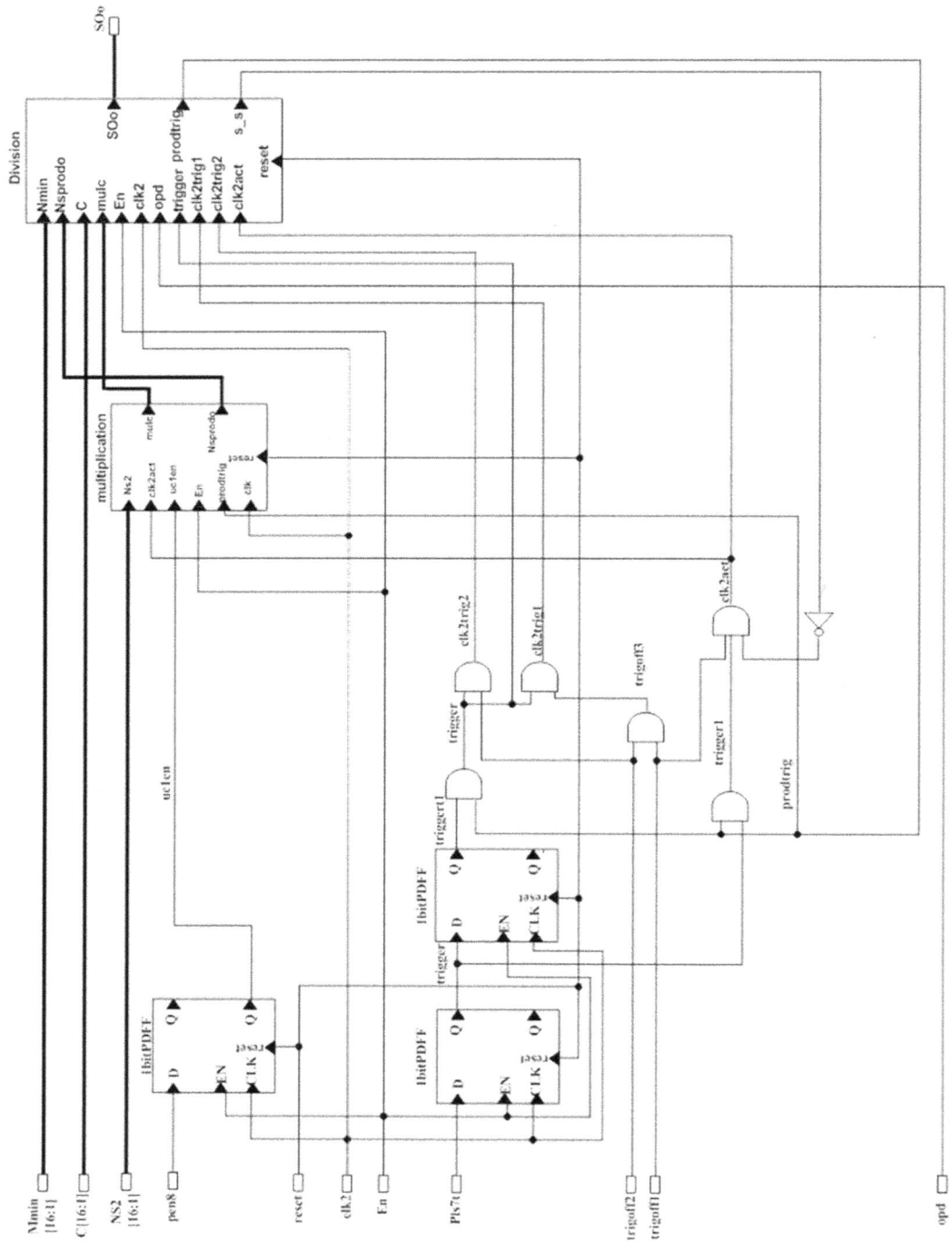

Fig. 1.36. The circuit of "*O(i)*" block.

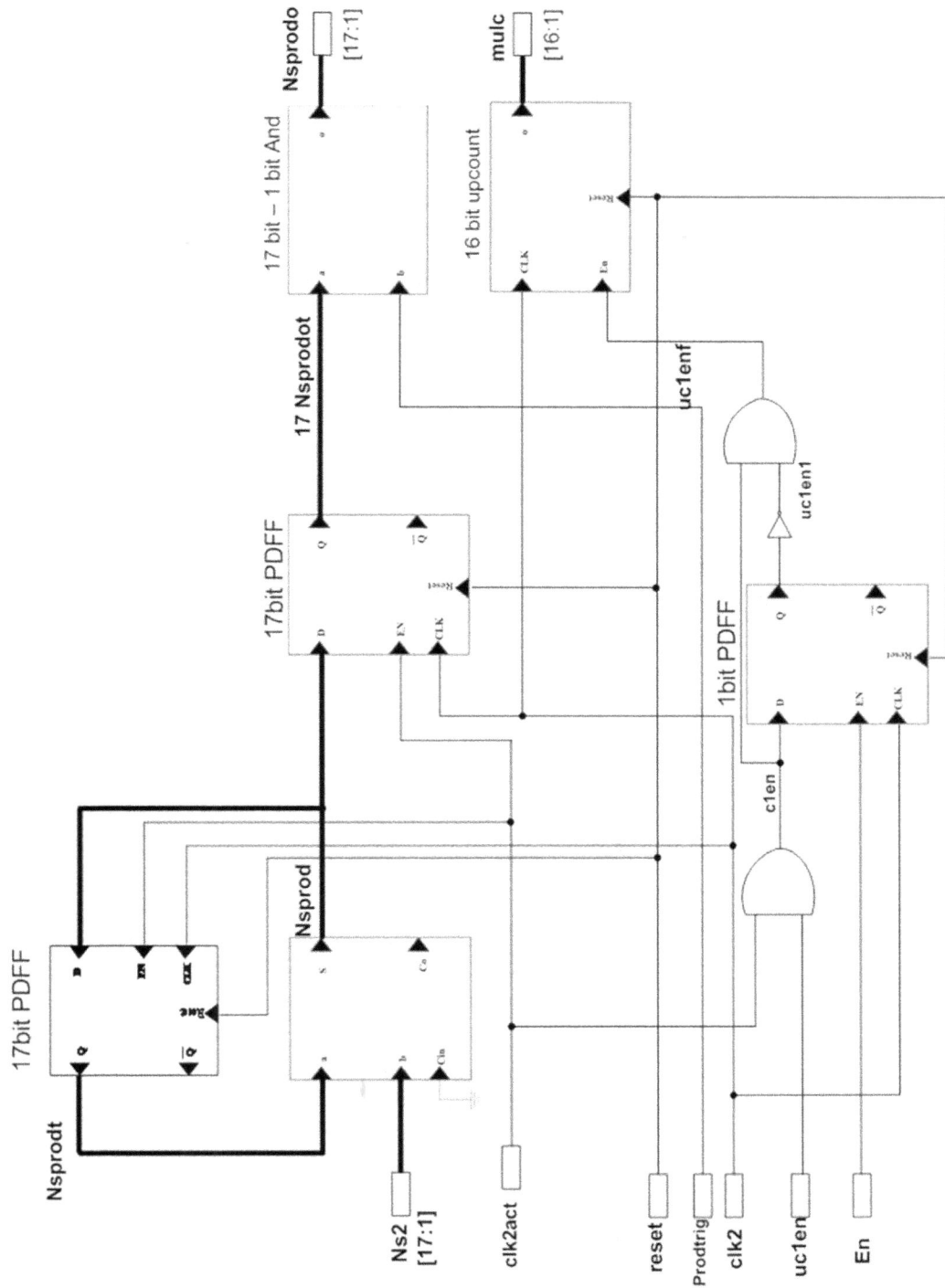

Fig. 1.37. Sub-circuit responsible for multiplication in the *O(i)* block.

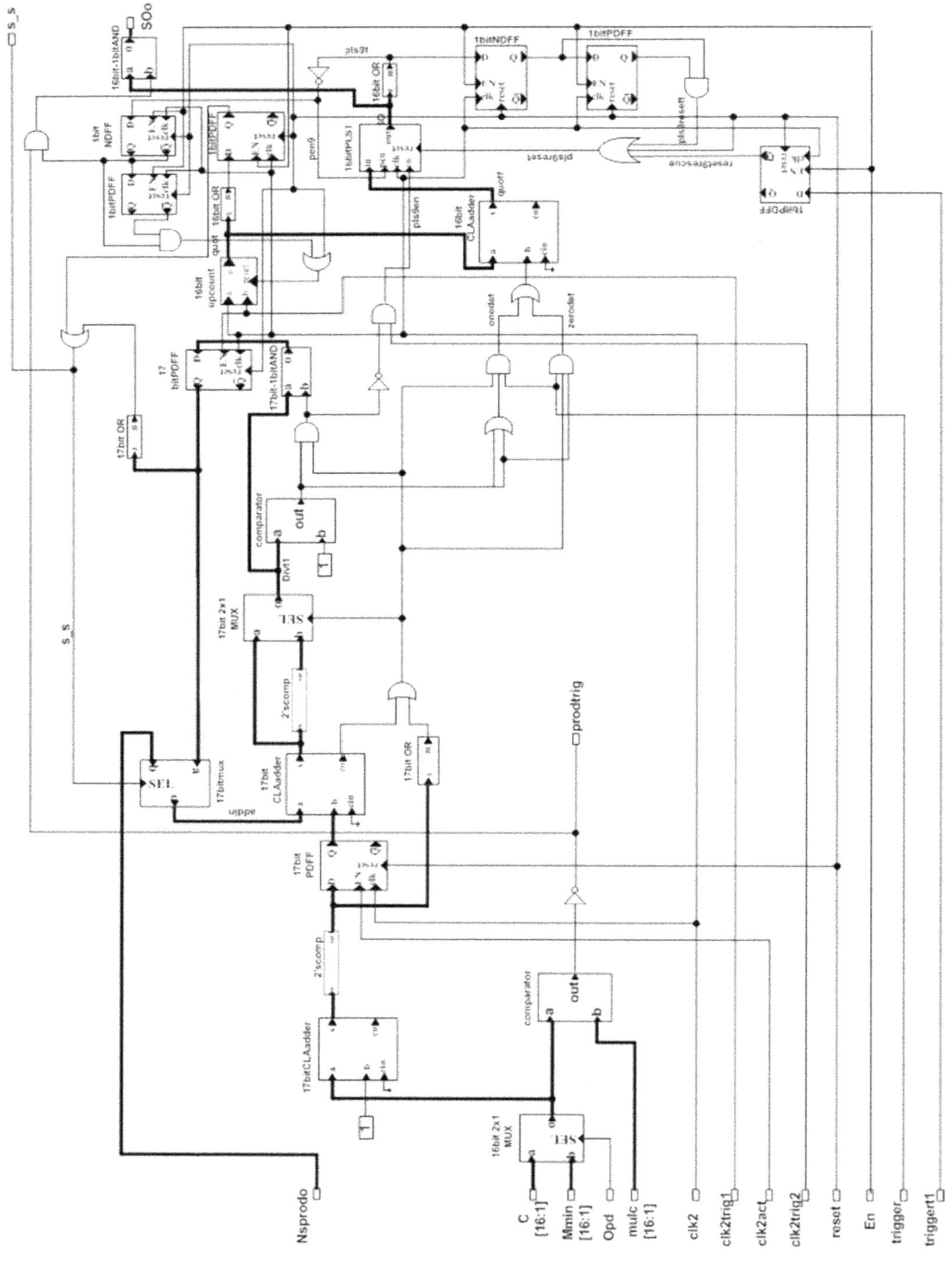

Fig. 1.38. Sub-circuit responsible for division in the *O(i)* block.

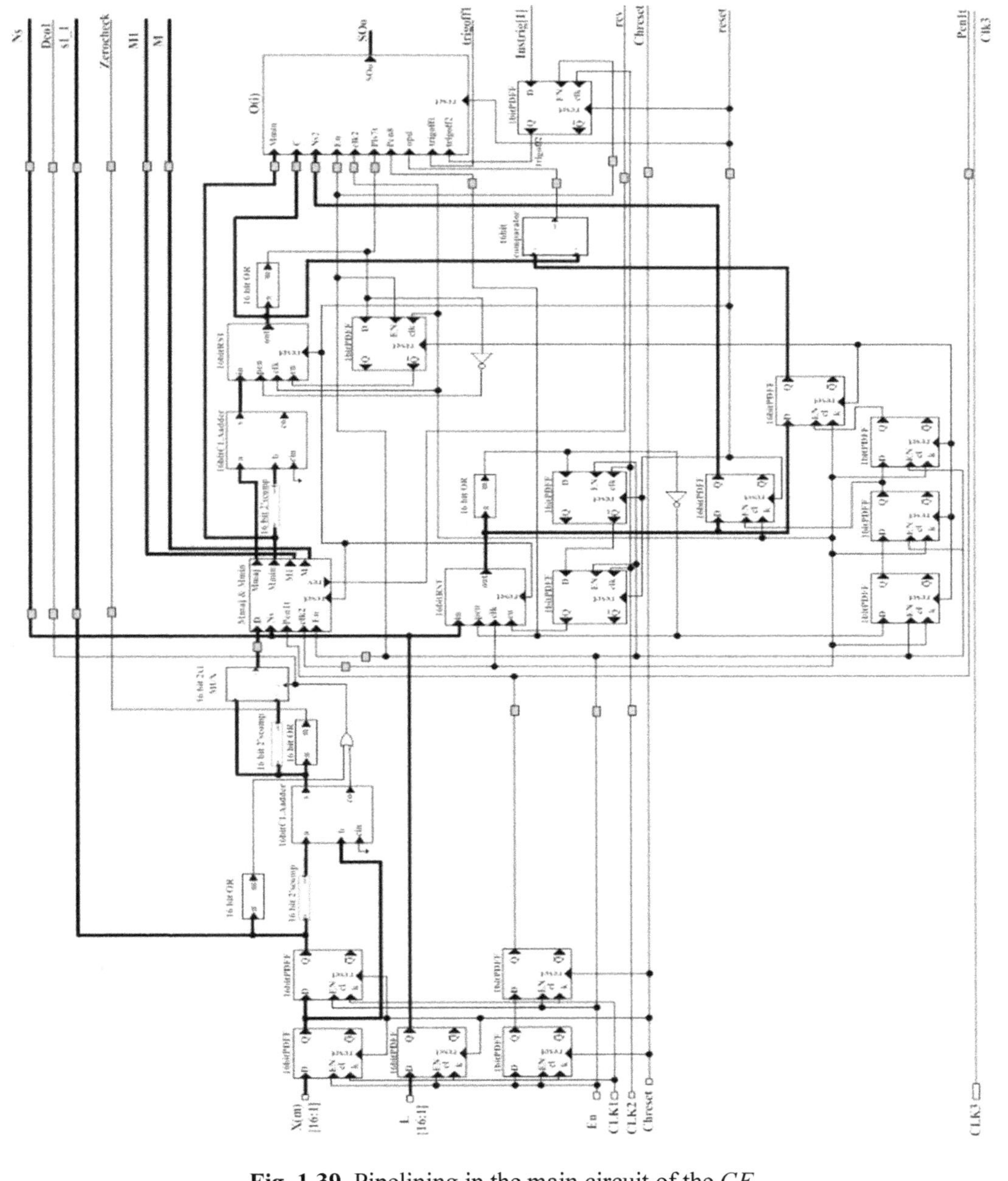

Fig. 1.39. Pipelining in the main circuit of the *CF*.

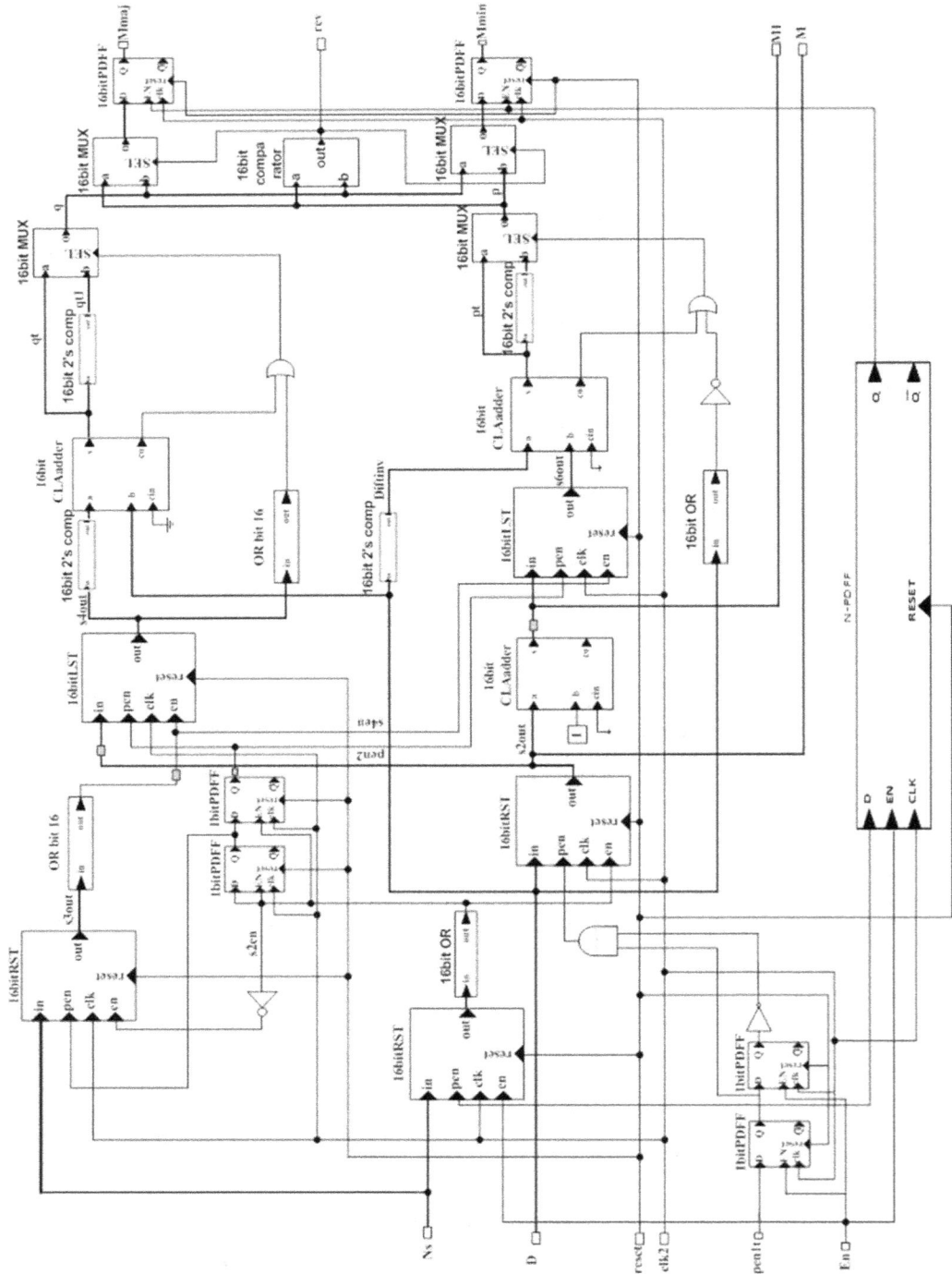

Fig. 1.40. Pipelining in the circuit of "M_{maj} and M_{min}" block.

1.4. Simulation Results and Comparison

Matlab is used to compare *CF* and *CFIR*. Modelsim tool by Mentor Graphics is used to simulate the HDL (Hardware Description Language) describing the circuit implementation using Verilog code [100]. Modelsim allows the designer to describe the large scale circuits into lines of code through different levels such as behavioral level, gate level implementation using VHDL or Verilog. After circuit validation, FPGA is used [101-102], to compare between *CF* and *CFIR*. This section is organized as follows. In Subsection 1.4.1, Matlab comparison is proposed. In Subsection 1.4.2, Modelsim simulation validation is presented. In Subsection 1.4.3, FPGA implementation is provided. In Subsection 1.4.4, performance analysis is discussed.

1.4.1. Matlab Comparison

CF and *CFIR* are compared using Matlab 2013. The input samples are represented with 16 bit fixed point representation. Different input sets (15 sets) are used to obtain the errors. The output sequences of both techniques are compared to the ideal linear interpolation sequence for each input set. The 15 sets are classified into three categories to meet the different conditions of *Shape1*, *Shape2* and equal successive inputs. The first category (from set 1 to set 5) follows *Shape1*. The second category (from set 6 to set 10) follows *Shape2*. The third category (from set 11 to set 15) is for equal successive inputs. Different interpolation factors are assumed. The percentage error for each input sequence for different values of interpolation factor [98-99] is shown in Fig. 1.41. The x-axis indicates the input sets from set 1 to set 10 for *Shape1* and *Shape2*. The y-axis indicates the percentage error of each set for *CF* and *CFIR*. *CF* has higher error than *CFIR*. However the maximum average error of *CF* for different sets is less than 2.6 %. The maximum error of *CFIR* for different sets is less than 0.02 %.

The input sets are listed in Tables 1.3 and 1.4 for *Shape1* and *Shape2*, respectively. The second and third columns indicate the input sets. The fourth and fifth columns indicate the error of *CF* and *CFIR*, respectively. The maximum average error of interpolation factor of 8, 16, 32, 64, 128, and 256 is 2.28 %, 2.59 %, 1.40 %, 1.11 %, 0.94 %, and 1.15 %, respectively.

For equal successive input sequences, the input is set to different values for interpolation factor of 8, 16, 32, 64, 128, and 256. The error is 0 % and 0.01 % for *CF* and *CFIR*, respectively. In next subsection, Modelsim simulation results are presented.

1.4.2. Modelsim Simulation Results

Full implementation for *CF* and *CFIR* techniques is developed using Verilog and is validated using Modelsim. The signal waveforms for *CF* and *CFIR* are shown in the following figures. The result of *CF* is shown in Fig. 1.42. The inputs for interpolation factor of 8 are, *X(m)*=2 and *X(m+1)*=14. The result of precomputer is shown in Fig. 1.43. The result of *S&A* unit is shown in Fig. 1.44. The result of 15 tap *CFIR* for interpolation factor of 8 is shown in Fig. 1.45.

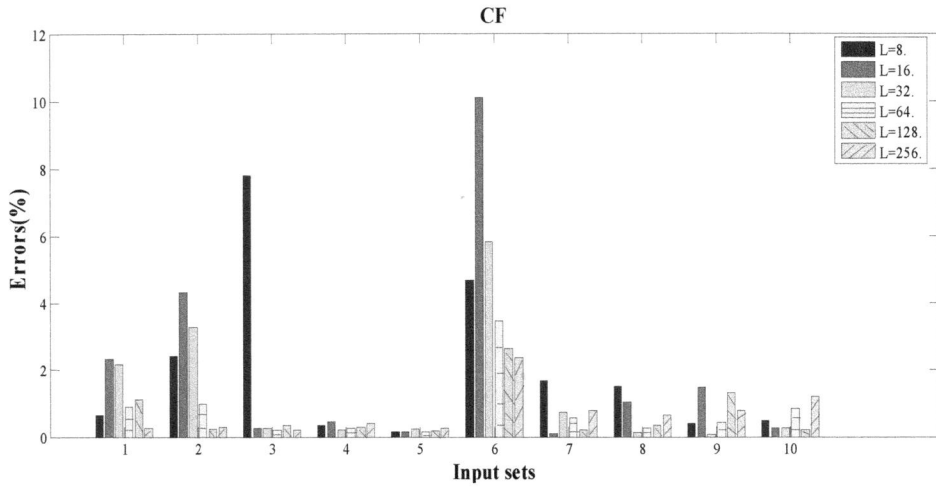

(a)

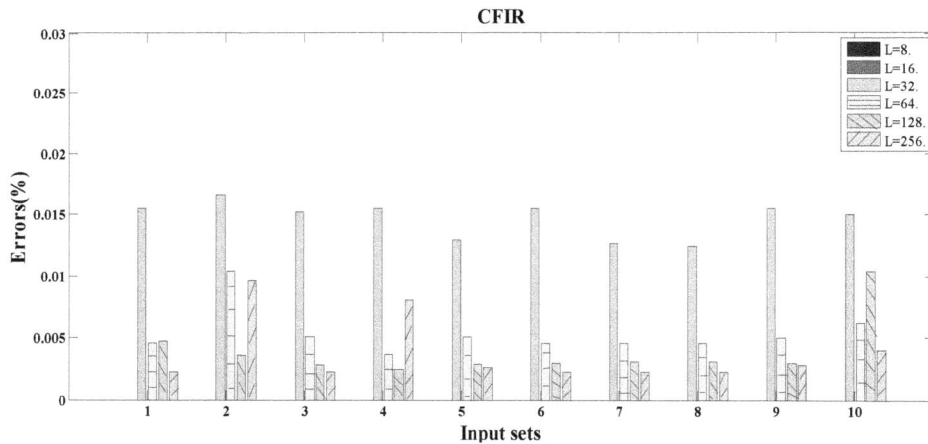

(b)

Fig. 1.41. Percentage errors for interpolation factor of 8, 16, 32, 64, 128, and 256:
(a) for CF; and (b) for CFIR.

To compare implementation of the two techniques on the gate level, the major gates of CF and *CFIR* are listed in Table 1.5. The number of gates of *CF* are much less than that of *CFIR*, which indicates significant reduction in area of the *CF*. The average reduction in the number of gates of interpolation factor of 8, 16, 32, 64, 128, and 256 is 86.9 %, 93.5 %, 96.7 %, 98.3 %, 99.1 %, 99.6 %, respectively.

Table 1.3. Interpolation accuracy for *Shape1*.

L	X(m)	X(m+1)	Error (%)	
			CF	*CFIR*
8	20	80	0.6506	0
	32	12	2.4370	0
	2	14	7.8103	0
	200	100	0.3456	0
	302	330	0.1583	0
	Average Error		2.2803	0
16	20	10	2.3289	0
	8	16	4.3291	0
	154	100	0.2699	0
	90	130	0.4596	0
	274	200	0.1438	0
	Average Error		1.5062	0
32	32	16	2.1657	0.0155
	2	50	3.2768	0.0166
	182	100	0.2567	0.0152
	228	114	0.2126	0.0155
	122	200	0.2355	0.0129
	Average error		1.2294	0.0151
64	30	60	0.8869	0.0046
	99	5	0.9873	0.0104
	110	272	0.2168	0.0051
	130	164	0.2556	0.0037
	150	372	0.1587	0.0051
	Average error		0.5010	0.0057
128	20	84	1.1211	0.0047
	120	312	0.2488	0.0036
	146	80	0.3365	0.0028
	158	96	0.2988	0.0025
	200	392	0.1752	0.0029
	Average error		0.4360	0.0033
256	128	256	0.2729	0.0023
	392	10	0.2972	0.0096
	252	126	0.2038	0.0023
	404	20	0.3943	0.0080
	100	226	0.2637	0.0026
	Average error		0.2863	0.0049

Table 1.4. Interpolation accuracy for *Shape2*.

L	X(m)	X(m+1)	Error (%)	
			CF	CFIR
8	20	10	4.6990	0
	30	60	1.6776	0
	99	5	1.5023	0
	110	272	0.4049	0
	130	164	0.4928	0
	Average Error		1.7553	0
16	2	14	10.101	0
	632	314	0.0900	0
	99	5	1.0450	0
	80	20	1.4858	0
	100	226	0.2587	0
	Average Error		2.5961	0
32	20	10	5.8276	0.0155
	30	60	0.7402	0.0126
	150	372	0.1295	0.0124
	632	314	0.0729	0.0155
	158	96	0.2664	0.0150
	Average error		1.4073	0.0142
64	20	10	3.4733	0.0046
	110	55	0. 5791	0. 0046
	65	130	0.2671	0. 0046
	50	120	0. 4306	0. 0050
	80	20	0.8480	0. 0062
	Average error		1.1196	0.005
128	20	10	2.6584	0.0030
	110	228	0.2141	0.0031
	208	100	0.3407	0.0031
	30	60	1.3069	0.0030
	392	10	0.2114	0.0104
	Average error		0.9463	0.00452
256	20	10	2.3659	0.0023
	110	55	0.7799	0.0023
	65	130	0.6438	0.0023
	50	120	0.7894	0.0028
	20	80	1.2003	0. 0040
	Average error		1.1558	0.0027

Table 1.5. Comparison between *CF* and *CFIR* through number of gates.

L	Gate Type	Number of Gates		Reduction (%)
		CF	*CFIR*	
8	AND	2570	24645	89.5
	OR	1169	8409	86.0
	NOT	1114	7536	85.2
	XOR	779	6027	87.0
	Average Reduction			86.9
16	AND	2570	50162	94.8
	OR	1169	16937	93.0
	NOT	1114	15472	92.7
	XOR	779	11819	93.4
	Average Reduction			93.5
32	AND	2570	100917	97.4
	OR	1169	33993	96.5
	NOT	1114	31344	96.4
	XOR	779	23403	96.6
	Average Reduction			96.7
64	AND	2570	202613	98.7
	OR	1169	68105	98.2
	NOT	1114	63088	98.2
	XOR	779	46571	98.3
	Average Reduction			98.3
128	AND	2570	406005	99.3
	OR	1169	136329	99.1
	NOT	1114	126576	99.1
	XOR	779	92907	99.1
	Average Reduction			99.1
256	AND	2570	812789	99.7
	OR	1169	272777	99.6
	NOT	1114	253552	99.6
	XOR	779	185579	99.6
	Average Reduction			99.6

1.4.2.1. *CF* Output

In Fig. 1.42, CF_F_ASA_tb is the module name in Verilog code. *In* is the input sample as *X(m)*. *L* is the interpolation factor that is used in the example. *C* is the difference between M_{maj} and M_{min} divided by 2. M_1 and M are the steps that are used to increment the samples. M_{maj} is the majority output samples which are the maximum of *r* and *s*. M_{min} is the minority output samples which are the minimum of *r* and *s*. *SOo* is the order of the minority samples that is replaced with the majority samples and the other way round. Y is the output samples.

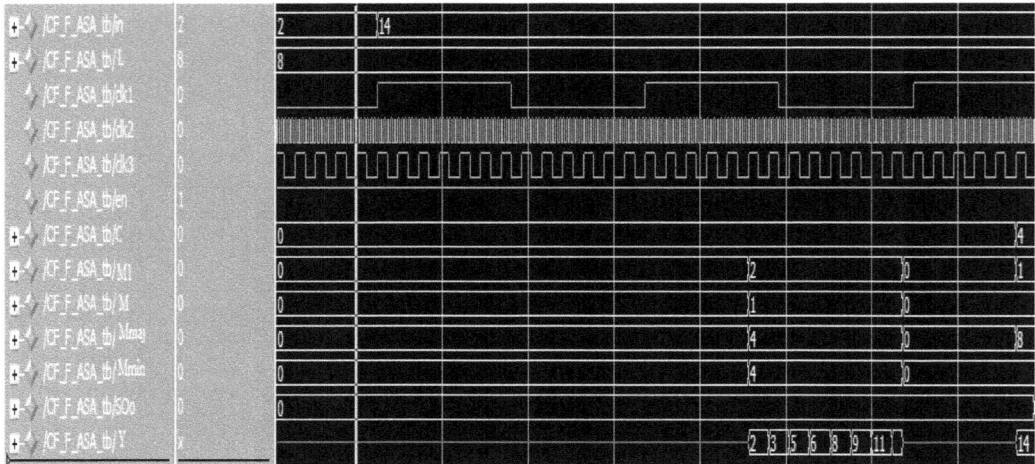

Fig. 1.42. The output of *CF* for *L*=8.

1.4.2.2. Validation for Functions of *CFIR*

Fig. 1.43 shows the output of the precomputer module in Verilog code. The input is *X*=2. The outputs are qm0=2, qm1=6, qm2=10, qm3=14, qm4=18, qm5=22, qm6=26, qm7=30, that are used to represent the alphabets *X, 3X, 5X, 7X, 9X, 11X, 13X, 15X*, respectively.

Fig. 1.44 shows the output of the *S&A* module in Verilog code. *Cin* is the input coefficient=1250. The inputs *m0=2, m1=6, m2=10, m3=14, m4=18, m5=22, m6=26, m7=30*, are the alphabets (the output of the precomputer). *Xsg* is the input sample sign=0. The output of the *S&A* is *XC*=2* 1250=2500. The output sign is sg=0 that means *XC* is positive number.

Fig. 1.45 shows the output of the fifteen tap *CFIR*, the module name of the Verilog code is FIR_5_tb. *X* is the input sample= [2, 0, 0, 0, 0, 0, 0, 0, 14]. The input coefficients are fractional numbers, so they are multiplied by 10^4, to be treated as integer numbers, so that the coefficients can be represented as integer binary numbers to be adapted with the system. Then the final output should be divided by 10^4. The input coefficients are *Cin1*=1250, *Cin2*= 2500, *Cin3*=3750, *Cin4*=5000, *Cin5*=6250, *Cin6*=7500, *Cin7*=8750, *Cin8*=10000, *Cin9*=8750, *Cin10*=7500, *Cin11*=6250, *Cin12*=5000, *Cin13*=3750, *Cin14*=2500, *Cin15*=1250. The output of the *CFIR* is *y1*= [2500, 5000, 7500, 10000, 12500, 15000, 17500, 20000, 35000, 50000, 65000, 80000, 95000, 110000, 125000, 140000, 122500, 105000, 87500, 70000, 52500, 35000, 17500]. The next subsection presentes FPGA implementation.

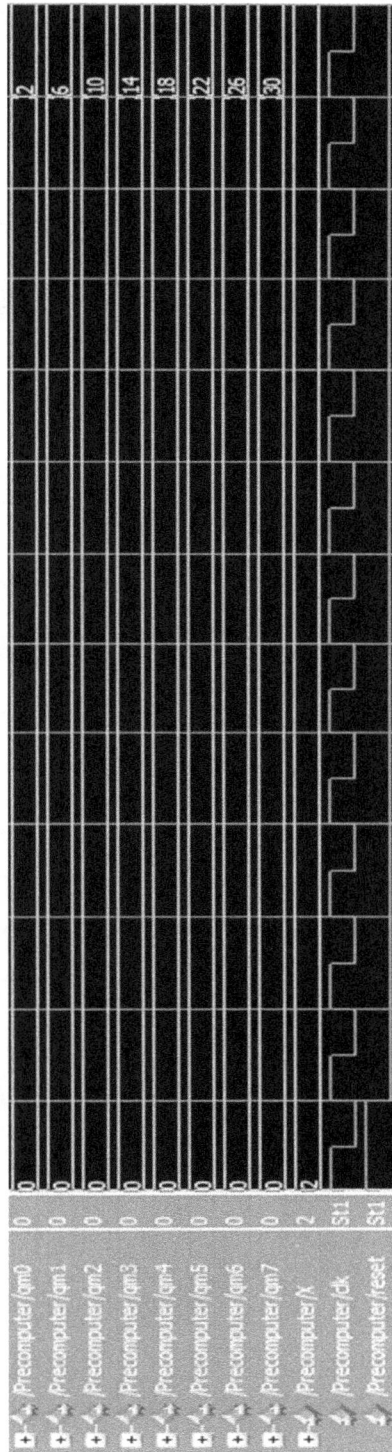

Fig. 1.43. The output of Precomputer.

Fig. 1.44. The output of S&A Unit.

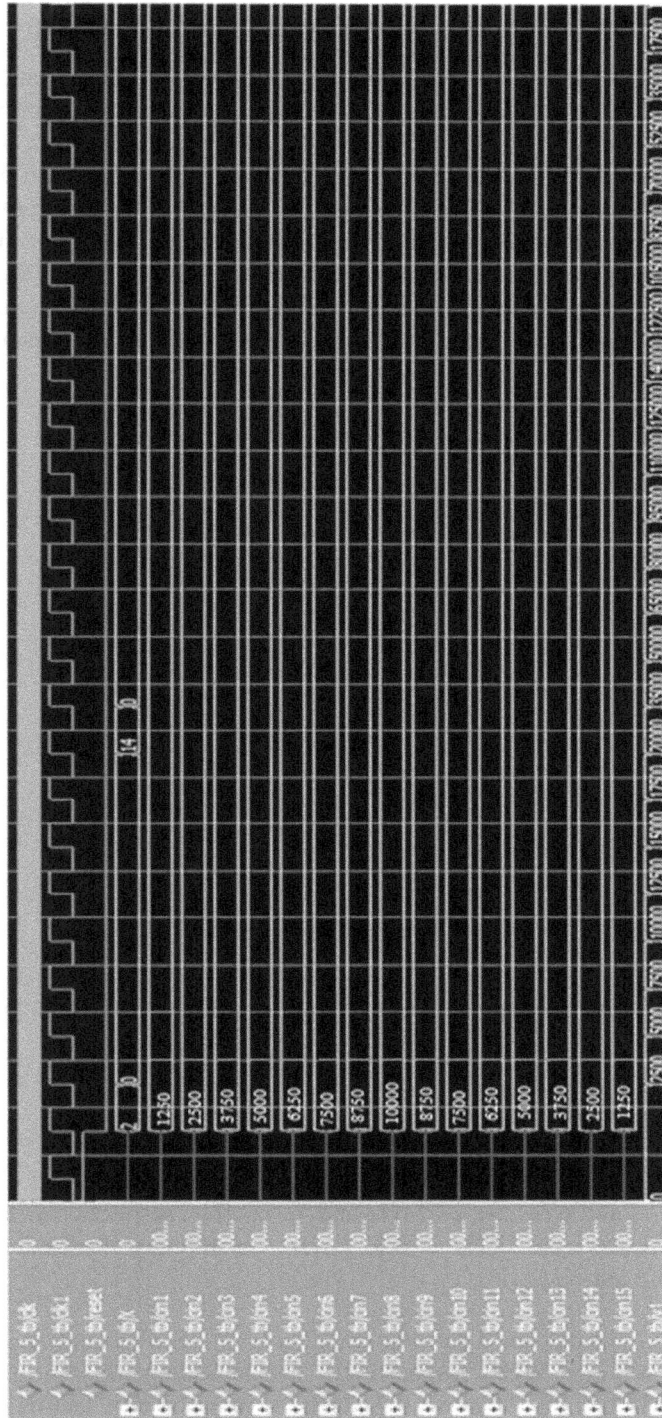

Fig. 1.45. The output of 15 tap *CFIR* for *L*=8.

1.4.3. FPGA Implementation

Xilinx Virtex5 FPGA XC5VTX240T is used to implement both *CF* and *CFIR*. *CF* uses 1 % of slice registers, 1 % of lookup table 1 % of the Memory, and occupy 1 % of slices. The maximum operating frequency is 0.92 MHz, 0.57 MHz, 0.32 MHz, and 0.2 MHz for interpolation factor of 8, 16, 32, and 64, respectively. The comparison between the two techniques as a percentage of FPGA is presented in Table 1.1.6. Average reduction of FPGA resources for interpolation factor of 8, 16, 32 and 64 is shown in Fig. 1.46. The average reduction in hardware implementation is 78.8 %, 89.5 %, 94.5 %, and 97.1 % for interpolation factor of 8, 16, 32, and 64, respectively. Note that the reduction increases as the interpolation factor increases. The next subsection presents performance analysis.

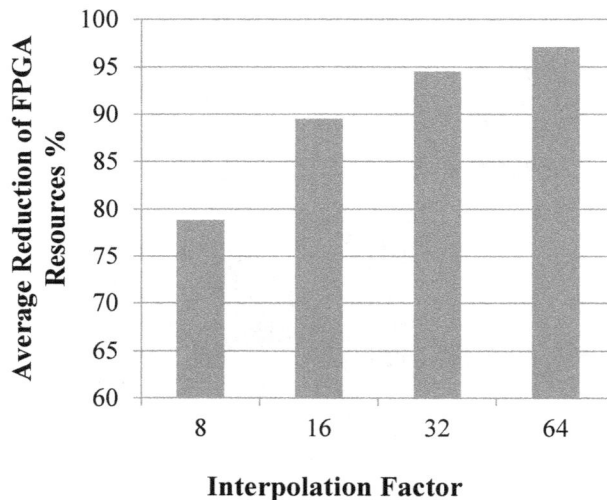

Fig. 1.46. Average reduction of FPGA resources for different interpolation factor of 8, 16, 32 and 64.

1.4.4. Performance Analysis

Xilinx Virtex5 FPGA XC5VTX240T is used. The accepted period of *Clk2* is 40 ns, so frequencies of *Clk2* and *Clk3* are 25 MHz and 4.16 MHz, respectively. For interpolation factor of 8, the frequency of *Clk1* is 0.32 MHz. To improve the frequency limitations pipelining is employed.

After the pipelining the accepted period of *Clk2* is 14 ns. The frequency of *Clk2* and *Clk3* are 71.42 MHz and 11.9 MHz, respectively. The maximum operating frequencies are 0.92 MHz, 0.57 MHz, 0.32 MHz and 0.2 MHz for interpolation factors of 8, 16, 32 and 64. The next section presents conclusions.

Table 1.6. Comparison between *CF* and *CFIR* to implement interpolation filter as a percentage of FPGA resources.

L	FPGA Resources	*CF*	*CFIR*	Reduction (%)
8	Memory	1 %	3 %	66.7
	Register	1 %	6 %	83.3
	LUT	1 %	4 %	75.0
	Slices	1 %	10 %	90.0
	Average Reduction			78.8
	Max Operating Frequency	0.92 MHz	1 MHz	8.0
16	Memory	1 %	6 %	83.3
	Register	1 %	12 %	91.7
	LUT	1 %	8 %	87.5
	Slices	1 %	22 %	95.5
	Average Reduction			89.5
	Max Operating Frequency	0.57 MHz	1 MHz	43.0
32	Memory	1 %	11 %	90.9
	Register	1 %	23 %	95.7
	LUT	1 %	16 %	93.8
	Slices	1 %	40 %	97.5
	Average Reduction			94.5
	Max Operating Frequency	0.32 MHz	1 MHz	68.0
64	Memory	1 %	22 %	95.5
	Register	1 %	45 %	97.8
	LUT	1 %	30 %	96.7
	Slices	1 %	67 %	98.5
	Average Reduction			97.1
	Max Operating Frequency	0.2 MHz	1 MHz	80.0

1.5. Conclusions

Interpolation filters are widely used in DSP. Interpolation process is performed by up-sampling followed by digital low pass filter. Digital low pass filter algorithm is either IIR or FIR. FIR is a non-recursive filter. Its output is the convolution sum of the input samples and the filter coefficient. *CFIR* is an advanced FIR filter using sharing multiplication technique. *CFIR* uses precomputer to perform the multiplication process according to the input samples. The precomputer outputs are shared through shift and add units (*S&A*). *S&A* units are used to select the correct multiplication output according to the coefficient value. Carry select adders adds the output of *S&A* units to get the output of the filter. A new multiplier free digital linear interpolation filter, the computational filter (*CF*), is presented. *CF* uses direct computation of the samples to guarantee uniform distribution of the interpolated samples. The output samples distribution is based on two shapes, *Shape1* and *Shape2*. No multiplier or divisor are needed to implement the filter, so its hardware is less complex and dissipates less power. The proposed filter dramatically reduces the hardware implementation as compared to the enhanced FIR (*CFIR*) technique. A detailed comparison between the *CF* and *CFIR* is provided. Insignificant reduction in the filter average precision (less than 3 %) is achieved for wide range of interpolation factor from 8 to 256. The Computational Filter average reduction in hardware implementation is 78.8 %, 89.5 %, 94.5 %, and 97.1 % for interpolation factor of 8, 16, 32, and 64, respectively. The maximum operating frequency for the CFIR is 1 MHz. The maximum operating frequency for the CF is 0.92 MHz, 0.57 MHz, 0.32 MHz, and 0.2 MHz for interpolation factor of 8, 16, 32, and 64, respectively.

References

[1]. S. R. Norsworthy, R. Schreier, G. C. Temes, Delta-Sigma Data Converters Theory, Design, Simulation, *IEEE Press*, 1997.

[2]. P. Kiss, J. Arias, D. Li, V. Boccuzzi, Stable high-order delta-sigma digital-to-analog converter, *IEEE Transactions on Circuits and Systems*, Vol. 51, No. 1, January 2004, pp. 200-205.

[3]. R. Schreier, An empirical study of high-order single-bit delta-sigma modulators, *IEEE Transactions on Circuits and Systems*, Vol. 40, No. 8, August 1993, pp. 461–466.

[4]. R. T. Baird, T. S. Fiez, Improved delta-sigma DAC linearity using data weighted averaging, in *Proceedings of the IEEE International Symposium on Circuits and Systems (ISCAS'95)*, Vol. 1, May 1995, pp. 13–16.

[5]. T. Zourntos, D. A. Johns, Variable structure compensation of delta sigma modulators: stability and performance, *IEEE Transactions on Circuits and Systems*, Vol. 49, No. 1, January 2002, pp. 41–53.

[6]. J. A. Cherry, W. M. Snelgrove, Approaches to simulating continuous-time delta sigma modulators, in *Proceedings of the IEEE International Symposium on Circuits and Systems (ISCAS'98)*, June 1998, Vol. 1, pp. 587–590.

[7]. B. Sklar, Digital Communication Fundamentals and Application, Second Edition, *Prentice Hall*, May 2004.

[8]. T. K. Moon, W. C. Stirling, Mathematical Methods and Algorithms for Signal Processing, PAP/CDR Edition, *Prentice Hall*, 2000.

[9]. R. G. Lyons, Understanding Digital Signal Processing, Third Edition, *Pearson Education*, November 2011.

[10]. R. E. Crochiere, L. R. Rabiner, Interpolation and decimation of digital signal – a tutorial review, *The Proceedings of the IEEE*, Vol. 69, No. 3, March 1981, pp. 300-331.

[11]. T. Kuo, A. Kwentus, A. N. Willson, A programmable interpolation filter for digital communications applications, in *Proceedings of the IEEE International Symposium on Circuits and Systems (ISCAS'98)*, May 1998, Vol. 2, pp. 97–100.

[12]. P. P. Vaidyanathan, Multirate digital filters, filter banks, polyphase networks, and applications: a tutorial, *The Proceedings of the IEEE*, Vol. 78, No. 1, January 1990, pp. 56-93.

[13]. B. Brandt, B. Wooley, A low-power, area-efficient digital filter for decimation and interpolation, *IEEE Journal of Solid-State Circuits*, Vol. 29, No. 6, June 1994, pp. 679 -687.

[14]. R. S. Balog, Topics in DSP: Interpolation & Delta Sigma Quantization, *Prentice Hall*, 1996.

[15]. J. G. Proakis, D. G. Manolakis, Digital Signal Processing: Principles, Algorithms, and Application, Fourth Edition, *Prentice Hall*, April 2006.

[16]. G. Bruun, Z-transform DFT filters and FFT's, *IEEE Transactions on Acoustics, Speech and Signal Processing*, Vol. 26, No. 1, February 1978, pp. 56-63.

[17]. E. O. Brigham, The Fast Fourier Transform and Its Applications, *Prentice Hall*, 1988.

[18]. D. R. Brillinger, Fourier Analysis of Stationary Processes, *The Proceedings of the IEEE*, Vol. 62, No. 12, December 1974, pp. 1628-1643.

[19]. T. Karp, N. J. Fliege, Modified DFT filter banks with perfect reconstruction, *IEEE Transactions on Circuits and Systems*, Vol. 46, No. 11, November 1999, pp. 1404-1414.

[20]. E. C. Ifeachor, B. W. Jervis, Digital Signal Processing: A Practical Approach, *Prentice Hall*, 2002.

[21]. M. Bhattacharya, T. Saramaki, All pass structure for multiplierless realization of recursive digital filter, in *Proceedings of the IEEE International Symposium on Circuit and Systems (ISCAS'03)*, Vol. 4, May 2003, pp. IV 237-IV.240.

[22]. G. K. Yeap, Practical Low Power Digital VLSI Design, *Springer*, August 1997.

[23]. P. Zhao, Z. Wang, Low power design of VLSI circuits and systems, in *Proceedings of the IEEE International Conference on ASIC (ASICON'09)*, October 2009, pp. 17-20.

[24]. L. Benini, E. Macii, G. De Micheli, Designing low power circuits: practical recipes, *The Proceedings of the IEEE Circuit and System Magazine*, Vol. 1, No. 1, 2001, pp. 6-25.

[25]. A. J. Chowdhury, M. S. Rizwan, S. J. Nibir, M. R. A. Siddique, A new leakage reduction method for ultra low power VLSI design for portable devices, in *Proceedings of the IEEE International Conference on Power Control and Embedded Systems (ICPCES'12)*, December 2012, pp.1-4.

[26]. S. R. S. Klavakolanu, M. K. Raju, F. Noorbasha, B. R. Kanth, A review report on low power VLSI systems analysis and modeling techniques, in *Proceedings of the IEEE International Conference on Signal Processing and Communication Engineering Systems (SPACES'15)*, January 2015, pp. 142-146.

[27]. K. Kaur, A. Noor, Strategies & methodologies for low power VLSI designs: a review, *International Journal of Advances in Engineering & Technology*, Vol. 1, No. 2, May 2011, pp. 159-165.

[28]. V. Sharma, J. K. Srivastava, Designing of low-power VLSI circuits using non-clocked logic style, *International Journal of Advancements in Research & Technology*, Vol. 1, No. 3, August 2012, pp. 1-5.

[29]. P. K. Hasmukh, A. S. Vipul, M. Durgamadhab, Estimation and optimization of power dissipation in CMOS VLSI circuit design: a review paper, *International Journal of Emerging Trends in Electrical and Electronics*, Vol. 1, No. 3, March 2013, pp. 14-21.

[30]. N. Zhu, W. L. Goh, W. Zhang, K. S. Yeo and Z. H. Kong, Design of low-power high-speed truncation-error-tolerant adder and its application in digital signal processing, *IEEE Transactions on Very Large Scale Integration Systems*, Vol. 18, No. 8, October 2009, pp. 1225-1229.

[31]. T. Solla, O. Vainio, Comparison of programmable FIR filter architectures for low power, in *Proceedings of the European Solid-State Device Research Conference (ESSDERC'02)*, September 2002, pp. 759-762.

[32]. S. Gao, J. Chi, C. Zhang, Quadric polynomial interpolation based on minimum local stretching energy, in *Proceedings of the IEEE International Conference on Computer and Information Technology Workshops (IPSN'08)*, July 2008, pp. 342-347.

[33]. J. Yuan, J. Lee, Narrow-band interference rejection in DS/CDMA systems using adaptive (QRD-LSL)-based nonlinear ACM interpolators, *IEEE Transactions on Vehicular Technology*, Vol. 52, No. 2, March 2003, pp. 374-379.

[34]. M. A. El-Moursy, A. G. Abdellatif, High-Performance Low-Power Digital Linear Interpolation Filter, in *Proceedings of the IEEE International Conference on Microelectronics (ICM'07)*, December 2007, pp. 57-60.

[35]. E. Hogenauer, An economical class of digital filters for decimation and interpolation, *IEEE Transaction on Acoustic, Speech and Signal Processing*, Vol. 29, No. 2, April 1981, pp. 155-162.

[36]. N. Jayant, S. Christensen, Effects of packet losses in wave form coded speech and improvements due to an odd-even sample-interpolation procedure, *IEEE Transactions on Communications*, Vol. 29, No. 2, February 1981, pp. 101-109.

[37]. J. C. Candy, G. C. Temes, *Oversampling Delta-Sigma Data Converters: Theory, Design, and Simulation, Wiley-IEEE Press*, August 1991.

[38]. Saeed V. Vaseghi, Advanced Digital Signal Processing and Noise Reduction, 4th Edition, *Wiley*, March 2009.

[39]. P. P. Vaidyanathan, Generalizations of the sampling theorem: seven decades after Nyquist, *IEEE Transactions on Circuits and Systems*, Vol. 48, No. 9, September 2001., pp. 1094-1109.

[40]. P. P. Vaidyanathan, S. M. Phoong, Discrete time signals which can be recovered from samples, in *Proceedings of the IEEE International Conference Acoustics, Speech and Signal Processing (ICASSP'95)*, May 1995, Vol. 2, pp. 1448-1451.

[41]. P. P. Vaidyanathan and S.M. Phoong, Reconstruction of Sequences from Nonuniform Samples, in *Proceedings of the IEEE International Symposium Circuits and Systems (ISCAS'95)*, Vol.1, May 1995, pp. 601-604.

[42]. Y. C. Eldar, A. V. Oppenheim, Filter bank reconstruction of bandlimited signals from nonuniform and generalized samples, *IEEE Transactions Signal Processing*, Vol. 48, No. 10, October 2000, pp. 2864-2875.

[43]. M. Unser, J. Zerubia, Generalized sampling without bandlimiting constraints, in *Proceedings of the IEEE International Conference Acoustics, Speech, and Signal Processing (ICASSP'97)*, Vol. 3, April 1997, pp. 2113-2116.

[44]. M. Unser, J. Zerubia, Generalized sampling: stability and performance analysis, *IEEE Transactions Signal Processing*, Vol. 45, No. 12, December 1997, pp. 2941-2950.

[45]. S.W. Smith, The Scientist and Engineer's Guide to Digital Signal Processing, *California Technical Publishing*, 1997.

[46]. A. Feuer, G. C. Goodwin, Sampling In Digital Signal Processing and Control, *Springer Science & Business Media*, 2012.

[47]. D. F. Elliott, Handbook of Digital Signal Processing: Engineering Applications, *Academic Press*, 2013.

[48]. P.P. Vaidyanathan, Multirate Systems and Filter Banks, *Prentice Hall*, 1993.

[49]. G. A. Nelson, L. L. Pfeifer, R. C. Wood, High-speed octave band digital filtering, *IEEE Transactions Audio and Electroacoustics*, Vol. 20, No. 1, March 1972, pp. 58-65.

[50]. T. A. Ramstad, Digital methods for conversion between arbitrary sampling frequencies, *IEEE Transactions on Acoustic, Speech Signal Processing*, Vol. 32, No. 3, June 1984, pp. 577-591.

[51]. A. W. Crooke, J. W. Craig, Digital filters for sample-rate reduction, *IEEE Transactions Audio and Electroacoustics*, Vol. 20, No. 4, October 1972, pp. 308-315.

[52]. R. E. Crochiere, L. R. Rabiner, Optimum FIR digital filter implementations for decimation, interpolation and narrow-band filtering, *IEEE Transactions on Acoustic, Speech Signal Processing*, Vol. 23, No. 5, October 1975, pp. 444-456.

[53]. R. E. Crochiere, L. R. Rabiner, further considerations in the design of decimators and interpolators, *IEEE Transactions on Acoustic, Speech Signal Processing*, Vol. 24, No. 4, August 1976, pp. 296-311.

[54]. T. Saramaki, A class of linear-phase FIR filters for decimation, interpolation and narrow-band filtering, *IEEE Transactions on Acoustic, Speech Signal Processing*, Vol. 32, No. 5, October 1984, pp. 1023-1036.

[55]. D. J. Goodman, M. J. Carey, Nine digital filters for decimation and interpolation, *IEEE Transactions on Acoustic, Speech Signal Processing*, Vol. 25, No. 2, April 1977, pp. 121-126.

[56]. F. Mintzer, B. Liu, Aliasing error in the design of multirate filters, *IEEE Transactions on Acoustic, Speech Signal Processing*, Vol. 26, No. 1, February 1978, pp. 76-87.

[57]. Z. Jing, A. Fam, A new structure for narrow transition band low pass digital filter design, *IEEE Transactions on Acoustic, Speech Signal Processing*, Vol. 32 , No. 2, April 1984, pp.362- 370.

[58]. F. Mintzer, Filter for distortion free two band multirate filter banks, *IEEE Transactions on Acoustic, Speech Signal Processing*, Vol. 33, No. 3, June 1985, pp. 626-630.

[59]. M. Narasimha, A. Peterson, On using the symmetry of FIR filters for digital interpolation, *IEEE Transactions on Acoustic, Speech Signal Processing*, Vol. 26, No. 3, June 1978, pp. 267-268.

[60]. M. G. Bellanger, G. Bonnerol, Premultiplication schemes for digital FIR filter with application to multirate filtering, *IEEE Transactions on Acoustic, Speech Signal Processing*, Vol. 26, No. 1, February 1978, pp. 50-55.

[61]. Fausto Pedro García Márquez, Noor Zaman, Digital Filters and Signal Processing, *InTech*, January 2013.

[62]. Web site, (https://users.dimi.uniud.it/~antonio.dangelo/MMS/materials/DigitalFilters.pdf).

[63]. L. Rabiner, R. W. Schafer, Recursive and nonrecursive realizations of digital filters designed by frequency sampling techniques, *IEEE Transactions on Audio and Electroacoustics*, Vol. 19, No. 3, September 1971, pp. 200-207.

[64]. H. D. Helms, Nonrecursive digital filters: design methods for achieving specifications on frequency response, *IEEE Transactions on Audio and Electroacoustics,* Vol. 16, No. 3, September 1968, pp. 336-342.

[65]. Dietrich Schlichthärle, Digital Filters: Basics and Design, Second Edition, *Springer*, April 2011.

[66]. Z. Milivojevic, Digital Filter Design, First Edition, *mikroElektronika*, 2009.

[67]. Wikipedia Web, (https://en.wikipedia.org/wiki/Digital_filter).

[68]. College of science and engineering Website:
http://www.ee.cityu.edu.hk/~hcso/ee5410_9.pdf

[69]. J. Rabaey, Low Power Design Essentials, *Springer*, 2009.

[70]. J. Rabaey and M. Pedram, Low Power Design Methodologies, Second Edition, *Kluwer Academic Publishers*, 1996.

[71]. M. H. Rashid, Power Electronics Handbook: Devices, Circuits and Application, Second Edition*, Academic Press*, July 2010.

[72]. P. E. Landman, Low-Power Architectural Design Methodologies, PhD thesis*, University of California, Berkeley*, 1994.

[73]. M. Mehendale, S.D. Stierleckar, G. Venkapesh, Low power realization of fir filters on programmable DSPs, *IEEE Transaction on Very Large Scale Integration Systems*, Vol. 6, No. 4, December 1998, pp. 546-553.

[74]. M. T. C. Lee, V. Tiwari, S. Malik, M. Fujita, Power analysis and minimization techniques for embedded DSP software, *IEEE Transactions on Very Large Scale Integration Systems,* Vol. 5, No. 1, March 1997, pp. 123-135.

[75]. D. A. Parker, K. K. Parhi, Low-area/power parallel FIR digital filter implementations, *Journal of VLSI Signal Processing Systems for Signal, Image and Video Technology,* Vol. 17, No. 1, September 1997, pp. 75-92.

[76]. M. Mehendale, S. D. Sherlekar, G. Venkatesh, Extensions to programmable DSP architectures for reduced power dissipation, in *Proceedings of the International Conference on VLSI Design (VLSID'98)*, January 1998, pp. 37-42.

[77]. F. A. Shah, H. Jamal and M. A. Khan, Reconfigurable low power FIR filter based on partitioned multipliers, in *Proceedings of the IEEE International Conference on Microelectronics (ICM'06)*, December 2006, pp. 87-90.

[78]. Henrik Ohlsson, Studies On Design and Implementation of Low-Complexity Digital Filters, *Linköpings University*, May 2005.

[79]. B. Amelifard, F. Fallah, M. Pedram, Closing the gap between carry select adder and ripple carry adder: a new class of low-power high-performance adders, in *Proceedings of the IEEE International Symposium on Quality Electronic Design (ISQED'05)*, March 2005, pp. 148-152.

[80]. A. Åslund, O. Gustafsson, H. Ohlsson, L. Wanhammar, Power analysis of high throughput pipelined carry-propagation adders, in *Proceedings of the IEEE NorChip Conference (NORCHIP'04)*, pp. 139–142, November 2004.

[81]. T. G. Noll, Carry save architectures for high-speed digital signal processing, *Journal VLSI Signal Processing Systems for Signal, Image and Video Technology*, Vol. 3, No. 1, June 1991, pp. 121–140.

[82]. M. Karlsson, Distributed Arithmetic: Design and Applications, Thesis No. 696, *Linköping University, Sweden*, 1998.

[83]. W. Jeong, K. Roy, Robust High-performance low power carry select adder, in *Proceedings of the Asia and South Pacific Design Automation Conference (ASP-DAC'03)*, January 2003, pp. 503-506.

[84]. O. Kwon, E. Swartzlander, K. Nowka, A fast hybrid carry-lookahead/ carry-select adder design, in *Proceedings of the Great Lakes symposium on VLSI (GLSVLSI'01)*, March 2001, pp. 149-152.

[85]. N. Sankarayya, K. Roy, D. Bhattacharya, Algorithms for low power high speed FIR filter realization using differential coefficients, *IEEE Transaction Circuits System II*, Vol. 44, No. 6, June 1997, pp. 488-497.

[86]. N. T. A. El-Kheir, M. S. El-Kharashi, M. A. El-Moursy, A low power programmable FIR filter using sharing multiplication technique, in *Proceedings of the IEEE International Conference on IC Design and Technology (ICICDT'12)*, June 2012, pp. 1-4.

[87]. J. Park, W. Jeong, H. Mahmoodi-Meimand, Y. Wang, H. Choo, K. Roy, Computation sharing programmable FIR filter for low-power and high- performance applications, *IEEE Journal of Solid-State Circuits*, Vol. 39, No. 2, February 2004, pp. 348-357.

[88]. K. Muhammad, Algorithmic and Architectural Techniques for Low-Power Digital Signal Processing, PhD thesis, *Purdue University*, 1999.

[89]. J. Park, H. Choo, K. Muhammad, S. Choi, Y. Im, K. Roy, Non-adaptive and adaptive filter implementation based on sharing multiplication, in *Proceedings of the IEEE International Conference on Acoustics, Speech, and Signal Processing (ICASSP'2000)*, Vol. 1, June 2000, pp. 460-463.

[90]. S. Khan, Digital Design of Signal Processing Systems a Practical Approach, First Edition, *John Wiley and Sons*, January 2011.

[91]. J. Rabaey, Digital Integrated Circuits a Design Perspective, Second Edition, *Prentice-Hall*, 2002.

[92]. B. Parhami, Computer Arithmetic: Algorithms and Hardware Design, Second Edition, *Oxford University Press*, December 2010.

[93]. N. H. E. Weste, D. Harris, CMOS VLSI Design: A Circuit and Systems Perspective, Fourth Edition, *Pearson Education*, 2011.

[94]. K. Johansson, Low Power and Low Complexity Shift-and-Add Based Computations, *Linköpings University*, 2008.

[95]. U. Meyer-Baese, Digital Signal Processing with Field Programmable Gate Arrays: Signals and Communication Technology, 3rd Edition, *Springer*, November 2007.

[96]. D. Harris, S. Harris, Digital Design and Computer Architecture, Second Edition, *Morgan Kaufmann*, August 2012.

[97]. R. W. Schafer, L. Rabiner, A digital signal processing approach to interpolation, *The Proceedings of the IEEE*, Vol. 61, No. 6, June 1973, pp. 692-702.

[98]. R. S. Abd El-Azeem, M. A. El-Moursy, A. M. Nassar, A. Gharib, N. T. Abou-El-Kheir, M. S. El-Kharashi, Direct computation for high performance interpolation filter, *Microelectronics Journal*, Vol. 51, May 2016, pp. 112-118.

[99]. R. S. Abd El-Azeem, M. A. El-Moursy, A. M. Nassar, A. Gharib, N. T. Abou-El-Kheir, M. S. El-Kharashi, High Performance Interpolation Filter Using Direct Computation, in *Proceedings of the IEEE International Design & Test Symposium (IDT'16)*, December 2016, pp. 121-124.

[100]. K. Ayob, Digital Filter Design for FPGA Engineers, *CreateSpace Independent Publishing Platform,* December 2014.

[101]. S. Kilts, Advanced FPGA Design: Architecture, Implementation, and Optimization, *Wiley-IEEE Press*, June 2007.

[102]. B. Readler, Verilog by Example: A Concise Introduction for FPGA Design, *Full Arc Press*, April 2011.

Chapter 2

Potential of Low-Voltage and Low-Energy MOS Devices in Coming Sensor Network Era

Yasuhisa Omura and Abhijit Mallik

2.1. Introduction

After 2001, the silicon-on-insulator (SOI) technology was seen as a well-known and well-understood technology for VLSI fabrication because IBM had commercialized SOI technology for PC and workstations [1]. The application of SOI Lubistors to ESD protection circuits for SOI integrated circuits (ICs) strongly contributed to the commercialization of SOI technology [2]. The SOI substrate itself and the material were reviewed to cover every stage of device scaling, and SOI devices are being applied far more often [1].

In the 21st century, various issues, such as global warming, local security issues, and the graying of society, demand solutions based on even more advanced technologies. It is clear that the electronics industries will have to contribute to society by implementing the original ideas of scientists and engineers. SOI materials and advanced architectures will play important roles in low-energy electronics [3]. Security issues of local society including health monitoring in particular, need maintenance-free reliable sensor networks without any chemical batteries if possible. Thus we should pay more attention to low-energy device and circuit technologies, not just low-power device technology and circuits [4-7].

This chapter addresses the potential and prospects of SOI materials and devices for low-energy signal processing and RF applications for the sensor network technology.

2.2. Potential of Silicon-on-Insulator Materials

The silicon-on-Sapphire (SOS) substrate was proposed in the 1960s for high-frequency devices [8]. Because of crystallographic and other technology issues, various silicon-on-

Yasuhisa Omura
Kansai University, Japan

SiO2 (SOI) substrates attracted more attention than SOS substrates [1]. After their successful production by IBM in 2001, new semiconductor-on-SiO$_2$ (SOI) substrates such as Ge-on-SiO2, SiGe-on-SiO$_2$, and compound material-on-SiO$_2$ (so-called "AOI") substrates [9] were proposed to realize high-speed digital ICs.

The worldwide trend in mobile communications changed the focus of technology in the 1990s. Many engineers paid attention to the potential of the SOI substrate because its high resistivity enables the low-loss transmission of RF signals [10]. This is utilized in recent 3D integration architectures [11].

2.3. Potential and Applications of SOI Devices

2.3.1. Fully-depleted SOI Devices

When the fully-depleted (FD) SOI MOSFET emerged in the 1990s [1, 12, 13], some companies intended to develop RF devices for telecommunications.

First, however, this chapter reviews its potential as a low-energy device. The impact of FD-SOI MOSFET scaling on low-energy performance has been examined [3]. Device simulation results of scaled single-gate FD-SOI MOSFETs are shown in Figs. 2.1 and 2.2, where L$_{min}$ is the minimum channel length and SS$_{max}$ is the maximum value of subthreshold swing.

The simulation results of I$_{ON}$ and I$_{OFF}$ satisfy the ITRS roadmap2003 with its target of 2013. When a high threshold voltage condition is assumed (V$_{TH}$=0.3 V), very low standby power dissipation is realized [3]. It is very noteworthy that the 15-nm channel single-gate FD-SOI MOSFET satisfies the roadmap request. The simulation results of I$_{ON}$ promise attractive performance from a 1 fJ/μm/device at V$_D$= 1V. Although the ITRS roadmap is not renewed, the business opportunity for FD-SOI ICs will grow because the concept of IoT requires FD-SOI devices.

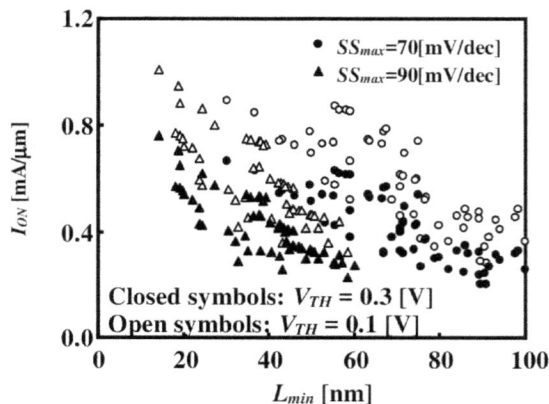

Fig. 2.1. Device simulation results of I_{ON} of FD-SOI MOSFETs at V_D = 1 V [3].

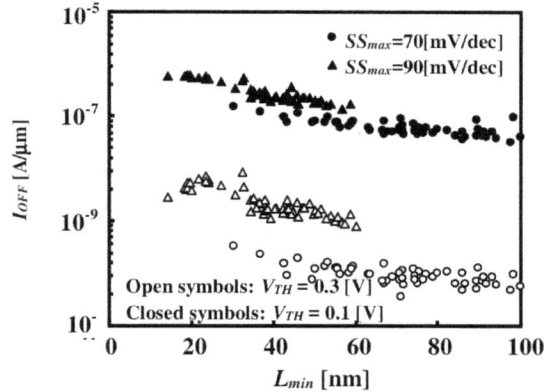

Fig. 2.2. Device simulation results of I_{OFF} of FD-SOI MOSFETs at $V_D = 1$ V [3].

Next, this chapter introduces a practical application of the FD-SOI MOSFET. The potential of FD SOI devices is examined from the viewpoint of new RF applications. As shown in Fig. 2.3, the Schenkel circuit for RF-ID chips needs many "diodes" [14]. Every diode can be replaced with an FD-SOI MOSFET as a "quasi-diode," which is shown in Fig. 2.4.

The RF performance of the quasi-diode made from a 0.4-μm gate FD-SOI MOSFET was examined in small-signal device simulations. Simulation results of boost-up efficiency of the quasi-diode are shown in Fig. 2.5 for various body doping conditions. It is found that appropriate body doping ($N_{A,ch}=1\times10^{17}$ cm^{-3}) promises Giga Hertz operation with high boost-up efficiency, which is a very interesting result. This yields circuits with higher efficiency and higher output voltage than possible with the bulk MOSFET [14].

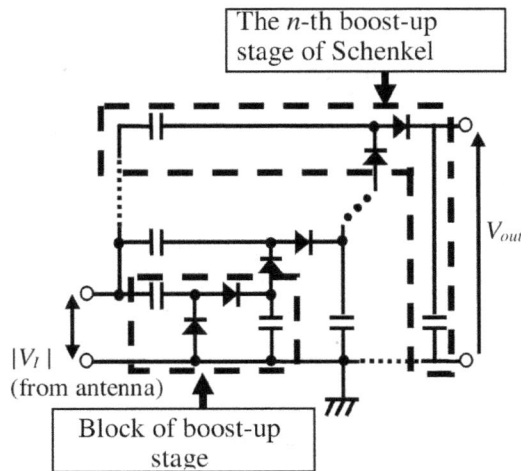

Fig. 2.3. Schematic of Schenkel circuit [14].

81

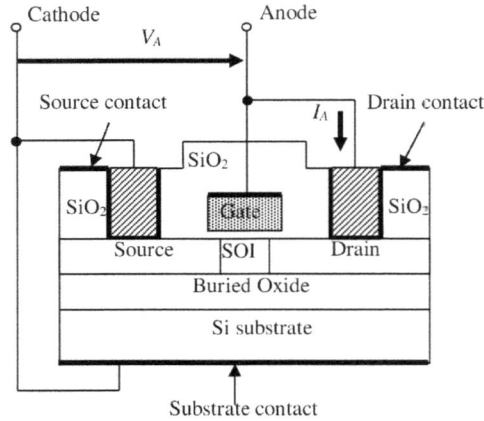

Fig. 2.4. Quasi-diode configuration by FD-SOI MOSFET [14].

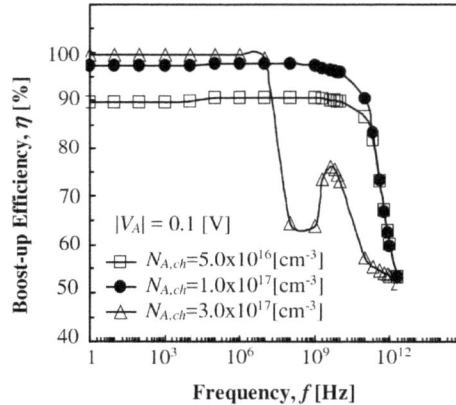

Fig. 2.5. Simulation results of boost-up efficiency of quasi-diode based on the FD-SOI MOSFET [14].

2.3.2. Cross-Current Tetrode SOI Devices

A particular interest has been to confirm if partially-depleted (PD) SOI devices can be utilized in low-power ICs [15]. Recently, the excellent potential of PD-SOI MOSFETs was demonstrated for low-energy applications [16]. Although PD-SOI MOSFETs themselves are not promising for future low-energy devices, its interesting aspects are introduced in this section.

Fig. 2.6 shows a three-dimensional view of the cross-current tetrode (XCT) SOI MOSFET. Fig. 2.7 shows simulation results of I_D-V_D characteristics of a 0.1-μm-gate XCT-SOI MOSFET calibrated by the experimental result of a 1-μm gate XCT-SOI MOSFET. Important aspects of the I_D-V_D characteristics of the 0.1-μm gate XCT-SOI

MOSFET are basically the same as those of the 1-μm gate device showing negative differential conductance in the drain current saturation state.

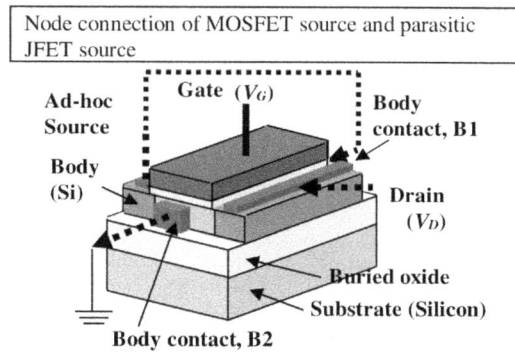

Fig. 2.6. Device structure of XCT-SOI MOSFET [15].

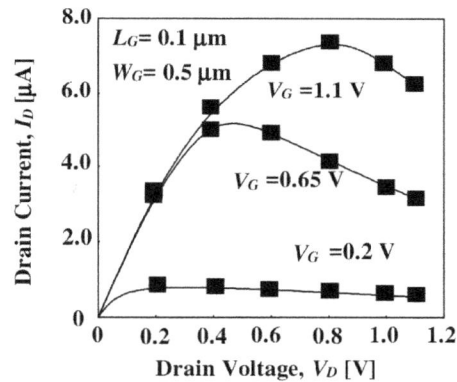

Fig. 2.7. Simulation results of I_D-V_D characteristics of 0.1-μm-gate XCT-SOI MOSFET [16].

SPICE simulations of EXOR-chain circuits were performed for the conventional PD-SOI CMOS and the XCT-SOI CMOS devices. The dissipated energy ratio per switching of both EXOR-chain circuits was calculated [17] (Fig. 2.8). It has been discovered that the energy dissipated by scaled XCT-SOI CMOS device decreases as the supply voltage (V_{DD}) rises; it is revealed that such low-energy operation of scaled XCT-SOI CMOS devices stems from the source potential floating effect (SPFE) [16, 17].

The XCT-SOI MOSFET was scaled down to 15 nm in order to investigate how scaling impacted the dissipated energy, and whether the switching delay time was improved. Simulation results of I_D-V_G characteristics are shown in Fig. 2.9 for two different junction profile conditions. Short-channel effects are well suppressed for the device with graded junctions [15, 16].

83

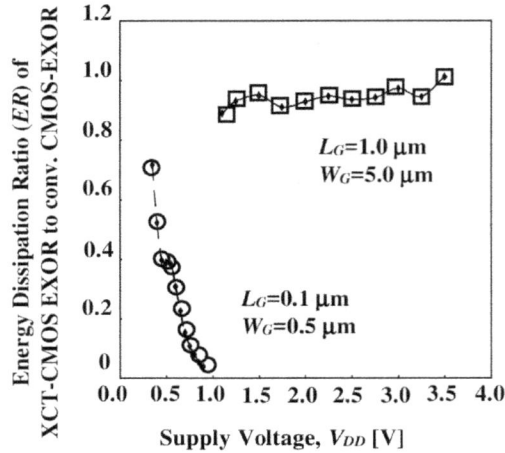

Fig. 2.8. Simulation results of Energy Dissipation Ratio per switching. XCT-SOI CMOS is compared with the conventional PD-SOI CMOS [17].

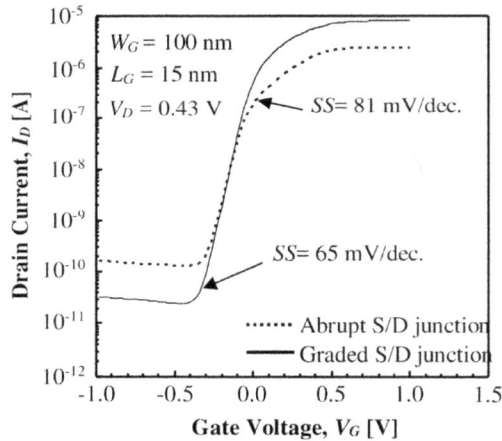

Fig. 2.9. Simulation results of I_D-V_G characteristics of 15-nm-gate XCT-SOI MOSFET [16].

The relation between the dissipated energy and the switching delay time of various devices is summarized in Fig. 2.10, where the switching energy of the XCT-SOI MOSFET is calculated without assuming SPFE. A reduction of about two-orders in the dissipated energy of XCT-SOI MOSFET is expected when SPFE is considered; "Expected zone" in Fig. 2.10 shows the lowest dissipated energy data predicted from the theoretical model.

Simulations demonstrated that scaled XCT-SOI CMOS devices should offer higher-frequency operation with much lower energy consumption than expected [3, 16]. It is expected that scaled XCT-SOI devices are very promising as very low energy ICs suitable as medical implants and intelligent sensor system chips.

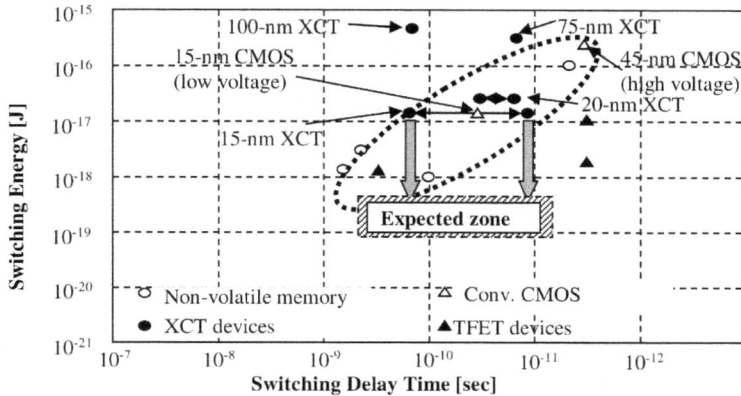

Fig. 2.10. Switching energy of various devices is compared as a function of switching delay time [16].

2.3.3. TBJ SOI MOSFET

A schematic cross-sectional view of the tunnel-barrier junction (TBJ) SOI MOSFET is shown in Fig. 2.11 [18, 19]. This device is similar to the single-electron transistor (SET), but it has a higher drivability than SET because the volume of the transistor body is much larger than that of SET. The drivability of the device is sensitive to the tunneling path length (thickness of insulating film) and the barrier height of insulating films covering source and drain electrodes because the resonant tunneling phenomenon is not utilized in this device's operation [19].

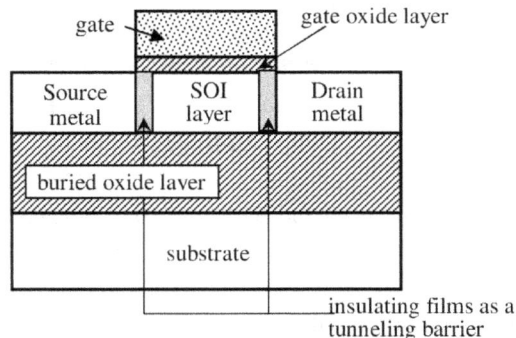

Fig. 2.11. Device structure of TBJ-SOI MOSFETS [19].

Typical simulation results of I_D-V_G and I_D-V_D characteristics of TBJ-SOI MOSFET are shown in Figs. 2.12 and 2.13. Subthreshold characteristics are similar to those of the conventional MOSFET, and the swing value is about 70 mV/dec at room temperature. I_D-V_D characteristics of the device are also similar to those of the conventional MOSFET, but its drivability is about 1/100 times that of the conventional MOSFET.

85

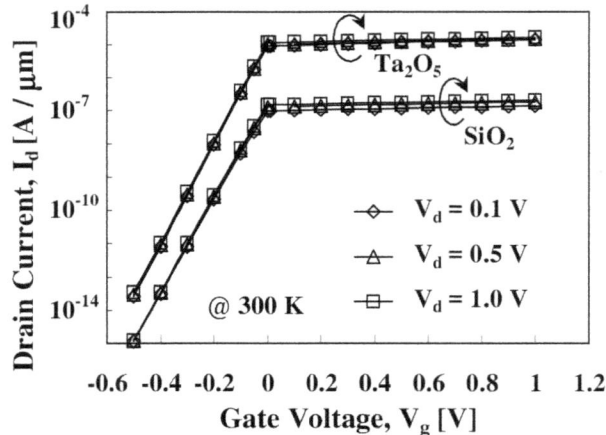

Fig. 2.12. Simulation results of I_D-V_G characteristics of 10-nm-channel TJB-SOI MOSFET with a 0.3 nm thick insulating film on the source and drain electrodes [19].

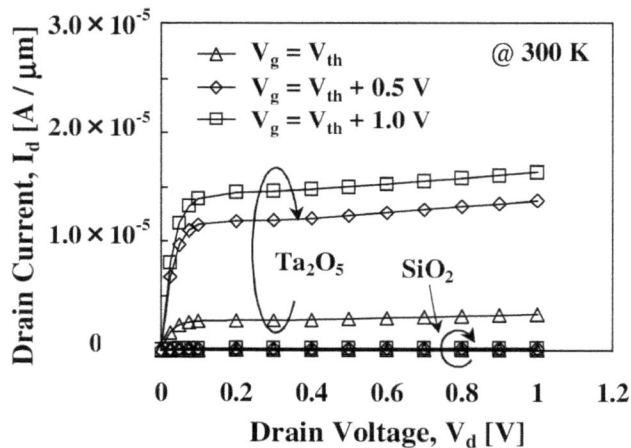

Fig. 2.13. Simulation results of I_D-V_D characteristics of 10-nm-channel TJB-SOI MOSFET with a 0.3 nm thick insulating film on the source and drain electrodes [19].

Calculated energy dissipation values of various devices are shown in Fig. 2.14 as a function of switching delay time. Simulations demonstrated that the scaled TBJ-SOI MOSFET should achieve higher-frequency operation with much lower energy use than anticipated [3, 19]. In addition, an IC chip having one hundred million TBJ-MOSFETs will work at 1 GHz with the power dissipation of $1 W/1$-cm^2 chip! This is not possible with conventional MOSFET devices.

The TBJ SOI MOSFET is also applicable to high-temperature operation tolerant devices by modifying the device structure [20]. This device will be also useful in low-energy device applications.

86

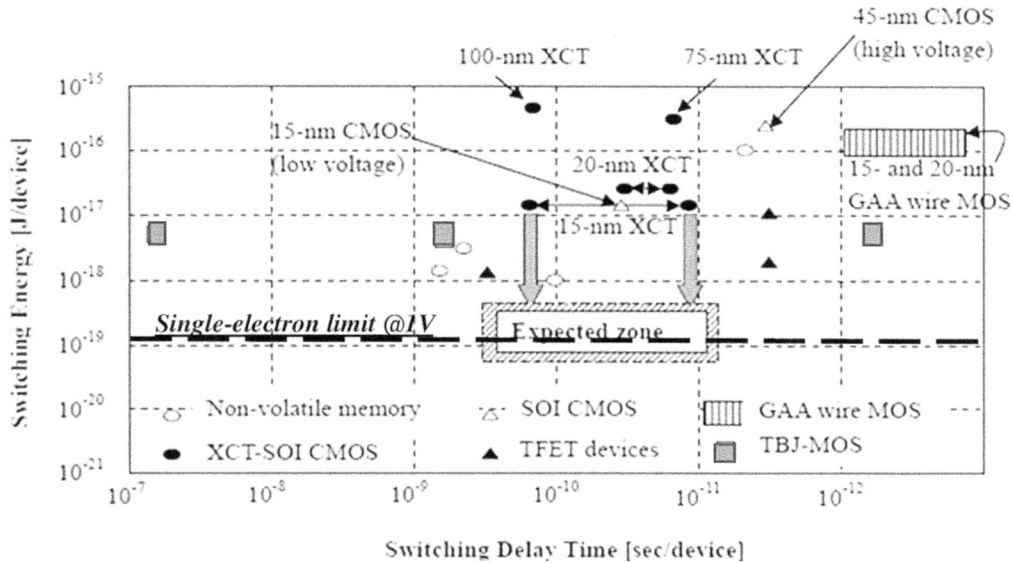

Fig. 2.14. Calculation results of switching energy as a function of a switching delay time for various devices [3]. Broken line reveals 'single-electron operation limit' at the supply voltage of 1 V.

2.3.4. Tunnel FET

Tunnel FETs are now attracting attention from the viewpoint of low-power and high-frequency device applications [21]. In the early stage of the device proposal, the lateral-field-induced tunnel FET (hereafter referred to as LTFET) was discussed experimentally and theoretically [22, 23]. This structure, however, failed to offer improvements in subthreshold swing and drivability [24, 25] because the lateral field causes carrier tunneling only near the MOS interface.

In contrast, it is anticipated that the vertical-field-induced tunnel FET (hereafter called VTFET), which employs carrier tunneling in line with the gate field, has a larger tunnel area along the MOS interface than the LTFET thereby improving its drivability considerably [26-28]. Hereafter we focus on the performance of VTFET only.

A possible VTFET structure is shown in Fig. 2.15. This device has a low bandgap Ge in source region to enhance the tunnel probability where electrons tunnel from the valence band of p-Ge to the conduction band of p-Ge near the HfO_2/p-Ge interface (Fig. 2.16). Simulated I_D-V_G characteristics, as shown in Fig. 2.17, demonstrate that a very steep subthreshold swing and a large on-current are achieved. Steep subthreshold is achieved by using: (i) a step in the Ge source region that eliminates lateral carrier tunneling, (ii) a small gate-channel overlap that decreases fringing-induced barrier lowering (FIBL), and (iii) a step in the Si channel region that reduces spreading of fringing field in the tunneling region [28]. Reduction in both FIBL and spreading of fringing electric field also results in improved on-current [28].

87

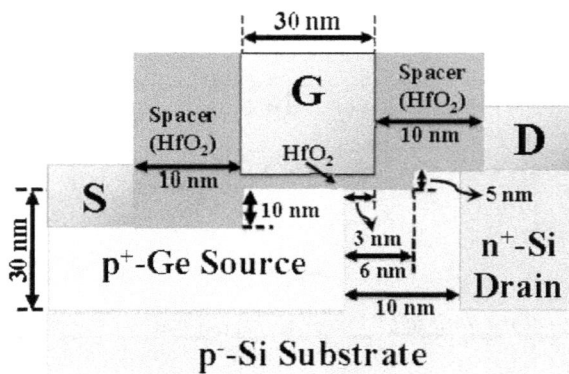

Fig. 2.15. A schematic device structure of vertical TFET. Here "GoGeS" structure is shown [28].

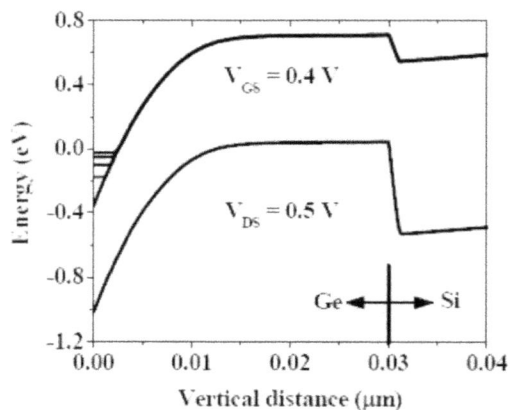

Fig. 2.16. Tunnel image of GoGeS VTFET [28].

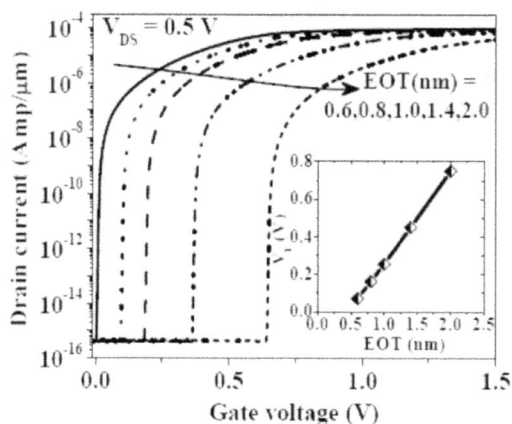

Fig. 2.17. Simulation results of I_D-V_G characteristics of GoGeS VTFET [28].

Another attractive Si VTFET architecture uses a vertically-stacked pn junction with a thin intrinsic channel under the gate region, as shown in Fig. 2.18 [29]. For the device in Fig. 2.16, the channel current is controlled by tunneling from the three dimensional valence band to the two dimensional surface subband. This tunnel process may not yield a large tunnel current [30]. On the other hand, three dimensional tunneling takes place in the device shown in Fig. 2.18. It is expected that the tunnel current for such devices would be larger than that of the conventional lateral double-gate TFET [31-33].

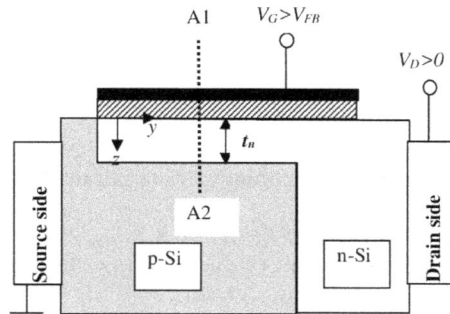

Fig. 2.18. A schematic device structure of vertical TFET with an n-Si/p-Si stack structure [29].

As VTFETs can be fabricated on SOI substrates and GOI substrates, both GoGeS VTFET and Si VTFET appear attractive for ICs comprising complementary VTFET circuits. Circuit performance will be better than that is possible with bulk TFET circuits because parasitic capacitances are reduced. Since a very thin semiconductor layer is not needed, fabrication cost is also reduced. Therefore complementary VTFET circuits on AOI substrates are highly attractive for future low-energy RF applications.

2.5. Conclusion

This chapter introduced the potential and prospects of various SOI devices including TFETs. Such advanced SOI devices should contribute to the advent of low-energy circuits, not only medical implant devices, but also sensor networks, mobile PC, and others. However, more promising SOI devices should be developed to support future low-energy device applications.

Acknowledgements

A part of this study was partially conducted by MEXT-Supported Program for the Strategic Research Foundation at Private Universities, 'Creation of 3D nano-micro structures and its applications to biomimetics and medicine', 2015-2019.

Publication of this study was financially supported by Kansai Research Foundation for Technology Promotion, 2017.

The author (Y. Omura) expresses his thanks to Mr. K. Yoshimoto (Presently, Sharp Corp., Japan), Mr. T. Tamura (Presently, Sharp Corp., Japan), Mr. H. Nakajima (Presently, Canon Corp., Japan), Mr. D. Sato (Presently, Panasonic Corp., Japan), Mr. O. Hayashi (Presently, Roam Corp. Japan), and Mr. D. Ino (Presently, Alps Tech. Corp., Japan) to their technical support when they were with Grad. School of Sci. and Eng., Kansai University.

References

[1]. J.-P. Colinge, Silicon-on-Insulator Technology: Materials to VLSI, Chapter 2, 3rd ed., *Kluwer Academic Pub.*, Massachusetts, 2004.

[2]. Y. Omura, SOI Lubistors, Chapter 1, *IEEE/Wiley & Sons,* 2013.

[3]. Y. Omura, A. Mallik, N. Matsuo, MOS Devices for Low-Energy Applications, Chapters 1, *IEEE/Wiley & Sons*, Singapore, 2016, pp. 37- 40.

[4]. S. Hanson, B. Zhai, K. Bernstein, D. Blaauw, A. Bryant, L. Chang, K. K. Das, W. Haensch, E. J. Nowak, D. M. Sylvester, Ultralow-voltage minimum-energy CMOS, *IBM J. Res. and Dev.,* Vol. 50, 2006, pp. 469-490.

[5]. A. P. Chandrakasan, D.C. Daly, D. F. Finchelstein, J. Kwong, Y. K. Ramadass, M. E. Sinangil, V. Sze, N. Verma, Technologies for ultradynamic voltage scaling, *Proc. of the IEEE*, Vol. 98, 2010, pp. 191-214.

[6]. L. Chang. D. J. Frank, R. K. Montoye, S. J. Koester, B. L. Ji, P. W. Coteus, R. H. Dennard, W. Haensch, Practical strategies for power-efficient computing technologies, *Proc. of the IEEE*, Vol. 98, 2010, pp. 215-236.

[7]. S. A. Vitale, P. W. Wyatt, N. Checka, J. Kedzierski, C. L. Keast, FDSOI process technology for subthreshold-operation ultralow-power electronics, *Proc. of the IEEE*, Vol. 98, 2010, pp. 333-342.

[8]. H. M. Manasevit, W. I. Simpson, Single-crystal silicon on a sapphire substrate, *J. Appl. Phys.*, Vol. 35, 1964, pp. 1349-1351.

[9]. M. Yokoyama, T. Yasuda, H. Takagi, H. Yamada, N. Fukuhara, M. Hata, M. Sugiyama, Y. Nakano, M. Takenaka, S. Takagi, Thin body III-V-semiconductor-on-insulator MOSFETs on Si fabricated using direct wafer bonding, *Appl. Phys, Exp.*, Vol. 2, 124501, 2009.

[10]. O. Rozeau, J. Jomaah, J. Boussey, Y. Omura, Comparison between high-dose and low-dose separation by implanted oxygen MOS transistors for low-power radio-frequency applications, *Jpn. J. Appl. Phys.*, Vol. 39, No. 4B, 2000, pp.2264 -2267.

[11]. 3D Integration for VLSI Systems, Chapter 8, Ed. by C.-S. Tan, K.-N. Chen, S. J. Koester, *Pan Stanford Pub.*, 2012.

[12]. Y. Omura, S. Nakashima, K. Izumi, T. Ishii, 0.1-mu m-Gate, ultrathin-film CMOS devices using SIMOX substrate with 80-nm-thick buried oxide layer, in *Proceeding of the IEEE Int. Electron Devices Meeting (IEDM'91)*, Tech. Dig., 1991, pp. 675-678.

[13]. Y. Omura, S. Nakashima, K. Izumi, T. Ishii, 0.1-mu m-Gate, ultrathin-film CMOS devices using SIMOX substrate with 80-nm-thick buried oxide layer, *IEEE Trans. on Electron Devices*, Vol. 40, No. 5, 1993, pp. 1019-1022.

[14]. Y. Omura, Y. Iida, Performance prospects of fully-depleted SOI MOSFET-based diodes applied to Schenkel circuit for RF-ID chips, *Sci. Res., J. Cir. and Syst.*, Vol. 4, No. 2, 2013, pp. 173-180.

[15]. Y. Azuma, Y. Yoshioka, Y. Omura, Cross-current SOI MOSFET model and important aspects of CMOS operations, *Ext. Abstr. Int. Conf. Solid State Devices and Mat.*, Tsukuba, Sept. 2007, pp. 460-461.

[16]. Y. Omura, D. Sato, Mechanisms of low-energy operation of XCT-SOI CMOS devices - prospect of sub-20-nm regime, *J. Low-Power Electron. and Appl.*, Vol. 4, 2014, pp. 14-25.

[17]. D. Ino, Y. Omura, D. Sato, Impact of dynamic body floating effect on low-energy operation of XCT-SOI CMOS devices with aim of sub-20-nm regime, *ECS Trans.*, Vol. 53, No. 5, 2013, pp. 75-84.

[18]. Y. Omura, A tunneling-barrier junction MOSFET on SOI substrates with a suppressed short-channel effect for the ultimate device structure, in *Proceedings of the 10th Int. Symp. on Silicon-on-Insulator Technology and Devices, The Electrochem. Soc.*, Washington D. C., 2001, PV2001-3, pp. 451-456.

[19]. H. Nakajima, A. Kawamura, K. Komiya, Y. Omura, Simulation models for silicon-on-insulator tunneling-barrier-junction metal-oxide-semiconductor field-effect transistor and performance perspective, *Jpn. J. Appl. Phys.*, Vol. 42, 2003, pp. 1206-1211.

[20]. Y. Omura, Proposal of high-temperature-operation tolerant (HTOT) SOI MOSFET and preliminary study on device performance evaluation, *J. Active and Passive Electronic Components Field-Effect Transistor, Hindawi Pub.*, Vol. 2011, 2011., pp. 1-8

[21]. A. C. Seabaugh, Z. Qin, Low-voltage tunnel transistors for beyond CMOS logic, *Proc. of the IEEE*, Vol. 98, 2010, pp. 2095-2110.

[22]. C. Sandow, J. Knoch, C. Urban, Q.-T. Zhao, S. Mantl, Impact of electrostatics and doping concentration on the performance of silicon tunnel field effect transistors, *Solid-State Electron.*, Vol. 53, 2009, pp. 1126-1129.

[23]. J. L. Padilla, F. Gámiz, A. Godoy, A simple approach to quantum confinement in tunneling field-effect transistors, *IEEE Electron Device Lett.*, Vol. 33, 2012, pp. 1342-1344.

[24]. R. Jhaveri, V. Nagavarapu, J. C. Woo, Effect of pocket doping and annealing schemes on the source-pocket tunnel field-effect transistor, *IEEE Trans. Electron Devices*, Vol. 58, 2011, pp. 80-86.

[25]. A. Revelant, A. Villalon, Y. Wu, A. Zaslavsky, C. L. Royer, H. Iwai, S. Cristoloveanu, Electron-hole bilayer TFET: experiments and comments, *IEEE Trans. Electron Devices*, Vol. 61, 2014, pp. 2674-2681.

[26]. A. Mallik, A. Chattopadhyay, Drain-dependence of tunnel field-effect transistor characteristics, *IEEE Trans. Electron Devices*, Vol. 58, 2011, pp. 4250-4257.

[27]. A. Mallik, A. Chattopadhyay, S. Guin, A. Karmakar, Impact of a spacer-drain overlap on the characteristics of a silicon tunnel field-effect transistor based on vertical tunneling, *IEEE Trans. Electron Devices*, Vol. 60, 2013, pp. 935-942.

[28]. A. Mallik, Avik Chattopadhyay, Y. Omura, A gate-on-germanium source (GoGeS) tunnel field-effect transistor enabling Sub-0.5-V operation, *Jpn. J. Appl. Phys.*, Vol. 53, 2014, pp. 104201-104208.

[29]. Y. Omura, Physics-based analytical model for gate-on-germanium source (GoGeS) TFET, *IEICE Technical Report*, SDM2014-97, 2014, pp. 7-12.

[30]. Y. Omura SOI Lubistors, 1st ed., *IEEE/Wiley & Sons*, 2013, Chapter. 11.

[31]. Y. Omura, A. Mallik, N. Matsuo, MOS Devices for Low-Energy Applications, Chapter 7, *IEEE/Wiley & Sons*, Singapore, 2016.

[32]. Sapan Agarwal, Eli Yablonovitch, Using dimensionality to achieve a sharp tunneling FET (TFET) turn-on, *Abstr. IEEE DRC*, 2011, pp. 199-200.

[33]. Yuan Taur, Jianzhi Wu, Jie Min, Dimensionality dependence of TFET performance down to 0.1 V supply voltage, *IEEE Trans. Electron Devices*, Vol. 63, 2016, pp. 877-880.

Chapter 3

Fabrication and Characterization of High Strength Electrodeposited Gold Toward High-Sensitive MEMS Inertial Sensors

Tso-Fu Mark Chang, Chun-Yi Chen, Hao-Chun Tang, Masaharu Yoshiba, Takashi Nagoshi, Daisuke Yamane, Katsuyuki Machida, Kazuya Masu and Masato Sone

3.1. Introduction

Gold has high chemical stability, corrosion resistance, electrical conductivity and density. These properties make gold promising materials to be used as structure materials in micro-electro-mechanical systems (MEMS) devices. Yamane et al. reported that the sensitivity of MEMS inertial sensors can be improved by replacing the conventional silicon-based components to gold-based materials [1-3], shown in Fig. 3.1. The high sensitivity is mostly contributed by the high density of gold (19.30 g/cm^3 at 298 K), which is much higher than that of silicon (2.33 g/cm^3 at 298 K).

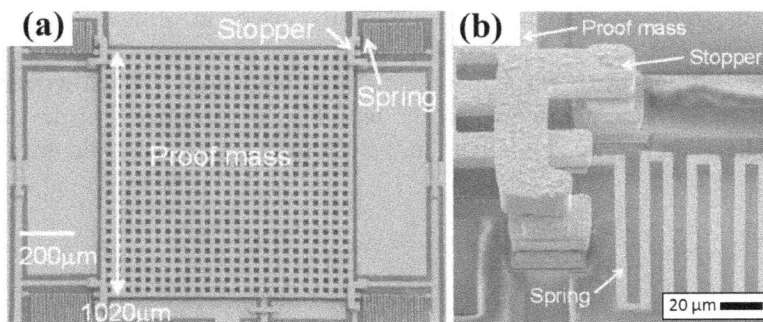

Fig. 3.1. SEM micrographs of a MEMS inertial sensor: (a) Chip view, and (b) close-up image.

Tso-Fu Mark Chang
Institute of Innovative Research, Tokyo Institute of Technology, Yokohama, 226-8503, Japan

However, mechanical strengths of gold are relatively low (i.e., yield strength of bulk gold: 50-200 MPa [4]) when compared with other metallic materials, which has been a concern in practical applications as the movable components in MEMS. Grain boundary strengthening mechanism is a typical method available to enhance the mechanical strength of metallic materials based on the Hall-Petch relationship [5]. As the grain size reaches submicron- or nano-scale, high strength can be obtained when compared to coarse-grained metals with the same chemical composition and phase constitution [6, 7].

Electrodeposition is often applied in fabrication of the electronic components because of the near-room-temperature operating environment, low-energy requirements, controllable deposition rates, low cost and simple scale-up with easily maintained equipment [8]. Fabrication of the MEMS components utilizing gold electrodeposition as a post-CMOS process enables integration of the CMOS and MEMS and demonstrates the potential for further reduction in size of the device and parasitic elements. Most importantly, grain size of the electrodeposited metal can be easily controlled by the electrodeposition parameters [9-13], which allows direct and simple realization of the grain boundary strengthening in electrodeposited metals.

On the other hand, the components used in MEMS devices are usually in micro-scale, and mechanical properties of metallic materials in micro-scale are different from those of the bulk materials. When size of the materials is reduced to micro-scale, an increase in the stresses is observed with a continuous reduction in size of the materials. This phenomenon is called size effect [4, 14]. The size effect indicates differences in the mechanism of deformation when comparing with the bulk materials. Thus, clarifying the micro-mechanical properties, such as yield strength, of the electrodeposited gold materials is important for the applications in MEMS.

3.2. Constant Current Electrodeposition

Cyanide-based gold electrolytes are commonly used in industrial level of gold electrodeposition, because stability of the gold ions in aqueous solution is always a concern, and stability of $Au(CN)_2^-$ in aqueous solution is high [15]. However, toxicity of cyanide solution is high, therefore, sulfite-based gold electrolyte is developed as an alternative. Chemical reactions involved in the electrodeposition of gold are shown in the following:

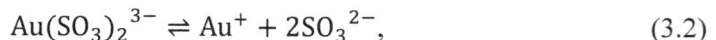

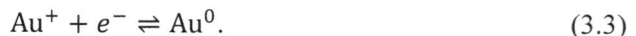

$$Au(CN)_2^- \rightleftharpoons Au^+ + 2CN^-, \tag{3.1}$$

$$Au(SO_3)_2^{3-} \rightleftharpoons Au^+ + 2SO_3^{2-}, \tag{3.2}$$

$$Au^+ + e^- \rightleftharpoons Au^0. \tag{3.3}$$

In this section, micro-mechanical properties of gold specimens prepared by constant current electrodeposition using cyanide-based and sulfite-based gold electrolytes are reported. The gold films were electrodeposited on Cu substrates. The cyanide-based electrolyte was purchased from Matex Japan Co. Ltd. (Matex Gold BOG-10), which contains 230 g/L of di-ammonium hydrogen citrate, 85 g/L of KCN, and 14.63 g/L of

K[Au(CN)$_2$] with pH of 5.0. Gold films electrodeposited with the cyanide-based electrolyte would be named as cyanide-gold in this chapter. The sulfite-based electrolyte was also provided by Matex Japan Co. Ltd. (Matex Gold NCA), which contains 50 g/L of Na$_2$SO$_3$, 50 of g/L (NH$_4$)$_2$SO$_3$, and 21.63 g/L of Na$_3$[Au(SO$_3$)$_2$] with pH of 8.0 and 5 % sodium gluconate. Gold films electrodeposited with the sulfite-based electrolyte would be named as sulfite-gold in this chapter. The current density used for the cyanide-based electrolyte was 4 mA/cm^2, and the temperature was 60 °C. The current density used for the sulfite-based electrolyte was 5 mA/cm^2, and the temperature was 50 °C.

For the micro-compression test, the micro-pillars were fabricated by focused ion beam (FIB, Hitachi: FB2100). Dimensions of the pillars were 10×10×20 μm^3 with a square cross-section. The pillar fabrication steps are illustrated in Fig. 3.2. At first, the specimen was polished to expose the interface between the electrodeposited gold and the Cu substrate as shown in Fig. 3.2(a). In order to minimize the tapering effect often observed in the FIB milling process, the ion-beam was irradiated from the side of the pillar at a tilt angle of 45±2° rather than at the commonly used orthogonal beam direction as shown in Figs. 3.2(b) and (c).

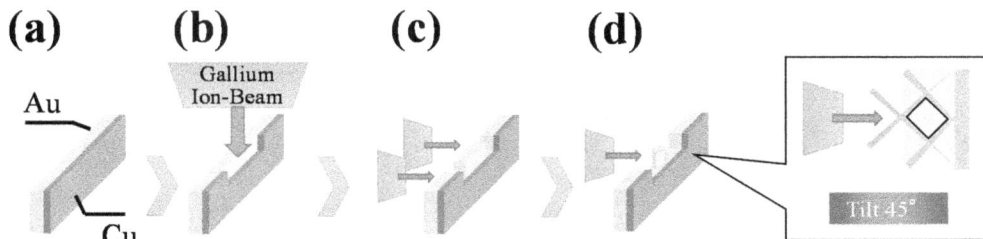

Fig. 3.2. Schematic images showing the micro-pillar fabrication steps. (a) Polished electrodeposited gold; (b) Fabrication of the pillar from edge of the specimen; (c) Size reduction of the pillar; and (d) Finishing with a low intensity beam at a tilt angle of ±2°.

The compression test was conducted using a test machine specially designed for micro-sized specimens equipped with a flat-ended diamond indenter at a constant displacement of 0.1 μm/s controlled by a piezo-electric actuator as shown in Fig. 3.3. The load resolution was 10 μN. An image showing the micro-pillar and the micro-indentor during the micro-compression test is shown in Fig. 3.4. Observation of the specimens was conducted using a scanning ion microscope (SIM) equipped in the FIB. The crystal structure was evaluated by X-ray diffraction (XRD, Ultima IV, Rigaku).

The electrodeposited gold films were confirmed to be pure gold from the XRD patterns shown in Fig. 3.5. Microstructure and deformation behavior of the pillars before and after the compression test were observed from the SIM images shown in Fig. 3.6. The ion channeling contrast shown in the SIM images indicates orientation changes of the grains on the surface. White and gray irregular strip-patterns having dimensions in several μm order were observed on both of the top and side-wall surfaces on the cyanide-gold pillar, and these patterns are attributed to the columnar textures with a direction perpendicular to the substrate/electrolyte interface. Cracks shown in Fig. 3.6 (b) indicate brittle fractures along the texture boundary, which suggest the brittle nature of the texture boundary. The grain size estimated by the XRD and the Scherrer equation was 17.6 nm, which is far smaller than size of the patterns observed in the SIM image. The Scherrer equation is shown in the following:

$$g = \frac{\lambda}{\beta \cos \theta}, \qquad (3.4)$$

where g, λ, β, and θ are the average grain size, X-ray wavelength (0.15418 nm), full width at half maximum in radians, and Bragg angle, respectively. While the contrast in the SIM image represents orientation against the beam direction, the boundaries observed in Figs. 3.6 (a) and (b) should be texture boundaries instead of grain boundaries. Camouflage patterns of the grain structure were observed on surface of the sulfite-gold pillar, as shown in the SIM image in Fig. 3.6 (c). The pillar showed broad shear crossing from the top front to the bottom after ca. 10 % of deformation, as shown in Fig. 3.6 (d), which is similar to the deformation of electrodeposited nanocrystalline Ni [16]. Grain size of the sulfite-gold was estimated to be 22.8 nm.

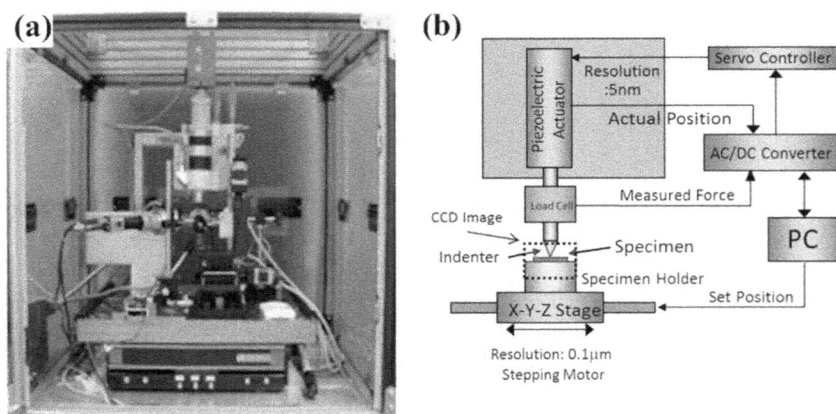

Fig. 3.3. (a) Photo, and (b) schematics of the micro-mechanical testing machine.

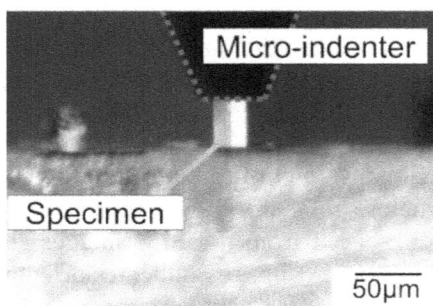

Fig. 3.4. A photo taken during the micro-compression test.

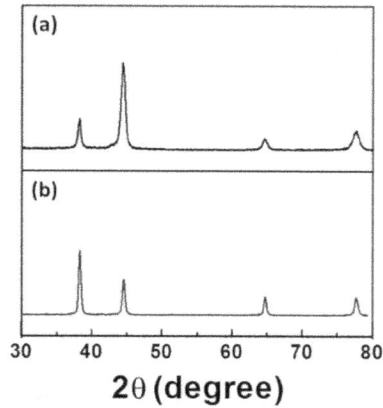

Fig. 3.5. XRD patterns of the gold film electrodeposited by (a) cyanide-based and (b) sulfite-based electrolyte.

Fig. 3.6. SIM images of the micro-pillar fabricated from the cynide-gold (a) before and (b) after the micro-compression test, and the micro-pillar fabricated from the sulfite-gold (c) before, and (d) after the test.

Fig. 3.7 shows compressive engineering stress-strain (SS) curves of the gold micro-pillars. Deformation behavior of the cyanide-gold pillar was different from the sulfite-gold pillar Yield drops observed in the cyanide-gold pillar may correspond to the cracks observed in Fig. 3.6(b). The cracks are suggested to be caused by the impurities derived from the bath containing cyanide ions, which lead to a decrease in the interfacial strength between the texture boundaries. The sulfite-gold pillar showed barrel-shape deformation, which is typically observed in polycrystalline specimens. Yield strengthes of the pillars were determined from the engineering stress-strain curves. By assuming volume conservation during the plastic deformation, the engineering stress σ is defined in the following equation:

$$\sigma = \frac{P}{S},$$ (3.5)

where P is the measured force, S is the cross-section area. The engineering strain ε can be calculated as shown the following equation:

$$\varepsilon = \frac{\Delta L}{L_0},$$ (3.6)

where ΔL is the displacement, L_0 is initial length of the pillar. Both of the P and the ΔL were acquired by the test machine, and S and L_0 were both obtained from the micrograph observations. The exact yield point is usually unclear in SS curves of micro-mechanical testing. Therefore, 0.2 % yield strength, which is the cross-point of the SS curve and 0.2 % offset line of the elastic deformation region, is determined to represent the yield strength. 0.2 % yield strength of the cyanide- and sulfite-gold pillar reached 650 and 600 MPa [13], respectively. The yield strength of the cyanide-gold was higher, which should be a result of the difference in the average grain size. Average grain size of the cyanide-gold is smaller than that of the sulfite-gold. On the other hand, both values were higher than the yield strength of bulk gold, which is 50-200 MPa. The strengthening is suggested to be results of sample size effect and grain boundary strengthening.

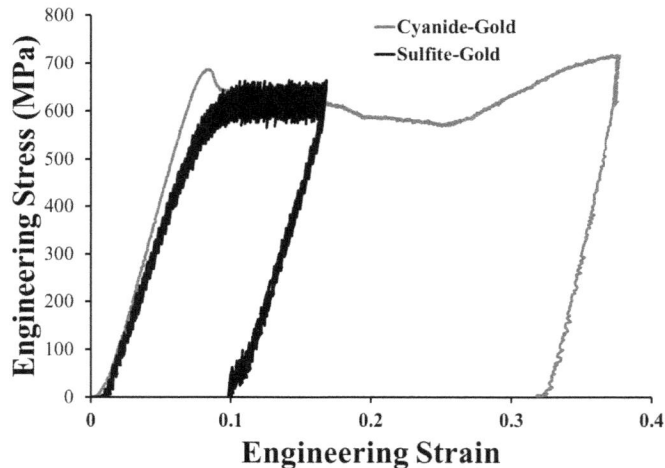

Fig. 3.7. Engineering stress-strain curves obtained from the micro-compression test of the cyanide- and sulfite-gold micro-pillars.

3.3. Electrodeposition with Supercritical CO_2

Grain size control is a fascinating feature in electrodeposition of metals by appropriately adjusting the deposition parameters such as the current density, pH value, plating bath composition, temperature, etc. [8]. For example, an increase in the current density can lead to a decrease in the grain size, and therefore, enhances the mechanical strength. However, impurities could be introduced into the electrodeposit because of side-reactions, and high hydrogen evolution caused by the high current density could also introduce defects into the electrodeposit. On the other hand, an alternative electrodeposition method employing a constant current density and a supercritical CO_2 (scCO_2) contained electrolyte was reported to be effective to give the grain refinement effect as demonstrated in electroplating of Ni and Cu [10,18]. Moreover, the specific properties of scCO_2 can promote transfer of materials into confined spaces and removal of H_2 gas bubbles away from the cathode, which allow application of high current density while keeping a defect- and pinhole-free plating at the same time [19-21]. In addition, co-deposition of carbon was reported in electroplating with an electrolyte containing scCO_2 at high pressure (EP-SCE) [22, 23]. The co-deposited carbon is effective to cause the grain refinement effect. Also, the carbon would precipitate at the grain boundary to suppress occurrence of the inverse Hall-Petch Relationship [22, 24] and reduce the grain boundary mobility [25], which is very effective in stabilizing the nanograin structures to maintain the high mechanical strength. With these advantages, an enhancement in mechanical strength of electrodeposited gold films is expected by introducing scCO_2 into the gold electrodeposition process to improve long-term reliability of the gold-based components used in MEMS devices.

In this section, micro-mechanical properties of gold films prepared by the EP-SCE, named as EP-SCE gold in the chapter, evaluated by micro-compression tests would be presented. For the EP-SCE, 20 vol.% of CO_2 with respect to the overall volume of the reaction chamber was used. The pressure was controlled at 10 MPa during the entire EP-SCE process. The electrolyte used was the commercially available sulfite-based electrolyte introduced in previous section. The temperature was fixed at 40 °C. The current density was 5 mA/cm^2. Dimensions of the micro-pillars were $10\times10\times20$ μm^3 with a square cross-section.

Fig. 3.8 shows SIM images of the $10\times10\times20$ μm^3 micro-pillar fabricated from the EP-SCE gold. The inconspicuous grain boundaries observed on the surface implied the grain size was finer than 1 μm [26]. After the micro-compression test, barrel-shape deformation was observed in the middle part of the pillar. The deformation following the slips of grain boundaries or formation of barrel-shape is well recognized as the typical characteristics of the deformation of poly- and nano-crystalline structures [27]. Average grain size of the EP-SCE gold was estimated to be 13 nm.

0.2 % yield strength of the EP-SCE pillar was 520 MPa and the compressive flow stress reached ~800 MPa as shown in Fig. 3.9. Again, the yield strength was much higher than that of the bulk gold. The significant enhancement of the mechanical properties in the EP-

SCE is attributed to the Hall-Petch relationship [5], that is, strength of polycrystalline metallic materials is enhanced following a decrease in the grain size.

Fig. 3.8. SIM images of the $10{\times}10{\times}20$ μm^3 micro-pillar fabricated from the EP-SCE gold (a) before, and (b) after the compression test.

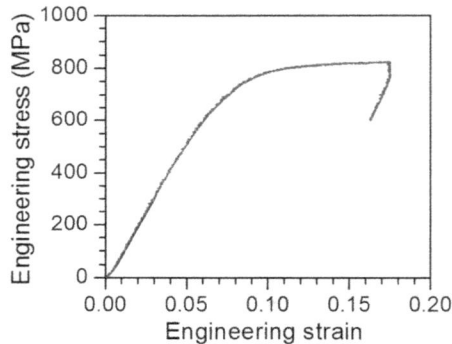

Fig. 3.9. Engineering SS curves of the micro-pillar fabricated from the EP-SCE gold.

3.4. Pulse Current Electrodeposition

Pulse current electrodeposition (PE) has been reported to be effective in fabricating Au materials with fine grains, high uniformity, and low porosity [28, 29]. Also, it is possible to control the composition and the film thickness by regulating the pulse amplitude and width. Most importantly, an increase in the nuclei density could be achieved to obtain electrodeposited films with finer grains. In this section, the pulse on-time (T_{on}) and the off-time (T_{off}) intervals were varied to clarify the effects on the crystal structure and the mechanical properties of the gold films, which are important for design of MEMS functional components.

The gold electrolyte used was a commercially the sulfite-based electrolyte provided by Matex Japan. Detail chemical composition of the electrolyte is described in the previous section. Surface area of the working electrode having contact with the electrolyte was 3.4 cm^2. The on-time current (I_{on}) and the off-time current (I_{off}) were fixed at 34 mA and 0 mA, respectively. The T_{on} and T_{off} were varied from 1 to 10 ms to study the effect on the microstructure. Detail information is given in Fig. 3.10.

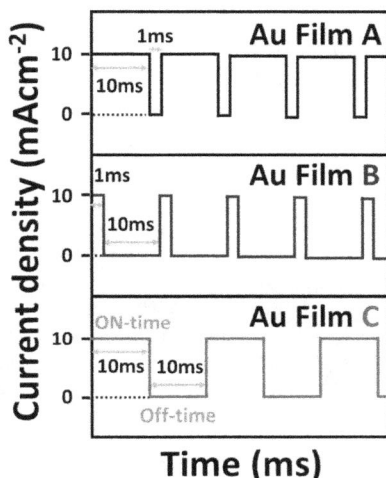

Fig. 3.10. Parameters of pulse-current waveform used in the PE-fabricated Au film A, B, and C.

SIM images of the three micro-pillars fabricated from the three PE gold films are shown in Fig. 3.11. Fig. 3.11 (d), (e), and (f) show side view of the pillar A, B, and C, respectively. The contrast observed in the SIM images represents the grain boundaries. Two types of grain structures were observed between Figs. 3.11(d) and (f). For the pillar A (Figs. 11 (a) and (d)), most of the grains were in micrometer order. Only a few grains having size less than 1 micrometer were observed. Grain structure of the pillar A was considered to be composed of non-homogeneous mixture of micro-grains and nano-grains. For the second type shown in the pillar C (Figs. 3.11(c) and (f)), there were no obvious grain/texture boundaries observed from the SIM image. The gold film was considered to be composed of homogeneous nano-grains, which no obvious boundary is observed because the grain size is too small and beyond the detection limit of the SIM. The pillar B (Figs. 3.11 (b) and (e)) showed a mixture of the two grain structure types. The nano-grains similar to those in the pillar C were located in the front corner as shown in Fig. 3.11 (b).

It was concluded that an interval of 10 ms for both the T_{on} and T_{off} is the optimized condition to achieve uniform nano-grains. During the T_{off}, the depleted surface concentration of gold ions on the cathode can be replenished by the gold ions diffuse from the bulk electrolyte, and the as-formed by-products by the side-reactions can diffuse away from the cathode surface. Then new nuclei are more likely to form because of the higher surface concentration of gold ions during the T_{on}. For the case of the pillar A, the T_{off} of 1 ms is too short, and it is insufficient for the gold ion replenishment. A low surface gold ion concentration leads to a large grain growth. Therefore, the plating condition is similar to the constant current plating, which leads to large grain structures. When the T_{on} is short enough, the gold ion consumed in each T_{on} period could be replenished during the T_{off} period. However, T_{on} should be long enough to allow fully charge of the electrical double layer. Otherwise, the reduction reaction cannot be initiated properly. On the other hand, when the T_{on} is too long, the plating condition would be close to the constant current plating conditions and lead to large grains and occurrence of side reactions.

Fig. 3.11. SIM images of the as-fabricated (a) pillar A, (b) B, and (c) C and side views of the as-fabricated (d) pillar A, (e) B, (f) C. In the area surrounded by red line, grains having size in several hundreds of nm and heterogeneous pattern were observed. In the area surrounded by blue line, no obvious grain/texture boundary was observed.

Scanning electron microscope (SEM) images of the pillar A, B, and C before and after the compression test are shown in Fig. 3.12. All the pillars showed broad shear crossing from the top to the bottom of the pillars, and protrusions on the pillar side faces were observed after the deformation. The formation of the protrusions was attributed to the grain boundary slip occurred during the compression test, because the pillars were all composed of poly-crystals.

The yield strength measured as a plastic strain of 0.2 % for the pillar A, B, and C are 400, 433, and 673 MPa, respectively. The differences in the yield strength are due to structural differences, i.e. grain size. As the grain size becomes smaller, the strength becomes higher due to the grain boundary strengthening mechanism well expressed as the Hall-Petch effect [5]. Relationship between the strength and the grain size is shown in the following equation:

$$\sigma = \sigma_0 + k_y d^{-1/2}, \tag{3.7}$$

where σ is the yield strength, σ_0 is a materials constant for the starting stress for dislocation movement (or the resistance of the lattice to dislocation motion), k_y is the strengthening coefficient (a constant specific to each material), and d is the average grain diameter.

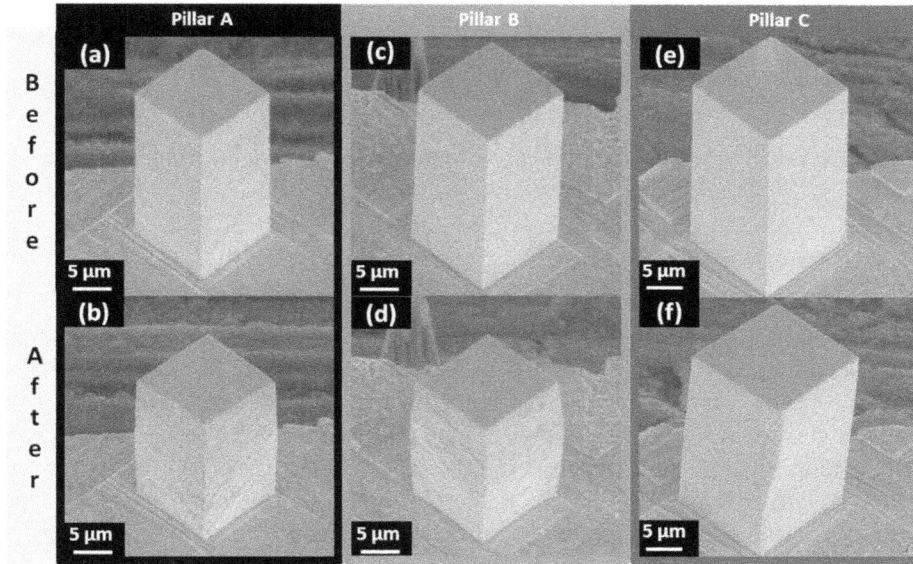

Fig. 3.12. SEM images of the (a) as-fabricated and (b) deformed pillar A; (c) as-fabricated and (d) deformed pillar B, and (e) as-fabricated, and (f) deformed pillar C.

Although the yield strength was only slightly higher for the pillar B when compared with the pillar A, the enhanced work hardening demonstrates the effect of the partially nano-grained region in the pillar B. In addition, the SEM images shown in Fig. 3.12(d) revealed smooth deformation morphology at the front corner of the pillar B, which indicates a resistance due to the deformation of the nano-grains. The fully nano-grained pillar C had a yield strength of 673 MPa as shown in Fig. 3.13. XRD diffraction patterns and the Scherrer equation were used to estimate average grain size of the film C. The estimated grain size was 10.4 nm, and it agrees well with the outstandingly high strength observed in this work. In summary, it can be deduced that the T_{on} of 10 ms and T_{off} of 10 ms are the optimum parameters of the PE, because average grain size of the film C is the finest, and the compressive yield strength of 673 MPa obtained is much higher than the other pillars. These results demonstrated that the PE is promising for MEMS applications since it demonstrated a simple method to control the grain size and the mechanical strength of the gold micro-components by regulating the T_{on} and T_{off}s.

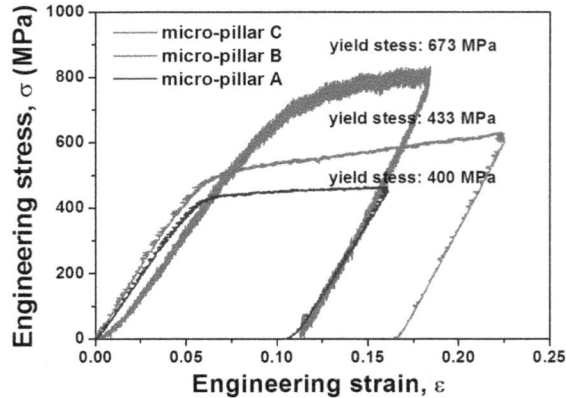

Fig. 3.13. Engineering stress-strain curves obtained from the micro-compression tests of the PE gold micro-pillars.

3.5. Conclusions

Micro-mechanical properties of pure gold electrodeposited by various electrodeposition methods for applications as movable structures in MEMS inertial sensors are presented in this chapter. A summary of the yield strengthes obtained from this work and form the literatures is shown in Fig. 3.14. The strengths obtained in this work well followed the Hall-Petch relationships and reached 673 MPa. To the best of our knowledge a compressive yield strength of 673 MPa is the highest value reported for electrodeposited pure gold. Also, the gold film prepared by the PE showed large work hardening which indicates high ductility and malleability of the material. In conclusion, the results provide essential information for development of MEMS devices applying gold-based materials as the component.

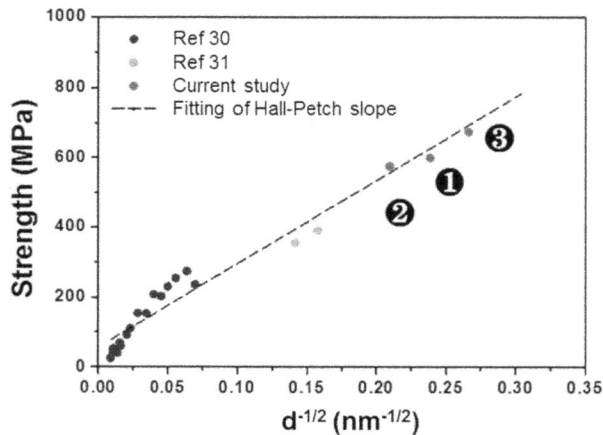

Fig. 3.14. Hall-Petch plot of gold pillars, where ❶ is the cyanide-gold, ❷ is the sulfite-gold, ❸ is the PE gold electrodeposited at the optimum condition.

Acknowledgements

This work is supported by CREST Project (#14531864) operated by the Japan science and Technology Agency (JST) and by the Grant-in-Aid for Scientific Research (S) (JSPS KAKENHI Grant number 26220907).

References

[1]. D. Yamane, T. Konishi, T. Matsushima, K. Machida, H. Toshiyoshi, K. Masu, Design of sub-1g microelectromechanical systems accelerometers, *Applied Physics Letters*, Vol. 104, Issue 7, 2014, pp. 074-102.

[2]. K. Machida, T. Konishi, D. Yamane, H. Toshiyoshi, K. Masu, Integrated CMOS-MEMS technology and its applications, *ECS Transactions*, Vol. 61, Issue 6, 2014, pp. 21-39.

[3]. D. Yamane, T. Konishi, H. Toshiyoshi, K. Masu, K. Machida, A Sub-1G MEMS Sensor, *ECS Transactions*, Vol. 66, Issue 5, 2015, pp. 131-138.

[4]. J. R. Greer, W. C. Oliver, W. D. Nix, Size dependence of mechanical properties of gold at the micron scale in the absence of strain gradients, *Acta Materialia*, Vol. 53, Issue 6, 2005, pp. 1821-1830.

[5]. N. J. Petch, The cleavage strength of polycrystals, *Journal of the Iron and Steel Institute*, Vol. 174, 1953, pp. 25-28.

[6]. Y. Wang, M. Chen, F. Zhou, E. Ma, High tensile ductility in a nanostructured metal, *Nature*, Vol. 419, 2002, pp. 912-915.

[7]. S. Cheng, E. Ma, Y. Wang, L. Kecskes, K. Youssef, C. Koch, U. Trociewitz, K. Han, Tensile properties of in situ consolidated nanocrystalline Cu, *Acta Materialia*, Vol. 53, Issue 5, 2005, pp.1521-1533.

[8]. M. Gad-el-Hak, The MEMS Handbook, *CRC Press*, 2005.

[9]. M. Sone, T. F. M. Chang, H. Uchiyama, Crystal growth by electrodeposition with supercritical carbon dioxide emulsion, Chapter 11, in Crystal Growth, (Ed. Sukarno Olavo Ferreira), *INTECH*, 2013, pp. 335-376.

[10]. T. F. M. Chang, M. Sone, A. Shibata, C. Ishiyama, Y. Higo, Bright nickel film deposited by supercritical carbon dioxide emulsion using an additive-free Watts bath, *Electrochimica Acta*, Vol. 55, Issue 22, 2010, pp. 6469-6475.

[11]. C. Y. Chen, M. Yoshiba, T. Nagoshi, T. F. M. Chang, D. Yamane, K. Machida, K. Masu, M. Sone, Pulse electroplating of ultra-fine grained Au films with high compressive strength, *Electrochemistry Communications*, Vol. 67, 2016, pp. 51-54.

[12]. H. C. Tang, C. Y. Chen, T. Nagoshi, T. F. M. Chang, D. Yamane, K. Machida, K. Masu, M. Sone, Enhancement of mechanical strength in Au films electroplated with supercritical carbon dioxide, *Electrochemistry Communications*, Vol. 72, 2016, pp. 126-130.

[13]. M. Yoshiba, C. Y. Chen, T. F. M. Chang, T. Nagoshi, D. Yamane, K. Machida, K. Masu, M. Sone, Brittle fracture of electrodeposited gold observed by micro-compression, *Materials Transactions*, Vol. 57, Issue 8, 2016, pp. 1257-1260.

[14]. M. D. Uchic, D. M. Dimiduk, J. N. Florando, W. D. Nix, Sample dimensions influence strength and crystal plasticity, *Science*, Vol. 305, Issue 5686, 2004, pp. 986-989.

[15]. P. Wilkinson, Understanding gold plating, *Gold Bulletin*, Vol. 19, Issue 3, 1986, pp. 75-81.

[16]. T. Nagoshi, M. Mutoh, T. F. M. Chang, T. Sato, M. Sone, Sample size effect of electrodeposited nickel with sub-10 nm grain size, *Materials Letters*, Vol. 117, 2014, pp. 256-259.

[17]. M. Stern, A. L. Geary, Electrochemical Polarization: I. A Theoretical Analysis of the Shape of Polarization Curves, *Journal of the Electrochemical Society*, Vol. 104, Issue 1, 1957, pp. 56-63.

[18]. T. F. M. Chang, T. Shimizu, C. Ishiyama, M. Sone, Effects of pressure on electroplating of copper using supercritical carbon dioxide emulsified electrolyte, *Thin Solid Films*, Vol. 529, 2013, pp. 25-28.

[19]. T. F. M. Chang, T. Tasaki, C. Ishiyama, M. Sone, Defect-free nickel micropillars fabricated at a high current density by application of a supercritical carbon dioxide emulsion, *Industrial & Engineering Chemistry Research*, Vol. 50, Issue 13, 2011, pp. 8080-8085.

[20]. J. Ke, W. Su, S. M. Howdle, M. W. George, D. Cook, M. Perdjon-Abel, P. N. Bartlett, W. Zhang, F. Cheng, W. Levason, G. Reid, J. Hyde, J. Wilson, D. C. Smith, K. Mallik, P. Sazio, Electrodeposition of metals from supercritical fluids, *Proceedings of the National Academy of Sciences of the United States of America*, Vol. 106, No. 35, 2009, pp. 14768-14772.

[21]. J. M. Blackburn, D. P. Long, A. Cabanas, J. J. Watkins, Deposition of conformal copper and nickel films from supercritical carbon dioxide, *Science*, Vol. 294, Issue 5540, 2001, pp. 141-145.

[22]. T. Nagoshi, T. F. M. Chang, S. Tatsuo, M. Sone, Mechanical properties of nickel fabricated by electroplating with supercritical CO2 emulsion evaluated by micro-compression test using non-tapered micro-sized pillar, *Microelectronic Engineering*, Vol. 110, 2013, pp. 270-273.

[23]. S. T. Chung, W. T. Tsai, Nanocrystalline Ni–C electrodeposits prepared in electrolytes containing supercritical carbon dioxide, *Journal of the Electrochemical Society*, Vol. 156, Issue 11, 2009, pp. D457–D461.

[24]. T. G. Nieh, J. Wadsworth, Hall-petch relation in nanocrystalline solids, *Scripta Metallurgica et Materialia*, Vol. 25, Issue 4, 1991, pp. 955-958.

[25]. T. Nagoshi, T. F. M. Chang, T. Sato, M. Sone, Effect of annealing on mechanical properties of nickel electrodeposited using supercritical CO_2 emulsion evaluated by micro-compression test, *Microelectronic Engineering*, Vol. 153, 2016, pp. 101-104.

[26]. J. M. Chabala, R. Levi-Setti, Y. L. Wang, Practical resolution limits of imaging microanalysis with a scanning ion microprobe, *Applied Surface Science*, Vol. 32, Issues 1-2, 1988, pp. 10-32.

[27]. M. Dietiker, S. Buzzi, G. Pigozzi, J. F. Löffler, R. Spolenak, Deformation behavior of gold nano-pillars prepared by nanoimprinting and focused ion-beam milling, *Acta Materialia*, Vol. 59, Issue 5, 2011, pp. 2180-2192.

[28]. J. Horkens, L. T. Romankiw, Pulsed potentiostatic deposition of gold from solutions of the Au (I) sulfite complex, *Journal of the Electrochemical Society*, Vol. 124, Issue 10, 1977, pp. 499-1505.

[29]. A. Ruffoni, D. Landolt, Pulse-plating of Au-Cu-Cd alloys: II. Theoretical modelling of alloy composition, *Electrochimica Acta*, Vol. 33, Issue 10, 1988, pp. 1281-1289.

[30]. D. Jia, K. T. Ramesh, E. Ma, Effects of nanocrystalline and ultrafine grain sizes on constitutive behavior and shear bands in iron, *Acta Materialia*, Vol. 51, Issue 12, 2003, pp. 3495-3509.

[31]. Z. Gan, Y. He, D. Liu, B. Zhang, L. Shen, Hall–Petch effect and strain gradient effect in the torsion of thin gold wires, *Scripta Materialia*, Vol. 87, 2014, pp. 4.

Chapter 4

On the Transimpedance Amplifiers in the Low Frequency Noise Characterization

Mattia Borgarino, Marco Salvaterra

4.1. Introduction

A Trans-Impedance Amplifier (TIA) is an electronic circuit used to amplify a current signal into a voltage signal. TIA is reported mainly applied in the design of optical transceiver front-ends [1-8], bio-sensors [9-16] and radiation detectors [17, 18]. As in the case of a generic amplifier, a couple of equivalent, input referred, partially correlated, voltage (e_N) and current (i_N) noise generators can describe the noise properties of a TIA, as shown in the following Fig. 4.1 [19].

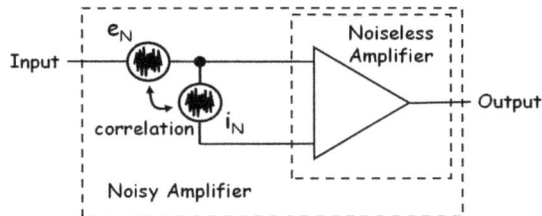

Fig. 4.1. Input referred equivalent noise sources for a generic amplifier.

Nevertheless, several papers on TIA reports only on the i_N noise generator, because, as in the case of the previously cited applications, the signal source behaves like a current source. Under these conditions, e_N is left floating and i_N is the only generator contributing to the noise at the TIA output.

In the past, the TIA demonstrated to be useful also in the field of the Low Frequency Noise (LFN) characterization of microelectronic devices [20-28]. In this application the noise source under test does not work perforce as a current generator, because the small-signal

Mattia Borgarino
Eng. Department "Enzo Ferrari", University of Modena and Reggio Emilia, Via Vivarelli, 10, 41125, Modena, Italy

impedance of the Device-Under-Test (DUT) may be not very high, as in the case of bipolar devices biased at moderate to high DC current [27]. Under these conditions, also e_N and its correlation with i_N contribute to the noise at the TIA output, leading to some concerns during the characterization. In particular, Section 4.2 is dedicated to the troubles may arise when a resistive feedback TIA is used to measure the power spectral density of the noise generated by two-terminal devices, such as resistors or diodes. In this case, a range of DUT small-signal impedance values makes unreliable the LFN measurement. Correlation measurements play an important role in the development of LFN compact models of two-port networks such as transistors. Section 4.3 is devoted to the measurement of the correlation between the noisy fluctuations in the base current and in the collector current of a bipolar transistor. Placing the TIA as close as possible to the DUT may be helpful in improving the measurement quality, because the noisy fluctuations generated by the DUT are very weak and can be aggressed by an interferer before the amplification. Section 4.4 addresses the issue of miniaturizing the TIA, with the idea in mind of placing it as close as possible to the DUT. Finally, Section 4.5 summarizes the previously carried out discussions by drawing some conclusions.

4.2. Characterizing a Noisy Two-terminal Device

The literature reports several approaches to design a TIA, as common-gate [17, 29], current-mode [30, 31], resistive feedback [17, 29, 30, 32], capacitive feedback [29, 32] and regulated cascode [17, 31, 33]. The following Fig. 4.2 sketches some example schematics of the cited solutions. All these circuits are single-ended but the literature report also pseudo-differential or differential solutions [34].

Resistive feedback TIA offers noise performance better than those offered by the common-gate [29, 32, 17] and current-mode [30] solutions, and comparable with those offered by the regulated cascode solution [17]. These better noise performances come with stability and voltage headroom issues [32] and a narrower bandwidth than that offered by the common-gate [32] and current mode [30] TIA. The capacitive feedback TIA offers low noise without stability issues [32] but its fabrication in an integrated circuit is not feasible in the case of low frequency signals, because of the too much large capacitors. The LFN characterization does not require large bandwidth, because the noise current fluctuations of interest fall typically in the 100 Hz - 100 kHz low frequency range. The resistive feedback TIA is therefore a good choice, because of its lower noise [30] and integrability.

Fig. 4.2 shows that the core of a resistive feedback TIA is a voltage amplifier. Fig. 4.3 shows two small-signal models of a resistive feedback TIA. Fig. 4.3a depicts the model obtained by replacing the voltage amplifier with its equivalent small-signal model constituted by an input resistor R_{IN}, an open-circuit voltage gain A_V, and two equivalent, input referred, partially correlated, voltage and current noise generators of power spectral densities $S_{V,AMP}$ and $S_{I,AMP}$, respectively. A Norton equivalent model renders the noisy DUT connected at the TIA input. R_{DUT} is the small-signal equivalent impedance of the DUT and $S_{I,DUT}$ is the power spectral density of its noise current fluctuations. Fig. 4.3b shows the model of the whole TIA in terms of its input impedance R_{TA}, gain G_{TA}, and input referred, partially correlated, voltage and current noise generators of power spectral densities $S_{V,TA}$ and $S_{I,TA}$, respectively. A Norton equivalent model still renders the noisy DUT at the TIA input. The double arrows between the couple of input referred, equivalent noise generators in Fig. 4.3 reminds that these generators, both for the voltage amplifier and the TIA, are partially correlated. In the following, this correlation, in the case of the TIA model in Fig. 4.3b, is taken into account by the cross-spectrum $S_{IV*,TA}$.

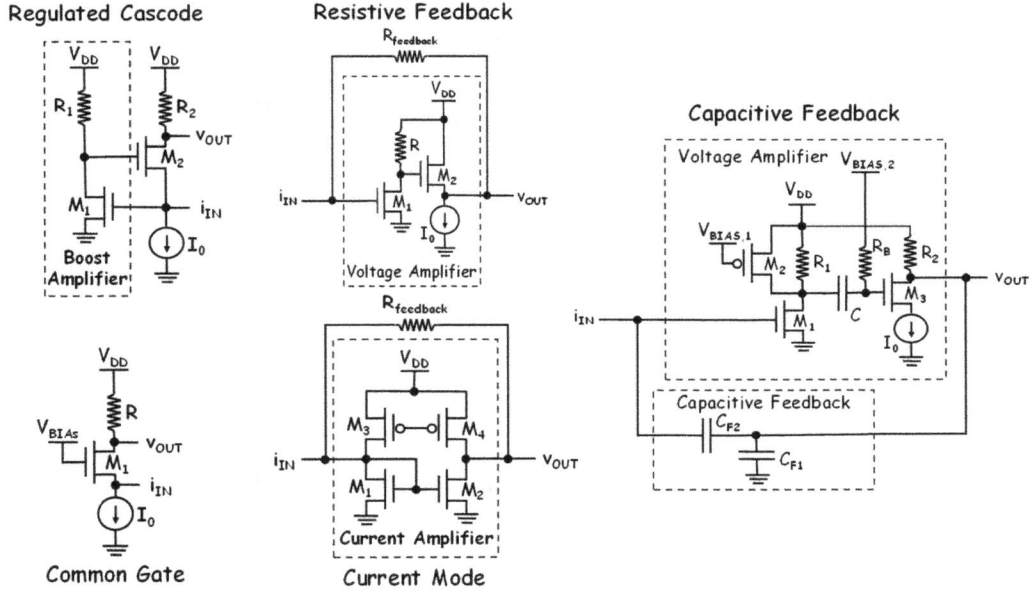

Fig. 4.2. Different kinds of TIA: Regulated cascode [17, 33], Resistive feedback [32, 31], Common gate [17, 29], Current_Mode [30, 31], Capacitive feedback [32, 29].

For the model in Fig. 4.3b it is straightforward to demonstrate [35] that the power spectral density of the noise voltage fluctuations at the output of the TIA S_{OUT} is given by:

$$S_{OUT} = G_{TA}^2 \frac{R_{DUT}^2 \left(S_{I,DUT} + S_{I,TA}\right) + 2R_{DUT} S_{IV*,TA} + S_{V,TA}}{\left(R_{DUT} + R_{TA}\right)^2}. \qquad (4.1)$$

Equation (4.1) was applied to the commercial EG&G5182 TIA, widely used by the scientific community working on the low frequency noise, to measure the thermal noise of a resistor [35]. Since in this case R_{DUT} is the resistance of a resistor, $S_{I,DUT}$ takes the following form:

$$S_{I,DUT} = \frac{4KT}{R_{DUT}}, \qquad (4.2)$$

where K is the Stephan-Boltzmann constant and T is the absolute temperature.

Fig. 4.4a shows that when the TIA is set for $G_{TA}=10^5$ Ω the agreement between the measured and computed S_{OUT} is good for several values of R_{DUT}. On the other hand, when the TIA is set up for a higher gain a discrepancy between the equation (4.1) and the experimental data emerges. Fig. 4.4b shows the experimental finding for $G_{TA}=10^7$ Ω.

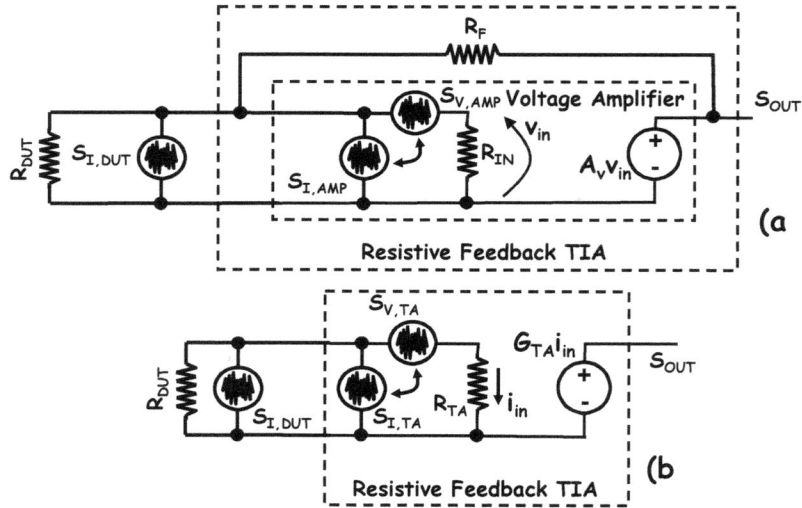

Fig. 4.3. Equivalent small-signal models for a resistive feedback TIA.

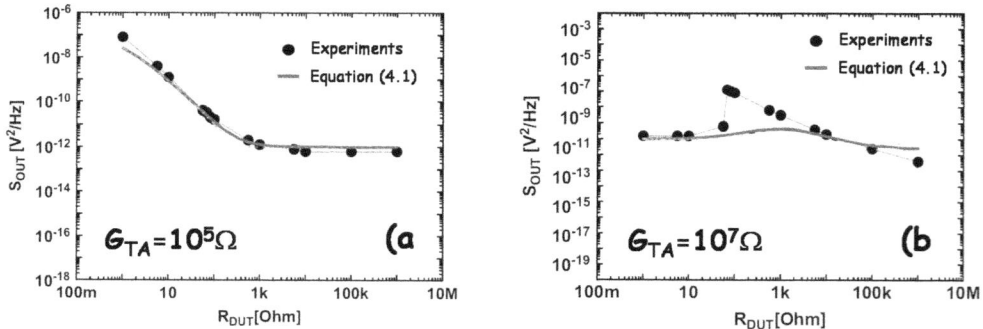

Fig. 4.4. Power spectral noise voltage at the output of the TIA for several value of R_{DUT}; data from [35].

The largest discrepancy occurs for an R_{DUT} of around 100 Ω. Similar behaviours appear for $G_{TA}=10^6$ Ω and $G_{TA}=10^8$ Ω, with the largest discrepancy taking place at around 10 Ω for $G_{TA}=10^6$ Ω and at around 1 kΩ for $G_{TA}=10^8$ Ω. By using the TIA model in Fig. 4.3a and accounting for the DUT noise only, it is possible to demonstrate [35] that S_{OUT} takes the following expression:

$$S_{OUT} = \frac{4KTA_V^2}{\left[\dfrac{1}{\sqrt{R_{DUT}}} + \sqrt{R_{DUT}} \left(\dfrac{1}{R_{IN}} + \dfrac{1-A_V}{R_F} \right) \right]^2}. \qquad (4.3)$$

The derivative of S_{OUT} with respect to R_{DUT} exhibits a maximum for the following value of R_{DUT} [35]:

$$R_{DUT,MAX} = \frac{R_{IN}R_F}{R_F + R_{IN}(A_V - 1)} \cong \frac{R_{IN}R_F}{R_F + R_{IN}A_V} \cong \frac{R_F}{A_V}. \qquad (4.4)$$

The approximations are introduced by considering first $A_V \gg 1$ and then $R_{IN}A_V \gg R_F$. Equation (4.4) is the expression of the R_{DUT} value for which the largest difference between equation (4.1) and the experimental data occurs. Since in the second approximation, R_F is the TIA gain, equation (4.4) can account for the previously described behavior in the discrepancy between equation (4.1) and experimental data. Equation (4.4) predicts indeed that when the G_{TA} increases (decreases) by one decade, the $R_{DUT,MAX}$ increases (decreases) by one decade. Equation (4.4) suggests that the discrepancy is a consequence of the finite value of the voltage amplier A_V. If A_V is infinite, the discrepancy disappears. In the case of a finite value of A_V, R_{DUT} may fall close to $R_{DUT,MAX}$, if the DUT exhibits a finite value of R_{DUT}. In order to minimize this issue, in a resistive feedback TIA aimed to LFN characterization the voltage amplifier at its core should exhibit a voltage gain as high as possible. It cannot therefore be the simple circuit depicted in Fig. 4.2 but a more complex circuit such a multi-stage operational amplifier.

4.3. Measuring the Correlation in Noisy Two-port Networks

The previous section addressed the use of one single TIA. A couple of TIA can useful measure the correlation between the base and collector current fluctuations in bipolar transistors [20, 21]. This coherence characterization leads to the correlation resistance [36], which is a helpful guideline to extract LFN compact models of transistors for purpose of analysis and design of microwave circuits working under large-signal conditions [26, 37-40] like oscillators [41, 42]. In Fig. 4.1, the two sources e_N and i_N are not, generally speaking, physically located where the model suggests. The model is indeed a mathematical representation of the noise properties of the amplifier, useful only to compute the voltage noise at the amplifier output, once the amplifier input is closed on a given resistor. Nevertheless, it may happen that these sources are physically located at the TIA input [21, 22]. In particular, in [21] a low noise voltage amplifier connected at the input of an EG&G5182 TIA demonstrated that e_N is physically located at the TIA input. This noise voltage source with a measured power spectral density of about 10^{-17} V^2/Hz was ascribed to the fluctuations of the virtual ground in the operational amplifier inside the TIA. Fig. 4.5 shows that, because of its physical location, this voltage source affects the measurements of the collector noise spectral density S_{IC}. This is an expected behavior, because the DUT is a bipolar transistor in a common-emitter configuration. It amplifies therefore the voltage noise on the base side into current noise on the collector side.

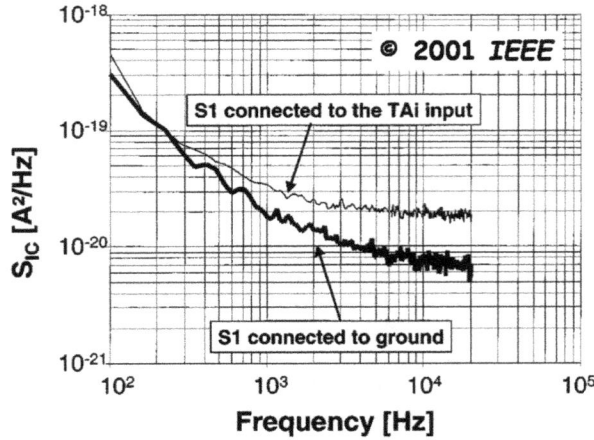

Fig. 4.5. Impact of the e_N noise voltage source on S_{IC} [21].

Consequently, this voltage source affects also the measurement of the correlation between base and collector current fluctuations S_{IBIC*}. A common-base Buffer Amplifier (BA) placed between the DUT base and the TIA input, as shown in the following Fig. 4.6, can solve this trouble. Thanks to this buffer, the virtual ground fluctuations of the TIA are unable to reach the DUT and therefore to affect the measurement of S_{IC} and of S_{IBIC*}. The effectiveness of this solution is demonstrated by comparing the full LFN characterization obtained by using a couple of TIA with the full LFN characterization obtained from another experimental set-up based on a multi-impedance technique similar to those employed for the measurement of the microwave noise. Fig. 4.7 compares the noise spectral densities of e_N (Fig. 4.7a) and i_N (Fig. 4.7b) together with their correlation C_{cor} (Fig. 4.7c) and the corresponding correlation resistance R_{cor} (Fig. 4.7d) obtained from a TIA based experimental set-up (light curves) and from a multi-impedance experimental set-up (heavy curves). The very good agreement demonstrates the effectiveness of the BA.

A voltage noise source physically located at the TIA input is not the only jeopardy for the measurement of S_{IBIC*}. For the experimental set-up in Fig. 4.6, it is possible to demonstrate, by neglecting the BA noise contribution, that the experimental cross-spectrum between collector and base noise current fluctuations $S_{IBIC*,EXP}$ is described by the following equation [28]:

$$S_{IBIC*EXP} = \frac{R_{DUT}}{R_{DUT} + R_{IN,BA}} \left[S_{IBIC*} - \beta \frac{R_{IN,BA}}{R_{DUT} + R_{IN,BA}} S_{IB} \right], \qquad (4.5)$$

where R_{DUT} is the small-signal resistance observed by looking into the DUT base, β is the common-emitter current gain of the DUT, $R_{IN,BA}$ is the BA input resistance, and S_{IB} is the power spectral density of the DUT base noise current fluctuations.

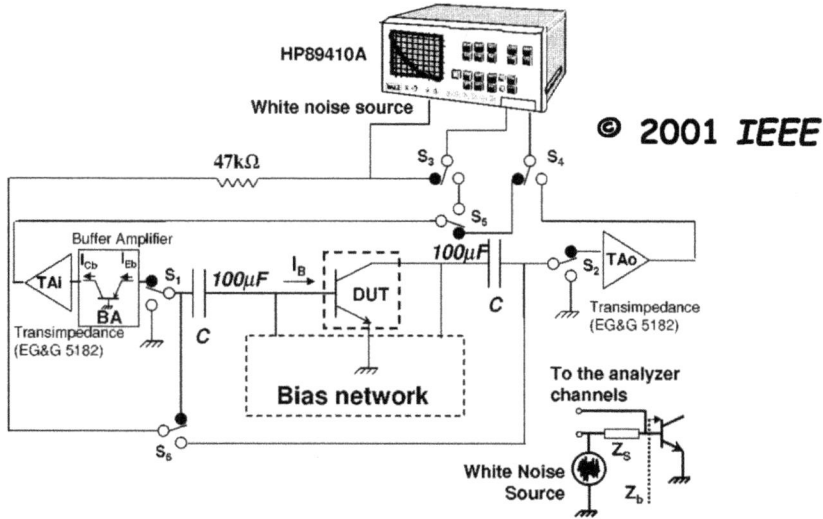

Fig. 4.6. Experimental set-up for the measurement of S_{IBIC*} by using a couple of TIA [21].

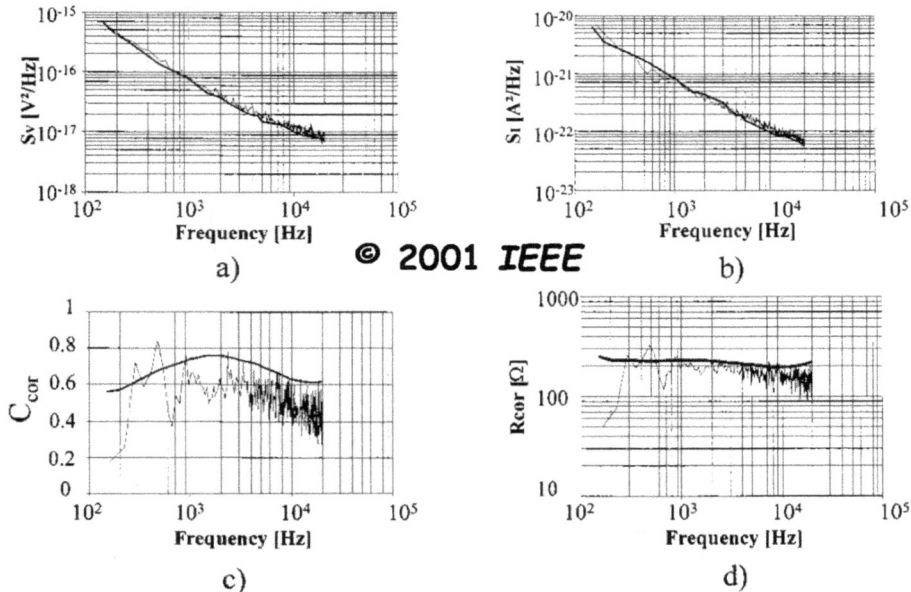

Fig. 4.7. Comparison of a full low-frequency characterization obtained using a TIA based set-up (light curves) and multi-impedance set-up (heavy curves) [21].

If the DUT exhibits low R_{DUT} and high S_{IB} Equation (4.5) shows that S_{IB} can corrupt the measurement of $S_{IBIC*,EXP}$. This is the typical situation when the DUT is a GaAs Heterojunction Bipolar Transistor (HBT). In a GaAs HBT R_{DUT} is indeed quite low, because of the quite high bias base current and S_{IB} is quite high, because the compound

113

semiconductors exhibit not negligible defect densities. This issue does not exist in the case of Si-based bipolar transistors or SiGe HBTs. These devices exhibit indeed a high R_{DUT}, because of the low bias base current, and a low S_{IB}, because silicon is a semiconductor with low defect densities.

The trouble in the case of GaAs HBTs can be overcome by a transformer placed between the DUT base and the BA input, as depicted in Fig. 4.8.

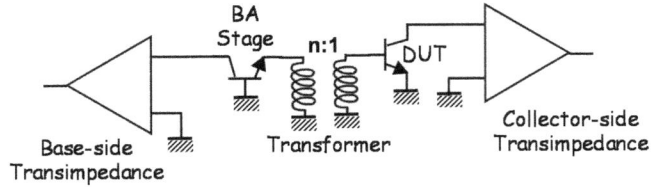

Fig. 4.8. TIA based experimental set-up improved by the introduction of a transformer.

Under the hypothesis that $n^2 R_{DUT} \gg R_{IN,BA}$ and by neglecting the BA noise contribution, it is possible to demonstrate that the new expression for $S_{IBIC*,EXP}$ is [28]:

$$S_{IBIC*,EXP} = -\frac{1}{n}\left[S_{IBIC*} - \beta \frac{R_{IN,BA}}{R_{DUT}} \frac{S_{IB}}{n^2} \right]. \qquad (4.6)$$

It is worth noticing that the contribution of S_{IB} appears reduced by n^2. The following Figs. 4.9 and 4.10 show the effect of the transformer on $S_{IBIC*,EXP}$ and on S_{IC}, respectively. In Fig. 4.10 it is interesting to remark that the higher the base bias current the larger the effect of the transformer. When the bias current increases, the R_{DUT} decreases and S_{IB} increases and thus the impact of S_{IB} on the measurement of S_{IC} increases. The effect of the transformer becomes therefore more relevant as the bias base current increases.

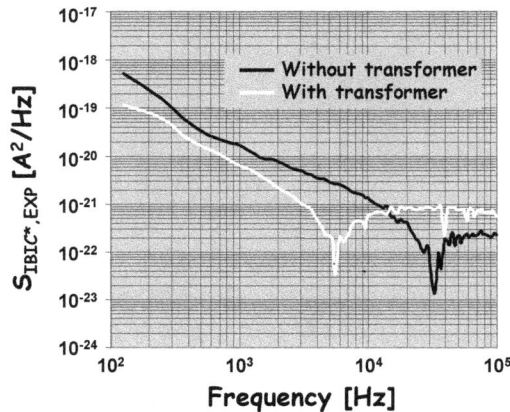

Fig. 4.9. Impact of the transformer on the measurement of S_{IBIC*}; data from [28].

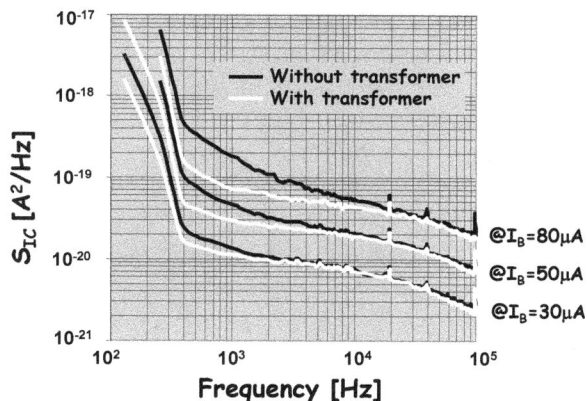

Fig. 4.10. Impact of the transformer on the measurement of S_{IC}; data from [28].

4.4. Miniaturizing the TIA

In an experimental set-up for the LFN characterization, the weak current fluctuations generated by the DUT are the signal of interest. Usually, the amplifiers used for these signals are commercial [20, 23-26] or home-made with discrete devices [43, 44]. In these cases, as the TIA is a large device, a cable connects the DUT and TIA. The unamplified LFN fluctuations generated by the DUT can be aggressed by interferers on this cable. In order to minimize this issue, the TIA should stay as close as possible to the DUT. A possible solution is to fabricate the TIA in the form of an integrated circuit, with the idea in mind of installating the TIA directly on the coplanar probes used to contact the DUT [45]. This paper [45] first suggests this use of an integrated TIA for the LFN characterization while other examples of integrated TIA in the literature address biosensor [29, 30] or radiation detection [17] applications. The TIA described in [45] covers the 100 kHz-800 MHz frequency range while most applications of the LFN characterization need data and/or models in a lower frequency range (e.g. 100 Hz-100 kHz). Fig. 4.11 shows the two stages building block diagram of an integrated TIA designed to cover the 100 Hz-100 kHz frequency range. As in [45], the first stage is a resistive feedback TIA. An operational amplifier provides the voltage amplification at its core, in agreement with the conclusions drawn in Section 4.2. A further operational amplifier is employed for the second stage, which serves as a voltage amplifier and as a buffer, making the input resistance of the whole TIA less dependent on the load. This may be useful when the load impedance is low, as in the case of a 50 Ω instrumentations channel.

The following Fig. 4.12 depicts the schematic of the operational amplifier. Its core is a telescopic cascode single-ended output Operational Transducer Amplifier (OTA). This solution offers a gain in the order of $(g_m r_o)^2$ [46] where g_m is the transconductance of the transistors in the input differential pairs and r_o is the small-signal resistance of the load p-channel transistors. The telescopic cascode OTA offers therefore the gain provided by the cascade of two traditional OTAs but with the advantage of avoiding DC coupling. Coupling capacitors cannot indeed be used, because the low frequencies (100 Hz -

115

100 kHz) would require too large integrated capacitors. The DC coupling is a trouble, because small variations in the DC bias point of the first stage may propagate along the amplification chain, jeopardizing the bias point of the last amplification stage. The telescopic cascode OTA is thus a practical solution, even if, on the other hand, it requires a large voltage headroom, because of the several transistors stacked between bias voltage (V_{DD}) and ground. Fig. 4.12 shows that a current mirror and a resistive voltage divider form the bias network for the telescopic cascode OTA, whose output stage is a common-source buffer. Finally, for sake of stability, an RC compensation network connect the two branches of the OTA.

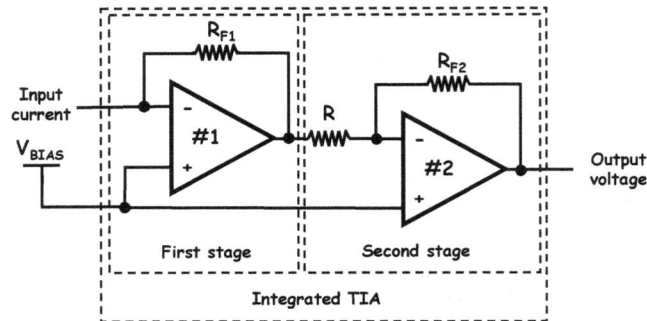

Fig. 4.11. Building block diagram of an integrated TIA working in the 100 Hz-100 kHz frequency range.

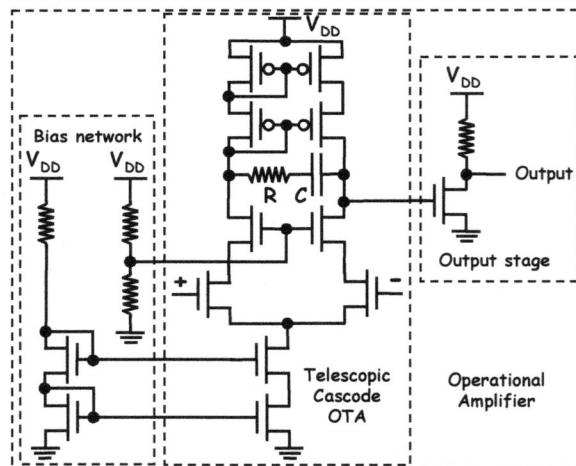

Fig. 4.12. Schematic of the operational amplifier used for the TIA in Fig. 4.11.

Implemented in a 350 nm CMOS technology by Austria Micro System, the operational amplifier in Fig. 4.12 exhibits a voltage gain of 65 dB and a bandwidth of about 10 kHz. 3.3 V were enough for a correct bias. By using this operational amplifier and by properly setting the resistors in the building block diagram in Fig. 4.11, a TIA with a G_{TA} of 120 dB, a bandwidth of about 200 kHz, and an input resistance of about 60 Ω at 1 kHz can be obtained. The following Fig. 4.13 shows the layout of the integrated circuit TIA, whose width and height are 1420 μm and 400 μm, respectively.

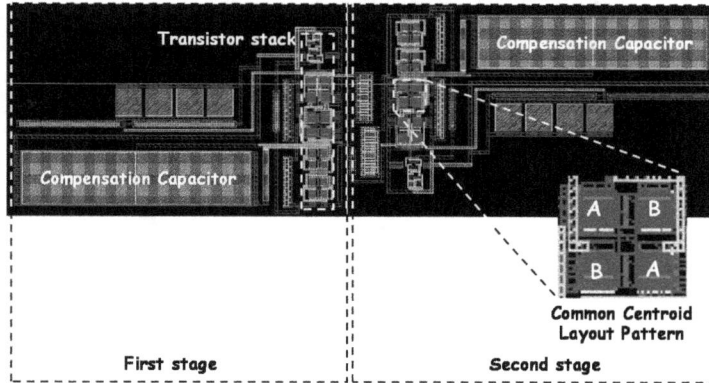

Fig. 4.13. Layout of the TIA in Fig. 4.11.

The two stages TIA architecture clearly appears in the layout. The Fig. 4.13 highlights the transistor stack of the telescopic cascode OTA. Because of the symmetry of the OTA circuit, the layout of the transistor couples exhibits a common centroid bidimensional pattern [47], in order to improve the matching. Finally, also the compensation capacitors are clearly visible. The power spectral density of the input referred equivalent voltage (current) noise source is about 10^{-15} V^2/Hz ($3 \cdot 10^{-25}$ A^2/Hz) at 1 kHz. The cross-spectrum of these two sources is about 10^{-20} W/Hz at 1 kHz. The sensitivity of the TIA was investigated by using an artificial DUT constituted by the parallel of a noiseless resistor and a white noise current generator. In this way, it was possible to change independently R_{DUT} and $S_{I,DUT}$. The following Fig. 4.14 shows the minimum correctly measurable $S_{I,DUT}$. The sensitivity is always very good when the R_{DUT} is very high (1 MΩ). This is the situation when the noise source behaves like a current generator. When the R_{DUT} is no longer a very high value, as in the previous Sections 4.2 and 4.3, some limitations may arise. Fig. 4.14 shows that, for a given frequency, the TIA sensitivity exhibits a decreasing trend when the R_{DUT} decreases. For a given R_{DUT}, a decreasing trend is also observable when the frequency increases. This is, because the TIA input resistance increases with the frequency. For the highest frequency of 10^5 Hz and the lowest R_{DUT} of 1 kΩ, the sensitivity is worse than 10^{-20} A^2/Hz. The discussions in Section 4.3 and 4.4 suggest that the limitations on measuring the power spectral density generated by a two-terminal device apply also to the measurement of the correlation in a noisy two ports network. The integrated TIA in Fig. 4.13 can therefore measure the correlation in Si bipolar transistors or in SiGe HBTs but in the case of GaAs HBTs the TIA should be improved by introducing a transformer as in Fig. 4.8. It is worth here noticing that the co-integration of the BA buffer with the TIA by using the same low cost technology it is not possible. The bipolar transistors available in a low cost commercial design kit does not indeed exhibit the noise and small-signal performance required by the transistor in the BA buffer. In [21] the discrete, high performance silicon bipolar transistor MAT02 by Analog Devices was indeed used. The BA buffer and the transformer are therefore off-chip.

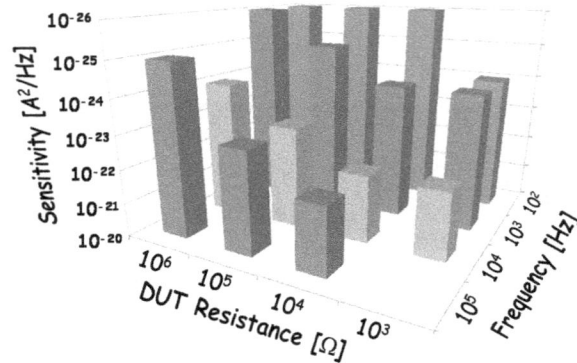

Fig. 4.14. Sensitivity versus frequency and R_{DUT} for the TIA in Fig. 4.13.

4.5. Conclusions

The present chapter reported on the troubles may arise when a transimpedance amplifier is applied to the low frequency noise characterization of microelectronics devices. After a short review on the possible approaches to design a transimpedance amplifier, the discussion focused on the issues may arise when a resistive feedback transimpedance amplifier is employed to characterize a noisy two-terminal device. The issues were described, discussed, and a possible solution was proposed. It was found that the higher is the voltage gain of the voltage amplifier at the core of the TIA, the largest is the range of the device-under-test small-signal resistance values where the LFN characterization is correct. The suggested tip is therefore that the voltage amplifier should be an operational amplifier. Afterwards, attention was paid to the issues concerning the use of a couple of transimpedance amplifiers to measure the correlation between the base and collector current noisy fluctuations in a bipolar transistor. Even in this case, the issues were presented, discussed, and a possible solution was proposed. It was found that the correctness of the LFN characterization can be improved by introducing a common base amplifier between the TIA and the base of the DUT. In addition, when the DUT is a GaAs-based bipolar transistor, the common-base amplifier should be cascaded with a transformer. Eventually, the chapter addressed the topic of miniaturizing the transimpedance amplifier, in order to place it as close as possible to the DUT. A possible circuit working in the 100 Hz-100 kHz frequency range was presented both at schematic and layout level together with indications about its limits.

Acknowledgements

The results reported in Section 4.4 were obtained in the frame of the project FAR2016 entitled "Miniaturized transimpedance amplifier for the low frequency noise characterization of microelectronics devices". M. Borgarino would like to thank Prof. S. Y. Yurish for having stimulated the writing of the present chapter.

References

[1]. O. Ghasemi, CS-based TIAs Using Inductive Feedback approach in 90 nm CMOS, in *Proceedings of the 28th IEEE Canadian Conference on Electrical and Computer Engineering (CCECE'15)*, Halifax, Canada, 3-6 May, 2015, pp. 1162-1167.

[2]. T. L. Nguyen, A. Izadi, G. Denoyer, SiGe BiCMOS technologies for high-speed and high-volume optical interconnect applications, in *Proceedings of the IEEE Bipolar/BiCMOS Circuits and Technology Meeting (BCTM'16)*, Miami, Florida, USA, 19-20 October, 2016, pp. 1-8.

[3]. K. Park, W.-S. Oh, A 40-Bb/s 310-fJ/b inverter-based CMOS optical receiver front-end, *IEEE Photonics Technology Letters*, Vol. 27, Issue 18, 2015, pp. 1931-1933.

[4]. Q. Pan, Y. Wang, Z. Hou, L. Si, Y. Lu, W.-H. Ki, P. Chiang, A 30-Gb/s 1.37-pJ/b CMOS receiver for optical inteconnects, *IEEE Journal of Lightwave Technology*, Vol. 33, Issue 4, 2015, pp. 778-786.

[5]. M. Atef, Transimpedance amplifier with a compression stage for wide dynamic range optical applications, *Microelectronics Journal*, Vol. 46, 2015, pp. 593-597.

[6]. F. Aznar, W. Gaberl, H. Zimmermann, A 0.18 um CMOS transimpedance amplifier with 26 dB dynamic range at 2.5 Gb/s, *Microelectronics Journal*, Vol. 42, 2011, pp. 1136-1142.

[7]. Y-J. Chen, M. du Plessis, An integrated 0.35 um CMOS optical receiver with clock and data recovery circuit, *Microelectronics Journal*, Vol. 37, 2006, pp. 985-992.

[8]. M. Seifouri, P. Amiri, M. Rakide, Design of broadband transimpedance amplifier for optical communication systems, *Microelectronics Journal*, Vol. 46, 2015, pp. 679-684.

[9]. J. Hu, Y-B. Kim, J. Ayers, A 65nm CMOS Ultra Low Power and Low Noise 131M Front-end Transimpedance Amplifier, in *Proceedings of the 23rd International Conference on System on Chip*, 27-29 September, 2010, Las Vegas, Nevada, USA, pp. 281-284.

[10]. V. Balasubramanian, P-F. Ruedi, Y. Temiz, A. Ferretti, C. Guiducci, C. C. Enz, A 0.18 um Biosensor Front-End Based on 1/f Noise, Distortion Cancelation and Chopper Stabilization Techniques, *IEEE Transactions on Biomedical Circuits and Systems*, Vol. 7, Issue 5, 2013, pp. 660-673.

[11]. K. A. Al Mamun, S. K. Islam, D. K. Hensley, N. McFarlane, A glucose biosensor using CMOS potentiostat and vertically aligned carbon nanofibers, *IEEE Transactions on Biomedical Circuits and Systems*, Vol. 10, Issue 4, 2016, pp. 807-816.

[12]. O. T. Inan, G. T. A. Kovacs, An 11 μW, two-electrode transimpedance biosignal amplifier with active current feedback stabilization, *IEEE Transactions on Biomedical Circuits and Systems*, Vol. 4, Issue 2, 2010, pp. 93-100.

[13]. K. Murari, R. Etienne-Cummings, N. V. Thakor, G. Cauwenberghs, A CMOS in-pixel CTIA high-sensitiviy fluorescence imager, *IEEE Transactions on Biomedical Circuits and Systems*, Vol. 5, Issue 5, 2011, pp. 449-458.

[14]. M. Tavakoli, L. Turicchia, R. Sarpeshkar, An ultra-low-power pulse oximeter implemented with an energy-efficient transimpedance amplifier, *IEEE Transactions on Biomedical Circuits and Systems*, Vol. 4, Issue 1, 2010, pp. 27-38.

[15]. H. F. Achigui, M. Sawanm, C. J. B. Fayomi, A monolithic based NIRS front-end wireless sensors, *Microelectronics Journal*, Vol. 39, 2008, pp. 1209-217.

[16]. V. Vijay, B. Raziyeh, S. Amir, D. Jelena, J, Muller, C. Yihui, H. Andreas, 32-Channel integrated electrical impedance sensors on a multi-functional neural microelectrode array platform, *Procedia Engineering*, Vol. 168, 2016, pp. 510-513.

[17]. L. B. Oliveira, C. M. Leitao, M. de Medeiros Silva, Noise performance of a regulated transimpedance amplifier for radiation detectors, *IEEE Transactions on Circuits and Systems, I: Regular Papers*, Vol. 59, Issue 9, 2012, pp. 1841-1848.

[18]. M. de Medeiros Silva, L. B. Oliveira, Regulated common-gate transimpedance amplifier designed to operate with a silicon photo-multiplier at the input, *IEEE Transactions on Circuits and Systems I: Regular Papers*, Vol. 61, Issue 3, 2012, pp. 725-735.

[19]. Z. Lu, K. S. Yeo, J. Ma, M. A. Do, W. M. Lim, X. Chen, Broad-band design techniques for transimpedance amplifiers, *IEEE Transactions on Circuits and Systems I: Regular Papers*, Vol. 54, Issue 3, 2007, pp. 590-600.

[20]. S. P. O. Bruce, L. K. J. Vandamme, A. Rydberg, Measurement of low-frequency base and collector current noise and coherence in sige heterojunction bipolar transistors using transimpedance amplifiers, *IEEE Transactions on Electron Devices*, Vol. 46, Issue 5, 1999, pp. 993-1000.

[21]. L. Bary, M. Borgarino, R. Plana, T. Parra, S. J. Kovacic, H. Lafontaine, J. Graffeuil, Transimpedance Amplifier-Based Full Low-Frequency Noise Characterization Setup for Si/SiGe HBTs, *IEEE Transactions on Electron Devices*, Vol. 48, Issue 4, 2001, pp. 767-773

[22]. S. Bruce, L. K. J. Vandamme, A. Rydberg, Improved correlation measurements using voltage and transimpednace amplifiers in low-frequency noise characterization of bipolar transistors, *IEEE Transactions on Electron Devices*, Vol. 47, Issue 9, 2000, pp. 1772-1773.

[23]. B. Lambert, N. Malbert, F. Verdier, N. Labat, A. Touboul, L. K. J. Vandamme, Low frequency gate noise in a diode-connected MESFET: measurements and modeling, *IEEE Transactions on Electron Devices*, Vol. 48, Issue 4, 2001, pp. 628-633.

[24]. M. Marin, M. J. Deen, M. de Murcia, P. Llinares, J. C. Vildeuil, Effects of body biasing on the low frequency noise of MOSFETs from a 130 nm CMOS technology, *IEE Proceedings - Circuits Devices Systems*, Vol. 151, Issue 2, 2004, pp. 95-101.

[25]. S. Okhonin, M. A. Py, B. Georgescu, H. Fischer, L. Risch, DC and low-frequency noise characteristics of SiGe P-Channel FET's designed for 0.13um technology, *IEEE Transactions on Electron Devices*, Vol. 46, Issue 7, 1999, pp. 1514-1517.

[26]. M. Borgarino, C. Florian, P. A. Traverso, F. Filicori, Microwave large-signal effects on the low-frequency noise characteristics of GaInP/GaAs HBTs, *IEEE Transactions on Electron Devices*, Vol. 53, Issue 10, 2006, pp. 2603-2609.

[27]. A. A. Lisboa de Souza, J. C. Nallatamby, M. Prigent, Low-frequency noise measurements of bipolar devices under high DC current density: whether transimpedance or voltage amplifiers, in *Proceedings of the 1st European Microwave Integrated Circuits Conference*, 2006, pp. 114-117.

[28]. M. Borgarino, Full direct low frequency noise characterization of GaAs heterojunction bipolar transistors, *Solid-State Electronics*, Vol. 49, 2005, pp. 1361-1369.

[29]. E. Kamrani, A. Chaddad, F. Lesage, M. Sawan, Integrated transimpedance amplifiers dedicated to low-noise and low-power biomedical applications, in *Proceedings of the 29th IEEE Southern Biomedical Engineering Conference*, Miami, Florida, USA; 3-5 May 2013, pp. 5-6.

[30]. A. Trabelsi, M. Boukadoum, Comparison of two CMOS front-end transimpedance amplifiers for optical biosensors, *IEEE Sensors Journal*, Vol. 13, Issue 2, 2013, pp. 657-663.

[31]. E. Sackinger, The transimpedance limit, *IEEE Transactions on Circuits and Systems I: Regular papers*, Vol. 57, Issue 8, 2010, pp. 1848-1856.

[32]. B. Razavi, A 622Mb/s 4.5pA/yHz CMOS transimpedance amplifier, in *Proceedings of the IEEE International Solid-State Circuits Conference*, San Francisco, California, USA, 9 February 2000, pp. 162-163.

[33]. S. M. Park, C. Toumazou, A packaged low-noise high-speed regulated cascade transimpedance amplifier using a 0.6um N-well CMOS technology, in *Proceedings of the 26th IEEE European Solid-State Circuits Conference*, 19-21 September 2000, Stockholm, Sweden, pp. 431-434.

[34]. K.-M. Lei, H. Heidari, P.-I. Mak, M.-K. Law, F. Maloberti, Exploring the noise limits of fully-differential micro-watt transimpedance amplifiers for Sub-pA/yHz sensitivity,

in *Proceedings of the IEEE 11ᵗʰ Conference on PhD Research in Microelectronics and Electronics*, 29th June-2nd July 2015, Glasgow, Scotland.

[35]. M. Borgarino, G. Betti Beneventi, V. Doga, P. Pavan, On the limitations of the transimpednace amplifiers as tools for the low frequency noise characterization, *Microelectronics Journal*, Vol. 45, 2014, pp. 152-158.

[36]. M. Borgarino, L. Bary, D. Vescovi, R. Menozzi, A. Monroy, M. Laurens, R. Plana, F. Fantini, J. Graffeuil, The correlation resistance for low-frequency noise compact modeling of Si/SiGe HBTs, *IEEE Transactions on Electron Devices*, Vol. 49, Issue, 5, 2002, pp. 863-870.

[37]. P. A. Traverso, C. Florian, M. Borgarino, F. Filicori, An empirical bipolar device nonlinear noise modeling approach for large-signal microwave circuit analysis, *IEEE Transactions on Microwave Theory and Techniques*, Vol. 54, Issue 12, 2006, pp. 4341-4352.

[38]. A. A. Lisboa de Souza, E. Dupouy, J.-C. Nallatamby, M. Prigent, J. Obregon, Experimental characterization of the cyclostationary low-frequency noise of microwave semiconductor devices under large signal operation, *International Journal of Microwave and Wireless Technologies*, Vol. 2, Issue 2, 2010, pp. 225-233.

[39]. P. A. Traverso, C. Florian, F. Filicori, A fully nonlinear compact modeling approach for high-frequency noise in large-signal operated microwave electron devices, *IEEE Transactions on Microwave Theory and Techniques*, Vol. 63, Issue 2, 2015, pp. 352-366.

[40]. A. Rodriguez-Testera, O. Mojon, M. Fernandez-Barciela, E. Sanchez, P. J. Tasker, Nonlinear HBT table-based model including low-frequency noise effects, *Electronics Letters*, Vol. 46, Issue 9, 2010, pp. 635-636.

[41]. C. Florian, P. A. Traverso, M. Borgarino, F. Filicori, A Non-Linear Noise Model of Bipolar Transistors for the Phase-Noise Performance Analysis of Microwave Oscillators, in *Proceedings of the IEEE International Microwave Symposium*, 11-16 June 2006, San Francisco, California, USA, pp. 659-662.

[42]. L. Bary, G. Cibiel, I. Telliez, J. Rayssac, A. Rennane, C. Boulanger, O. Llopis, M. Borgarino, R. Plana, J. Graffeuil, Low frequency noise characterization and modeling of microwave bipolar devices: application to the design of low phase noise oscillator, in *Proceedings of the IEEE Radio Frequency Integrated Circuits Symposium*, 2-4 June 2002, Seattle, Washington, USA, pp. 359-362.

[43]. C. Ciofi, F. Crupi, C. Pace, G. Scandurra, How to enlarge the bandwidth without increasing the noise in OP-AMP-based transimpedance amplifier, *IEEE Transactions on Instrumentation and Measurement*, Vol. 55, Issue 3, 2006, pp. 814-819.

[44]. P. Magnone, P. A. Traverso, G. Barletta, C. Fiegna, Experimental characterization of low-frequency noise in power MOSFETs for defectiveness modelling and technology, *Measurement*, Vol. 52, 2014, pp. 47-54.

[45]. K. Ohmori, R. Hasunuma, S. Yamamoto, Y. Tamura, H. Jiang, N. Nishihara, K. Masu, K. Yamada, Application of low noise TIA ICs for novel sensing of MOSFET noise up to the GHz region, in *Proceedings of the IEEE Symposium on VLSI Circuits*, 10-13 June 2013, Kyoto, Japan, pp. C40-C41.

[46]. B. Razavi, Design of Analog CMOS Integrated Circuits, *McGrawHill*, 2001.

[47]. A. Hastings, The Art of Analog Layout, *Prentice Hall*, 2001.

Chapter 5

Integrated Low-Power Gating Scan Cell for Test Power Minimization

Mahshid Mojtabavi Naeini, Sreedharan Baskara Dass, Chia Yee Ooi, Tomokazu Yoneda and Michiko Inoue

5.1. Introduction

While operational frequency increase is becoming an inevitable part of the current high density integrated circuits, it has brought up new challenges to test engineers. The major limiting factor on concurrent testing of cores in System-On-Chip (SOC) is the inherent excessive power consumption. High test power consumption not only increases test cost, but also may cause reliability hazards or even instant damages [1]. The excessive average power consumption, in the first place, produces extra heat in CUT (Circuit-Under-Test) which has inevitable role in appearing hot spots, degradation of performance, circuit premature destruction, functional failures and as a result, circuit reliability degradation. The main mechanisms which put the circuit at the risk of structural degradations are corrosion (oxiding of conductors), electro-migration (molecular migration of the conductor structure toward the electronic flow), hot-carrier-induced defects, or dielectric breakdown (loss of insulation of the dielectric barrier) [2]. Note that excessive average power effect on temperature increase may turn to more severe consequences such as temperature variations. The temperature variations may induce timing variations during test and, in some cases, may lead to test-induced yield loss [3]. In addition, intensive heat generated by high switching activity during the test process has negative influences on the circuit packaging cost to make higher level of temperature tolerable. However, it is worth to note that, on top of the elevated average power consumption during test application that can cause severe problems in the CUT, excessive peak power violation is another area which requires particular attention. One of the problems that can arise from elevated peak power dissipation during test application is noise induced by probes (probes used for wafer level testing) which generally have higher inductances than the power and ground pins embedded on the chip. Therefore, increasing switching activity during test time can cause power supply noise which leads to voltage glitches and finally die failure. Thus, this

Mahshid Mojtabavi Naeini
Malaysia-Japan International Institute of Technology, Universiti Teknologi Malaysia

results in unwanted yield loss. It is widely reported that IR-drop, ground bounce and cross-talk that are also known as noise phenomena, appear as a result of higher imposed switching activity during the test phase. Therefore, test power reduction has been turned to one of the hot topics for current test engineers. Scan-based testing is a structural Design-For-Testability (DFT) technique and due to its strong characteristics in controllability and observability, it has become the most widely employed techniques in both Built-In-Self-Test (BIST) and non-BIST testing schemes. Unfortunately, scan-based architectures are very expensive in terms of power as each scan test pattern contributes to a shift operation with high power consumption

Scan test power consumption can fall into two major sub-categories: shift power consumption and capture power consumption. Shift power consumption is due to the transitions occurring in scan cells when the adjacent bits in test vector have different values. These transitions not only cause switching activities in scan cells but also propagation of switching activities to the combinational logic through scan cells outputs. Often, capture power consumption is referred to transitions that happen within scan cells when they have different values before and after capture. However, in some literatures, transitions in combinational part in launch cycles also have been considered as capture power. In this paper, we consider the whole capture clock cycle including both launch and capture clock edges for evaluating peak power consumption. The remainder of the paper is organized as follows. In Section 5.2, the existing methods in literature on reducing shift switching activity in combinational logic and scan chain are reviewed. Section 5.3 outlines the motivation and novelty of the proposed gating scan architectures. Sections 5.4 and 5.5 elaborate the proposed ILPG and CPILPG scan cell structures and their operating principles, respectively. In Section 5.6, the application of the proposed gating scan cells in partial gating is discussed. Experimental results and analysis have been presented in Section 5.7. Finally, Section 5.8 draws the conclusion.

5.2. Literature Review

In recent years, enormous research studies have been conducted to reduce power in scan-based testing architecture. For scan-based designs in particular, there are two sources of power consumption during shift mode. The switching activities happen in the scan chain caused by scan-ripple and the switching activities in the combinational logic due to propagation of rippling transitions in the scan chain. Switching activity in the combinational part contributes to a large portion of the total switching activity in the circuit [4]. The shift power reduction strategies can mainly fall into two categories: 1) ATPG-based approach; and 2) DFT-based approach.

ATPG-based technique is a widespread approach that generally focuses on power reduction by incorporating power optimized strategies in test pattern generation phase. The main approaches in this category that attempt to minimize the switching activity in the scan chain and consequently in combinational part through reducing the transitions in the scan chain can be classified as power-aware test pattern generation [5-7], test vector reordering [8-10], scan cell reordering [11-14] and X-filling [15-19]. Test vector reordering and equivalently scan cell reordering methods reduce toggling in scan chain based on the correlation between successive test patterns. However, these methods suffer from high computational time due to solving hard problems. Furthermore, they may raise the scan routing congestion problem and higher silicon area cost. In X-filling techniques, the don't care bits in test cube are re-assigned in such a manner that causes less transitions during scan shifting. However, test cube bit re-assignment techniques usually cannot achieve the same amount of power reduction as DFT-based techniques do [18]. In [20, 21], ATPG-based algorithms have been presented to generate primary input patterns such that they suppress the redundant transitions in combinational part originated from scan chain during shifting. However, the primary input control method in [20, 21] is not sufficient enough to reduce spurious switching activities in combinational part since, in large industrial designs, internal signal lines are mostly controllable through scan cells rather than primary inputs. Finally, often utilizing power optimizations strategies in ATPG algorithms prolongs test application time [22]. In addition, they are not able to eliminate all redundant transitions in combinational part during shift mode of operation.

The DFT-based approaches are less intrusive to the designs, and are easy to be integrated with different embedded compression techniques [23]. An approach known as scan partitioning splits the scan chain into multiple partitions and activates only one partition at any time interval [24-27]. A segment regrouping algorithm has been proposed by Yamato *et al.* in [27] to identify an optimal combination of scan segments to be clocked simultaneously and further instantaneous shift power reduction in scan chain has been achieved. There are two main drawbacks to the scan chain partitioning approaches. First, they may cause data dependence problems. Second, they require complicated control circuits for activation of proper partition at the right time; thus, they have high penalty in terms of area. Although significant achievement in test power reduction has been gained due to scan chain reordering and partitioning techniques, none of the solutions within these categories have completely eliminated spurious switching activities in the combinational logic completely during shift mode. Clock gating is one of the well-known power reduction techniques in the wide range of DFT-based methods [28-32]. The main drawback of this method rises from clock skew problems in normal mode of operation. The approach in [31] divides the scan chain into two partitions (odd and even scan cells) and uses two separate clocks for each partition. At any time instant, only one partition is active so that the peak and average power during scan period have been reduced by a factor of two. The major disadvantage of this approach is that by increasing the scan cells derived by each clock control signal, the elevated delay may make it difficult to meet the timing closure. Scan chain modification was proposed in [22] by inserting extra logic gates (such as INV, XOR, XNOR) to minimize scan chain transitions during shifting. This approach suffers from high area overhead due to additional logic gates and great computation complexity to find optimal position for logic gates to be inserted. More importantly, it adds more delay to the shift path that slows down the shift process, thus increasing test application time. Gating logic insertion is a straight-forward method to isolate the switching activities occurring in scan chain from propagating to combinational part during data shift-in/out process. This method is less intrusive to the original designs, and is independent of test set. However, performance degradation due to the delay that gating logic imposes on the critical paths is the main concern. Furthermore, depending on the added logic, high penalty in terms of area may be unavoidable. In [33, 34], authors reported gating the stimulus paths of flip-flops to a constant logic "1" or "0" by utilizing gating logic (NOR [33], transmission gate (TG) together with a pull up or a pull down transistor [34]) at the scan cells output, and thus eliminating spurious switching in logic gates. However, depending on the logic on the scan cell output, they still may not be able to block all switching from propagating to the combinational logic in the beginning of shift cycle if the logic value on scan cell output before shift mode differs from the gating logic value. Thus, the unblocked transient still can propagate to the deeper level of the combinational logic, causing many transitions at circuit internal lines before reaching the steady state. Approaches in [35, 36] experimented the gating logic (MUX [35], extra inverter-base latch [36]) to hold the scan output at the previous logic and completely blocks any redundant switching activity in the combinational logic during shift mode. An enhanced scan structure was presented in [4] for delay fault testing. In this method, each scan cell has been augmented into an AND-NOR-based hold latch at the scan cell output to ensure that the scan stimulus paths remain unchanged during shift session. The hold latch has been implemented by a cross-coupled of NOR and two AND gates to hold the scan cells' stimulus paths. However, these approaches suffer from large overhead in terms of area and propagation delay. The first level supply (FLS) gating scheme proposed by Bhunia *et al.* [37] have inserted a common pull-down gating transistor to the gates at the first level of logic which are typically connected to the scan chain outputs. In order to prevent the outputs of the first level gates from being floating, they have added another pull-up transistor to force the output to VDD. FLS scheme has less overhead in terms of area, propagation delay or even switching activity in gating logic compared with other gating schemes; however, it is unable to block all transients in the combinational logic in the beginning of shift cycle because the output of the first level gates are forced to a fixed value, which is similar to NOR and TG gating approaches. Bhunia *et al.* gave an alternative solution in [38] for Enhanced Scan in delay fault testing named First Level Hold (FLH) which gates the rippling transition to the combinational logic during shift mode. Recently, shift power reduction through scan cell modification has been proposed [39-42]. In Jump scan architecture [39], scan cell structure has been modified such that an extra multiplexer and D-latch has been applied to the scan cell architecture. The shift power, therefore, is reduced by halving scan clock frequency since it shifts two bits of scan pattern per clock cycle. This modification, however, causes performance degradation of scan cell during normal mode of operation, in addition to high area cost.

Suhag *et al.* [40] proposed a modified scan flip-flop for gating critical paths to eliminate performance degradation caused by gating logics. Fig. 5.1 demonstrates a detailed architecture of this scan element. The presented modified scan cell ties the data propagation line from scan cell to combinational logic to constant logic value '1'. Therefore, it is not able to cut off all the redundant switching activity in the combinational logic for the same abovementioned reason about freezing the scan cell output to a constant logic during shifting. Nineteen (19) extra transistors in the scan cell compared to the original structure result in significant power consumption inside the scan cell while having severe impacts on area overhead. In addition, a modified scan cell has been presented in [41] for delay fault testing application. As it has been shown in Fig. 5.2, it bypasses the slave latch with an alternative low cost dynamic latch in scan shifting path. Therefore, it can successfully eliminate all transitions to the combinational logic. However, there is no improvement in the level of power consumption inside the scan cell and as a result in scan chain.

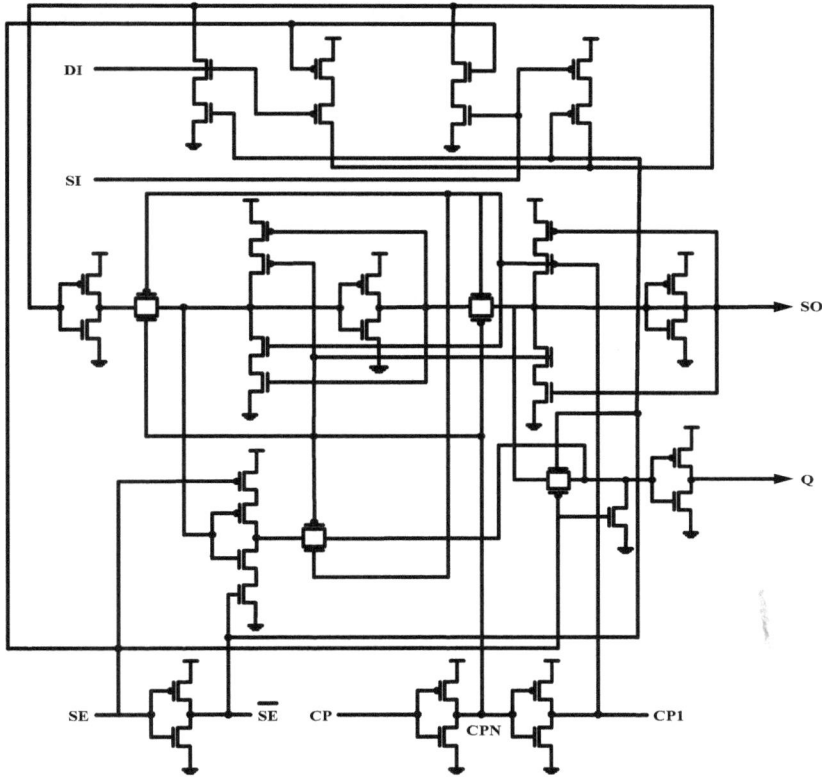

Fig. 5.1. Modified scan cell for critical path.

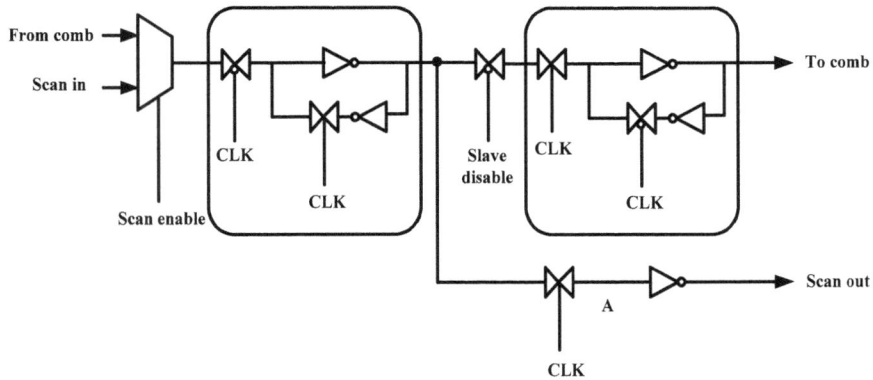

Fig. 5.2. Modified scan flip-flop for low power delay fault testing [41].

Jayagowri *et al.* [42] proposed a double edge trigger scan cell called DMOLDET scan cell that is able to reduce power consumption in clock line by halving the clock frequency. Two gating schemes as demonstrated in Figs. 5.3 (a) and (b) are presented to gate functional output of DMOLDET scan flip-flop to combinational logic during shift mode.

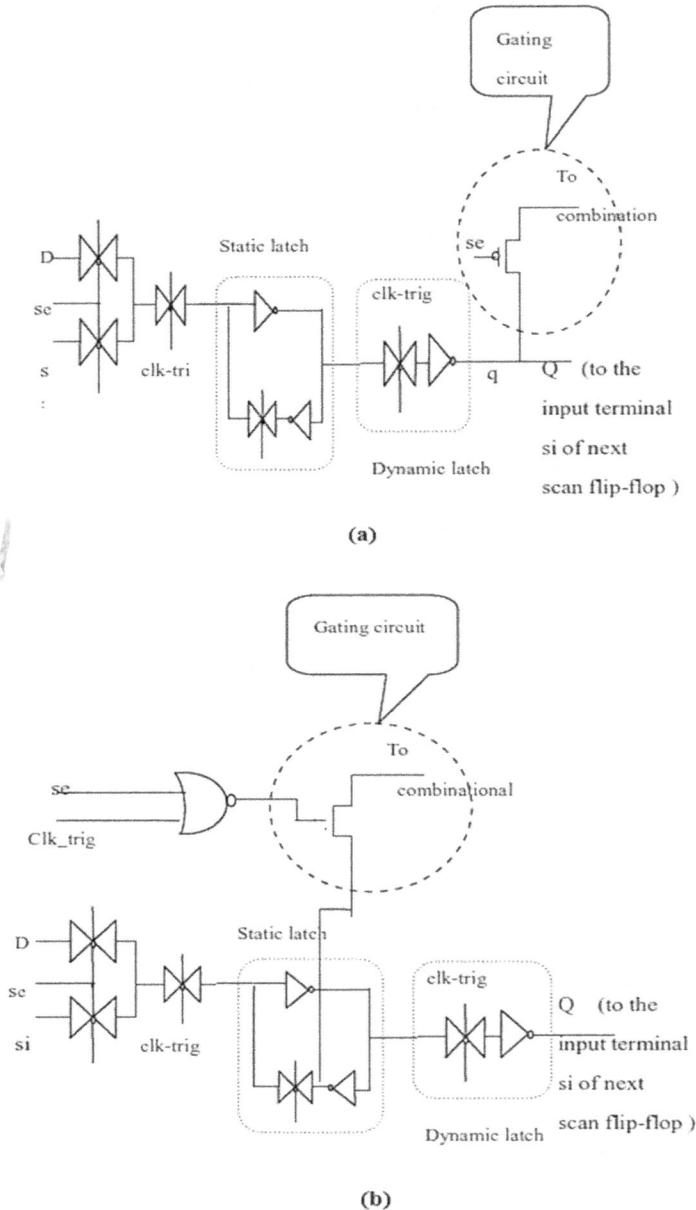

(a)

(b)

Fig. 5.3. Switch level circuit diagrams of the gating techniques (excluding clock driver circuits). (a) PMOS gated DMOLDET scan flip-flop; (b) NMOS and NOR gated DMOLDET scan flip-flop [42].

First gating scheme includes a single pass transistor controlled by shift enable signal and second scheme includes a single pass transistor controlled by a NOR gate that has been utilized on the scan cell output to combinational circuit in order to block redundant transition from scan cell to combinational logic. The main drawback of this design is that scan cell stimulus path to combinational part is driven by a single pass transistor and is remained floating. Regarding [37], the voltage of a floated output is determined by the leakage balance between the pull-up PMOS and pull-down NMOS network of the gate. Crosstalk noise or transient effects due to soft error also can easily change the voltage of floating output. Therefore, the floating output in this architecture can cease many transitions in combinational circuit during shifting.

In the past few years, partial gating techniques [23, 43-47] have been receiving more attention as a replacement for full gating approaches to alleviate the full gating penalties in area overhead and performance degradation. By proper selection of scan cells on non-critical paths to be gated and their gating values, they try to maximize shift power reduction with acceptable performance degradation. Hardware implementations of the gated scan cell in partial gating method presented in [23] is depicted in Figs. 5.4 (a) and (b).

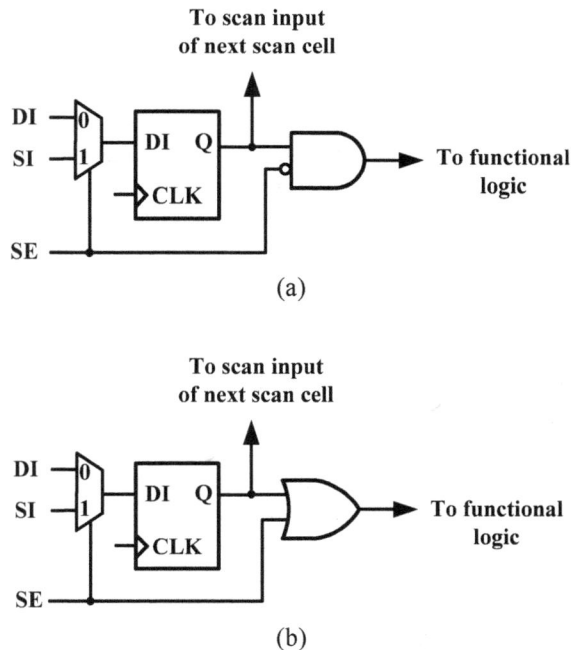

Fig. 5.4. Hardware implementation of the gated scan cell in partial gating method [23]: (a) Gating to logic "0" by ANDB. (b) Gating to logic "1" by OR.

The partial gating method in [43] tries to control performance degradation by inserting multiplexer as gating logic only on non-critical scan cells as shown in Fig. 5.5.

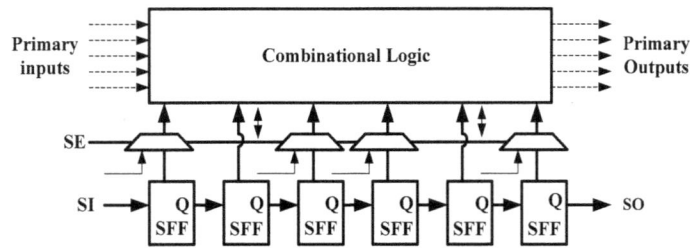

Fig. 5.5. Partial gating architecture using multiplexer [43].

Most recently reported partial gating methods such as those in [23, 48-50] gate a sub-set of scan cells not only during shift mode but also during capture mode in order to reduce peak power besides shift power reduction. However, in large industrial designs, scan cells have large fan out cones. Thus, un-gated scan cells in partial gating method can still cause a great amount of switching activities in the combinational logic. The missing part of both existing full gating and partial gating methods is that they have limited their approaches only to power reduction in combinational circuits during shift cycles without highlighting much that shift power consumption is also related to the total amount of power consumed in scan chain besides combinational logic.

5.3. Motivation

In scan-based testing, the major portion of the power and energy is dissipated during shift process as has been outlined in [33] since a large portion of the test application time includes shift cycles especially for large industrial designs with long scan chains. Aforementioned gating methods mainly have tried to reduce the level of switching activity in the combinational part. However, only a few schemes can be found in the literature that have concentrated on reducing the level of switching activity inside the scan cell.

The missing part of both existing full gating and partial gating methods is the highlight of the relation of shift power consumption to the total amount of power consumed in scan chain besides combinational logic. Moreover, in both existing full gating and partial gating techniques, significant power is consumed in the gating elements themselves, which causes the peak power to increase from 5 % to 60 % in all the benchmark circuits when the gating overhead was considered [41]. This is due to the large switching activity occurring in the gating logic when scan mode changes to capture mode or vice versa [40]. Varieties of previous works have successfully reduced power consumption in both combinational logic and scan chain by using separate methods for each section. In [51], FLS has been utilized with a scan partitioning method which activates a part of scan cell at a time to reduce power consumption in scan chain. Finally, a scan chain partitioning technique has been combined with scan cell modification for low power scan operation [52].

To the best of our knowledge, the current literature does not offer an integrated solution targeting for power reduction in combinational part as well as inside the scan cell itself that is the main novelty of our proposed scheme. Therefore, the proposed design is able to reduce total power consumption during both shift (test) mode and functional mode of operation. Finally, an important point to note is that, although power issue is one of the major concerns in testing of modern VLSI circuits, other parameters such as test application time, area overhead, and fault coverage needs especial attention to be devoted. In this paper, we focus on shift power reduction in both combinational part and scan chain while maintaining the capture peak power under reasonable threshold. The proposed scan cells are positive edge triggered.

5.4. Integrated Low Power Gating (ILPG) Scan Cell

Motivated by the limitation of the existing gating methods, a novel low power gating (ILPG) scan architecture for shift power reduction considering both scan chain and combinational part has been proposed. The proposed gating scan cell covers the following features:

A. Gating redundant transitions from scan cell to combinational logic during shift mode.

B. Reducing switching activity inside the scan cell during shift mode.

The rippling transitions through shift session cause great switching activities in the scan chain. The propagation of this switching activity into the combinational part contributes to large redundant transitions in the circuit lines. In order to suppress the scan chain transitions from propagating during shift cycles, we have proposed a scan cell which contains a modified slave latch augmented by a gating logic.

For constructing the gating logic, we have utilized a transmission gate and an inverter to gate the scan output to the combinational logic. It uses the transmission gate to cut off the connection between the inverted scan cell output $\bar{Q}$ (used for stitching the scan cells) and the output Q of the scan cells during shift mode. As a result, the switching activities on $\bar{Q}$ during shift mode does not affect the scan cell output Q which is used for driving the combinational logic. High resistance offered by an inactive transmission gate reduces the leakage current in the transmission gate during shift mode and response capture cycle since the transmission gate is idling in these intervals. In addition, transmission gate is a strong driver to feed the gating logic inverter and pseudo primary inputs during normal/capture mode.

In order to totally prevent the unnecessary transitions to the combinational logic during shifting, a state preserving logic has been proposed. It is a feedback structure that refreshes the scan output Q with the previous logic state. The two pull-up and pull-down sleep transistors controlled by *SE* signal are active during shift mode which cause the state preserving logic fixes the scan output logic to the same previous logic. However, unlike gating logic, this section is transparent in normal/capture mode of operation since the sleep transistors are inactive. Figs. 5.6 (a) to (c) indicate the transistor-level implementation of proposed ILPG scan cell and the data paths through shift and normal/capture mode of operation, respectively.

During **shift mode** of operation, the value of *SE* signal is '1' that causes transmission gate in modified slave latch to be switched off. This results in blocking clock signal and shifting test patterns from propagating to the modified slave latch and activates the refresh inverter feedback to maintain the logic value on the scan cell *Q* output. Although during this mode, two pull up/down sleep transistors are active, the stacking effect still cause active leakage reduction in both proposed scan cells during shift. This is because the leakage of a two-transistor stack is an order of magnitude less than the leakage in a single transistor [53]. An analysis of the subthreshold leakage through a stack of n transistors is given in [54].

During **normal/capture mode** of operation, the value of *SE* signal is '0', this causes the transmission gate in modified slave latch to be switched to transfer the *DI* data from combinational logic that has passed through master latch output and capture it on scan cell *Q* output. Therefore, during this mode, scan cell is clocked exactly the same as conventional scan cell. At this time, refresh feedback inverter (enclosed in a red square in Fig. 5.6 (c)) is switched off since the set of pull up/down sleep transistors that are controlled by *SE* signal are inactive. This will cut off the power rail and results in standby leakage power reduction in the refresh feedback inverter since the main source of leakage power is subthreshold current flowing from VDD to GND [55].

According to [55], subthreshold leakage current flowing through a stack of transistors connected in series will decrease when more than one transistor in the stack is turned off. This effect is known as the stacking effect. Note that, one way to increase the efficiency of stack effect in reducing leakage power as outlined in [56] is to apply leakage control transistor(s) (low V_{th}) in series to the gates as sleep transistor(s). However, in our proposed scan cells, we have employed typical transistors in the technology library for implementing all transistors including sleep transistors in order to evaluate the proposed scan cells in the worst case scenario and also to conduct fair comparisons over conventional scan cell and other existing techniques.

It is worth a mention that, when clock signal is low, the voltage value on $\overline{Q}$ is determined based on loading capacitance at the transmission gate. Therefore, by changing clock edge from positive to negative, this node realizes some voltage changes due to leakage discharge of the loading capacitance. However, in a similar trend with proposed scan cell schemes in [41, 42], in the proposed ILPG scan cell the amount of change in voltage value at $\overline{Q}$ is minor and very close to the exact voltage value (VDD or GND). Thus, the proposed ILPG scan cell is still able to deliver the right binary logic value through scan cell output $\overline{Q}$. Furthermore, $\overline{Q}$ is a uni-fanout driver since it is allocated as a scan output to merely derive the next scan cell in the scan chain. The voltage instability of the derived scan data through $\overline{Q}$ at clock negative edge is thoroughly stabilized through the CMOS feedback inverter ring of the master latch of the successive scan cell and the right binary logic is stored in the next scan cell. The scan cell output *Q*, however, does not experience the above mentioned instability at negative edge of clock since $\overline{Q}$ drives *Q* through a CMOS inverter which is full swing logic and stabilizes the voltage level.

(a)

(b)

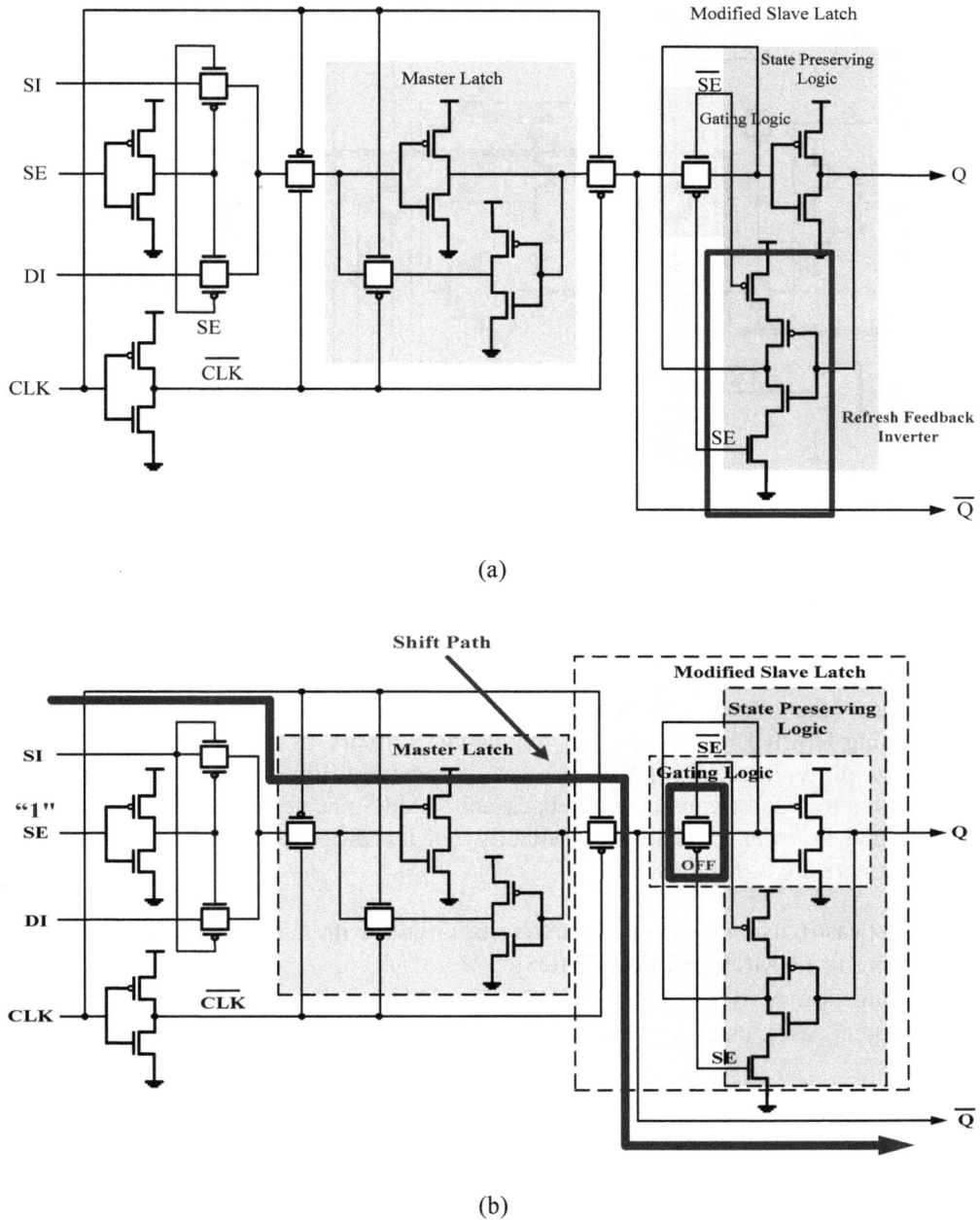

Fig. 5.6 (a, b). Transistor-level implementation of proposed scan cell: (a) Proposed ILPG scan cell with gating and state preserving logic; (b) Data path in shift mode.

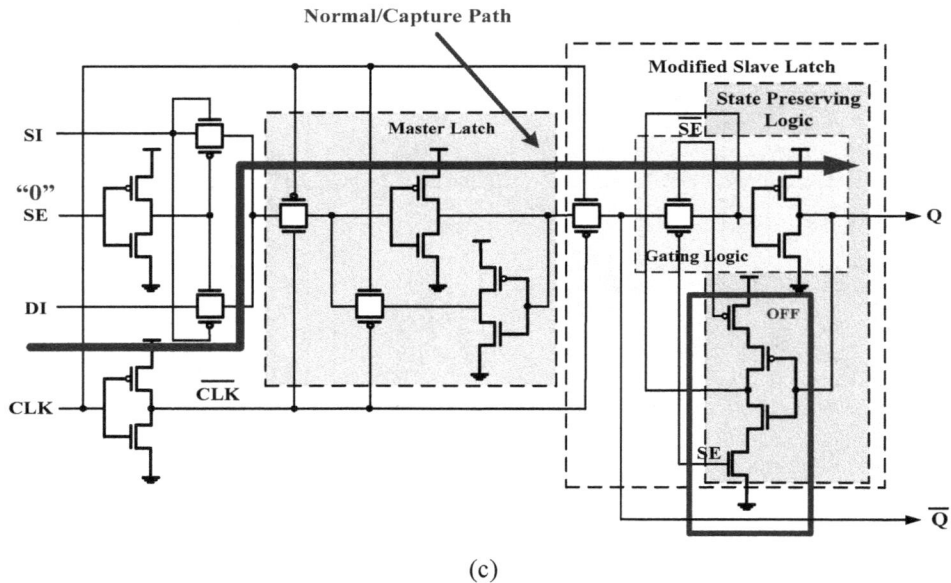

(c)

Fig. 5.6 (c, Continued). Transistor-level implementation of proposed scan cell: (c) Data path in normal/capture mode.

Fig. 5.7 shows the proposed ILPG scan cell Q and $\overline{Q}$ output waveforms in various process corners using HSPICE in Synopsys 90nm standard library by applying random patterns with the supply voltage of 1.0 Volt at room temperature. Five different process corners were adopted to precisely evaluate the back-annotated Spice netlist of the proposed ILPG scan cell and in order to assure the reliability of the proposed scheme. The employed process corners are as follows:

- FF (best case): using the best case extracted circuit, with fast NMOS and fast PMOS;
- FS: using fast NMOS and slow PMOS;
- SF: using slow NMOS and fast PMOS;
- TT (typical case): using a typical extracted circuit, with typical NMOS and typical PMOS;
- SS (worst case): using the worst case extracted circuit, with slow NMOS and slow PMOS.

As it is mentioned earlier, one of the concerns regarding scan gating is high capture peak and average power that is due to extra power consumption in gating logics during capture mode of operation.

In general, switching activities related to capture mode take place when (i) circuit mode changes from capture mode to shift mode; or (ii) circuit mode changes from shift mode to capture mode. The following text elaborates how the internal structure of the proposed ILPG scan cell reduces the power dissipation during these mode changes.

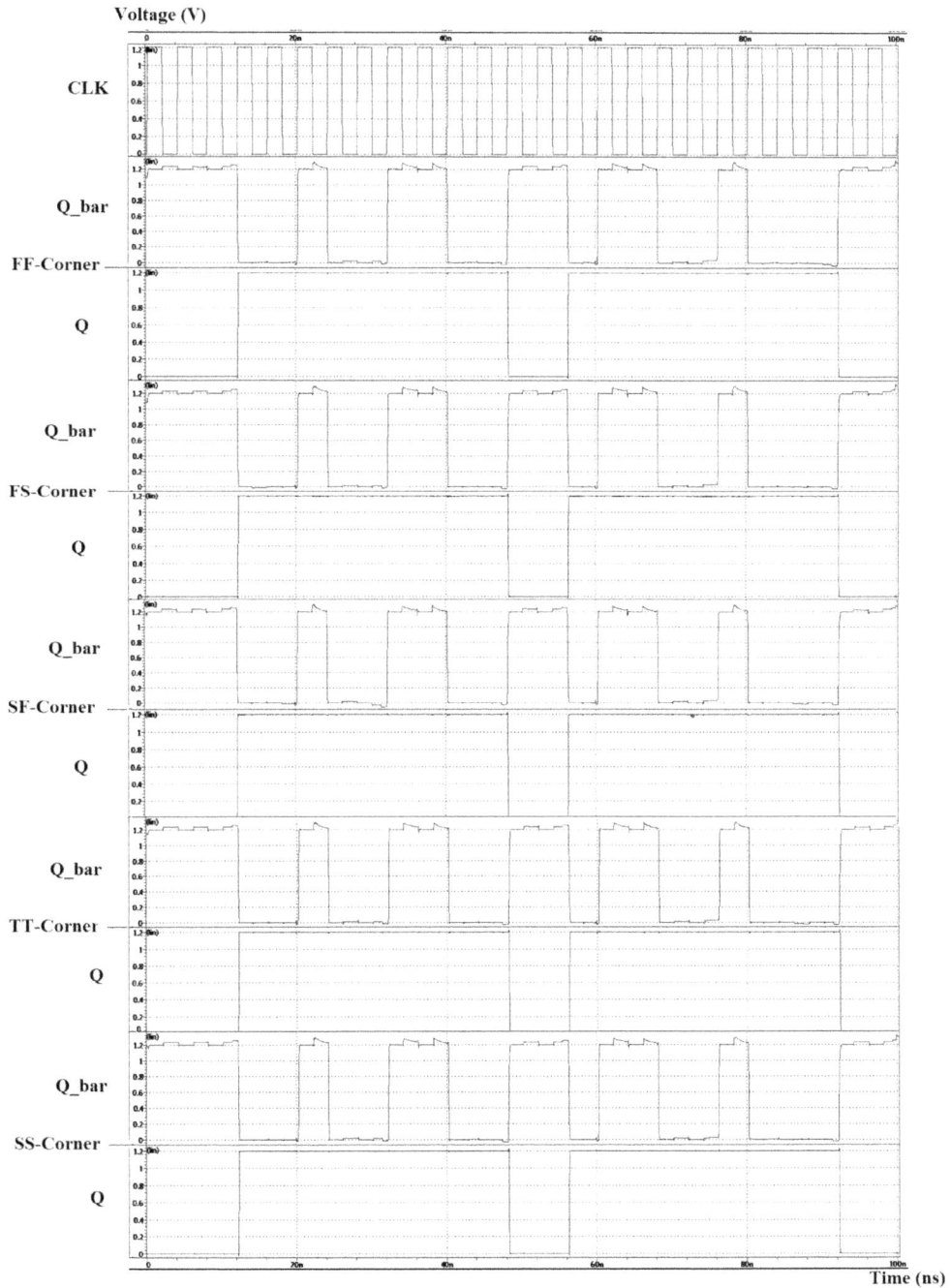

Fig. 5.7. Post-layout simulated waveform of the proposed ILPG scan cell outputs using five different process corners at 250 MHz clock frequency.

During testing when circuit mode changes from capture to shift mode, those scan gating approaches [33-34] that try to tie the scan functional output to constant logic '0' or '1' will fail to block all the spurious switching from propagating to the combinational logic. This is because the previously captured value on this path may be different from the constant logic value which the scan gating approach is trying to tie this path to and this will cause extra switching activity on the flip-flop output that will be propagating into the combinational logic. However, this problem has been solved in some approaches including our proposed state preserving scan cell that maintain the previously captured value on the flip-flop output and they do not cause the extra switching activity. Note that the difference between our proposed approach and other holding approaches [4, 35-36, 38] that are able to maintain the previously captured logic value during shift mode is in the scan cell internal structure. For more clarification, please consider Fig. 5.8 as an example scenario.

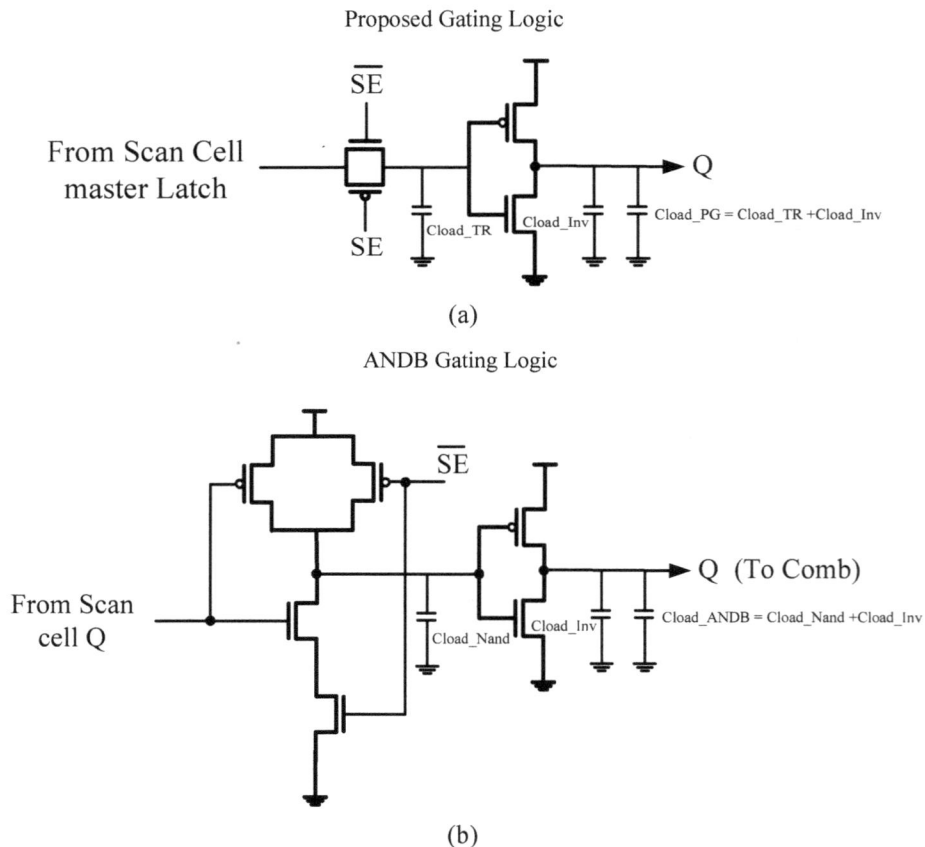

Fig. 5.8. Transistor-level implementation of gating logics. (a) Our proposed gating Logic; (b) ANDB gating logic.

During testing when circuit mode changes from shift mode to capture mode, all scan gating approaches including our proposed methods cause switching activity depending on the gating logic applied to the functional output path. The missing part of previous gating works is that they ignored the impact of gating elements on capture power [49]. There are two sources that cause such significant switching activity as follows:

I. When shift mode changes to capture mode, the value that is supposed to be captured may be different from the value that flip-flop output is holding at gating logic. This will cause switching activity on the scan cell's functional path.

II. The second source is switching activity inside the gating elements (ANDB, OR, MUX etc.). All gating elements are controlled by *SE* signal. As a result, changing the value from '1' to '0' on this signal line causes switching activity in all gating elements.

The two above mentioned sources will cause significant switching activity at the beginning of each capture cycle. That is why capture peak power usually happens at the beginning of each capture cycle.

Similar to existing gating approaches, our proposed method is not able to make any difference in the source of switching activity explained in part I. However, it is able to reduce the second source of switching activity (part II). In the proposed state preserving scan cell, gating logics participating in gating action are a transmission gate and an inverter, as it is demonstrated in Fig. 5.8 (a), located inside the scan cell as a part of the modified slave latch. Fig. 5.8 (b) shows ANDB gating element as an example of one the most common gating techniques that is applied as a redundant logic to scan cell functional output and used to hold scan functional path on logic '0'.

Internal gating logic is one of the reasons that the proposed scan cell does not elevate capture power significantly. Another reason is that the output load capacitance of our gating logic is smaller than gating logic such as ANDB and OR. This is because the main source of switching power in a gate is charging and discharging of the gate output load capacitance. Therefore, the amount of switching power differs from one gate to another based on their complication and implementation technology.

As it can be seen in Figs. 5.8 (a) and (b), the only capacitances elements that make differences in the proposed gating approach and ANDB gating output load include $C_{load\ TR}$ and $C_{load\ Nand}$, respectively. With the same transistor size and technology implementation for both gating logics, $C_{load\ TR}$ has less capacitance compared to $C_{load\ Nand}$ since it is derived by less number of transistors compared to C_{load_Nand}. Therefore, it results in less switching power during data capture.

The capability of the proposed state preserving in standby leakage reduction can alleviate the effect of state preserving on peak power during normal/capture mode. The transmission gate, pull-up and pull-down sleep transistors are driven by shift enable signal *SE* so no extra control signal is required. Sharing the pull-up and pull-down transistors of the state preserving logic among all the scan cells can moderate the scan chain area overhead.

137

The proposed ILPG scan cell works as flip-flop because the inserted state preserving logic together with gating logic function as modified slave latch which enables power reduction.

Due to the shift path with less complexity, the scan cell speed has been accelerated during shifting which consequently alleviate setup time violations. However, likewise other scan cells, hold time still remains as a matter of criteria. The reduced area and propagation delay due to the removal of two inverters and a transmission gate in the scan structure can moderate the area and delay overhead imposed by the augmented gating logic and also state preserving logic.

Another drawback of scan data gating approaches is their high penalty in terms of propagation delay and area overhead. Note that although the proposed ILPG scan cell is still able to deliver improvements in the propagation delay on the launch data path with lower area overhead compared with most existing gating schemes, it still increases this delay arch. In our case, this is because the proposed modified slave latch is not derived directly by clock signal line thereby resulting in higher *CLK-to-launch* propagation delay compared to conventional scan cell. The elevated launch propagation delay may be tolerable based on the timing closure of the path which ILPG scan cell is inserted. However, depending on the path it may still cause timing closure violation in case that the proposed ILPG is inserted on the critical paths with skew margin less than imposed delay. To address this issue, we have proposed Critical Path Integrated Low Power Gating (CPILPG) scan cell which is able to totally eliminate redundant transition in combinational part during shifting while it can improve gating penalty in terms of launch propagation delay. Therefore, it is a more suitable candidate for scan cells on the critical paths.

5.5. Critical Path Integrated Low Power Gating (CPILPG) Scan Cell

In order to eliminate the timing violation problems on critical paths, a new low power gating scan cell called Integrated Low Power Gating (CPILPG) scan cell for critical path is proposed. The critical path in scan-based design is the longest path between any two scan elements through combinational logics. The maximum delay between any two scan cells, also known as critical path delay, is used to determine the maximum clock frequency at which a design can work, as described by equation (5.1).

$$f_{Max} = \frac{1}{(T_{clk-Q} + T_{logic} + T_{setup} + T_{routing} + T_{skew})}, \tag{5.1}$$

where T_{clk-Q} represents time from clock arrival until data arrives at scan cell Q output. T_{logic} is propagation delay through logic between scan cells. $T_{routing}$ is routing delay between scan cells. T_{setup} gives the minimum time that data must arrive at D input of scan cell before the next rising edge of clock. Finally, T_{skew} is the propagation delay of clock between the launch scan cell and the capture scan cell.

According to equation (5.1), reducing T_{clk-Q} and T_{skew} has a direct impact on the maximum clock frequency and consequently, it leads to CUT's overall performance enhancement. The proposed CPILPG scan cell is able to deliver lower T_{clk-Q} during normal/capture mode which results in lower T_{skew}. Thus, it allows higher possible clock operating frequency and consequently overall performance can be improved. The transistor-level structure of the proposed CPILPG scan cell is illustrated by Fig. 5.9 (a).

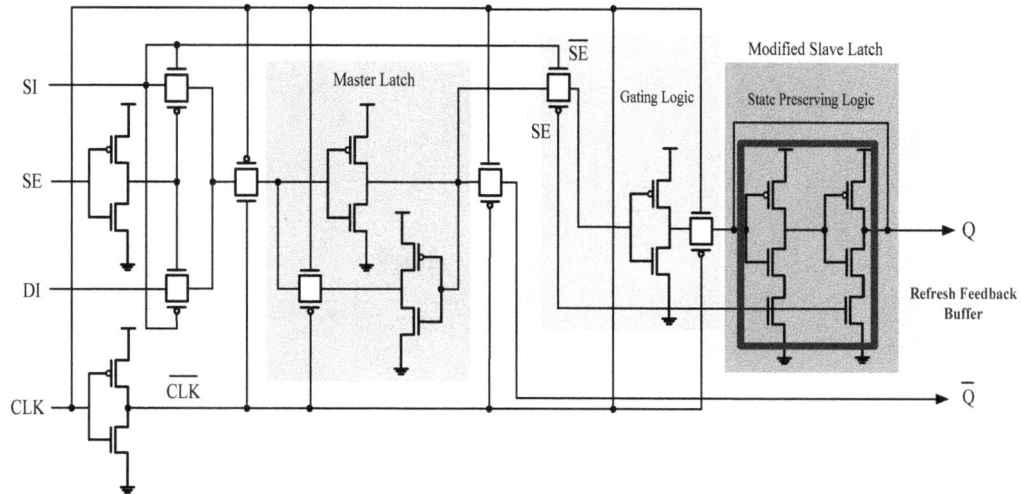

(a)

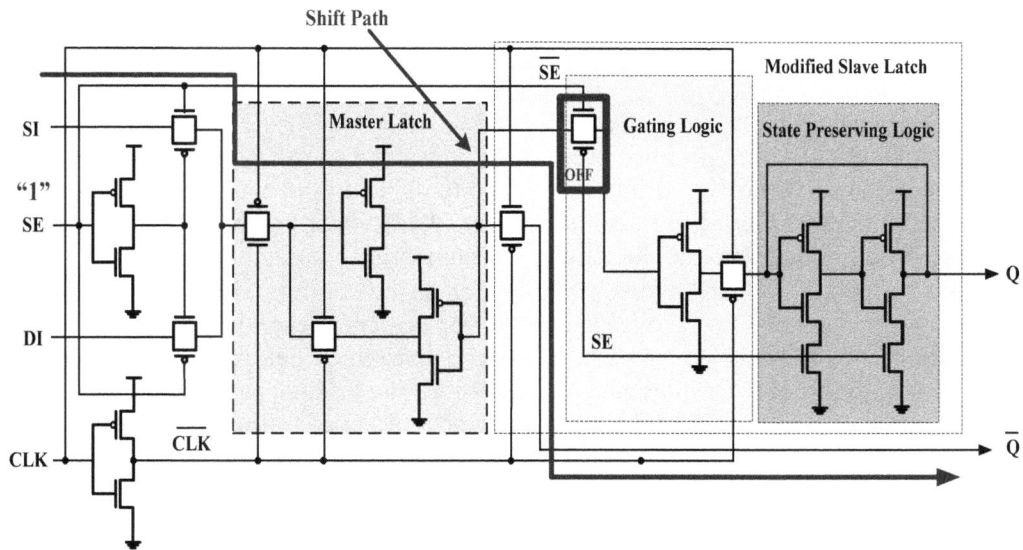

(b)

Fig. 5.9 (a, b). Transistor-level implementation of proposed scan cell for critical path:
(a) Proposed CPILPG scan cell with gating and state preserving logic;
(b) Data path in shift mode.

139

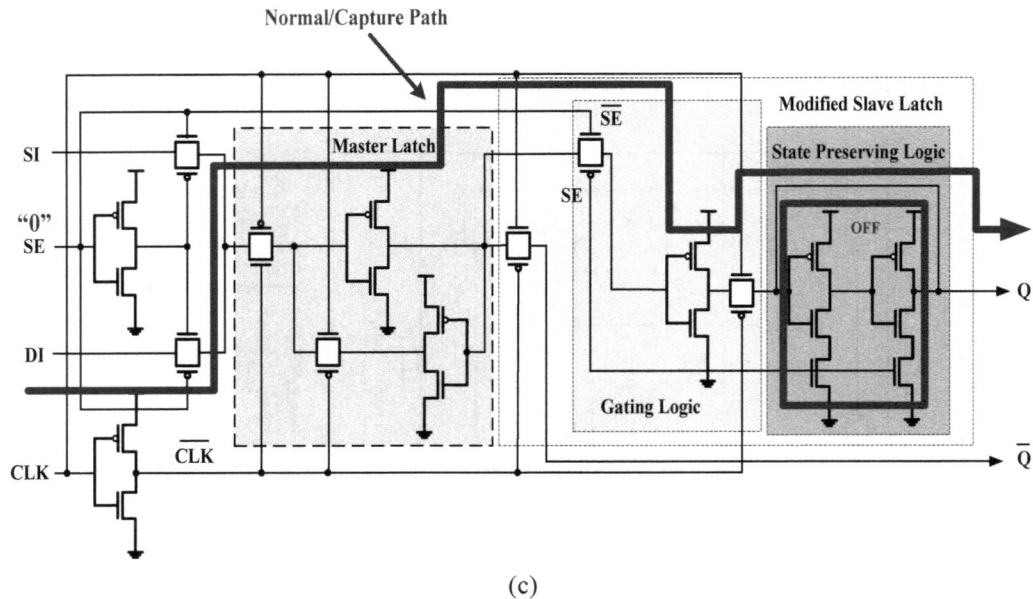

Fig. 5.9 (c, Continued). Transistor-level implementation of proposed scan cell for critical path: (c) Data path in normal/capture mode.

The fundamental principal regarding the proposed CPILPG scan cell is almost the same as ILPG scan cell; except that the gating and state preserving logic structures have been re-modified in order to provide higher speed during normal/capture mode of operation. The complete scan cell operation is elaborated as follows. During shift mode (*SE=1*), similar to ILPG scan cell, a transmission gate controlled by *SE* signal incorporates to prevent the scan cell transition from propagation to the combinational circuit. In order to ensure the hold capability of the output, inverters feedback structure in state preserving logic are active and hold/refresh the logic value on output *Q*, continuously. The shift path through scan cell $\bar{Q}$ output has been highlighted in bold line in Fig. 5.9 (b). During normal/capture mode (*SE=0*), the feedback inverters in state preserving logic are off, thus being bypassed by the data launch path. Again as discussed earlier, thanks to the stacking effect, the two pull-down sleep transistors can reduce the leakage current in the in-active state preserving logic. Using high V_{th} sleep transistors can even further reduce the standby leakage power; however, it will impose more delay penalty on the scan cell output *Q*.

In order to speed up the scan cell operation during normal/capture mode, a transmission gate derived by clock has been located at the end of the launch path as it is observable in Fig. 5.9 (c). This will improve scan cell speed in two ways. First, it reduces clock driving propagation delay, T_{clk-Q} during data launch operation. Second, the charge/discharge of the output *Q* is done through the transmission gate that ensures output *Q* reaches to the steady state at higher speed. This is due to the fact that transmission gate swing voltage is lower than V_{dd} (equal to $V_{dd}-V_{th}$) for logic "1" and higher than V_{ss} (equal to V_{th}) for logic "0". However, this slight decrease in swing voltage can be easily compensated at the first level of CMOS circuits in the combinational part.

5.6. Partial Gating Application

A combination of ILPG scan cells on non-critical delay paths and CPILPG scan cells on critical paths can be efficiently utilized in the application of partial gating in order to deliver higher overall performance. The main advantages of such a hybrid configuration include:

- Both ILPG and CPILPG scan cells have minimum impact on the performance degradation due to smaller launch and capture propagation delay compared to existing gating methods. Therefore, more scan cells can be gated using the ILPG scan cell without timing closure violation. At the same time, CPILPG scan cell can be used to gate more scan cells on the critical paths without causing any critical path timing violation. Thus, higher power reduction is achievable at the cost of neglectable performance loss.
- The proposed scan architectures introduce a new short shift path that improves both shift and capture propagation delays as well as power consumption in scan chain during shift mode. This makes shifting at higher frequency possible in those cases that maximum shift frequency has been bounded by maximum allowable power consumption. Therefore, they can efficiently improve test application time over existing gating solutions.

5.7. Post-Layout Simulation Results and Analysis

In order to evaluate the proposed ILPG and CPILPG scan cells, full-custom layout design was first carried out using Synopsys Custom Designer and Laker. Fig. 5.10 shows the full-custom layout design for the proposed gating scan cells.

Next, in order to verify the proposed scan cells in a realistic environment, a set of post-layout simulations has been conducted on parasitic-included Spice netlist using HSPICE in Synopsys 90 nm CMOS technology with supply voltage of 1.2 Volt at clock frequency 250 MHz and at room temperature. Input buffers were inserted at the scan cell inputs while the output load capacitance of 100fF has been considered for each participated scan cell in comparison.

The simulation results then are compared to Conventional Scan Cell and Modified Scan Cell [41], TG Scan Cell [34], Modified Scan Cell for critical path [40], ICMux Scan Cell [43], ANDB Scan Cell and OR Scan Cell [23] in terms of power, propagation delay, PDP and area overhead. Whereas the proposed scan cells with state preserving and gating logic is a compact architecture, scan cells with ICMUX [43] TG [34], ANDB and OR [23] are considered as an entity composed of their corresponding conventional scan cells and gating logics, respectively.

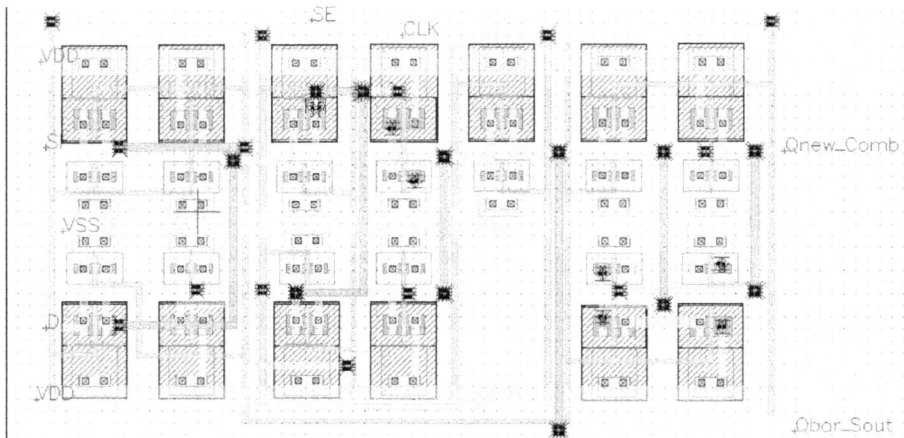

(a)

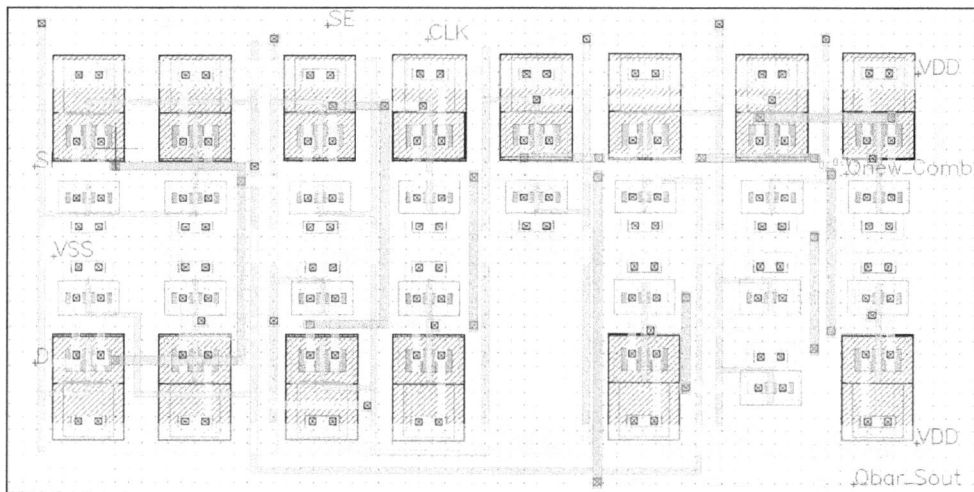

(b)

Fig. 5.10. Full-custom layout design of the proposed scan cells: (a) proposed ILPG scan cell; (b) proposed CPILPG scan cell in 90 nm CMOS technology using Laker.

Power consumption and propagation delay were observed by applying randomly generated waveforms. The signal waveforms of each scan cell to combinational logic along with the input stimuli signals has been shown for all compared scan cells in Fig. 5.11.

It is clearly observable that, unlike conventional scan cell, the proposed ILPG scan cell and modified scan cell have no transitions in the combinational logic during shift session as the scan cell output "*Qnew_Comb*" remains the same while the scan-enable (*SE*) signal is high. Note that "*Qnew_Comb*" is the scan cell output connecting to the combinational part.

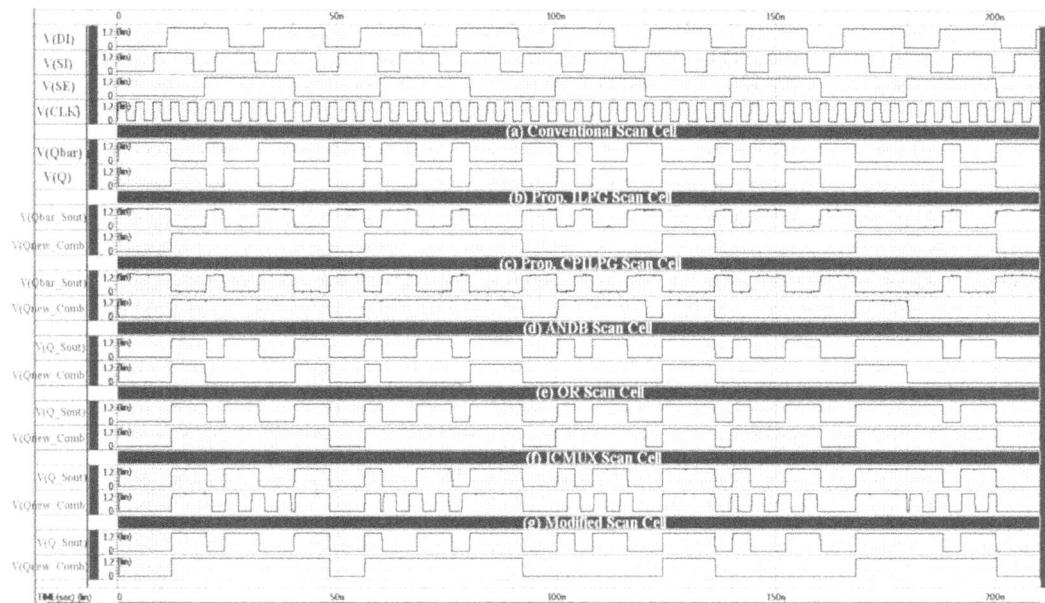

Fig. 5.11. Post-layout signal waveform sequences at 250 MHz clock frequency: (a) Conventional Scan Cell; (b) Proposed ILPG Scan Cell; (c) Proposed CPILPG Scan Cell; (d) ANDB Scan Cell; (e) OR Scan Cell; (f) ICMUX Scan Cell; (g) Modified Scan Cell.

5.7.1. Test Average Power Consumption

To measure the total power consumption, we applied random vectors to the scan cells inputs and calculated the total power consumption for each scan cell during the test (Shift and capture) and functional mode of operations.

Table 5.1 demonstrates the percentage of power reduction achieved by the proposed ILPG and CPILPG scan cells over each scan cells from previous works.

Table 5.1. Comparison of total average power consumption during whole test application time.

Percentage of Total Power Consumption Reduction		
	Prop. ILPG Scan Cell	Prop. CPILPG Scan Cell
% Imp. over Conventional Scan Cell	15.12 %	-9.09 %
% Imp. over Modified Scan Cell	22.01 %	-0.23 %
% Imp. over TG Scan Cell	17.35 %	-6.22 %
% Imp. over ICMUX Scan Cell	28.98 %	6.63 %
% Imp. over ANDB Scan Cell	29.47 %	7.26 %
% Imp. over OR Scan Cell	39.15 %	19.99 %

As proposed by previous works shown in Table 5.1, ILPG scan cell can reduce power consumption inside the scan cell by 15.12 % compared with conventional scan cell and up to 39.15 % compared to OR scan cell. The proposed CPILPG scan cell imposes

9.09 % penalty on scan cell's total power consumption compared with conventional scan cell. The most likely explanation about this is that, contrary to ILPG scan cell, CPILPG scan cell is able to reduce scan cell power consumption merely during test mode rather than functional mode as well.

5.7.2. Shift Power Consumption

Table 5.2 shows the percentage of average power reduction for each related scan cell over conventional scan cell during five successive shift cycles. It can be seen that among all compared scan cells, only the proposed ILPG and CPILPG scan cells are able to improve shift power consumption over the original shift power dissipation in conventional scan cell by 22.16 % and 6.67 %, respectively. This is expected since in both proposed scan cells, shift data is transferred through the $\overline{Q}$ output of scan cell which leads to a shorter shift path compared to other scan cells.

Leakage power comparisons during shifting are listed in Table 5.3. As discussed earlier, employing stacking effect in refresh feedback inverter results in average leakage power improvements of 22.16 % over conventional scan cell.

Table 5.2. Comparison of shift average power consumption of each scan cell over conventional scan cell during shift mode.

Percentage of Shift Average Power Consumption Reduction						
	Shift Cycle#1	Shift Cycle#2	Shift Cycle#3	Shift Cycle#4	Shift Cycle#5	Average
ANDB Scan Cell	-17.73 %	-17.43 %	-16.27 %	-16.77 %	-19.36 %	**-17.51 %**
OR Scan Cell	-50.88 %	-29.63 %	-52.92 %	-50.36 %	-52.94 %	**-47.34 %**
ICMux Scan Cell	-17.37 %	-22.10 %	-24.44 %	-22.56 %	-15.36 %	**-20.36 %**
TG Scan Cell	-7.85 %	-0.57 %	-11.22 %	-3.25 %	-10.67 %	**-6.71 %**
Modified Scan Cell	4.86 %	9.52 %	-29.82 %	5.62 %	-0.34 %	**-2.03 %**
Prop. ILPG Scan Cell	21.92 %	24.67 %	23.07 %	24.80 %	16.37 %	**22.16 %**
Prop. CPILPG Scan Cell	10.58 %	13.97 %	-18.10 %	14.29 %	12.61 %	**6.67 %**

Table 5.3. Comparison of shift Leakage power consumption of each scan cell over conventional scan cell during shift mode.

Percentage of Shift Leakage Power Consumption Reduction						
	Shift Cycle#1	Shift Cycle#2	Shift Cycle#3	Shift Cycle#4	Shift Cycle#5	Average
ANDB Scan Cell	-22.00 %	-23.29 %	-20.70 %	-24.15 %	-13.78 %	**-20.79 %**
OR Scan Cell	-58.67 %	-60.02 %	-54.63 %	-49.57 %	-38.24 %	**-52.14 %**
ICMux Scan Cell	-21.33 %	-14.54 %	-16.46 %	-22.38 %	-7.14 %	**-16.39 %**
TG Scan Cell	-6.15 %	-5.78 %	-1.44 %	-5.13 %	-1.65 %	**-4.00 %**
Modified Scan Cell	12.89 %	12.99 %	-43.51 %	5.15 %	12.10 %	**-0.47 %**
Prop. ILPG Scan Cell	28.72 %	29.33 %	28.55 %	28.21 %	24.90 %	**27.94 %**
Prop. CPILPG Scan Cell	17.50 %	17.59 %	-25.73 %	16.68 %	-8.22 %	**3.56 %**

The most likely reason that CPILPG scan cell is not as effective as ILPG scan cell in improving leakage power over conventional scan cell is that the refresh feedback buffer in CPILPG scan cell only contains pull-down sleep transistors compared to a pull-up and a pull-down sleep transistors in feedback refresh inverter of ILPG scan cell. However, implementing CPILPG scan cell with both pull-up and pull-down transistors will add delay on the scan cell launch output which has conflict with CPILPG design goal.

5.7.3. Capture Power Consumption

The main objective of our proposed state preserving scan cell is shift power reduction while keeping capture (peak and average) power as close to other existing gating techniques as possible. Comparisons of average normal/capture power consumption for four capture cycles are presented in Table 5.4. The simulations have been carried out by HSPICE using Synopsys standard cell library in 90 nm technology.

Table 5.4. Comparison of capture average power consumption of each scan cell over conventional scan cell.

Percentage of Capture Average Power Consumption Reduction					
	Capture Cycle#1	Capture Cycle#2	Capture Cycle#3	Capture Cycle#4	Average
ANDB Scan Cell	-28.68 %	-13.17 %	-21.24 %	-25.50 %	**-22.36 %**
OR Scan Cell	-37.03 %	-36.46 %	-49.76 %	-31.40 %	**-39.08 %**
ICMux Scan Cell	-23.51 %	-4.67 %	-20.52 %	-17.77 %	**-17.05 %**
TG Scan Cell	-9.21 %	0.05 %	8.82 %	-1.92 %	**-0.92 %**
Modified Scan Cell	-15.87 %	-12.84 %	-14.15 %	-14.44 %	**-14.37 %**
Prop. ILPG Scan Cell	-2.56 %	0.77 %	4.15 %	3.78 %	**1.45 %**
Prop. CPILPG Scan Cell	-18.02 %	-5.33 %	-11.94 %	-17.21 %	**-9.21 %**

As mentioned earlier, the elevated peak power beyond the chip power budget caused by gating logic toggling between gating (shift) mode and transparent (normal/capture) mode has a direct impact on chip reliability.

Generally, restricting the time window to just one clock cycle in power measurement is not realistic enough since the power consumption within one clock cycle may not be large enough to elevate the temperature over the thermal capacity limit of the chip [1]. On the other hand, the excessive peak power through one normal/capture cycle is not strong enough to elevate the circuit temperature beyond the thermal threshold of the chip.

In order to show the impact of our proposed scan cells on the peak power, we have compared the percentage of peak power increase in the proposed gating scan cells and existing scan cells over conventional scan cell through five consecutive normal/capture cycles (each including one clock period). As it is demonstrated in Fig. 5.12, the proposed ILPG scan cell achieves improvement in peak power over conventional scan cell in almost all observed capture cycles up to 13.73 % in capture cycle #3.

CPILPG scan cell, however, increases original peak power in most of capture cycles. Yet, it shows better results in peak power overhead compared with ANDB in all capture cycles and in comparison with OR and modified scan cell in most cases.

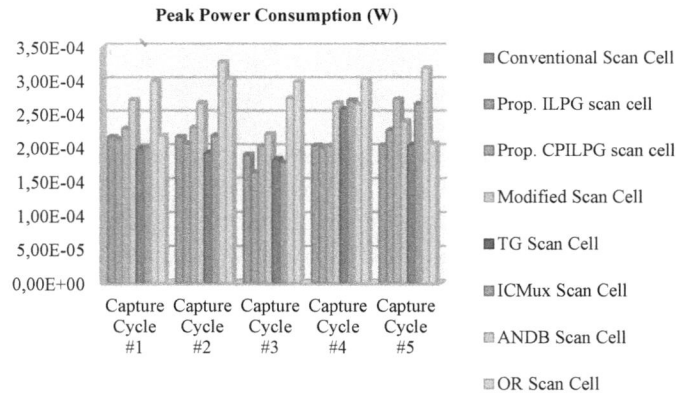

Fig. 5.12. Peak power comparisons during normal/capture mode.

Note that, in highly integrated industrial designs, the number of scan cells located on critical paths is significantly less than those on non-critical paths. Therefore, peak power increased by a small number of CPILPG scan cells is not high enough to violate circuit peak power threshold.

Another point to note is that ICMUX scan cell shows significantly lower peak power in the last two capture cycles. This is expected because ICMux scan cell's power consumption highly depends on the selected input control vector that determines the freezing logic on the scan cell output. In those capture cycles whereby the input control vector is identical to the logic value on the scan cell output when *SE* signal changes to '0', no redundant transition occurs. Therefore, the level of average and peak power is dramatically lower than the case when the freezing logic differs from the logic on scan cell output. Similar scenario can be observed for OR and ANDB scan cells.

It is illustrated in Table 5.5 that the proposed CPILPG scan cell is not as successful as the proposed ILPG scan cell in decreasing stand by leakage power.

The reason is that since CPILPG scan cell is proposed for the critical paths, each of the two inverters in refresh feedback in the modified slave latch are implemented with one pull down sleep transistor in order to reduce the delay that refresh feedback add to the data capture path. On the other word, using one sleep transistor to cut off standby subthreshold current is not as efficient as the set of pull up/down transistors, but it will impose less delay on the capture path due to smaller output capacitance.

However, it is noteworthy that CPILPG scan cell has been designed to improve the proposed ILPG scan cell launch delay on non-critical paths. In highly integrated industrial designs, the number of scan cells located on critical paths is significantly less than those on non-critical paths. Therefore, the leakage overhead imposed by the small number of CPILPG scan cells does not affect the total normal/capture leakage power seriously.

Table 5.5. Comparisons of normal/capture leakage power improvements over conventional scan cell in 90nm Synopsys technology.

Percentage of Normal/Capture Leakage Power Consumption Reduction					
	Capture Cycle#1	Capture Cycle#2	Capture Cycle#3	Capture Cycle#4	Capture Cycle#5	Average
ANDBScan Cell	-42.50 %	-39.89 %	-23.57 %	-32.53 %	-28.26 %	**-33.58 %**
OR Scan Cell	5.54 %	-46.59 %	-52.69 %	-69.54 %	-49.30 %	**-43.68 %**
ICMux Scan Cell	-7.71 %	-20.89 %	-12.53 %	-16.38 %	-21.42 %	**-15.99 %**
TG Scan Cell	14.78 %	-5.90 %	-4.13 %	-22.05 %	-15.80 %	**-6.90 %**
Modified Scan Cell	-14.99 %	-20.95 %	-23.02 %	-27.30 %	-32.76 %	**-23.64 %**
Prop. ILPG Scan Cell	17.31 %	5.02 %	2.55 %	-5.83 %	-5.59 %	**2.65 %**
Prop. CPILPG Scan Cell	-4.80 %	-6.23 %	-13.12 %	-13.67 %	-29.32 %	**-12.85 %**

5.7.4. Propagation Delay and Power-Delay-Product (PDP)

Various delay arches for all compared scan cells have been investigated and presented in Table 5.6 to show the impact of different existing gating strategies on gating scan cell propagation delay. Rows 1-3 are related to the percentage of improvement achieved by each scan cell in terms of shift, capture and launch operation over conventional scan cell. It can be seen that the proposed ILPG scan cell delivers improvements on shift delay and capture delay over conventional scan cell by 1.86 % and 0.58 %, respectively. Note that the achieved improvement in shift delay is measured within a single scan cell. However, the real impact can be observed in large industrial designs with hundred thousands of scan cells in scan chain. It contributes to speed up test application time via decreasing shifting time in testing process that involves huge number of test patterns. The delay overhead on the ILPG scan cell data path to combinational logic is 4.64 % through normal/capture mode of operation. This has been improved in CPILPG scan cell as it shows only 1.95 % penalty on the launch delay variable.

Rows 4-5 represent propagation delay on the clock signal in shift (*CLK-to-shift output*) and launch (*CLK-to-Qnew*) based on the percentage of simulated clock period. According to this sub-section, the portion of time needed for propagating data through shift path after rising edge of clock is 0.6 % of clock period for ILPG scan cell and 0.69 % of clock period for CPILPG scan cell. This corresponds to 51.42 % improvement for ILPG scan cell and 43.94 % improvement for CPILPG scan cell in *CLK-to-shift output* over conventional scan cell.

The proposed ILPG scan cell increases *CLK-to-Qnew* (launch output) from 1.25 % of clock period in conventional scan cell to 2.24 %. However, CPILPG scan cell achieves 25.44 % improvement in *CLK-to-Qnew* (launch output) over ILPG scan cell, which is the best result among other alternative gating scan cells. Finally, the last row in Table 5.6, depicts the exact amount of Power-Delay-Product (PDP) for each studied scan cell.

Table 5.6. Comparison of propagation delay and Power-Delay-Product (PDP).

		Percentage of Propagation Delay Reduction over Conventional Scan Cell						
		ANDB Scan Cell	**OR** Scan Cell	**ICMux** Scan Cell	**TG** Scan Cell	**Modified** Scan Cell	**Prop. ILPG** Scan Cell	**Prop.CPILP G** Scan Cell
1	SI to shift output (Shift Delay) (s)	-0.37 %	-1.08 %	-1.24 %	0.71 %	1.08 %	1.86 %	1.66 %
2	DI to shift output (Capture Delay) (s)	-1.71 %	-1.76 %	-2.55 %	1.77 %	1.08 %	0.58 %	2.59 %
3	DI to Qnew (Launch Delay) (s)	-5.96 %	-5.83 %	-5.78 %	-2.35 %	-2.11 %	-4.64 %	-1.95 %
		Percentage of propagation delay over Clock period						
		Conventional Scan Cell	**ANDB** Scan Cell	**OR** Scan Cell	**ICMux** Scan Cell	**TG** Scan Cell	**Modified** Scan Cell	**Prop. ILPG** Scan Cell	**Prop.CPILP G** Scan Cell
4	CLK-to-shift output	1.24 %	1.65 %	1.55 %	1.53 %	1.46 %	0.86 %	0.60 %	0.69 %
5	CLK-to-Qnew (launch) output	1.25 %	2.50 %	2.49 %	2.01 %	1.75 %	1.70 %	2.24 %	1.67 %
		Power-Delay-Product (PDP)							
6	PDP	20.10E-17	48.08E-17	55.66E-17	38.45E-17	28.86E-17	29.69E-17	30.45E-17	29.85E-17

Note that we assume minimum transistor size for implementing sleep transistors in state preserving logics (both ILPG and CPILPG scan cells). However, further improvement on delay overhead is feasible for both proposed scan cells. As it is pointed out in [38], sleep transistors with larger size for scan cells in the critical path can be used to further reduce the delay penalty. This increases the area overhead but does not affect the switching power of the gates.

5.7.5. Comparative Case Analysis

In this section, a comparative analysis on all scan cell parameters between the proposed CPILPG and the modified scan cell for critical path in [40] is discussed. The work in [40] proposed a high performance gating scan cell for critical paths to alleviate the delay penalty of existing gating schemes and to enhance scan cell overall performance. The advantage of our proposed CPILPG scan cell is shown in Table 5.7 in terms of percentage of improvements over the modified scan cell for critical path [40]. According to Table 5.7, CPILPG scan cell outperforms modified scan cell for critical path not only in propagation delay characteristics but also in scan cell power consumption by 70.51 %. The proposed CPILPG scan cell also shows 49.90 % improvement in PDP.

Although the two scan cells shows almost the same propagation delay on launch output (*DI* to *Qnew*), 16.49 % improvement on *CLK-to-Qnew* has been obtained for our proposed CPILPG compared with modified scan cell for critical path. It is noteworthy that the scan cell speed on transferring data is mostly determined based on the time interval between clock trigger and changing value on the launch output which has been presented by *CLK-to-Qnew*.

Table 5.7. Prop. CPILPG scan cell percentage of improvements over modified scan cell for critical path.

Percentage of Improvements on Scan Cell Characteristic			
	Prop. CPILPG Scan Cell	**Modified Scan Cell for critical path**	**% Imp.**
Total Power Consumption	4.3569E-06	7.4291E-06	70.51 %
Ave. Shift Power Consumption	3.9882E-06	7.9023E-06	49.53 %
SI-to-shift output (Shift Delay) (s)	1.3265E-09	1.3865E-09	4.32 %
DI-to-So shift output (Capture Delay) (s)	8.2826E-10	8.8948E-10	6.88 %
DI-to-Qnew (launch output) (Launch Delay (s)	8.6698E-10	8.8021E-10	1.50 %
CLK-to-shift output Delay(s)	2.7956E-11	8.6071E-11	67.51 %
CLK-to-Qnew (launch output) Delay (s)	**6.6982E-11**	**8.0216E-11**	**16.49 %**
PDP	29.853E-17	59.593E-17	49.90 %

5.7.6. Area Overhead

To evaluate the area overhead imposed by the proposed structures, we calculated the actual size of the layout design. Table 5.8 demonstrates the number of transistors along with the layout area for each gating scan cell design. The routing overhead has not been considered in area overhead comparisons. However, the routing overhead associated with the proposed gating structure is not high since no additional control signal is required. Correspondingly, the area overhead for other scan cell designs has been measured.

Table 5.8. Layout-aware area overhead comparison of scan cell designs.

	No. of Transistors	**Layout Area ($\mu m2$)**
Conventional Scan Cell	24	116.39
ANDB Scan Cell	30	158.48
OR Scan Cell	30	157.32
ICMux Scan Cell	28	150.76
TG Scan Cell	27	137.46
Modified Scan Cell	30	156.816
Prop. ILPG Scan Cell	26	130.84
Prop. CPILPG Scan Cell	30	146.95
Modified Scan Cell for critical path	43	248.41

Next, some large ITC'99 benchmark circuits are employed in order to explore the impact of the proposed gating scan cells on the overall area overhead of the investigated benchmark circuits. Furthermore, full gating architecture has been considered for all benchmark circuits to demonstrate the worst case of area penalty. We have considered the actual scan chain size in each benchmark circuit as a metric for calculating the area overhead in each case. Since the compared scan cells address area overhead only on the scan chain structure, and they do not have any impact on increasing the combinational part area. Table 5.9 summarizes the characteristic of the selected benchmark circuits.

Followed by the circuit name, the number of primary inputs, primary outputs, the number of gates and the number of flip-flops has been shown for each circuit.

Table 5.9. ITC'99 Benchmark circuits' characteristics.

Benchmark Circuit	# PI	# PO	# Gates	# Flip-Flops
b14	32	54	10,098	245
b15	36	70	8,922	449
b17	37	97	32,326	1,415
b18	36	23	114,621	3,320
b19	21	30	231,320	6,642
b21	32	22	20,571	490
b22	32	22	29,951	735

Fig. 5.13 indicates the comparison in the size of the scan chain for each benchmark circuit. The calculation results shows ILPG scan cell incurs area overhead of 12.41 % over original scan design in case of full gating. While this metric is up to 36.16 % and 35.16 % for ANDB scan cell and OR scan Cell, respectively. In order to alleviate the active area taken by the proposed ILPG scan cell, the pull-up/down sleep transistors in state preserving logic can be shared between all scan cells in full gating architecture. In partial gating architecture, only those scan cells that are on non-critical paths share the sleep transistors. We have considered full gating scan in our experiment and have assumed that for random input patterns nearly, half of the scan cells do not switch and the pull up and pull down transistors are exercised for almost half of the scan cells at the same time. Thereby, the channel width for shared pull-up and pull-down transistors are chosen to be as follows.

$$W_{Shared\_pull-up} = 0.5 * FF * (5 * W_{min(Pmos)}), \qquad (5.2)$$

$$W_{Shared\_pull-down} = 0.5 * FF * (5 * W_{min\,(Nmos)}), \qquad (5.3)$$

where *FF* is the number of scan cells. The active area for shared pull-up and pull-down transistors is calculated based on total transistor active area (*W*L* for a transistor) and is denoted by *Area$_{Pull-up}$* and *Area$_{Pull-down}$*, respectively. Next, similar to previous experiment, the actual size of the proposed design excluding the pull-up and pull-down sleep transistors has been extracted from the layout design. Based on this discussion, the actual size of benchmark circuits fully gated by the proposed shared IPLG scan cells is given by:

$$Area_{Benchmark\_circuit} = (Area_{PGscan\_cell} * FF) + Area_{Pull-up} + Area_{Pull-down}. \qquad (5.4)$$

It is observable from Fig. 5.13 that the alternative scan chain using modified scan cell for critical path [40] significantly increase the area overhead by 113.49 % over conventional scan structure.

The proposed CPILPG scan cell increases area overhead of 26.25 % over conventional scan cell for all benchmark circuits. Yet it still has less penalty in terms of area compared with other scan cells except TG scan cell. However, in reality, area overhead addressed by CPILPG is far less than this amount since it is only practiced for the scan cells on critical paths. Note that alike the proposed ILPG scan cell, the two pull-down sleep transistors in the CPILPG scan cell state preserving logic can be shared between all scan cells on critical paths which will lead to further area penalty compensation.

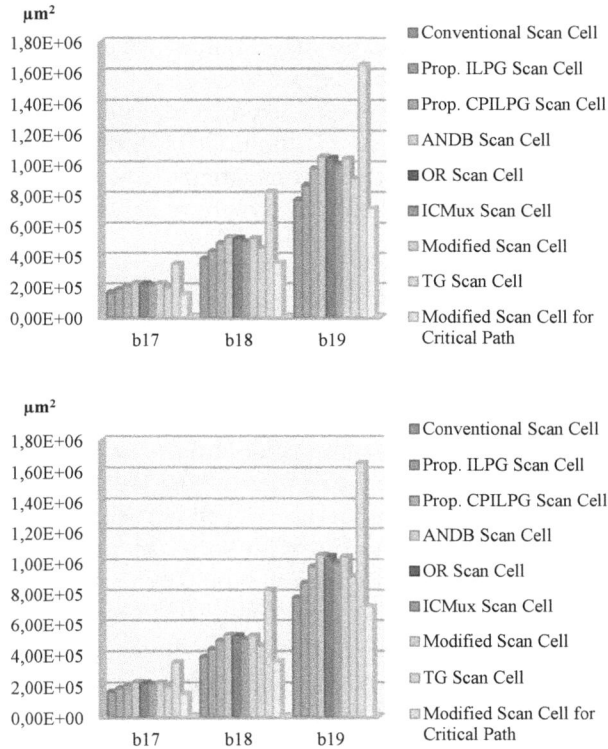

Fig. 5.13. Full gating area overhead comparisons in ITC'99 benchmark circuits.

5.7.7. Test Quality

The efficiency of test power reduction policies are not only evaluated based on their effectiveness in reducing power consumption, but also according to their impact on test quality.

The proposed gating scan cells offer a new shift path through $\bar{Q}$ that necessitates adaptive shift process through scan chain. However, since the functional operation remains intact, no impact on test quality and original fault coverage will be issued. Adaptive test patterns generation for stuck-at faults as well as other fault models such as transition faults can be carried out fully automated by ATPG based on the new scan chain structure. A comprehensive description on adaptive shift-in/out process is discussed and can be found in [57].

Note that in the proposed scan cells, *SE* signal controls gating operation of modified slave latch. Consider a scenario in the proposed scan cell that *DI* input value is '0', *SI* input value is '1' and *SE* signal value is '1' (shift mode). In a fault free condition, ATPG expects logic '0' on $\bar{Q}$ output and no switching on *Q* output. Now let us assume a stuck-at-0 fault on *SE* signal, in that case ATPG receives logic value of '1' instead of logic value '0' on the $\bar{Q}$ output and stuck-at-0 fault will be diagnosed. In addition, the fact that the proposed scan cell maintains the previous value on the *Q* output during shift gives ATPG more flexibility due to an extra output to check for fault diagnosis during testing. This is because any switching during shift mode on *Q* output can also be realized as faulty situation. For instance, let the preserved logic value on *Q* output is '1'. In that case stuck-at-0 fault on *SE* signal will results in *DI* logic value ('0') propagating to *Q* output and changes the value from '1' to '0'. Consequently, ATPG realizes switching on *Q* output and considers it as faulty condition.

5.7.8. Technology Adaptability

The structure of the proposed scan cells does not allow using *Q* output as shift output. This is because the modified slave latch blocks any input to *Q* output and hold the previous value during shift mode. Therefore, if synthesis tool tries to use this output as shift output, the process of pattern shifting will fail. Hence, evaluating the proposed scan cell requires $\bar{Q}$ synthesis (synthesis using only $\bar{Q}$ output of scan cells as scan output in scan insertion). In this case, DFT synthesis should be run with some adaptive changes compared to traditional method. The adaptive configuration should be able to make synthesis tool, which in our experiments is Synopsys Design Compiler, to run scan insertion only using $\bar{Q}$ output of scan cells. Fortunately, this is feasible by slight modification in technology library including keeping signal type description as test-scan-out-inverted for $\bar{Q}$ output and remove signal type test-scan-out for *Q* output of scan cells. This way, Design Compiler will be capable to dedicate $\bar{Q}$ output of scan cells as the only output in constructing scan chain.

Before following up this section with power estimation results, first, we discuss the overhead resulted from $\bar{Q}$ synthesis compared to traditional synthesis. We have run both traditional synthesis and $\bar{Q}$ synthesis on two ITC'99 benchmark circuits using Synopsys Design Compiler in 90nm technology. The statistics of the two evaluated ITC'99 benchmark circuits are summarized in Table 5.10.

Table 5.10. Statistics of ITC'99 benchmark circuits.

Benchmark Circuit	# PI	# PO	# Inverters	# Buffers	# Logics	# Scan Cells	# Total Gates
b14	32	54	1518	188	6728	245	8679
b20	36	70	3537	332	14748	490	19107

According to Design Compiler reports, no significant difference in any of the relevant synthesis metrics including leaf cell count, buffer/inverter count, number of sequential cells, total number of nets and critical path slack is observed. The comparison of circuit statistics in cell count, area, net and timing for each synthesis method is shown in Table 5.11.

Table 5.11. Comparison of cell count and area overhead for traditional synthesis and modified synthesis.

	Traditional Synthesis		$\bar{Q}$ **Synthesis**		
	Cell Count		**Cell Count**		**Overhead**
	Leaf Cell Count	2735	Leaf Cell Count	2791	2.04 %
	Buf/Inv Cell Count	387	Buf/Inv Cell Count	403	4.13 %
	Buf Cell Count	99	Buf Cell Count	77	-22.22 %
	Inv Cell Count	288	Inv Cell Count	326	13.19 %
	Comb. Cell Count	2520	Comb. Cell Count	2576	2.22 %
	Seq. Cell Count	215	Seq. Cell Count	215	0.00 %
b14	**Area (nm)**		**Area (nm)**		
	Buf/Inv Area	3732.480097	Buf/Inv Area	3820.953699	2.37 %
	Net Area	2597.927745	Net Area	2617.367485	0.74 %
	Cell Area	31622.860749	Cell Area	32776.704158	3.64 %
	Design Area	34220.788494	Design Area	35394.071643	3.42 %
	Net		**Net**		
	Total No. of Nets	3204	Total No. of Nets	3453	7.77 %
	Timing (ns)		**Timing (ns)**		
	CriticalPath Slack	0.13	Critical Path Slack	0.22	
	Traditional Synthesis		$\bar{Q}$ **Synthesis**		
	Cell Count		**Cell Count**		**Overhead**
	Leaf Cell Count	5570	Leaf Cell Count	5528	2.04 %
	Buf/Inv Cell Count	1203	Buf/Inv Cell Count	1207	4.13 %
	Buf Cell Count	468	Buf Cell Count	372	-22.22 %
	Inv Cell Count	735	Inv Cell Count	835	13.19 %
	Comb. Cell Count	5084	Comb. Cell Count	5039	2.22 %
	Seq. Cell Count	486	Seq. Cell Count	486	0.00 %
b14	**Area (nm)**		**Area (nm)**		
	Buf/Inv Area	9342.259381	Buf/Inv Area	8999.424189	2.37 %
	Net Area	6200.305513	Net Area	6308.002928	0.74 %
	Cell Area	65023.487747	Cell Area	65431.756798	3.64 %
	Design Area	71223.793260	Design Area	71739.759726	3.42 %
	Net		**Net**		
	Total No. of Nets	6342	Total No. of Nets	6490	7.77 %
	Timing (ns)		**Timing (ns)**		
	CriticalPath Slack	0.02	Critical Path Slack	0.00	

One of the main concerns regarding $\bar{Q}$ synthesis is the increase in net fan outs. Since the timing properties of the proposed scan cells with $\bar{Q}$ scan out differ from traditional scan cells, for some (not all) parts of the design, Design Compiler may need to insert buffers while at other parts, it removes buffers.

In real world designs, the exact number of buffers inserted during synthesis is not a relevant issue. What is important is the ratio of buffers to the total number of leaf cells. This ratio for both the abovementioned benchmark circuits in $\bar{Q}$ synthesis is less than

15 % (2.75 % for b14 and 6.72 % for b20). This shows that designs with the proposed $\bar{Q}$ synthesis do not have any logical design rule problem such as maximum fan out, maximum capacitance and maximum transition violations which are critical metrics for fabrication.

In the next phase of experiments, automatic test pattern generation for stuck-at fault model was conducted using Synopsys Tetramax in 90nm technology. Table 5.12 summarizes the fault coverage achieved by traditional synthesis for circuit with no gating technique, circuit with scan functional output hold at '0' and circuit with scan functional output hold at '0'.

The last column in Table 5.12 demonstrates fault coverage achieved by $\bar{Q}$ synthesis (using only $\bar{Q}$ as scan shift output in scan insertion) for the proposed state preserving hold scan.

Table 5.12. Test Comparisons of total number of faults, internal/ basic scan test patterns and fault coverage.

	No gating (Original)			Hold at '0'		
	#Total Faults	# Internal/Basic Scan patterns	Fault Coverage	#Total Faults	# Internal/Basic Scan patterns	Fault Coverage
b14	11662	592	**99.96%**	12441	595	**98.19%**
b20	25354	539	**99.65%**	25777	538	**98.67%**
	Hold at '1'			Prop. State Preserve Hold		
	#Total Faults	# Internal/Basic Scan patterns	Fault Coverage	#Total Faults	# Internal/Basic Scan patterns	Fault Coverage
b14	12441	610	**98.19%**	12441	610	**97.96%**
b20	25777	524	**98.68%**	25777	513	**98.18%**

To show the effectiveness of the proposed method in test power and more specifically shift power reduction, we have conducted four experiments for the two abovementioned ITC'99 benchmark circuits.

The first experiment has been conducted with traditional synthesis and scan insertion using both Q and $\bar{Q}$. In this experiment, no gating technique has been applied. The second and third experiments used traditional synthesis, in addition to OR scan gating and ANDB scan gating [23] were carried out for holding the scan stimulus path at logic '0' and '1', respectively. The last experiment is dedicated to the proposed scan cells evaluation with the aim of $\bar{Q}$ synthesis and scan insertion. ATPG then has been conducted to generate test patterns for stuck-at fault model using Synopsys Tetramax.

To get the most accurate estimation on power estimation, we conducted vector dependent power analysis. Synopsys VCS simulation tool was employed to simulate the Verilog test bench and to dump value change dump (VCD) file that was later used by Synopsys Prime Time for power estimations. Power analysis experiments were carried out in two modes:

1. Event-based power calculation that calculates power consumption for specified shift time intervals; and 2. Average power calculation that measure average power consumption through the whole test application time. The result obtained through each power analysis procedure is presented in Table 5.13 and Table 5.14.

Relevant to actual design requirements, there are three major constraints for all types of IC design: Timing, power and area (size). The goal is to reduce power consumption while keeping the design size low and meeting performance (timing) demands. Thus, any method or technique that improves one design metric (such as power consumption) must not degrade the other design metrics (such as design timing and size) below a certain threshold which is design dependent.

Table 5.13. Comparison of improvements of proposed state preserving approach in shift power consumption over other methods.

Shift Power Consumption Comparisons								
		(Original) No Gating	Imp over %	Hold at '0'	Imp over %	Hold at '1'	Imp over %	Prop. State Preserve Hold
b14	Combinational Internal Power (W)	3.204e-05	80.02	2.454e-05	73.92	3.038e-05	78.93	6.399e-06
	Combinational Switching Power (W)	9.825e-06	90.94	4.445e-06	79.99	1.059e-05	91.60	8.892e-07
	Combinational Leakage Power (W)	7.978e-05	-0.97	8.220e-05	1.99	8.074e-05	0.22	8.056e-05
	Combinational Total Power (W)	1.216e-04	27.75	1.112e-04	20.99	1.217e-04	27.81	8.785e-05
b20	Combinational Internal Power (W)	7.101e-05	71.17	5.552e-05	63.13	6.941e-05	70.50	2.047e-05
	Combinational Switching Power (W)	2.164e-05	73.47	1.197e-05	52.03	2.458e-05	76.64	5.741e-06
	Combinational Leakage Power (W)	1.639e-04	2.56	1.614e-04	1.05	1.629e-04	1.96	1.597e-04
	Combinational Total Power (W)	2.565e-04	27.52	2.289e-04	18.78	2.569e-04	27.63	1.859e-04

Table 5.14. Comparison of improvements of proposed state preserving approach in total average test power consumption over other methods.

Shift Power Consumption Comparisons								
		(Original) No Gating	Imp over %	Hold at '0'	Imp over %	Hold at '1'	Imp over %	Prop. State Preserve Hold
b14	Combinational Internal Power (W)	3.214e-05	29.77	3.593e-05	37.18	4.393e-05	48.62	2.257e-05
	Combinational Switching Power (W)	1.141e-05	27.80	1.246e-05	33.89	1.791e-05	54.00	8.237e-06
	Combinational Leakage Power (W)	8.266e-05	-1.20	8.309e-05	-0.68	8.196e-05	-2.07	8.366e-05
	Combinational Total Power (W)	1.262e-04	9.27	1.315e-04	12.92	1.438e-04	20.37	1.145e-04
b20	Combinational Internal Power (W)	7.353e-05	25.56	8.921e-05	38.65	1.011e-04	45.86	5.473e-05
	Combinational Switching Power (W)	2.446e-05	23.63	2.832e-05	34.03	3.674e-05	49.15	1.868e-05
	Combinational Leakage Power (W)	1.670e-04	2.39	1.636e-04	0.36	1.636e-04	0.36	1.630e-04

As shown in the reports, testing on the two designs reveals that the proposed technique is able to reduce power consumption (both dynamic and leakage) during shift cycles and the whole test application with no impact on functional mode operation. In addition, the proposed gating method is able to reduce side effects of scan gating which are high capture peak and average power and maintain them below a certain level. However, it is still necessary that power-aware test pattern generation and/or test schedule techniques to be employed in conjunction with the proposed gating scheme in order to reduce the excessive capture power consumption caused by entire switching activities in all gating scan cells during this mode. Therefore, for designs that require low power consumption during test phase, employing this technique will produce the desired results with no impact on design quality including timing and area.

5.8. Summary and Conclusion

In this paper, we have proposed a cost-efficient and robust integrated DFT-based solution for power reduction in combinational part as well as scan chain by low power gating scan cell. The proposed scheme eliminates the need for low power ATPG efforts and it does not resort to any X-filling, vector or scan cell reordering techniques. The proposed ILPG scan cell is simple, yet effective in terms of total power reduction during both test (shift) and functional mode of operations. Another scan cell called CPILPG scan cell is proposed to reduce the gating penalty in terms of launch propagation delay during normal/capture mode besides power reduction in combinational logic through data gating during shift mode. Employing CPILPG scan cell on critical paths provides the possibility for more scan cells to be gated in the application of partial gating due to CPILPG's higher ability to meet critical path timing. The proposed scan cells improve shift propagation delay by employing shorter shift path compared with scan cell original shift path that causes further power reduction in scan chain. Therefore, it allows shifting at higher frequency possible, which leads to test application time reduction where shift power dissipation is a threshold bound on the shift frequency. Both proposed scan cells can be employed effectively in full gating method as well as partial gating method. The proposed schemes have small impact on peak power and do not need any hardware configuration outside the scan cells. They do not face any routing problem as no extra control signal has been employed. No degradation on test development (quality, coverage etc.) ensues as the scan cells modifications have no impact on functional operation. Furthermore, no penalty in terms of time complexity is issued as the proposed approaches do not get involved in any problem solving algorithm. The proposed schemes are adaptively addressed with low-power ATPG-based techniques such as power-aware test pattern generation, test vector reordering, X-filling etc., to suppress excessive gating capture power in both BIST-based and non BIST-based testing. Finally, other DFT-based power reduction strategies like scan chain partitioning and scan reordering can be effectively applied to the proposed low-power gating method for further test power reduction.

Acknowledgements

This research is partly sponsored by UTM Research University Grant vote no. 4F504.

References

[1]. P. Girard, Survey of low-power testing of VLSI circuits, *IEEE Design & Test of Computers,* Vol. 19, No. 3, 2002, pp. 82-92.

[2]. J. Altet, A. Rubio, Thermal Testing of Integrated Circuits, *Springer*, Boston, 2002.

[3]. A. Bosio, L. Dilillo, P. Girard, A. Todri, A. Virazel, Why and how controlling power consumption during test: a survey, in *Proceedings of the IEEE 21st Asian Test Symp.(ATS'12)*, Niigata, Japan, 19–22 November 2012, pp. 221–226.

[4]. M. L. Bushnell, V.D. Agrawal, Essentials of Electronic Testing for Digital memory, and Mixed-Signal VLSI Circuits, *Kluwer*, Boston, MA, 2000.

[5]. S. Wang, S. K. Gupta, ATPG for heat dissipation minimization during test application, *IEEE Transaction on Computers,* Vol. 47, No. 2, 1998, pp. 256-262.

[6]. V. Devanathan, C. Ravikumar, V. Kamakoti, Glitch-aware pattern generation and optimization framework for power-safe scan test, in *Proceedings of the VLSI Test Symp. (VTS'07)*, Berkeley, CA, USA, 6-10 May 2007, pp. 167-172.

[7]. M.-F. Wu, K.-S. Hu, J.-L. Huang, An efficient peak power reduction technique for scan testing, in *Proceedings of the IEEE 16th Asian Test Symp. (ATS'07)*, Beijing, China, 9-11 October 2007, pp. 111-114.

[8]. V. Dabholkar, S. Chakravarty, I. Pomeranz, S. M. Reddy, Technique for minimizing power dissipation in scan and combinational circuits during test application, *IEEE Transaction on Computer-Aided Design of Integrated Circuits and Systems,* Vol.17, No. 12, 1998, pp. 1325-1333.

[9]. P. Girard, L. Guiller, C. Landrault, S. Pravossoudovitch, A test vector ordering technique for switching activity reduction during test application, in *Proceedings of the IEEE Great Lakes Symp. on VLSI (GLSVLSI'99)*, Michigan, USA, 4-6 March 1999, pp. 24-27.

[10]. J. Tudu, E. Larsson, V. Singh, V. Agrawal, on minimization of peak power for scan circuit during test, in *Proceedings of the IEEE European Test Symp (ETS'09)*, Seville, Spain, 25-29 May 2009, pp.25-30.

[11]. Ghosh, S. Bhunia, K. Roy, A low complexity scan reordering algorithm for low power test-per-scan BIST, in *Proceedings of the International Conference on VLSI Design (VLSID'04)*, Mumbai, India, 5–9 January 2004, pp. 883–888.

[12]. M. Bellos, D. Bakalis and D. Nikolos, Scan cell ordering for low power BIST, in *Proceedings of the IEEE Computer Society Annual Symposium on VLSI (ISVLSI'04)*, Lafayette, LA, USA, 19–20 February 2004, pp. 281–284.

[13]. W. D. Tseng, Scan chain ordering technique for switching activity reduction during scan test, *IEE Proc.-Comp. Digit. Tech.,* Vol. 152, No. 5, 2005, pp. 609–617.

[14]. -K. Baek, I. Kim, J.-T. Kim, Y.-H. Kim, H. B. Min, J.-H. Lee, A dynamic scan chain reordering for low power VLSI testing, in *Proceedings of the International Conference on Information Technology Convergence and Services (ITCS'10)*, Cebu, Philippines, 11–13 August 2010, pp. 1–4.

[15]. K. Butler, J. Saxena, T. Fryars, G. Hetherington, A. Jain, and J. Lewis, Minimizing power consumption in scan testing: pattern generation and DFT techniques, in *Proceedings of the IEEE International Test Conference (ITC'04)*, Washington, DC, USA, 26-28 October 2004, pp. 355-364.

[16]. S. Remersaro, X. Lin, Z. Zhang, S. Reddy, I. Pomeranz, J. Rajski, Preferred Fill: A scalable method to reduce capture power for scan based designs, in *Proceedings of the IEEE International Test Conference (ITC'06)*, Santa Clara, California, USA, 22-27 October 2006, pp. 1-10.

[17]. X. Wen, K. Miyase, T. Suzuki, Y. Yamato, S. Kajihara, L.-T. Wang, K. K. Saluja, A highly-guided X-filling method for effective low-capture-power scan test generation, in *Proceedings*

of the IEEE International Conference on Computer Design (ICCD'06), San Jose, California, USA, 1-4 October 2006, pp. 251-258.

[18]. J. Li, Q. Xu, Y. Hu, X. Li, X-filling for simultaneous shift-and-capture-power reduction in at-speed scan-based testing, *IEEE Transaction on Very Large Scaled Integration (VLSI) Systems,* Vol. 18, No. 7, 2010, pp. 1081-1092.

[19]. S. Balatsouka, V. Tenentes, X. Kavousianos, K. Chakrabarty, Defect aware X-filling for low-power scan testing, in *Proceedings of the IEEE/ACM Design, Automation and Test in Europe Conference and Exhibition (DATE'10),* Dresden, Germany, 8-12 March 2010, pp. 873-878.

[20]. N. Nicolici, B. M. Al-Hashimi, A. C. Williams, Minimization of power dissipation during test application in full-scan sequential circuits using primary input freezing, *IEE Proc.-Comp. Digit. Tech.,* Vol. 147, No. 5, 2000, pp. 313-322.

[21]. T.-C. Huang, K.-J. Lee, Reduction of power consumption in scan-based circuits during test application by an input control technique, *IEEE Transaction on Computer-Aided Design of Integration Circuits and Systems,* Vol.20, No. 7, 2001, pp. 911-917.

[22]. O. Sinanoglu, I. Bayraktaroglu, A. Orailoglu, Test power reduction through minimization of scan chain transitions, in *Proceedings of the 20th IEEE VLSI Test Symposium (VTS'02),* Washington, DC, USA, 28 April- 2 May 2002, pp. 166-171.

[23]. X. Lin, Y. Huang, Scan shift power reduction by freezing power sensitive scan cells, *J. Electron Test, Springer,* Vol. 24, No. 4, 2008, pp. 327–334.

[24]. P. Rosinger, B. M. Al-Hashimi, N. Nicolici, Scan architecture with mutually exclusive scan segment activation for shift and capture power reduction, *IEEE Transaction on Computer-Aided Design of Integrated Circuits and Systems,* Vol. 23, No. 7, 2004, pp. 1142–1153.

[25]. S. Almukhaizim, O. Sinanoglu, Dynamic scan chain partitioning for reducing peak shift power during test, *IEEE Transaction on Computer-Aided Design of Integrated Circuits and Systems,* Vol. 28, No. 2, 2009, pp. 298–302.

[26]. Xiang, D. Hu, Q. Xu, A. Orailoglu, Low power scan testing for test data compression using a routing-driven scan architecture, *IEEE Transaction on Computer-Aided Design of Integrated Circuits and Systems,* Vol. 28, No. 7, 2009, pp. 1101–1105.

[27]. Y. Yamato, X. Wen, M. A. Kochte, K Miyase, S. Kajihara, L.-T. Wang, A novel scan segmentation design method for avoiding shift timing failure in scan testing, in *Proceedings of the IEEE International Test Conference (ITC'11),* Anaheim, California, USA, 20-22 September 2011, pp. 1-8.

[28]. C. Chen, C. Kang, M. Sarrafzadeh, Activity-sensitive clock tree construction for low power, in *Proceedings of the International Symposium on Low Power Electronics and Design (ISLPED'02),* Monterey, CA, USA, 12–14 August 2002, pp. 279-282.

[29]. A. Farrahi, C. Chen, A. Srivastava, G. Tallez, M. Sarrafzadeh, Activity-driven clock design, *IEEE Transaction on Computer-Aided Design of Integrated Circuits and Systems,* Vol. 20, No. 6, 2001, pp.705-714.

[30]. W. Shen, Y. Cai, X. Hong, J. Hu, An effective gated clock tree design based on activity and register aware placement, *IEEE Transactions on Very Large Scaled Integration (VLSI) Systems,* Vol. 18, No. 12, 2010, pp. 1639-1648.

[31]. Y. Bonhomme, P. Girard, L. Guiller, C. Landrault, S. Pravossoudovitch, A gated clock scheme for low power scan testing of logic ICs or embedded cores, in *Proceedings of the 10th IEEE Asian Test Symposium (ATS'01),* Washington, DC, USA, 19-21 November 2001, pp. 253-258.

[32]. J. C. Rau, C. L. Wu, P. H. Wu, An efficient algorithm to selectively gate scan cells for capture power reduction, *Tamkang Journal of Science and Engineering,* Vol. 14, No. 1, 2011, pp. 39-48.

[33]. S. Gerstendörfer, H. J. Wunderlich, Minimized power consumption for scan-based BIST, in *Proceedings of the IEEE International Test Conference (ITC'99),* Atlantic City, NJ, USA, 28–30 September 1999, pp. 77–84.

[34]. S. P. Khatri, S. K. Ganeshan, A modified Scan-D flip flop to reduce test power, in *Proceedings of the 15th IEEE International Test Synthesis Workshop (ITSW'08),* Santa Barbara, CA, USA, 7–9 April 2008, pp. 1–3.

[35]. X. Zhang, K. Roy, Power Reduction in test-per-scan BIST, in *Proceedings of the International Online Testing Workshop,* Palma De Mallorca, Spain, 3-5 July 2000, pp. 133-138.

[36]. N. Parimi, X. Sun, Design of A low power D flip-flop for test-per-scan circuits, In *Proceedings of the IEEE Canadian Conference on Electrical and Computer Engineering (CCECE'04),* Niagara Falls, ON, Canada, 2–5 May 2004, pp. 777–780.

[37]. S. Bhunia, H. Mahmoodi, D. Ghosh, S. Mukhopadhyay, K. Roy, Low-power scan design using first level supply gating, *IEEE Transactions On Very Large Scaled Integration (VLSI) Systems,* Vol. 13, No. 3, 2005, pp. 384-395.

[38]. S. Bhunia, H. Mahmoodi, D. Ghosh, S. Mukhopadhyay, K. Roy, Arbitrary two-pattern delay testing using a low-overhead supply gating technique, *J. Electron Test, Springer,* Vol. 24, No. 6, 2008, pp. 577-590.

[39]. M.-H. Chiu, J. C.-M. Li, Jump scan: a DFT technique for low power testing, in *Proceedings of the 23rd IEEE VLSI Test Symposium (VTS'05),* Palm Springs, California, USA, 1-5 May 2005, pp. 277-282.

[40]. A. K. Suhag, S. Ahlawat, V. Shivastava, N. Singh, Elimination of output gating performance overhead for critical paths in scan test, *International Journal of Circuit and Architecture Design,* Vol. 1, No. 1, 2013, pp. 62-72.

[41]. A. Mishra, N. Sinha, Satdev, V. Singh, S. Chakravarty, A. D. Singh, Modified scan flip-flop for low power testing, in *Proceedings of the 19th IEEE Asian Test Symposium (ATS'10),* Shanghai, China, 1–4 December 2010, pp. 367-370.

[42]. R. Jayagowri, K. S. Gurumurthy, Implementation of gating technique with modified scan flip-flop for low power testing of VLSI chips, *Progress in VLSI Design and Test Lecture notes in Computer Science, Springer-Verlag Berlin Heidelberg,* Vol. 7373, 2012, pp. 52-58.

[43]. S. Sharifi, J. Jaffari, M. Hosseinabady, A. A. Kusha, Z. Navabi, Simultaneous reduction of dynamic and static power in scan structures, in *Proceedings of the Design, Automation and Test in Europe Conference and Exhibition (DATE'05)*, Munich, Germany, 7-11 March 2005, pp. 846–851.

[44]. M. Elshoukry, M. Tehranipour, C. P. Ravikumar, A critical-path-aware partial gating approach for test power reduction, *ACM Transactions on Design Automation of Electronic Systems,* Vol. 12, No. 2, 2007, pp. 1-22.

[45]. R. Sankaralingam, N. A. Touba, Inserting test points to control peak power during scan testing, in *Proceedings of the 17th IEEE International Symposium on Defect and Fault Tolerance in VLSI Systems (DFT'2),* Vancouver, BC, Canada, 6-8 November 2002, pp. 138-146.

[46]. X. Kavousianos, D. Bakalis, D. Nikolos, Efficient partial scan cell gating for low-power scan-based testing, *ACM Transactions on Design Automation of Electronic Systems,* Vol. 14, No. 2, 2009, pp. 1–15.

[47]. D. Jayaraman, R. Sethuram, S. Tragoudas, Gating internal nodes to reduce power during scan shift, in *Proceedings of the ACM International Conference on Great Lakes Symposium on VLSI (GLSVLSI'10),* Rhode Island, USA, 16–18 May 2010, pp. 79-84 .

[48]. X. Lin, J. Rajski, Test power reduction by blocking scan cell outputs, in *Proceedings of the 17th IEEE Asian Test Symposium (ATS'08),* Hokkaido, Japan, 24–27 November 2008, pp. 329–336.

[49]. W. Zhao, M. Tehranipour, S. Chakarvarty, Power-safe test application using an effective gating approach considering current limits, in *Proceedings of the 29th IEEE VLSI Test Symposium (VTS'11),* Dana Point, CA, USA, 1-5 May 2011, pp. 160-165.

[50]. Y. T. Lin, J. L. Huang, X. Wen, A transition isolation scan cell design for low shift and capture power, in *Proceedings of the 21ᵗʰ IEEE Asian Test Symposium (ATS'12)*, Niigata, Japan, 19-22 November 2012, pp. 107-112.

[51]. S. Bhunia, H. Mahmoodi, D. Ghosh, K. Roy, Power reduction in test-per-scan BIST with supply gating and efficient scan partitioning, in *Proceedings of the 6ᵗʰ International Symposium on Quality Electronic Design (ISQED'05)*, Washington, DC, USA, 21-23 March 2005, pp. 453-458.

[52]. E. Arvaniti, Y. Tsiatouhas, Low-power scan testing: a scan chain partitioning and scan hold based technique, *J. Electron Test,* Vol. 30, No. 3, 2014, pp. 329-341.

[53]. Y. Ye, S. Borkar, V. De, New Technique for standby leakage reduction in high-performance circuits, *Proceedings of the Symposium on VLSI Circuits, Digest of Technical Papers,* 11-13 June 1998, pp. 40-41.

[54]. Z. Chen, M. Johnson, L. Wei, K. Roy, Estimation of standby leakage power in CMOS circuits considering accurate modelling of transistor stacks, in *Proceedings of the International Symposium on Low Power Electronics and Design (ISLPED'98)*, Monterey, CA, USA, 10-12 August 1998, pp. 239-244.

[55]. K. Roy, S. Mukhopadhyay, H. Mahmoodi, Leakage current mechanism and leakage reduction techniques in deep-submicrometer CMOS circuits, *Proceedings of the IEEE*, Vol. 91, No. 2, 2003, pp. 305-327.

[56]. M. C. Johnson, D. Somasekhar, K. Roy, Leakage control with efficient use of transistor stacks in single threshold CMOS, in *Proceedings of the 36ᵗʰ ACM/IEEE Design Automation Conference (DAC'99)*, New Orleans, LA, USA, June 21-25 1999, pp. 442-445.

[57]. M. M. Naeini, C. Y. Ooi, A novel scan architecture for low power scan-based testing, *VLSI Design,* Vol. 2015, No. 3, 2015, pp. 1-13.

Chapter 6

Digital Predistortion for Linearization of Wireless Transmitters

Ahmad Rahati Belabad

6.1. Introduction

Power amplifiers, as essential parts, are utilized in wireless communication systems. A great amount of power is consumed in a transmitter by the radio frequency power amplifier, hence this element is known as a power-hungry block [1]. Although power amplifiers are inherently nonlinear circuits, they have been aimed at amplifying communication signals linearly [2]. The nonlinear nature of this element produces in-band and out-of-band distortions. These distortions lead to adjacent channel interface and increase in error vector magnitude (EVM) [3]. In addition, many types of the signal with diverse digital modulations such as code division multiple access (CDMA) for the third generation and orthogonal frequency-division multiple access (OFDMA) for the fourth generation have been introduced to improve spectrum efficiency and data transmission rate [4].

Because the signals have high peak to average power ratio (OFDM~10dB) and a large back-off is required from the nonlinear region of the power amplifier for linear amplification, transmission of the non-constant envelope signals using linear power amplifiers has poor efficiency [5]. Thus, the power added efficiency (PAE) of power amplifiers is greatly diminished as a result of the large back-off so as to achieve sufficient linearity. Consequently, there is always a trade-off between linearity and efficiency as designing RF power amplifiers. In order to remove the compromise in the design of power amplifiers, several linearization techniques, such as feedback, feedforward, and predistortion, can be employed to improve the linearity of power amplifiers without sacrificing efficiency [6]. Among the linearization techniques, the predistortion method has good performance and low cost [7]. Inserting a nonlinear model which has reverse

Ahmad Rahati Belabad
Department of Electrical Engineering, Amirkabir University of Technology, Tehran, Iran

transfer characteristic of the power amplifier before the power amplifier is the principal idea of linearization in the predistortion method.

The predistortion can be divided into two types in accordance with the frequency in which it is implemented: 1. Analog Predistortion (APD) 2. Digital Predistortion (DPD).

The analog predistortion (ADP) works directly on the input signal of the power amplifier. The implementation of the analog predistortion is very simple but the predistortion works at a high frequency; hence it has low performance and limited adaptability [8]. There are many analog predistortion circuits that have simple and cost-effective structures [9-14]. The analog predistortion has several advantages in comparison with the digital predistortion, such as low cost and simple structure, but its capacity to improve linearity is less than that of the digital predistortion. Generally, the performance of the analog predistortion procedure is not comparable to the digital predistortion and most analog predistortion structures are exploited in a narrow band application. As an example, Mincheol Seo has proposed analog predistorter based on a Schottky diode but the improvement in spectral regrowth is very low [15]. One of the linearization methods always utilized in modern communication systems is the digital predistortion method [6, 7] and [16, 17].

Digital predistortion is one of the most effective methods of linearization, with low cost and high flexibility due to digital hardware implementation. The digital implementation greatly increases the accurate linearization capability [18]. The early linearization methods in literature utilized a memoryless structure in which predistortion compensated for the instantaneous nonlinear behavior of the power amplifier [19]. This behavior is characterized by amplitude modulation/amplitude modulation (AM/AM) and amplitude modulation/phase modulation (AM/PM) terms. For instance, in [20], Yunsung Cho has employed lookup table (LUT) rather than a memoryless model, but the LUT could not efficiently eliminate distortion in an advanced wireless transmitter. However, when the bandwidth is augmented in most advanced systems, the memory effect of the power amplifier is not negligible. In order to obtain the best performance of DPD, the memory effect must be considered [21]. A proper model for accurate modeling of nonlinear dynamic systems is the Volterra series [22]. However, this model has many coefficients and the number of coefficients grows very quickly when the order of nonlinearity and memory depth increases; hence, the computational complexity of the model grows.

Lei Guan [23] has recently employed the Volterra series in order to model predistortion, however the complexity of model as well as the resource consumption increase. In order to overcome the complexity of the Volterra series, many models originating from Volterra series such as Wiener model [24], Hammerstein model [18] and memory polynomial model [25] have been proposed. Wiener and Hammerstein models, which are called two-box models, are appropriate for applications with low bandwidth. In [3], Jungwan Moon has proposed an Enhanced Hammerstein model, but this model and identification of coefficients are very complex. Memory effects are well considered in the memory polynomial model, and the coefficients of the model can be easily extracted using the least square method.

6.2. Conventional Linearization Methods of Power Amplifiers

In the advanced telecommunications systems, new modulation architectures such as high spectral efficiency quadrature amplitude modulation (QAM) have been more developed. However, because these modulations generate a nonconstant envelope signal, they are highly sensitive to the effects of the non-linear power amplifier in order to maintain the quality of the transmitted signal. Linearization techniques are the best solution to improve the performance of the nonlinear power amplifier and at the same time to maintain the efficiency of power amplifiers near the saturation area. In fact, by this linearization method, the high-efficiency power amplifier which is operated near the saturation area is linearized and then the harmonic distortion of the power amplifier is reduced. Due to the strong performance limitations in many modern communication systems, the linearization of the power amplifier has become a basic requirement. This means that the utilization of the linearization methods in conjunction with the power amplifier reduces distortion at the output of the power amplifier. At results, the power amplifier satisfies the regulation of the telecommunications standards.

Linearization method can be called the procedure that reduces or even remove in-band and out of band distortions in the power amplifier. There are many methods for linearization such that most of them usually add one or more elements or system to the power amplifier. The linearization enables this possibility that power amplifier can produce higher output power and obtain higher efficiency for a fixed amount of distortion [26]. By applying the linearization method, the high-efficiency power amplifier is obtained without linear operation reduction.

Although linearization techniques at the system level compared to linearization techniques at the circuit level establish more reduction in the intermodulation components, system-level linearization method requires more cost and larger size. Therefore, system-level linearization method is appropriate for professional equipment such as the communication base stations.

System-level linearization techniques can be divided into two types according to their performance against distortion: linearization with the aim of removing the distortion at the output and linearization with the aim of avoiding the distortion at the output. Linearization techniques can be generally divided into three groups in terms of structure: feedback, feedforward, and predistortion [27, 28]. In the many resources, two approaches of the envelope elimination and restoration (EER) and (LINC) are expressed as ways to increase efficiency.

6.2.1. Feedforward

Nonlinear power amplifier produces the output signal that it can be assumed as a summation of a linear coefficient of the input signal and an error signal. In this approach, the error is calculated and is decreased by an appropriate factor from the power amplifier output signal in order to create the signal without distortion [29] and [30]. Fig. 6.1 shows the block diagram of feed forward method. In this method, power amplifier output signal

V_M is attenuated by a factor of $1/A$v and then the signal of V_N is generated. The signal V_N is subtracted from the input signal. The signal V_M can be introduced as the following:

$$V_M = A_v V_{in} + V_D,\qquad(6.1)$$

where the variable V_D denotes to the distortion. At the node N, The signal is introduced as follows:

$$V_N = V_{in} + \frac{V_D}{A_V}.\qquad(6.2)$$

In this case $V_P = V_D + A_v$, $V_Q = V_D$ and the result will be $V_{Out} = A_v V_{in}$.

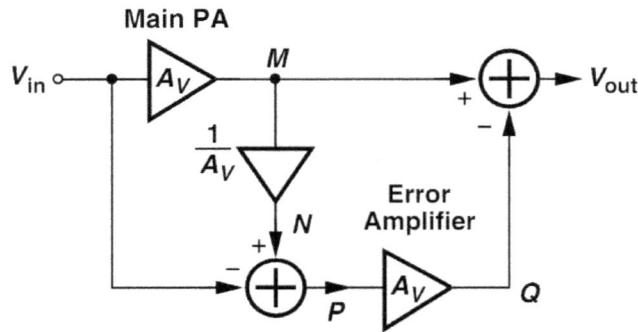

Fig. 6.1. Feedforward linearization structure.

In fact, two power amplifiers (i.e., main PA and Error Amplifier) have large phase shift at high frequencies, so the signal V_D will not be totally removed. For this reason, one stage delay Δ_1 is employed to compensate the phase shift of the main power amplifier as it is shown in Fig. 6.2. Also, another delay stage is used to compensate phase shift of the error amplifier.

Feed forward structure is inherently stable structure due to the lack of the feedback loop provided that two of its power amplifier are stable. This measure is the main advantage of the feedforward method. However, feed forward structure has several problems that have limited its performance. The first problem with this method is analog delay element. When the analog delay element to be a passive device, it is lossy elements. As the same way, When the analog delay element to be an active device, it generates distortion. The second problem is the loss of the second subtractor such that it decreases the efficiency the system. The third disadvantage is that the performance of the linearization method is depended on gain and phase matching of the received signal to each subtraction block.

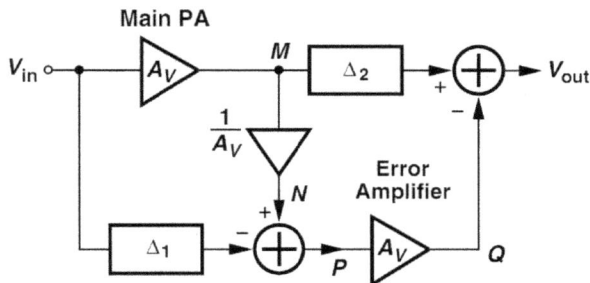

Fig. 6.2. The modified feed forward linearization structure.

6.2.2. Feedback

The basic structure of the power amplifier linearization by the feedback method is shown in Fig. 6.3 [31].

In this structure, the variables of the $x(t)$, A, $-1/K$ and $d(t)$ are the input signal, power amplifier gain, feedback loop gain, and distortion, respectively. The distortion $d(t)$ is inserted after the power amplifier. The system output is specified as follows:

$$y(t) = \frac{AK}{K+A}x(t) + \frac{K}{A+K}d(t) \qquad (6.3)$$

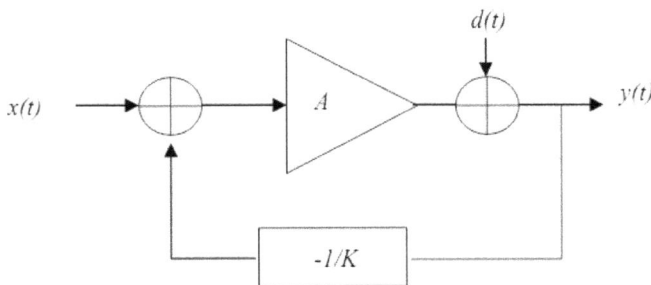

Fig. 6.3. Power Amplifier Linearization of the feedback method.

If we assume that the power amplifier gain is much more than feedback loop gain (i.e., A>>K), equation (3) is simplified as follows.

$$y(t) = Kx(t) + \frac{K}{A}d(t) \qquad (6.4)$$

It is obvious that the signal gain is reduced from A to K and the amount of distortion is highly decreased to the value of the K/A. Therefore, the power amplifier which has feedback requires more input voltage in order to produce the same power output that the power amplifier without feedback is generated.

However, feedback linearization method is simple in terms of implementation, this method when applied to the RF power amplifier systems have several drawbacks. The main problem is power amplifier gain reduction and stability problem.

6.2.3. Predistortion

Predistortion method is the very simple idea of linearization. Instead of applying input directly to the amplifier power, the input signal is applied to the predistortion block. Predistortion block is a nonlinear system which counteracts the nonlinearity of the power amplifier. In ideal, predistortion block has a behavior that it is the converse behavior of the power amplifier. Cascading the predistortion module and the power amplifier makes the all the system linear. This nonlinear model should be such that cascading power amplifier and predistortion creates a linear response.

A power amplifier together with the predistortion is shown in Fig 6.4 [32]. As seen in Fig. 6.4, the power amplifier has compressing input/output characteristic and the input/output characteristic of predistortion has an expanding behavior. This expanding characteristic can compensate for the nonlinearity of the power amplifier and theoretically, the relationship between the input and output of the system (PA+DPD) is highly linear.

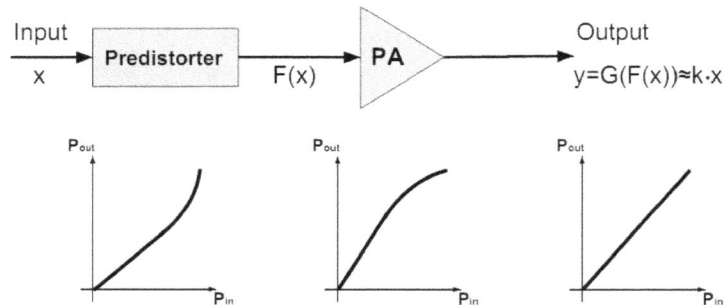

Fig. 6.4. The concept of predistortion linearization method.

In Fig. 6.4, the function of F(.) as a predistortion module and G(.) as a power amplifier have been marked. However, constructing complementary behavior of the power amplifier is not an easy task. Predistortion can be implemented in both analog and digital domain. Although we focus on digital predistortion, several analog methods are investigated in the next section.

Without the power amplifier, the power consumption of this technique is usually lower than that of the feedforward method. Also, this linearization method presents the better stability and bandwidth compared to the feedback method due to its open loop structure. Modern mobile wireless systems utilize digital modulation techniques to improve the efficiency of the system. Digital modulation techniques require high linearity power amplifier. Amplifier linearization techniques used in wireless systems require low size and high-efficiency structure. Although feed forward linearization technique obtains a significant improvement in removing distortion, the system requires a lot of circuits. In addition, when feedback linearization techniques are used in the radio frequency, it has a stability problem. Because of the stability and simple structure, predistortion linearization technique is suitable especially for integrated circuits. Some references offer two methods of envelope elimination and restoration (EER) and linear amplification using nonlinear component (LINC) as a linearization method. However, these methods have been suggested as a way to increase efficiency in some references. In the following, a brief description is carried out by the methods.

6.2.4. Envelope Elimination and Restoration (EER)

In this method, RF modulated signal is transmitted in both directions by passing the signal from the power divider [33] as depicted in Fig. 6.5. In the first path, the signal is applied to the limiter and as a result, a constant envelope phase modulated signal is generated. This signal is then amplified by a nonlinear power amplifier with high efficiency such as class E power amplifiers. However, in a parallel direction, the signal is applied to the envelope detector such that its amplitude information has been extracted and amplified by an envelope power amplifier. Finally, both signals are applied to an amplifier in order to reconstruct the envelope information. The two signals containing amplitude and phase information are in-phase (I) and quadrature Q signals. The signal containing the amplitude information control the DC bias of final power amplifier that its input is the signal containing the phase information. Finally, the generated signal is the boosted main signal without distortion.

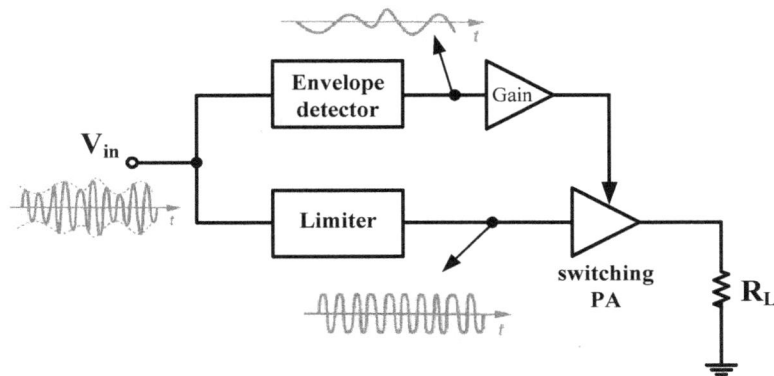

Fig. 6.5. The schematic of the envelope elimination and restoration method.

The main advantage of the envelope elimination and restoration method is a measure that the power amplifier can always be operated as switched mode power amplifier. Therefore, high efficiency is achieved without loss of linearity. But, the problem with this method is a time difference of gain the phase of two paths in Fig. 6.5 such that this difference should be less than an acceptable value [34]. Each of the paths has a different operating frequency. Also, this method is limited to the signal with average envelope variations.

Because envelope severe variation causes the final power amplifier and envelope power amplifier to be saturated and then the distortion will be generated at the output.

6.2.5. Linear Amplification using Nonlinear Components (LINC)

In this procedure, nonlinear elements are combined and used as a structure such that overall system operates linearly [35]. For this purpose, the bandpass signal $v_{in}(t) = \alpha(t)\cos[w_c t + \varphi(t)]$ can be shown as a summation of two signals with constant envelope but modulated phase as follows:

$$v_1(t) = 0.5V_0 \sin\left[w_c t + \varphi(t) + \theta(t)\right], \tag{6.5}$$

$$v_1(t) = -0.5V_0 \sin\left[w_c t + \varphi(t) - \theta(t)\right] \tag{6.6}$$

Which $\theta(t) = \sin^{-1}(\dfrac{\alpha(t)}{V_0})$ and V_0 is constant coefficient. So, if the signal v_1 and v_2 are generated, amplified by nonlinear stages and then two amplified signal components are added, the output includes the same phase and envelope information which the input had at first. Obtaining the signal v_1 and v_2 is very complex task because their phases should be modulated with $\theta(t)$ such that it is nonlinear function of $a(t)$. Fig. 6.6 shows the block diagram of the structure of LINC. In this structure, the input signal is divided by an AM to PM modulator to the two signals with constant envelope. Since the envelope of the two signal is constant, the nonlinear power amplifier with high efficiency can be used in order to amplify the signal power to a certain value. Finally, two signals after amplification are added by a combiner in order to comprise a linear factor of the input signal.

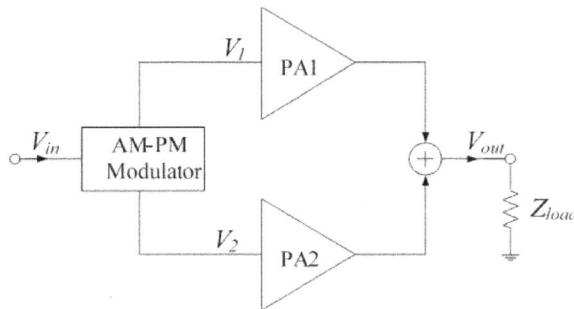

Fig.6.6. The schematic of linear amplification using the nonlinear component (LINC).

The main problem with this approach is that gain and phase balance of the two paths should be carefully controlled in order to create ideal linearization. But, one of the most important benefits of this approach is the use of the nonlinear power amplifier at the two paths [36]. Practical designs are not as simple as the mentioned equation. First, separation of the complex signal is very important in order to convert input envelope modulation to phase modulation in accordance with $V_1(t)$ and $V_2(t)$. Second, the combiner is required to establish low losses and perfect isolation between the two ports in order to prevent magnetic coupling of the two paths. Finally, this method is sensitive to component changes due to temperature variation and aging such that they may change the signal separator efficiency. Also it is possible that gain variations of the two power amplifier PA_1 and PA_2 are not similar to each other.

6.3. Behavioral Modeling of Digital Predistortion and Power Amplifiers

The idea of predistortion is applying the nonlinearity which is complementary of the nonlinearity of the power amplifier so that cascading the predistortion module and the power amplifier causes a linear and behavior response. In order to have a linear system, behavioral modeling is very necessary and important in order to predict power amplifier nonlinearity [37].

In cases where internal power amplifier circuit is not available such as Traveling-Wave-Tube (TWT) amplifiers, or when circuit simulation at the system level is considered, the behavioral or experimental model is better than the physical model. In this model, it is assumed that there is no information about the internal structure of the power amplifier (i.e., black box method) and we have only a series of conclusions and observations between input and output of the power amplifier. Usually, these models are known as the behavioral model. The accuracy of this method is highly sensitive to the choice of models and model parameter extraction methods. Behavioral models are divided into two categories as follows:

1. Memoryless model;
2. Models with memory.

6.3.1. Memoryless Model

The memoryless model depends only on its momentary input and in this model, there is no memory of previous inputs which can impact over the output signal. In the other words, there is a one by one mapping between input and output signals. The memoryless model usually divides distortions into two parts: amplitude to amplitude distortion (AM/AM) and amplitude to phase distortion (AM/PM). Amplitude to amplitude distortion (AM/AM) function is utilized to model the saturated output signal which is generated due to the gain compression. Amplitude to phase modulation (AM/PM) is utilized to model the signal phase shift.

6.3.1.1. Saleh Model

Saleh model is a simple nonlinear model based on two-parameter equations. In fact, this model emulates AM/AM and AM/PM distortion with non-linear equations in which there are only two unknown parameters [38]. The parameters of the model are easily obtained using simulation and calculation.

We assume that the input signal is in the form of the following equation:

$$x(t) = r_x(t) \cdot \cos\left[\omega_0(t) + \theta(t)\right], \tag{6.7}$$

which ω_0, $r_x(t)$ and $\theta(t)$ are carrier signal frequency, amplitude modulated envelope and phase modulated envelope, respectively. Due to the effects of power amplifiers nonlinearity and using the definition of the AM/AM and AM/PM modulation, the output can be represented in the following format:

$$y(t) = r_y\left[r_x(t)\right] \cdot \cos\left[\omega_0(t) + \theta(t) + \varphi_y\left[r_x(t)\right]\right], \tag{6.8}$$

where $r_y[r_x(t)]$ denotes AM/AM distortion and $\varphi_y[rx(t)]$ is AM/PM distortion.

Saleh model for the two above terms specifies two formula with two parameters as Eq. (6.9) and Eq. (6.10) as follows:

$$r_y\left[r_x(t)\right] = \frac{\alpha_r \cdot r_x(t)}{1 + \beta_r\left[r_x(t)\right]^2}, \tag{6.9}$$

$$\varphi_y\left[r_x(t)\right] = \frac{\alpha_\varphi \cdot r_x^2(t)}{1 + \beta_\varphi\left[r_x(t)\right]^2}. \tag{6.10}$$

The unknown parameters of the above equation are extracted by mapping each of the formulas with AM/AM and AM/PM curves of the power amplifier. when the signal of $r_x(t)$ is increased, the functions of the $r_y[r_x(t)]$ and $\varphi_y[r_x(t)]$ are limited to $1/r_x(t)$ and constant value, respectively.

6.3.1.2. Taylor Series Model

Taylor series model is one of the simplest and most practical forms of power amplifier modeling [39]. This method is based on the idea that the output of the system can be represented as follows:

$$V_{out}(t) = \sum_{n=1}^{\infty} a_n V_{in}^n(t), \tag{6.11}$$

where the coefficients of a_n are constant.

When the effects of the nonlinearity to be a week, this model has good accuracy for predicting output and intermodulation components. One advantage of this approach is this measure that the term of each intermodulation component and its relation is totally obvious. For example, third and fifth intermodulation are characterized by the coefficients of a_3 and a_5, respectively. But, the main disadvantage of this method is when the effects of power amplifier nonlinearity to be severe. Therefore, we can not rely only on the several first terms for power amplifier modeling and then this model leads to complexity of calculation. Taylor series standard model is expressed by Eq. 11 in which only AM/AM distortion has been included. But because of the distortion of AM-PM in most power amplifiers, Taylor's series complex model is used for power amplifier modeling.

$$V_{out}(t) = (\alpha_n + j\beta_n) \cdot V_{in}^n(t).$$ (6.12)

In recent years, Taylor's series complex model has been a lot used for simulation and predicting power amplifier behavior in most communications applications.

6.3.2. Models with Memory

The output of the models with memory depends on not only the input current but also part of the previous input that is affected the output of the power amplifier. These systems with memory are named so-called dynamic systems. For accurate modeling, many systems also depend on the values of previous outputs. Memory effects can be observed as changes in the frequency domain on the power amplifier transfer function. Previous static models without memory do not depend on the frequency and these are desirable for a power amplifier that is triggered by a narrowband signal. For new modulation methods, broadband signals are used to achieve higher data rates. Memory effects are much more evident by greater bandwidth signals. For this reason, memory effects are an important measure for linearization of the power amplifier using the predistortion. In the following, models with memory are described to observe how these models emulate the memory effects of the power amplifier.

6.3.2.1. Volterra Series

Generally, whatever the model of the power amplifier has more coefficients, it has more accuracy. The most accurate amplifier models with a memory effect are Volterra series and this type of model can be considered as a generalized Taylor series [40]-[41]. Volterra series discrete form is as follows:

$$y[n] = \sum_{l_1}^{L-1} h_1[l_1].x[n-l_1] + \sum_{k=2}^{k} \sum_{l_1=0}^{L-1} ... \sum_{l_k=0}^{L-1} h_i[l_1,...,l_k].x[n-l_k]...x[n-l_k], \quad (6.13)$$

where $h_i[*]$ denotes to discrete Volterra kernel of order k–th and L refer to limited memory length. The Volterra model has usually more accuracy than other power amplifier models with memory. Anyway, the number of Volterra model coefficients increases exponentially with the length of memory and kernel orders. Using a very high number of coefficients not only lead to an increase in the cost calculations but also reduces the stability of the entire system. Because of the weakness of Volterra model with the large memory and high-order kernel, it is rarely utilized in a real application. Often, only selected Volterra kernels are used to create a simple version of Volterra model. The models which are discussed following, such as the Wiener and Hammerstein models, are simplified Volterra model.

6.3.2.2. Wiener Model

As shown in Fig. 6.7, Wiener model is a two-box model that it consists of a linear FIR filter with memoryless nonlinearity function such that memoryless nonlinearity function can be considered as a power series like a Taylor series [42].

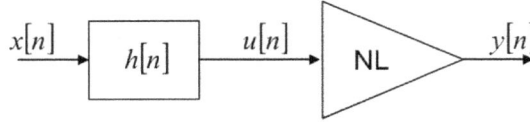

Fig. 6.7. Power amplifier Wiener model.

FIR filter and the nonlinear function displays dynamic nonlinearity and static nonlinearity, respectively. The nonlinear function can be implemented using a lookup table based on the curve AM /AM and AM/PM curves of the power amplifier.

The equation of two subsystems is determined as follows:

$$u[n] = \sum_{q=0}^{Q-1} \gamma_q \cdot x[n - l_q], \qquad (6.14)$$

$$y[n] = \sum_{k=0}^{K} \alpha_k \cdot u[n] \cdot |u[n]|^{k-1}. \qquad (6.15)$$

Where the variable of α_k denotes to complex coefficients of the power amplifier memoryless model. γ_q and l_q specify the filter coefficients and delay.

By replacing Eq. (6.14) to Eq. (6.15) the following equation is generated:

$$y[n] = \sum_{k=1}^{K} \alpha_k \cdot \left[\sum_{q=0}^{Q-1} \gamma_q \cdot x[n - l_q] \right] \cdot \left| \sum_{q=0}^{Q-1} \gamma_q \cdot x[n - l_q] \right|^{k-1}. \qquad (6.16)$$

Wiener model is one of the simplest ways to incorporate the effects of memory and nonlinearity. Anyway, this model has the limited performance for modeling most power amplifiers [43].

6.3.2.3. Hammerstein Model

Hammerstein model is a two-box model such that it contains memoryless nonlinear function and linear FIR filter. The block diagram of the Hammerstein model is shown in Fig. 6.8. The FIR filter emulates dynamic nonlinearity and nonlinear function part shows static nonlinearity such as the Wiener model which was described in the last section [44].

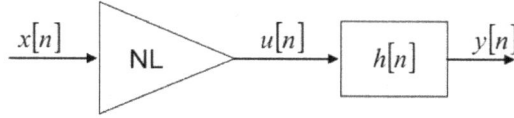

Fig. 6.8. Hammerstein model.

Two sub-models of Hammerstein model are defined as follows:

$$u[n] = \sum_{k=1}^{K} \alpha_k \cdot x[n] \cdot |x[n]|^{k-1}, \qquad (6.17)$$

$$y[n] = \sum_{p=0}^{P-1} \beta_p \cdot x[n - l_p]. \qquad (6.18)$$

The relationship between the input and output of the model is described as follows:

$$y[n] = \sum_{p=0}^{P-1} \beta_p \cdot \sum_{k=1}^{K} \alpha_k \cdot x[n - l_p] \cdot |x[n - l_p]|^{k-1}, \qquad (6.19)$$

where the variable of α_k denote to complex coefficients of the memoryless power amplifier model. Similar to the Wiener model, the Hammerstein model has very simple memory nonlinearity which is leading to limited performance for the predistortion.

6.3.2.4. Wiener- Hammerstein Model

Wiener-Hammerstein model is a combination of Wiener and Hammerstein models [45]. This model is a three-box model such that lookup table block is placed between the two FIR filters. This model is shown in Fig. 6.9. Static nonlinear function output F(.) is defined as follows:

$$u_{WH}(n) = F[x_{WH}(n)] = \sum_{i=1}^{N} b_i \cdot x_{WH}(n) \cdot |x_{WH}(n)|^{i-1}, \qquad (6.20)$$

where the variables of N and b_i denote the nonlinear order and coefficients F(.), respectively.

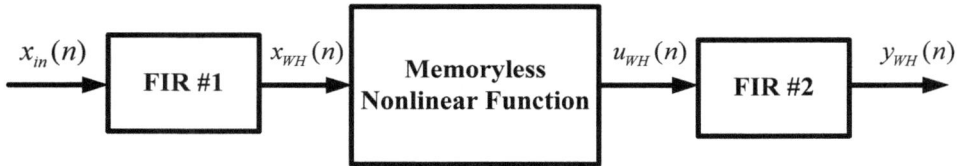

Fig. 6.9. Wiener-Hammerstein model.

The outputs of the two FIR filter (i.e., $X_{WH}(n)$ and $y_{WH}(n)$) is specified as follows:

$$x_{WH}(n) = \sum_{m_1=0}^{M_1} h_{m_1} \cdot x_{in}(n - m_1),$$ (6.21)

$$y_{WH}(n) = \sum_{m_2=0}^{M_2} k_{m_2} \cdot u_{WH}(n - m_2).$$ (6.22)

$M1$ and $M2$ are the memory depths of two FIR filter and h_{m1} and k_{m2} are the filter coefficients, respectively.

6.4. Memory Polynomial Model and Model Identification Procedure

In this part as an example, the memory polynomial model has been proposed in order to model both the power amplifier and the digital predistortion. Kim and konstantinou proposed the memory polynomial model several years ago [46]. Owing to the complexity and accuracy of trade-off, this model is one of the most popular models for behavior modeling of predistortion. When the coefficients of Volterra series change to diagonal terms (i.e. removing all cross terms), the memory polynomial model is generated. The model is a two summation formula with two parameters: nonlinear order and memory depth. Therefore, this model has great flexibility. The baseband complex output signal (y_{MP}) of the memory polynomial model as a function of baseband complex input signal (x) can be described by the following equation:

$$y_{MP}(n) = \sum_{m=0}^{M} \sum_{k=1}^{K} \alpha_{mk} \cdot x(n-m) \cdot \left| x(n-m) \right|^{k-1},$$ (6.23)

where α_{mk} is the coefficients of the model and K and M refer to nonlinear order and memory depth, respectively. If we rewrite Eq. 6.23 as a generic formulation, the equation is written in this form:

$$y_{MP}(n) = \varphi_{MP}(n) \cdot A,$$ (6.24)

where $\varphi_{MP}(n)$ and A are defined as follows:

$$\varphi_{MP}(n) = \begin{bmatrix} x(n) \\ \vdots \\ x(n) \cdot \left| x(n) \right|^{k-1} \\ x(n-1) \\ \vdots \\ x(n-1) \cdot \left| x(n-1) \right|^{k-1} \\ \vdots \\ x(n-M) \cdot \left| x(n-M) \right|^{k-1} \end{bmatrix},$$ (6.25)

$$A = \begin{bmatrix} \alpha_{01} & \cdots & \alpha_{0k} & \alpha_{11} & \cdots & \alpha_{1k} & \cdots & \alpha_{MK} \end{bmatrix}^T, \tag{6.26}$$

where $\begin{bmatrix} \cdot \end{bmatrix}^T$ refers to transpose operator. According to Eq. 6.23, the dimension of this model is defined by nonlinear order and memory depth; thus, the memory polynomial model has two degrees of freedom. The block diagram of the memory polynomial model is depicted in Fig. 6.10. As illustrated in Fig. 6.10, the memory polynomial model can be understood as a combination of $(M+1))$ polynomial functions. Each of them has been applied to the delayed complex input sample x(n).

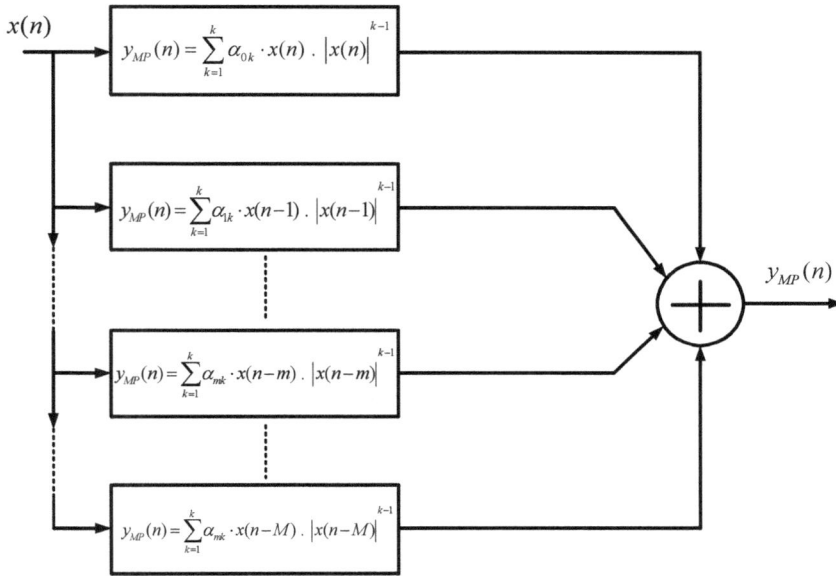

Fig. 6.10. The block diagram of the memory polynomial model.

For a set of N samples, this vector representation can be rewritten in matrix format as follows:

$$y = X \cdot A, \tag{6.27}$$

where y is the vector of N samples of the output signal and given by:

$$y = \begin{bmatrix} y(n) \\ y(n-1) \\ \cdots \\ \cdots \\ \cdots \\ y(n-N+1) \end{bmatrix}^T. \tag{6.28}$$

175

Also, X is a matrix whose rows are delayed versions of $\varphi_{MP}(n)$. It is defined as follows:

$$X = \begin{bmatrix} \varphi_{MP}(n) & \varphi_{MP}(n-1) & \cdots & \varphi_{MP}(n-N+1) \end{bmatrix}^{T}, \qquad (6.29)$$

$$X = \begin{bmatrix} x(n) & \cdots & x(n) \cdot |x(n)|^{K-1} & x(n-1) & \cdots & x(n-1) \cdot |x(n-1)|^{K-1} \\ x(n-1) & \cdots & x(n-1) \cdot |x(n-1)|^{K-1} & x(n-2) & \cdots & x(n-2) \cdot |x(n-2)|^{K-1} \\ \vdots & \cdots & \vdots & \vdots & \cdots & \vdots \\ x(n-N+1) & \cdots & x(n-N+1) \cdot |x(n-N+1)|^{K-1} & x(n-N) & \cdots & x(n-N) \cdot |x(n-N)|^{K-1} \end{bmatrix}$$

$$\begin{bmatrix} \cdots & x(n-M) \cdot |x(n-M)|^{K-1} \\ \cdots & x(n-M-1) \cdot |x(n-M-1)|^{K-1} \\ & \vdots \\ \cdots & x(n-M-N+1) \cdot |x(n-M-N+1)|^{K-1} \end{bmatrix}^{T}. \qquad (6.30)$$

If the matrix X was invertible, the coefficients identification would be given by:

$$A = X^{-1} \cdot y. \qquad (6.31)$$

In order to perform coefficients extraction, an approximate solution can be achieved by minimizing the mean squared error, e, given by:

$$e = \|y - XA\|^{2}. \qquad (6.32)$$

To solve this problem, one method which is usually utilized is the method of computing the pseudo-inverse of the matrix X as follows [47]-[48]:

$$pinv(X) = (X^{T}X)^{-1}X^{T}. \qquad (6.33)$$

Then, the coefficients are calculated using:

$$A = pinv(X) \cdot y. \qquad (6.34)$$

By minimizing the error using least square (LS) method, this method allows for proper identification of the model coefficients.

6.5. Digital Predistortion

The predistortion can be designed by using a lookup table (LUT) or predistortion models. In the first case, the LUT is a simple memory that maps the input to the predistorted output. This predistorted output is directly applied to the input of the power amplifier as an input. In the second case, the behavioral modeling is utilized in the predistortion model. The behavioral modeling is basically similar to the LUT method. But in the use of the predistortion model, the output of the predistortion is calculated for each input sample. These two implementations have different advantages and disadvantages.

When the more resolution of digital predistortion is required, the size of the LUT will be larger and then a lot of memory space is consumed. Anyway, the implementation of the model in the hardware usually requires the more processing time compared to the LUT model and as a result, the more power is consumed in the use of the behavioral model in the predistortion.

To create LUT values and model coefficients, the power amplifier specification should be investigated. Nowadays, most methods that are employed for linearizing the power amplifier have a feedback structure. Feedback makes the overall system (including amplifier power and pre-distortion) is operated as an adaptive. When the power amplifier specification is changed, this adaptive structure causes that predistortion is operated with high performance. Changes in the power amplifier characteristics are due to temperature fluctuations, power supply variation, aging.

Nowadays, with the impressive growth in DSP techniques, most linearization methods based on digital predistortion are completely implemented in baseband domain [49]. An overall system made of the power amplifier and digital predistortion based linearization method has been depicted in Fig. 6.11. In this system, first, the generated baseband signal is distorted using the predistortion block, and then the distorted signal is passed through a digital to analog converter (DAC) in order to convert it to an analog signal. Similarly, the frequency of the analog signal is converted to radio frequency using a modulator composed of a mixer, local oscillator, and combiner, and then the RF signal is applied to the power amplifier in order to amplify the transmission signal.

In order to extract and update the coefficients of the digital predistortion model, as illustrated in Fig. 6.11, a small portion of the transmitted signal is taken by a coupler. The frequency of the return signal is down-converted to the modulator and the resultant signal is transformed to a digital signal using an analog to digital converter (ADC).

The digital signal is applied to the adaptation algorithm block which takes a sample of input and output data and updates the coefficients of the predistortion block in order to achieve an effective linearization method. The feedback path which is shown in Fig. 6.11 works only in the initial system setup or whenever the characteristics of the system have wide variation. Therefore, the design and implementation of a digital predistortion system include two parts: 1) The digital predistortion block. In this chapter, the use of memory polynomial model has been proposed to model it. The input signals I and Q should be distorted by applying the digital predistortion block; 2) The adaptation algorithm block utilized for updating the coefficients of the predistortion block.

In the next section, the digital predistortion block based on the memory polynomial model is designed for the linearization of the power amplifier with memory effects. The coefficients of the predistortion block are identified using the indirect learning structure. Compared to Hammerstein based predistortion, the digital predistortion function constructed by memory polynomial has more terms. But this predistortion has high robustness and its coefficients are easily identified using the least square algorithm.

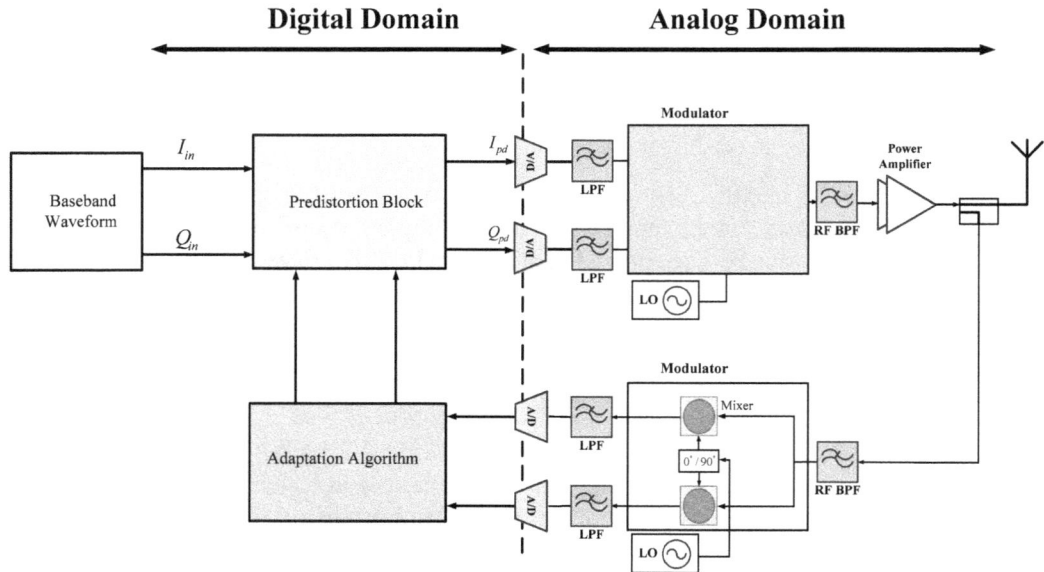

Fig. 6.11. The schematic of the digital predistortion block and power amplifier.

6.6. Performance Assessment of the Digital Predistortion

The flowchart summarizing the evaluation steps of power amplifier linearization using memory polynomial model based digital predistortion and indirect learning structure is presented in Fig. 6.12.

The evaluation steps are classified into four parts. At first, input/output data of the power amplifier are gathered using a simulation of the power amplifier in the Matlab software. Next, one behavioral model is proposed for the power amplifier model and the coefficients of the model are identified. Then in step III, coefficients identification of the digital predistorter based on the memory polynomial using the indirect learning structure is carried out. Finally, the overall of the system (PA+DPD) is simulated. A detailed description of each stage is described as follows.

6.6.1. Providing Power Amplifier Data Set

The data required for modeling the power amplifier are the input and output data (in-phase and quadrature) of the power amplifier. To provide the input and output data of the power amplifier, the real model of the power amplifier has been utilized. This model is like a text file with the extension of .s2d and it is imported into Matlab Simulink block of the power amplifier. The .s2d file contains large signal S-parameters and describes nonlinearity using gain compression description, third-order intercept point, 1-dB gain compression, saturated power and gain compression.

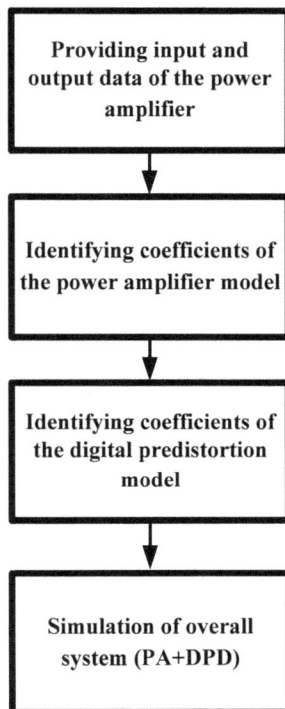

```
┌─────────────────────────┐
│   Providing input and    │
│  output data of the power│
│        amplifier         │
└─────────────────────────┘
            │
            ▼
┌─────────────────────────┐
│  Identifying coefficients of│
│ the power amplifier model│
└─────────────────────────┘
            │
            ▼
┌─────────────────────────┐
│  Identifying coefficients of│
│  the digital predistortion│
│          model           │
└─────────────────────────┘
            │
            ▼
┌─────────────────────────┐
│   Simulation of overall  │
│    system (PA+DPD)       │
└─────────────────────────┘
```

Fig. 6.12. Simulations steps of the power amplifier and digital predistortion.

The transmitter system shown in Fig. 6.13 is simulated for providing the input and output data of the power amplifier. The system works as follows: first, one and zero bits are randomly generated using the Random Integer block. Then, these random numbers are mapped onto a 16 QAM structure using a QAM modulator block. The QAM signal is passed through a transmitter raised cosine filter and then the output signal of the raised cosine filter is applied to the power amplifier. In the raised cosine filter, each symbol is multiplied by a sinc function.

The power amplifier has an input and output matching network specified in Fig. 6.13 as input and output ports. In the output of the power amplifier, input modules have been conversely placed. First, the output signal of the power amplifier is applied to a receiver raised cosine filter, after which it is passed through a QAM demodulator. Finally, at the output of the system, one and zero bits are observed. The power amplifier operates at a 2.1 GHz center frequency and has bandwidth as high as 15 MHz.

Constellation diagram at node In1 with 2048 samples is shown in Fig. 6.14. As clearly shown in Fig. 6.4, all of the data without any distortion have been placed onto the 16 QAM points. The maximum amplitude of in-phase and quadrature signal is ± 1.

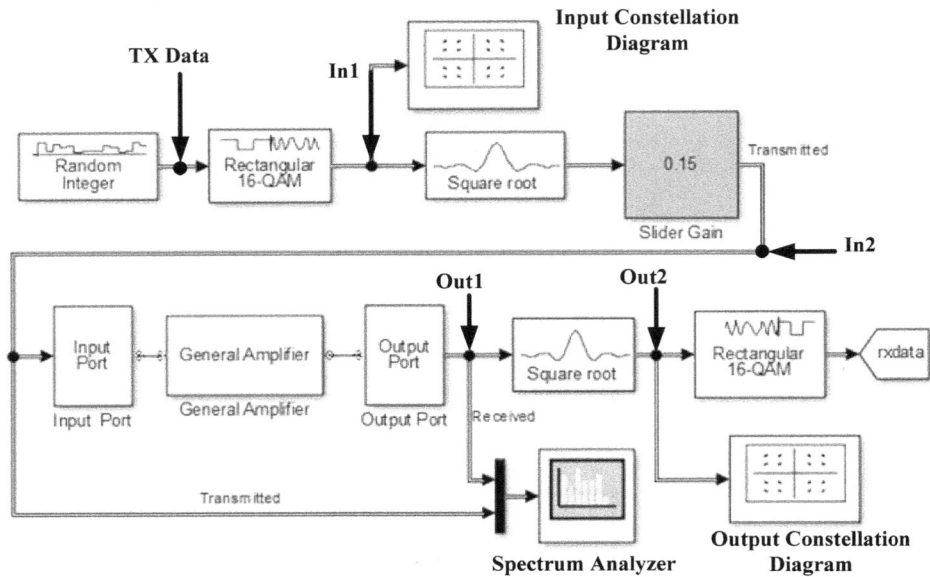

Fig. 6.13. Simulated transmitter system.

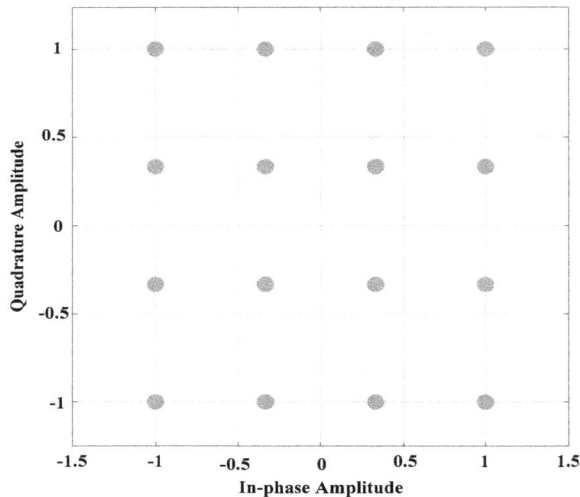

Fig. 6.14. Constellation diagram of input data at node In1.

Constellation diagram of the power amplifier output at node In2 with 2048 samples is depicted in Fig. 6.15. As obviously depicted in Fig. 6.15, all of the output data have been distorted because of AM/AM and AM/PM distortion of the power amplifier, and these signals have deviated from their original points. The constellation diagram which was shown in Fig. 6.14 is converted to the constellation diagram of Fig. 6.15 due to nonlinear distortion of the power amplifier.

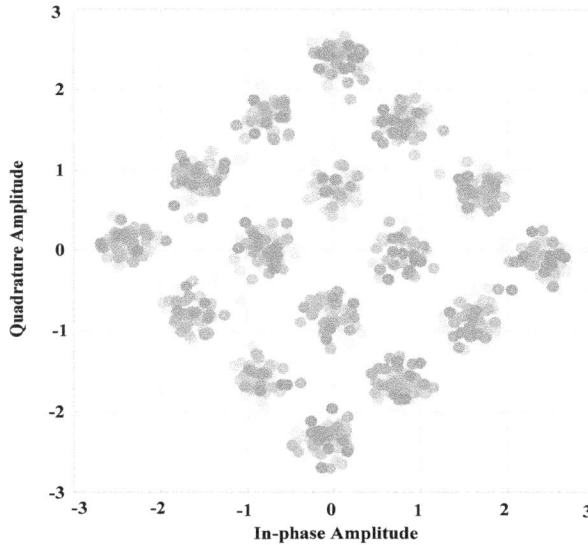

Fig. 6.15. Constellation diagram of output data at node out2.

The power spectrum density curves of the input and output of the power amplifier have been depicted in Fig. 6.16 at the operation frequency of 2.1 GHz with a bandwidth of 15 MHz. These two spectrum curves have been generated using the power amplifier simulation with .s2d file in the Matlab software. It is obvious that the power amplifier causes spectral regrowth phenomena and the side lobes of the spectrum curves have been grown.

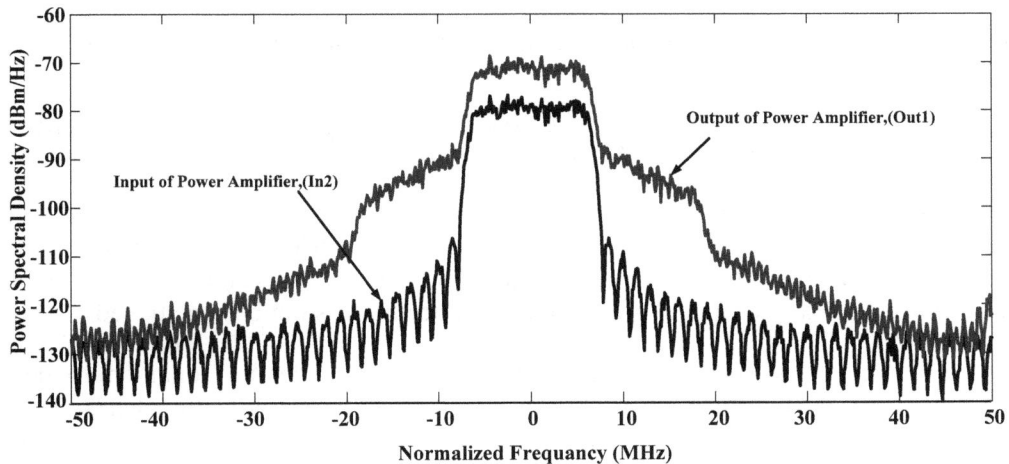

Fig. 6.16. Simulated input and output spectrum of the power amplifier at the center frequency of 2.1 GHz with a bandwidth of 15 MHz.

The adjacent power ratio (ACPR) is utilized in order to quantify the nonlinearity of PAs driven by modulated signals in the frequency domain. This is a significant linearity parameter since the power which is generated by the nonlinear distortions in the adjacent channels cannot be eliminated by filtering. Therefore, the power generated in the adjacent channels is considered as an unwanted emission that needs to be minimized and controlled.

6.6.2. Identifying Coefficients of the Power Amplifier Model Using the Least Square Algorithm

In this subsection, the coefficients of the power amplifier memory polynomial model have been obtained using the least square algorithm described in the previous part. The use of memory polynomial model has been proposed to model power amplifiers with memory effects. The characterization of the power amplifier nonlinear behavior using modulated signals consists of acquiring the input and output baseband waveforms of the power amplifier.

In other words, in order to obtain the coefficients of the power amplifier memory polynomial model, input/output in-phase and quadrature signals of the power amplifier is necessary. For this reason, input and output baseband signals with 500 samples have been acquired and input and output in-phase signals with 500 samples have been illustrated in Fig. 6.17 and Fig. 6.18, respectively. As illustrated in Fig. 6.17, The in-phase input signal of the power amplifier has a non-constant envelope and remarkable peak to average power ratio (PAPR) due to its complex modulation. The in-phase input signal of the power amplifier is boosted by the power amplifier and the resultant signal is shown in Fig. 6.18. It is obvious that the output signal in some samples has been saturated because of inherent nature of the power amplifier.

The relationship between the input and output of power amplifiers are as follows:

$$y(n) = \sum_{k=1}^{k}\sum_{q=0}^{Q} a_{kq} z(n-q)\left|z(n-q)\right|^{k-1}. \tag{6.35}$$

The nonlinearity order k and memory depth q have been swept with normalized mean square error (NMSE) criterion. When the amount of nonlinearity order and memory depth to be 5, the NMSE will be as minimum as possible. If the proposed amounts are greater than 5, the complexity of model grows and better accuracy is not been achieved.

For this reason in this section, the value of the nonlinearity order $k=5$ and memory depth $q=5$ have been chosen. A transfer function of nonlinear power amplifier leads to odd-order intermodulation products. These third-order intermodulation products cause distortion. For this reason, only the odd-order terms in the memory polynomial are considered. The effect of spectra generated by the even-order terms on the passband is negligible.

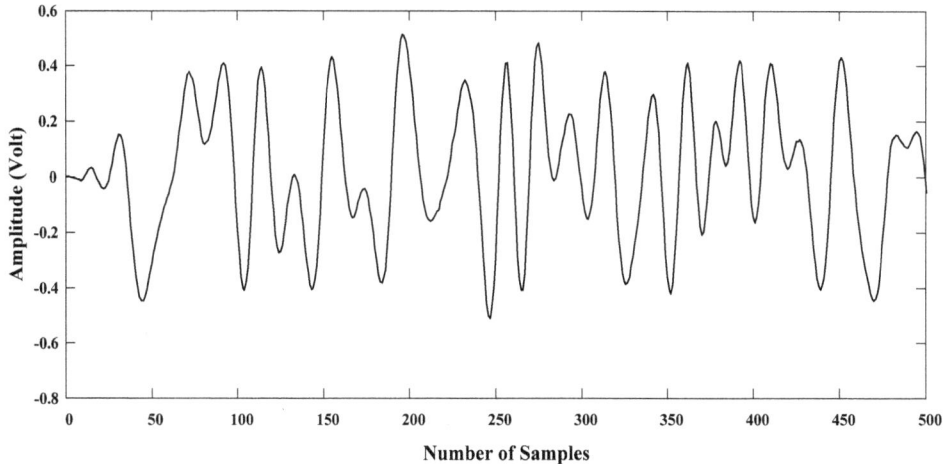

Fig. 6.17. Input in-phase signal of the power amplifier.

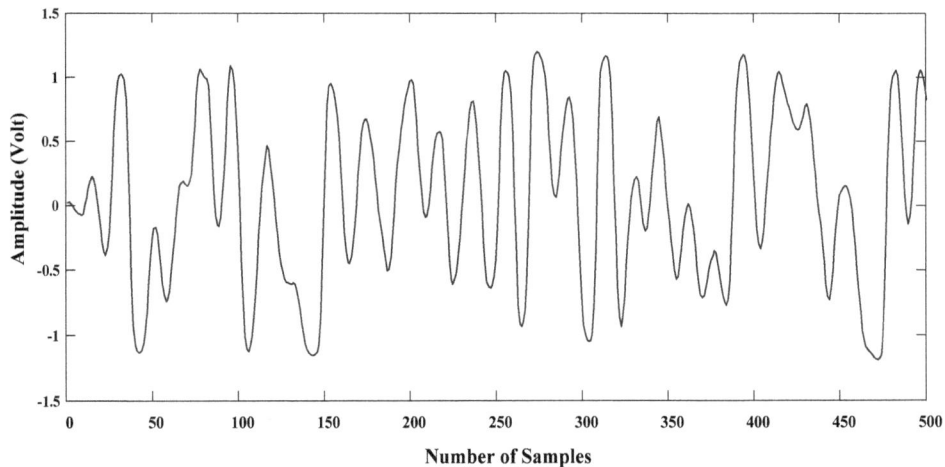

Fig. 6.18. Output in-phase signal of the power amplifier.

Using input/output in-phase and quadrature data and the least square method, we can map simulated data to the power amplifier memory polynomial model; hence, the coefficients of the power amplifier memory polynomial model are extracted. The coefficients of the power amplifier model were clearly identified using the least square algorithm. Therefore, the power amplifier memory polynomial model with the coefficients accurately extracted has been simulated. The input and output spectrum curves of the power amplifier modeled by memory polynomial are shown in Fig. 6.19. It is obvious from Fig. 6.19 that this power amplifier output spectrum is similar to the power amplifier output spectrum that was simulated with .s2d files in the Matlab software. The least square algorithm has demonstrated that the coefficient of the power amplifier memory polynomial model has been accurately generated.

183

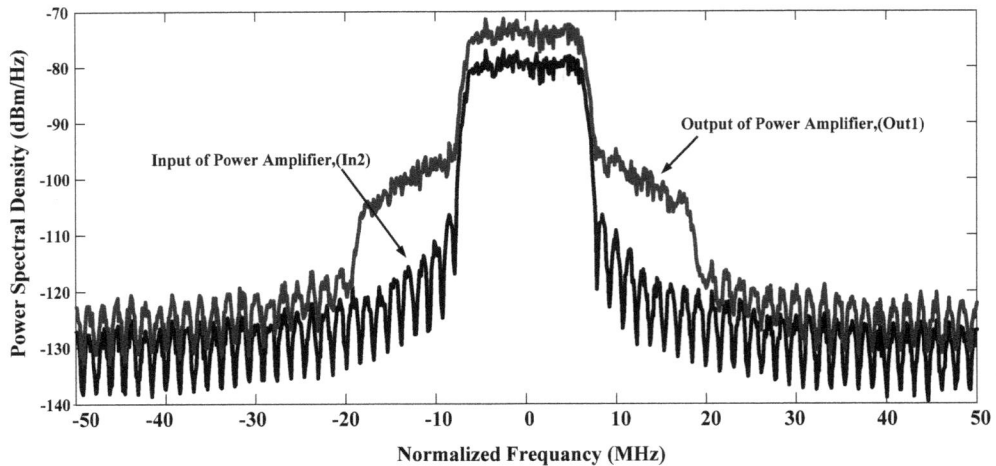

Fig. 6.19. Input and output spectrum of modeled power amplifier at the center frequency of 2.4 GHz with a bandwidth of 15 MHz.

6.6.3. Identifying Coefficients of the Digital Predistortion Model

In order to linearize the power amplifier and eliminate its nonlinearity, the predistortion block which is accurately modeled by the memory polynomial model is placed upstream of the power amplifier. Indeed, the behavior of the predistortion block is ideally the inverse characteristics of the power amplifier. In this step, the coefficients of the digital predistortion modeled by the memory polynomial are obtained using the indirect learning structure. First, the output signal is divided into the gain of the power amplifier and the attenuated signal is utilized as the input of digital predistortion block modeled by the memory polynomial. Afterward, the input data applied to the power amplifier is considered as an output signal of the predistortion model. The coefficient of digital predistortion memory polynomial model has accurately been identified using the mentioned set of input/output data in conjunction with the least square algorithm.

The memory polynomial model shown in the previous section is employed as a digital predistorter and the memory polynomial model is utilized as a power amplifier model. The indirect learning structure employed for coefficients identification of the proposed predistortion has been illustrated in Fig. 6.20.

Actually, the indirect learning structure is a careful method in which a learning loop is closed around the power amplifier [49]. The input and output signals of the power amplifier, which was described in the previous section, are utilized for extraction of the predistorter model. When the parameters of the predistortion in the learning loop are identified, this predistortion is then directly copied and employed as the predistortion block. The complex baseband input of the proposed predistorter is denoted by $x(n)$; the output of the predistorter and input of the power amplifier are denoted by u(n), and the complex baseband output of the power amplifier is denoted by $y(n)$. To identify the coefficients of the proposed model, the predistorter block located in the upstream of the power amplifier is disconnected and the indirect learning loop is closed. The indirect learning loop includes the power amplifier, predistorter block, and estimation of parameters block.

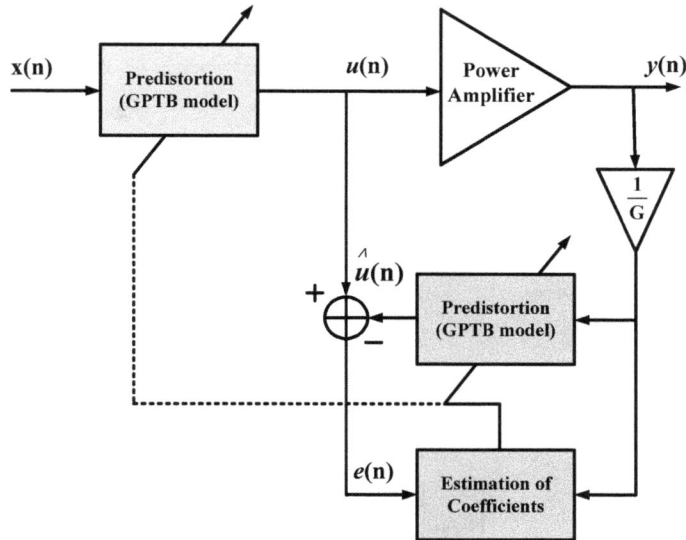

Fig. 6.20. Indirect learning structure.

When the input of the power amplifier and the output of the predistorter block in the feedback path are equal (and consequently an error term, i.e. e(n)= $u(n) - \hat{u}(n) \simeq 0$, is close to zero), the input and output of the system will be linear (i.e. y(n)=Gx(n)).

The coefficients of the proposed model have been identified by this structure. In this approach, when the error of energy $\left\| e(n) \right\|^2$ is minimum, the algorithm for finding the coefficients of the predistortion is fully converged. The identification method of the predistorter model is similar to the procedure that was clearly described in the previous part by replacing the input of the predistorter with *y(n)/G*, where G is a small signal gain of the power amplifier and the output is $\hat{u}(n)$.

6.6.4. Simulation of the Transmitter (PA+DPD)

The schematic of the simulated transmitter with digital predistortion has been depicted in Fig. 6.21. The digital predistortion based on the memory polynomial model is placed between the raised cosine filter block and modulator. In-phase and quadrature (I and Q) signals are applied to the predistortion block, and the signals are passed through the proposed predistorter model. When these signals are applied to the predistorter, they experience the static and dynamic nonlinearity of the model. Finally, the predistorted

185

signals are applied to the modulator. The memory polynomial model, which was clearly described in previous part, has been employed for the predistorter block. The nonlinearity order, k, and memory depth, q, of the proposed model have been considered with normalized mean square error (NMSE) criterion. Using indirect learning structure, the coefficients of the proposed model are accurately identified.

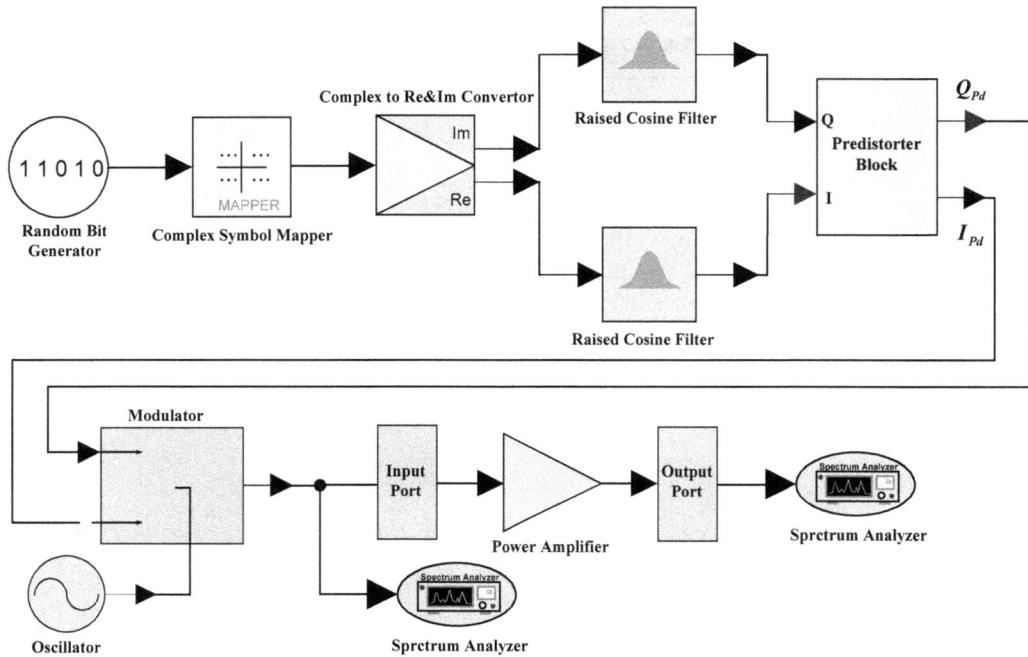

Fig. 6.21. Simulated transmitter system with digital predistortion.

Fig. 6.22 illustrates the input signal spectrum of the power amplifier X(n), the output signal spectrum of the nonlinear power amplifier and the output signal spectrum of the overall system (power amplifier + digital predistortion). It is very clear that the digital predistortion linearization method has compensated for the distortion of the power amplifier, and the spectral regrowth of the power amplifier is totally removed. The measure of adjacent channel power ratio (ACPR) is presented in Table 1. It is very apparent that the digital predistortion linearization method has compensated for the distortion of the power amplifier, and the spectral regrowth of the power amplifier is totally removed. As a result, the output spectrum of the overall system is exactly the same as the input signal, and the system operates linearly.

The measure of adjacent channel power ratio (ACPR) is presented in Table 6.1. It is very clear that the ACPR of the power amplifier with digital predistortion has been improved to about 15 dBc.

186

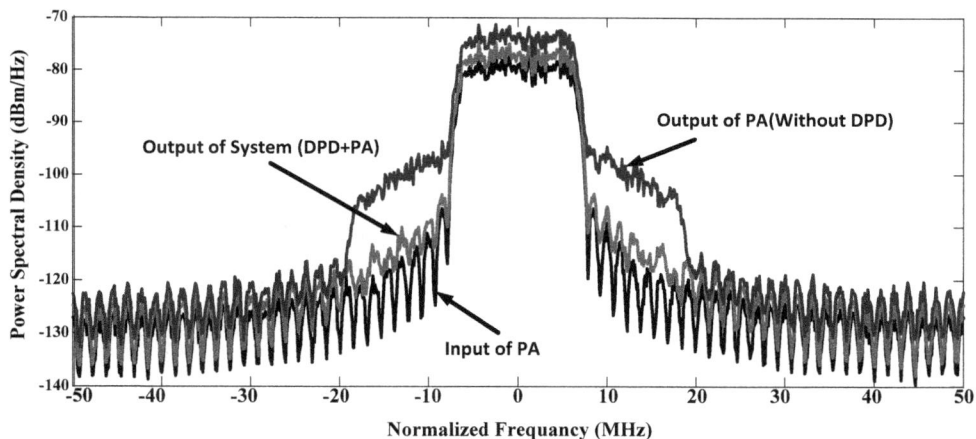

Fig. 6.22. Power amplifier input spectrum, power amplifier output spectrum without digital predistortion and system output spectrum which consists of the power amplifier and DPD.

Table 6.1. ACPR for a power amplifier with and without Digital predistortion.

	ACPR (dBc)	
	+/-15 MHz	+/-20 MHz
Without DPD	-25	-25
With DPD	-40	-40

6.7. Conclusions

A digital predistortion structure has been described in this chapter. In this structure, the memory polynomial model was utilized for both the power amplifier and digital predistortion modeling. The memory depth and nonlinear order of the memory polynomial model were carefully chosen to have the greatest impact on non-linear effect reduction. The coefficients of the power amplifier model were identified using the least square algorithm and those of the digital predistortion model were extracted using the indirect learning structure. Finally, the overall system, including the power amplifier and digital predistortion, was simulated. The spectrum of the overall system output showed that spectral regrowth due to the nonlinearity of the power amplifier was totally removed. It demonstrated that ACPR has improved to about 15 dB using digital predistortion linearization method. This proposed digital predistortion system can be easily implemented in real digital circuits such as DSP and FPGAs.

References

[1]. B. Razavi, RF Microelectronics, Second Edition, *Prentice Hall*, Upper Saddle River, NJ, 2011.

[2]. Alireza Saberkari, Saman Ziabakhsh, Herminio Martinez, Eduard Alarcón, Active inductor-based tunable impedance matching network for RF power amplifier application, *Integration, the VLSI Journal*, Vol. 52, 2016, pp. 301-308.

[3]. Junghwan Moon, Bumman Kim, enhanced hammerstein behavioral model for broadband wireless transmitters, *IEEE Transactions on Microwave Theory and Techniques*, Vol. 59, No. 4, February 2011, pp. 924-933.

[4]. Ahmad Rahati Belabad, Nasser Masoumi, Shahin J. Ashtiani, A fully integrated 2.4 GHz CMOS high power amplifier using parallel class A&B power amplifier and power-combining transformer for WiMAX application, *AEU-International Journal of Electronics and Communications*, Vol. 67, No. 12, December 2013, pp. 1030–1037.

[5]. Debopriyo Chowdhury, Christopher D. Hul, Ofir B. Degan, Yanjie Wang, Ali M. Niknejad, A fully integrated dual-mode highly linear 2.4 GHz CMOS power amplifier for 4G WiMAX applications, *IEEE Journal of Solid-State Circuits*, Vol. 44, No. 12, December 2009, pp. 3393-3402

[6]. Hua Qian, Hao Huang, Saijie Yao, A general adaptive digital predistortion architecture for stand-alone RF power amplifiers, *IEEE Transactions on Broadcasting*, Vol. 59, No. 3, August 2013, pp. 528-538.

[7]. Ahmad Rahati Belabad, Eiman Iranpour, Saeed Sharifian, FPGA Implementation of a Hammerstein based digital predistorter for linearizing RF power amplifiers with memory effects, *Amirkabir International Journal of Science & Research*, Vol. 47, No. 2, 2015, pp. 9-17.

[8]. Roman Marsalek, Contributions to the power amplifier linearization using digital baseband adaptive predistortion. Ph. D. Thesis, *Brno University of Technology*, 2003.

[9]. S. Jung, H. Park, M. Kim, G. Ahn, J. Van, H. Hwangbo, C. Park, S. Park, Y. Yang, A new envelope predistorter with envelope delay taps for memory effect compensation, *IEEE Transactions on Microwave Theory and Techniques*, Vol. 55, No. 1, January 2007, pp. 52-59.

[10]. K. Yamauchi, K. Mori, M. Nakayama, Y. Mitsui, and T. Takagi, A microwave miniaturized linearizer using a parallel diode, in *Proceedings of the IEEE MTT-S Int. Microw. Symp. Dig.*, June 1997, pp. 1199-1202.

[11]. K. Lim, G. Ahn, S. Jung, H. Park, M. Kim, J. Van, H. Cho, J. Jeong, C. Park, Y. Yang, A 60 watt multi-carrier WCDMA power amplifier using an RF predistorter, *IEEE Trans. Circuits Syst. II, Exp. Briefs*, Vol. 59, No. 4, Apr. 2009, pp. 265–269.

[12]. J. Yi, Y. Yang, M. G. Park, W. W. Kang, B. Kim, Analog predistortion linearizer for high-power RF amplifiers, *IEEE Trans. Microw. Theory Tech.*, Vol. 48, No. 12, December 2000, pp. 2709-2713.

[13]. T. Nojima, T. Kon No, Cuber predistortion linearizer for relay equipment in 800 MHz band land mobile telephone system, *IEEE Trans. Veh. Tech.*, Vol. 34, No. 4, Nov. 1985, pp. 169-177.

[14]. J. Cha, J. Yi, J. Kim, B. Kim, Optimum design of a predistortion RF power amplifier for multicarrier WCDMA applications, *IEEE Trans. Microw. Theory Tech.*, Vol. 52, No. 2, February 2004, pp. 655-663.

[15]. Mincheol Seo, Kyungwon Kim, Minsu Kim, Hyungchul Kim, Jeongbae Jeon, Myung-Kyu Park, Hyojoon Lim, Young Yang, Ultrabroadband Linear Power Amplifier Using a Frequency-Selective Analog Predistorter, *IEEE Transactions on Circuits and Systems II*, Vol. 58, No. 5, May 2011, pp. 264-268.

[16]. Mayada Younes, Fadhel M. Ghannouchi, An accurate predistorter based on a feedforward Hammerstein structure, *IEEE Transactions on Broadcasting*, Vol. 58, No. 3 May 2012, pp. 454-461.

[17]. Atefeh Salimi, Rasoul Dehghani, Abdolreza Nabavi, A digital predistortion assisted hybrid supply modulator for envelope tracking power amplifiers, *Integration, the VLSI Journal*, Vol. 52, 2016, pp. 282-290.

[18]. Taijun Liu, S. Boumaiza, F. M. Ghannouchi, Augmented Hammerstein predistorter for linearization of broadband wireless transmitters, *IEEE Transactions on Microwave Theory and Techniques*, Vol. 54, No. 4, June 2006, pp. 1340 - 1349.

[19]. K. J. Muhonen, M. Kavehrad, R. Krishnamoorthy, Look-up table techniques for adaptive digital predistortion: A development and comparison, *IEEE Trans. Veh. Tech.*, Vol. 49, No. 9, September 2000, pp. 1995-2002.

[20]. Yunsung Cho, Juyeon Lee, Sangsu Jin, Byungjoon Park, Junghwan Moon, Joo Seung Kim, Bumman Kim, Fully integrated CMOS saturated power amplifier with simple digital predistortion, *IEEE Microwave and Wireless Components Letters*, Vol. 24, No. 8, August 2014, pp. 533-535.

[21]. J. S. Kenney, W. Woo, L. Ding, R. Raich, H. Ku, G. T. Zhou, The impact of memory effects on predistortion linearization of RF power amplifiers, in *Proceedings of the 8th Int. Microw. Opt. Tech. Symp. (ISMOT'01)*, June 2001, pp. 189-193.

[22]. M. Schetzen, The Volterra & Wiener Theories of Nonlinear Systems, *Krieger Publishing Co.*, Melbourne, Fl., USA, 2006

[23]. Lei Guan, Anding Zhu, Low-cost FPGA implementation of Volterra series-based digital predistorter for RF power amplifiers, *IEEE Transactions on Microwave Theory and Techniques*, Vol. 58, No. 4, March 2010, pp. 866-872.

[24]. Taijun Liu, S. Boumaiza, F. M. Ghannouchi, Deembedding static nonlinearities and accurately identifying and modeling memory effects in wide-band RF transmitters, *IEEE Transactions on Microwave Theory and Techniques*, Vol. 53, No. 11 November 2005., pp. 3578-3587.

[25]. Lei Ding, G. Tong Zhou, Dennis R. Morgan, Zhengxiang Ma, J. Stevenson Kenney, Jaehyeong Kim, Charles R. Giardina, A robust digital baseband predistorter constructed using memory polynomials, *IEEE Transactions on Communications*, Vol. 52, No. 1, January 2004, pp. 159-165.

[26]. A. Katz, Linearization: reducing distortion in power amplifiers, *IEEE Microwave Magazine*, Vol. 2, No. 4, December 2001, pp. 37-49.

[27]. Wangmyong Woo, Hybrid Digital/RF envelope predistortion linearization for high power amplifiers in wireless communication systems, PhD Thesis, The Academic Faculty, *Georgia Institute of Technology*, Atlanta, USA, April 2005.

[28]. Taijun Liu, Slim Boumaiza, Fadhel M. Ghannouchi, Augmented Hammerstein predistorter for linearization of broad-band wireless transmitters, *IEEE Transactions on Microwave Theory and Techniques*, Vol. 54, No. 4, June 2006, pp. 1340-1349.

[29]. H. Seidel, A microwave feed-forward experiment, *The Bell System Technical Journal*, Vol. 50, No. 9, November 1971, pp. 2879-2916.

[30]. E. E. Eid, F. M. Ghannouchi, F. Beauregard, Optimal feedforward linearization system design, *Microwave J.*, November 1995. pp. 78-86.

[31]. Ming Xiao, Novel predistortion techniques for RF power amplifiers, PhD Thesis, School of Electronic, Electrical and Computer Engineering, *The University of Birmingham*, United Kingdom, October 2009.

[32]. Erik Andersson, Linearization of power amplifier using digital predistortion, implementation on FPGA, Student Thesis, Department of Electrical Engineering, Electronics System, *Linköpings Universitet*, Sweden, 2014.

[33]. Ildu Kim, Young Yun Woo, Junghwan Moon, Jungjoon Kim, Bumman Kim, High-efficiency hybrid EER transmitter using optimized power amplifier, *IEEE Transactions on Microwave Theory and Techniques*, Vol. 56, No. 11, November 2008, pp. 2582-2593.

[34]. Chi-Tsan Chen, Tzyy-Sheng-Horng, Kang-Chun Peng, Chien-Jung Li, High-gain and high-efficiency EER/Polar transmitters using injection-locked oscillators, *IEEE Transactions on Microwave Theory and Techniques*, Vol. 60, No. 12, November 2012, pp. 4117-4128.

[35]. B. Stengel, W. R. Eisenstadt, LINC power amplifier combiner method efficiency optimization, *IEEE Transactions on Vehicular Technology*, Vol. 49, No. 1, January 2000, pp. 229-234.

[36]. Jingshi Yao, S. I. Long, Power amplifier selection for LINC applications, *IEEE Transactions on Circuits and Systems II*, Vol. 53, No. 8, August 2006, pp. 763-767.

[37]. Fadhel M. Ghannouchi, Oualid Hammi, Behavioral modeling and predistortion, *IEEE Microwave Magazine*, Vol. 10, No. 7, December 2009, pp. 52-64.

[38]. A. A. M. Saleh, Frequency-independent and frequency-dependent nonlinear models of TWT amplifiers, *IEEE Transactions on Communications*, Vol. 29, No. 1, November 1981, pp. 1715-1720.

[39]. P. B. Kenington, High-Linearity RF Amplifier Design, *Artech House Publishers*, Sept. 2000.

[40]. A. Zhu, J. C. Pedro, and T. J. Brazil, Dynamic deviation reduction-based Volterra behavioral modeling of RF power amplifiers, *IEEE Transactions on Microwave Theory and Techniques*, Vol. 54, No. 12, December 2006, pp. 4323-4332.

[41]. C. Eun, E. J. Powers, A new Volterra predistorter based on the indirect learning architecture, *IEEE Transactions on Signal Processing*, Vol. 45, No. 1, January 1997, pp. 223-227.

[42]. M. Schetzen, Nonlinear system modeling based on the Wiener theory, *Proceedings of the IEEE*, Vol. 69, No. 12, December 1981, pp. 1557-1573.

[43]. Gozde Erdogdu, Linearization of RF power amplifier by using memory polynomial digital predistortion technique, MSD Thesis, *Middle East Technical University*, Ankara, Turkey, June 2012.

[44]. L. Ding, R. Raich, G. T. Zhou, A Hammerstein predistortion linearization design based on the indirect learning architecture, in *Proceedings of the IEEE International Conference on Acoustics, Speech, and Signal Processing (ICASSP '02)*, Orlando, Fl., USA, 2002.

[45]. F. Taringou, O. Hamm, B. Srinivasan, R. Malhame, F. M. Ghannouchi, Behaviour modeling of wideband RF transmitters using Hammerstein-Wiener models, I*ET Circuits, Devices & Systems*, Vol. 4, No. 4, July 2010, pp. 282-290.

[46]. J. Kim, K. Konstantinou, Digital predistortion of wideband signals based on power amplifier model with memory, *Electronics Letters*, Vol. 37, No. 23, November 2001, pp. 1417-1418.

[47]. R. Penrose, A generalized inverse for matrices, *Proceedings of the Cambridge Philosophical Society*, 1955, pp. 406-413.

[48]. J. Stoer, R. Bulirsch, Introduction to Numerical Analysis, 3rd ed., *Springer-Verlag New York*, New York, USA, 2002.

[49]. Ahmad Rahati Belabad, Seyed Ahmad Motamedi, Saeed Sharifian, An adaptive digital predistortion for compensating nonlinear distortions in RF power amplifier with memory effects, *Integration, the VLSI Journal*, Vol. 57, 2017, p. 184–191.

Chapter 7

An Overview of Diffusion Barriers in Cu Interconnection

Y. Meng, K. W. Xu and F. Ma

7.1. Introduction

7.1.1. Cu Metallization

Nowadays, integrated circuits (ICs) are moving towards smaller technology nodes and, copper (Cu) has been widely adopted in the interconnection due to its low resistivity and excellent electro-migration resistance [1-3]. However, the inter-diffusion and reaction between Cu and Si substrates still restrict the reliability of devices and have become a crucial issue in the field of microelectronics [4-6]. The most effective way to suppress the inter-diffusion is to add a diffusion barrier between Cu interconnects and Si substrates. On the one hand, it will prevent Cu atoms from diffusing into Si substrates, on the other hand, it can enhance the adhesion strength between the two layers [7-10]. Therefore, the selection of barrier material and the optimum preparation method become the research focus.

7.1.2. Requirements for Diffusion Barriers

According to the diffusion equation: $D=D_0 \cdot \exp(-Q/kT)$, smaller diffusion coefficient corresponds to higher activation energy. As described by $E_a=A \cdot T_m+B$, in which A and B are constant, high melting point means high activation energy [11]. The noble transition metals possess low resistivity, high melting point and weak chemical reactivity, and they are believed to be the best candidates for diffusion barriers. Commonly, the materials used as barrier layers must meet the following requirements [12-16]: (1) Low resistivity, otherwise, the RC of the system will be increased; (2) Excellent thermal stability; (3) Good adhesion with the two layers, which will improve the electromigration resistance but

F. Ma
State Key Laboratory for Mechanical Behavior of Materials, Xi'an Jiaotong University, Xi'an, Shaanxi, China

191

reduce the interfacial resistance; (4) The barrier must be thin enough and has a uniform thickness.

7.2. Preparation and Characterization for Diffusion Barrier

7.2.1. Preparation Methods

With the development of chip integration and the shrinkage of feature size of devices, the thickness of barrier layer is reduced down to nanometer scale, and the diffusion barrier performances directly affect the reliability of Cu interconnection. Therefore, the preparation of barrier layer is one of the key technologies for Cu metallization. Especially, the barrier layer in Cu/porous low k dielectric configuration should be: (1) Dense without pinhole; (2) Covering interconnect bottom and sidewalls; (3) Low surface roughness. Therefore, fabrication of dense, smooth, ultra-thin and uniform barrier layer becomes one of the research focuses in Cu interconnection.

Fabrication of Cu interconnects is based on the dual damascene technology, the dielectric layer is firstly deposited, and then etched into trenches, finally the barrier layer and Cu plating layer are deposited subsequently. So the compatibility for the deposition methods must be considered. The commonly used preparation methods are physical vapor deposition (PVD), chemical vapor deposition (CVD), atomic layer deposition (ALD) and electrochemical deposition (ECD).

7.2.1.1. Physical Vapor Deposition

Physical vapor deposition includes evaporation plating, sputtering deposition, plasma deposition, ion plating and molecular beam epitaxy (MBE), etc. Magnetron sputtering (MS) is one of the most commonly used methods and has the potential to be further optimized to deposit diffusion barriers in the 22 nm technology nodes. Magnetron sputtering has several advantages, such as, low substrate temperature, high sputtering rate, moreover, the targets with high melting point can usually be deposited with good adhesion, controllable atomic ratio, high purity, and good step coverage. However, as the feature size of devices is reduced, and the aspect ratio (>5:1) is increased, the step coverage of the sputtered diffusion barrier is poor.

7.2.1.2. Chemical Vapor Deposition

In chemical vapor deposition, gas phase reactions or gaseous decomposition commonly takes place. Good step coverage is the major advantage of CVD, as compared to PVD. There are several forms of CVD, such as, atmospheric pressure CVD (APCVD), low-pressure CVD (LPCVD), plasma-enhanced CVD (PECVD), metalorganic CVD (MOCVD) and laser-enhanced CVD (LECVD). However, the thin films fabricated by CVD are limited owing to the gas precursor, reaction products and chemical reaction thermodynamics. Moreover, the deposition temperature is commonly so high that the substrate will be affected and, C, O and Cl atoms might be doped into the thin films, resulting in high resistivity and poor diffusion barrier performance. PECVD has low deposition temperature, fast deposition rate, and exhibits the potential to be adopted for the fabrication of diffusion barriers.

7.2.1.3. Electrochemical Deposition

Step coverage and control of the barrier performance are the major problems for barrier layer deposition. It can be overcome by electrochemical deposition (ECD), which has been used to prepare CoWP alloy [17] and NiWP alloy [18]. However, ECD can only be adopted to deposit conductive thin films on the activated surface, but not directly on dielectric layers. So the application potential is small.

7.2.1.4. Atomic Layer Deposition

Atomic layer deposition is a new method developed to prepare thin films in recent years [19]. ALD has the following advantages: (1) the as-deposited films, such as Ru [20], TaN/Ru [21] and MnO_x [22] are ultrathin; (2) the film thickness and the composition can be controlled by changing the reaction cycles; (3) low deposition temperature, <400 °C is adopted; (4) the thin films are conformal with the high aspect-ratio structures and good step coverage. Jill et al. [23] prepared 1.5 nm thick WN diffusion barrier by using ALD, and found that the failure temperature is 600 °C; 3 nm TiO_2 diffusion barrier prepared by ALD can block the inter-diffusion of Cu atoms at 650 °C [24]. The low deposition rate, low efficiency, and high production cost limit the application of ALD into production in the short term.

7.2.1.5. Self-Forming Approach

The concept of self-forming diffusion barriers was proposed by Murarka [25]. In this process, Cu(M) alloy (M = Zr, Mg, Mn, W, etc.) thin films are prepared on ILD or Si substrates, and then annealed at high temperature, promoting the alloying elements precipitating into the interface between Cu(M) alloy and ILD or Si substrate, and reacting with the residual oxygen in the thin film, as a result, an ultra-thin stable phase as a reliable barriers formed [26]. Fei et al. [27] deposited 100 nm thick Cu-C (4.8 at. % C) and Cu-V (4.1 at. % V) alloy films on SiO_2/Si substrate by using magnetron sputtering, and a 10 nm thin layer is formed at the interface between the alloy film and the substrate after annealed at 500 °C. The failure temperature of Cu-C and Cu-V is 700 °C and 600 °C, respectively. Liu et al. [28] reported that a self-formed double layer of $ZrGe_x$/Cu_3Ge and ZrO_x($ZrSi_yO_x$)/Cu_3Ge appeared after thermal annealing of Cu (Ge, Zr) alloy films, and Cu (Ge, Zr) alloy thin films keep good thermal stability up to 650 °C. The self-formed barrier layer usually possesses the reduced film thickness, the decreased resistivity, and enhanced adhesion strength. However, the preparation process is not easy to control, and a great number of research works are needed in the near future.

7.2.2. Characterization Methods

7.2.2.1. Chemical Properties

X-ray photoelectron spectroscopy (XPS) is one of the most popular surface analysis methods, and is used to analyze the chemical composition and the combining state. Chemical information can be interpreted by simple peak shift and the atomic ratio can be quantified by analyzing the peak area after background subtraction. XPS was commonly measured by using Al Kα radiation, the applied voltage and current were 12 kV and 6 mA, respectively. The measurements were done on surface and after etching for 10 s by 2 keV Ar$^+$ with the etching rate of 0.28 nm/s.

7.2.2.2. Phase Structure

The phase structures of the as-deposited and annealed diffusion barriers are characterized by X-ray diffraction (XRD). According to the Scherrer's Equation, the grain size is measured from the full width at half maximum (FWHM) of the diffraction peaks. XRD is also used to detect the Cu_3Si phase and thus to prove whether inter-diffusion between Cu and Si takes place. As for the ultra-thin films, grazing incident X-ray diffraction (GIXRD, XRD-7000, Shimadzu) with Cu Kα radiation at an incident angle of 1° is commonly adopted. The tube current and voltage were 40 mA and 40 kV, respectively. The measurement angle is scanned in the range of 30-80° with a scanning step of 0.02° and a scanning speed of 5°/min.

7.2.2.3. Microtopography

Scanning electron microscopy (SEM) is used to observe the surface morphology of Cu thin films after annealing at different temperatures, and the existence of Cu_3Si grains on thin films suggests the failure of diffusion barrier. In addition, surface morphology also affects the resistivity of the barrier layer. So SEM has been widely used to analyze the surface morphology of the films.

Transmission electron microscopy (TEM) is used to observe the interface morphology and to analyze the chemical composition distribution. Specifically, bright-field TEM images can be used to determine the film thickness and interface morphology, and electron diffraction is used for phase analysis. The EDS (Energy Dispersive Spectroscopy) attachment is used to analyze the chemical composition distribution across interface, particularly, nano-beam EDS analysis can provide elemental compositions in nanometer level. High-Resolution TEM is employed to characterize the cross-section morphology at sub-nanometer scale. The TEM sample is prepared in the following steps: the thin films on Si substrates are cut into square pieces 2 mm×2 mm in size, and then glued together face to face, followed by thermal annealing in an incubator at 120 °C for 2 h. After that, the samples are mechanically grinded from the cross-section direction using multi-granularity sandpapers until the cross-section thickness is reduced down to several micrometers. Finally, ion polishing is done to further reduce the thickness in the center region down to about 50 nm so that the electron beam can penetrate [29].

7.2.2.4. Electrical Properties

Four point probe (FPP) method is commonly applied to measure the sheet resistance and resistivity of the semiconductors. Since the sheet resistance is sensitive to the microstructure of Cu interconnects, the failure temperature of the barrier layer can be judged according to the jump on the Rs curve. Before the sheet resistance measurement, the samples are cut into 1 cm × 1 cm pieces. Five measurements are conducted for each Cu/barrier thin film, and the average value is taken. The functional properties of the diffusion barrier can be evaluated by analyzing with a metal-oxide-semiconductor (MOS) capacitor composed of a thin dielectric layer between Cu thin films and Si substrate.

7.3. Cu Alloy Diffusion Barriers

The diffusion activation energies in pure Cu thin films are much lower than those in bulk counterparts, but cannot fully reflect the excellent resistance to electrical migration. Given that most of the migration along the grain boundaries (GBs) and interface, thus the Cu interconnects can be alloyed with another element with low solid solubility so that it can precipitate at GBs, improving the resistance to electrical migration of Cu. In addition, the increased resistivity is not apparent. The alloying elements can be Al, Mg, Ag, Mo, Cr, Zr, Mn, Ru, etc. [30-34].

7.3.1. Microstructure and Properties of Cu Alloy Thin Films

The surface morphology of Cu thin films is affected by the alloying elements. Fig. 7.1 presents the AFM images of the as-deposited Cu-Zr films and Cu films [35]. The surface roughness (Rms) of Cu-Zr alloy film and pure Cu film are 7.239 nm and 23.883 nm, respectively. So the surface of Cu-Zr thin film (Fig. 7.1a) is smoother than that of Cu film (Fig. 7.1b) [35, 36]. Wang et al. reported that the as-deposited Cu film has poor texture while Cu(Zr) and Cu(Cr) thin films exhibit much finer columnar structure [33]. Consequently, the doped Zr and Cr atoms suppress the grain growth in Cu films and, moreover, act as the nucleus seeds for Cu grains resulting in more grains during thermal annealing. Besides, Zr and Cr atoms with high kinetic energy also bombard on the Cu film promoting the surface flattening [37].

The alloying elements might cause lattice distortion in Cu films, and thus the XRD peaks shift. As compared to pure Cu thin films, the XRD peaks of the as-deposited Cu-Cr thin films shift toward a lower angle as a result of expanded lattice due to the doped Cr (Fig. 7.2) [2]. However, upon annealing, the XRD peaks moves to higher angles, suggesting the segregation of Cr alloy. Similar results were also observed in Cu-Zr, Cu-Ru, Cu-C, Cu-Ta, Cu-Mo and Cu-V alloy and so on [2, 38-41]. There is no other peaks corresponding to any Cu based compound, due to the small alloying content [42]. Strong Cu(111) texture slows down the chemical reaction between the Cu seed layer and diffusion barrier [43]. Damayanti et al. demonstrated that Ru might promote the formation of Cu(111) texture (Fig. 7.3), which provides higher electro-migration resistance than Cu (200) [44].

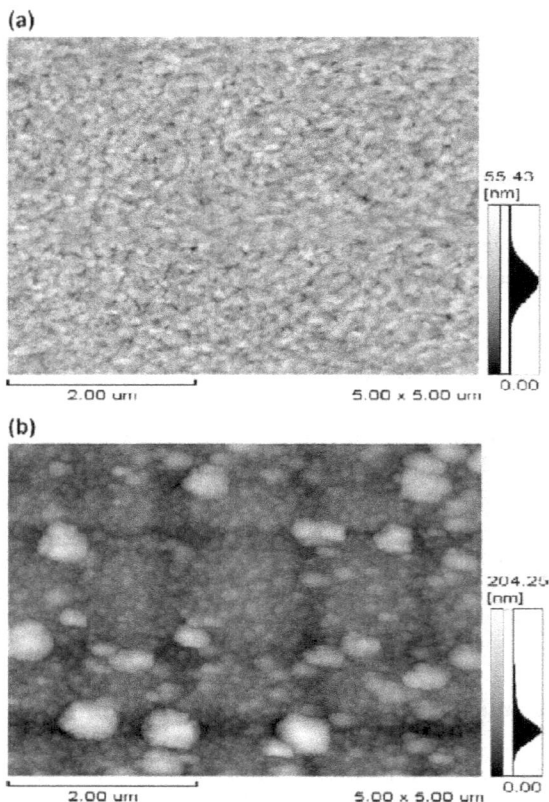

Fig. 7.1. AFM images of the as-deposited thin films: (a) Cu-Zr, and (b) Cu [35].

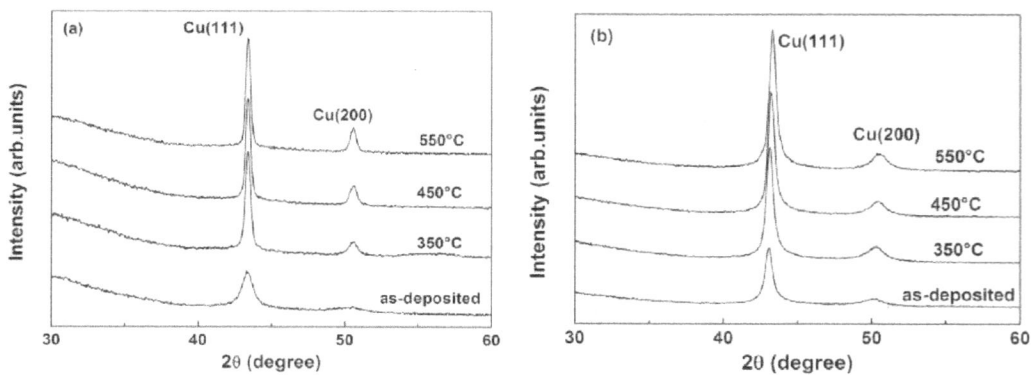

Fig. 7.2. XRD patterns of (a) Cu/SiO$_2$/Si, and (b) Cu(4.5 at% Cr)/SiO$_2$/Si samples before and after annealing at the temperature in the range of 350 °C-500 °C for 1 h [2].

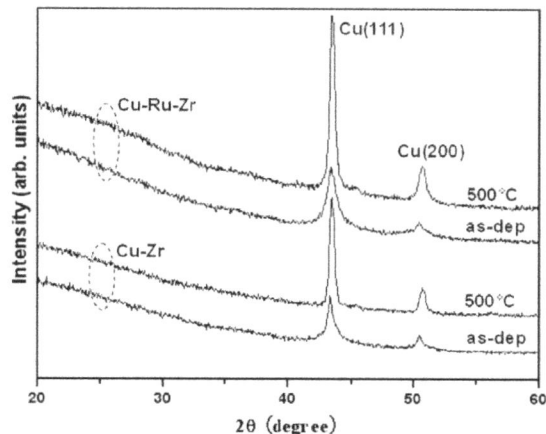

Fig. 7.3. XRD patterns of the Cu-Ru-Zr/SiO$_2$/Si and Cu-Zr/SiO$_2$/Si samples before and after annealing at 500 °C [42].

The as-deposited Cu-Zr alloy thin films possess higher resistivity than that of pure Cu films owing to the enhanced scattering by GBs. After thermal annealing, the resistivity will decrease due to the grain growth and the defect elimination in thin films. Barmak et al. [45] reported that the resistivity of as-deposited and annealed Cu film is 2.5 μΩ·cm and 1.7 μΩ·cm, respectively. But the resistivity of as-deposited and annealed Cu alloy thin films is 44.2 μΩ·cm and 2.0 μΩ·cm, respectively.

7.3.2. Thermal Stability of Cu Alloy/Si Systems

Thermal stability of Cu alloy/Si systems was substantially improved by adding small amounts of alloying elements. Fig. 7.4 shows the XRD patterns of the Cu/TaN/Si and Cu-Zr/TaN/Si structure after annealed at 800 °C for 60 min [35] Cu$_3$Si and TaSi$_2$ compound were observed, suggesting the failure of the structure. But the Cu-Zr barrier was still kept stable since no new phase appeared. So the thermal stability might be improved by alloying Zr atoms. Essentially, Zr precipitates at the Cu-Zr/TaN interface, and acts as a new diffusion barrier to prevent Cu and Si diffusion.

It was reported that Cu can directly react with Ge resulting in Cu$_3$Ge phase with a low resistivity of 5.5 μΩ·cm at 150 °C [46, 47]. It can also be applied as a diffusion barrier in Cu interconnection. Moreover, Cu$_3$Ge was well adhered onto SiO$_2$ and stable against oxidation in air up to 520 °C [48]. But Cu$_3$Ge can react with Si substrate at 400 °C. Thus the thermal stability of Cu$_3$Ge needs to be improved. The above results demonstrated the excellent properties of Cu-Zr alloy, so Cu(Ge, Zr) alloy was proposed and the barrier performance was studied.

Fig. 7.5 shows the XRD patterns of Cu/Cu(Ge, Zr)/Si (Group A) and Cu/Cu(Ge)/Si (Group B). After annealed at 400 °C, Cu$_3$Ge emerged in both system but no Cu-Si phase appeared. Particularly, it was even maintained up to 650 °C for the Cu/Cu(Ge, Zr)/Si system. In contrast, Cu_3Si phase appeared at 450 °C (Fig. 7.5 b), implying that some Cu atoms migrate into the Cu_3Ge/Si interface [28]. When the annealing temperature was elevated up to 500 °C, the peaks of Cu_3Si phase become stronger, indicating severe diffusion between Cu and Si in Group B (Fig. 7.5 b).

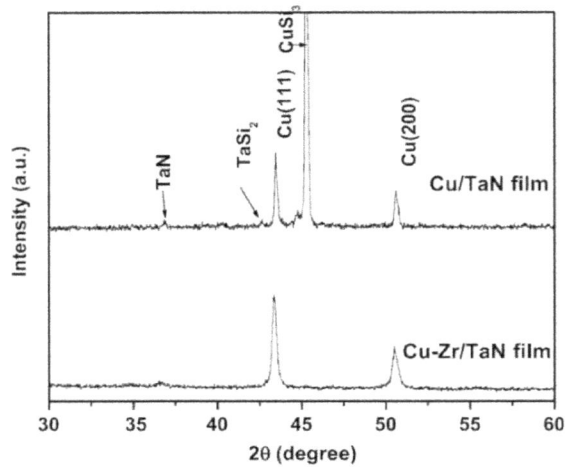

Fig. 7.4. XRD patterns of Cu films and Cu-Zr alloy films annealed at 800 °C [35].

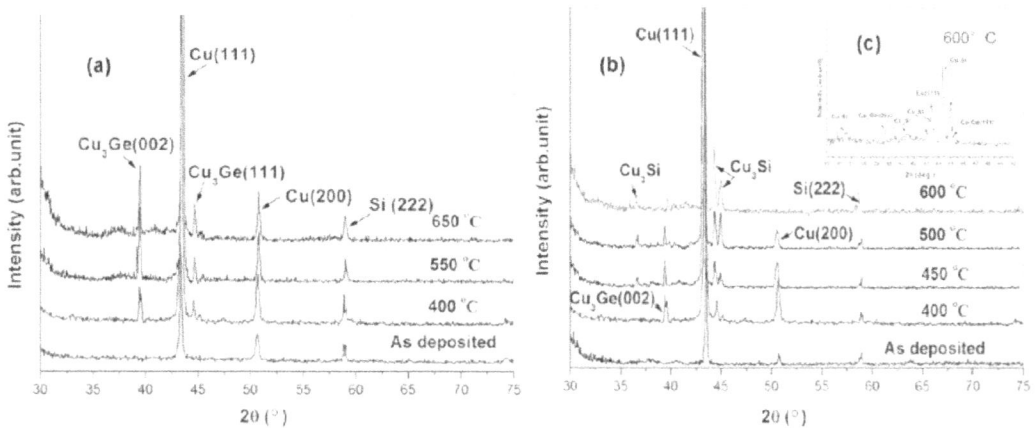

Fig. 7.5. GIXRD patterns of the film stacks as-deposited and annealed for 1 h: (*a*) sample of group A; (*b*) sample of group B, and inset (*c*) is the XRD pattern of group B annealed at 600 °C [28].

Electrical property of Cu alloy samples was measured by a MOS capacitor structure, Cu(Ge) and Cu(Ge, Zr) films were used as the gate electrodes [28]. Fig. 7.6 displays the current-voltage (I-V) curves. Upon annealing at 500 °C, the leakage current of $Cu(Ge)/SiO_2/Si$ system at low electrical field was increased rapidly, indicating massive diffusion of Cu into SiO_2. However, the I-V curve of Cu(Ge, Zr) MOS system still keeps steady even after annealed at 650 °C, suggesting higher thermal stability of Cu(Ge, Zr)/SiO$_2$/Si system, consistent with the XRD patterns. This can be attributed to the self-formed ZrGe$_x$ or ZrO$_x$/ZrSi$_x$O$_y$ at the Cu(Ge, Zr)/SiO$_2$ interface, as indicated by the HRTEM image and EDS spectrum (Fig. 7.8).

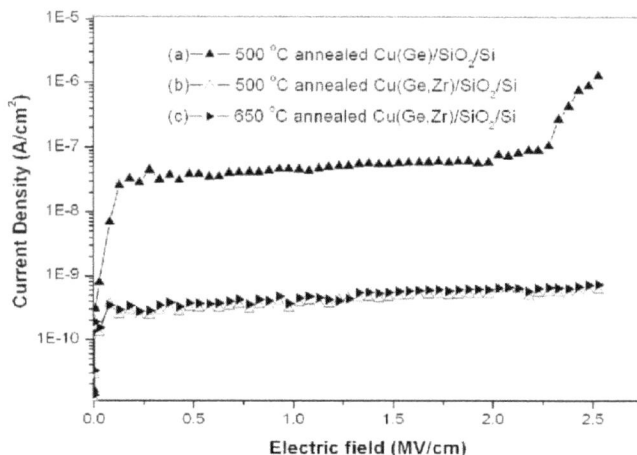

Fig. 7.6. Leakage current densities of (a) Cu(Ge)/SiO$_2$/Si MOS structure annealed at 500 °C; (b), (c) Cu(Ge, Zr)/SiO$_2$/Si MOS structure annealed at 500 °C and 650 °C, respectively [28].

Wang et al. prepared Cu-Al alloy with different Al concentration and the barrier performances were studied [49]. Fig. 7.7 shows the Cross-sectional SEM images of the samples after thermal annealing. The Cu-1.8Al alloy thin films exhibited superior thermal stability and low resistivity after annealed at 500 °C for 1 h or at 400 °C for 7 h. But triangular-shaped Cu$_3$Si compound was observed at the interface between Cu-0.9Al alloy film and Si substrate after annealed at 500 °C for 1 h, indicating film failure. So Al doping can help to improve the barrier performance.

7.3.3. Self-Forming Diffusion Barriers

In terms of Cu alloy diffusion barrier, a self-formed barrier always appears, and thus improves the thermal stability of Cu alloy films. It might meet the requirement in the interconnect technology node down to 14 nm [26, 50]. In such a case, Cu (X = Al, Mg, Mn, Zr, Ti, Ru and so on) alloying thin films were integrated, and MnSi$_y$O$_x$, TiO$_x$, ZrO$_x$, MgO and Al$_y$O$_x$ self-formed barrier layers with a controllable thickness were produced during post-plating annealing [26, 51].

ZrO$_x$/ZrSi$_x$O$_y$ self-forming layer at the Cu$_3$Ge/Si interface was observed in the annealed Cu(Ge, Zr) alloy films by the cross-sectional HRTEM and EDS mapping (Fig. 7.8). The self-forming reactions can be divided in the following steps: (1) Ge atoms reacts with Cu to form polycrystalline Cu$_3$Ge films; (2) Zr atoms are precipitated in the Cu$_3$Ge GBs and the Cu$_3$Ge/Si interface at the annealing temperature of 400 °C, and reacts with excess Ge atoms resulting in $ZrGe_2$ phase at 450 °C or above (Fig. 7.8a); (3) $ZrGe_2$ layer reacts with the original native silicon oxide layer leading to $ZrSi_xO_y$ layer at 500 °C or above (Figs. 7.8 d and 7.8 e). The incorporated Zr changes the boundary structure of the Cu/Cu(Ge, Zr)/Si stacks, and the amorphous $ZrO_x/ZrSi_xO_y$ layer and Cu_3Ge layer exhibit superior barrier properties [28].

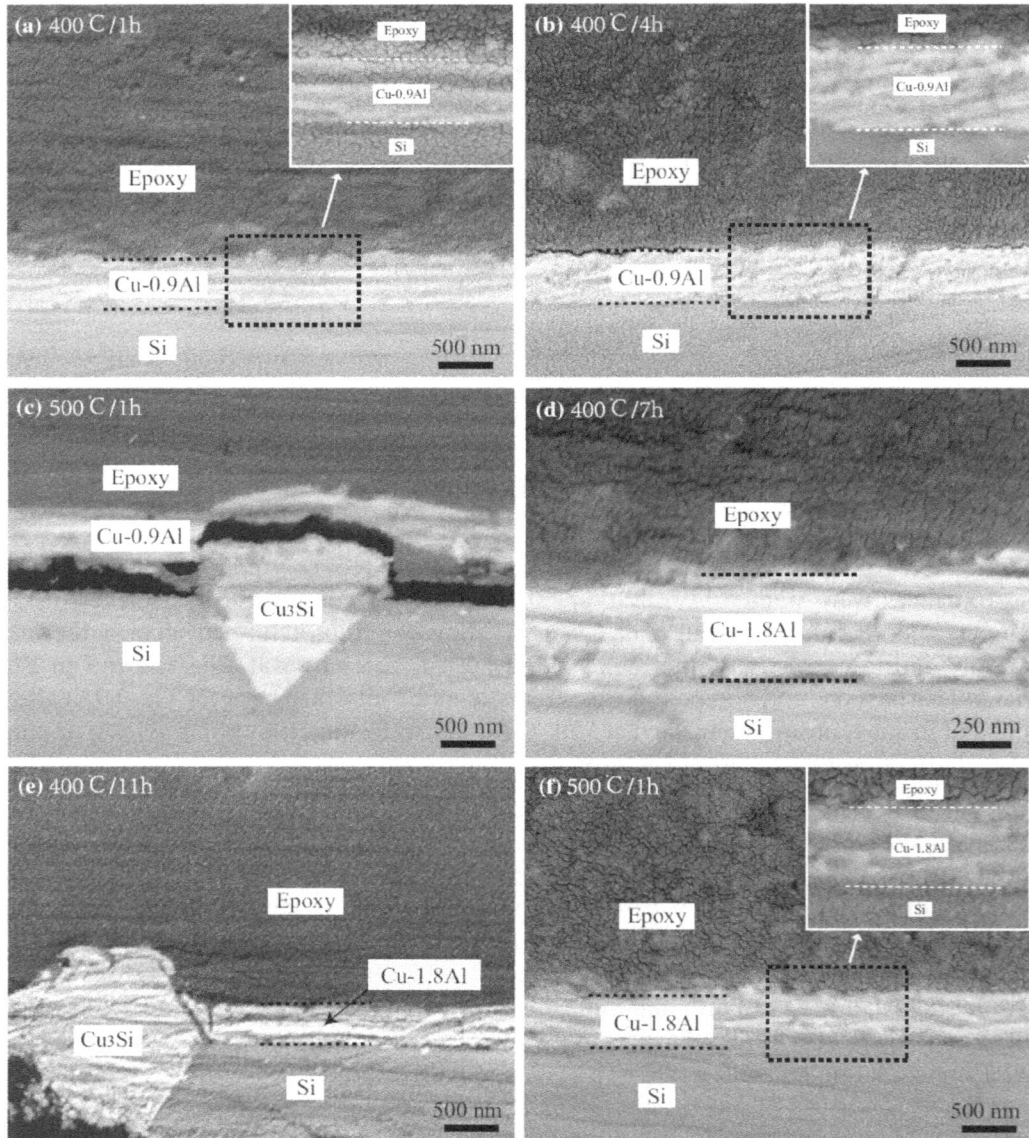

Fig. 7.7. Cross-sectional SEM images of Cu-0.9Al thin film annealed at (a) 400 °C for 1 h; (b) 400 °C for 4 h, and (c) 500 °C for 1 h, and Cu-1.8Al thin film annealed at (d) 400 °C for 7 h; (e) 400 °C for 11 h, and (f) 500 °C for 1 h [49].

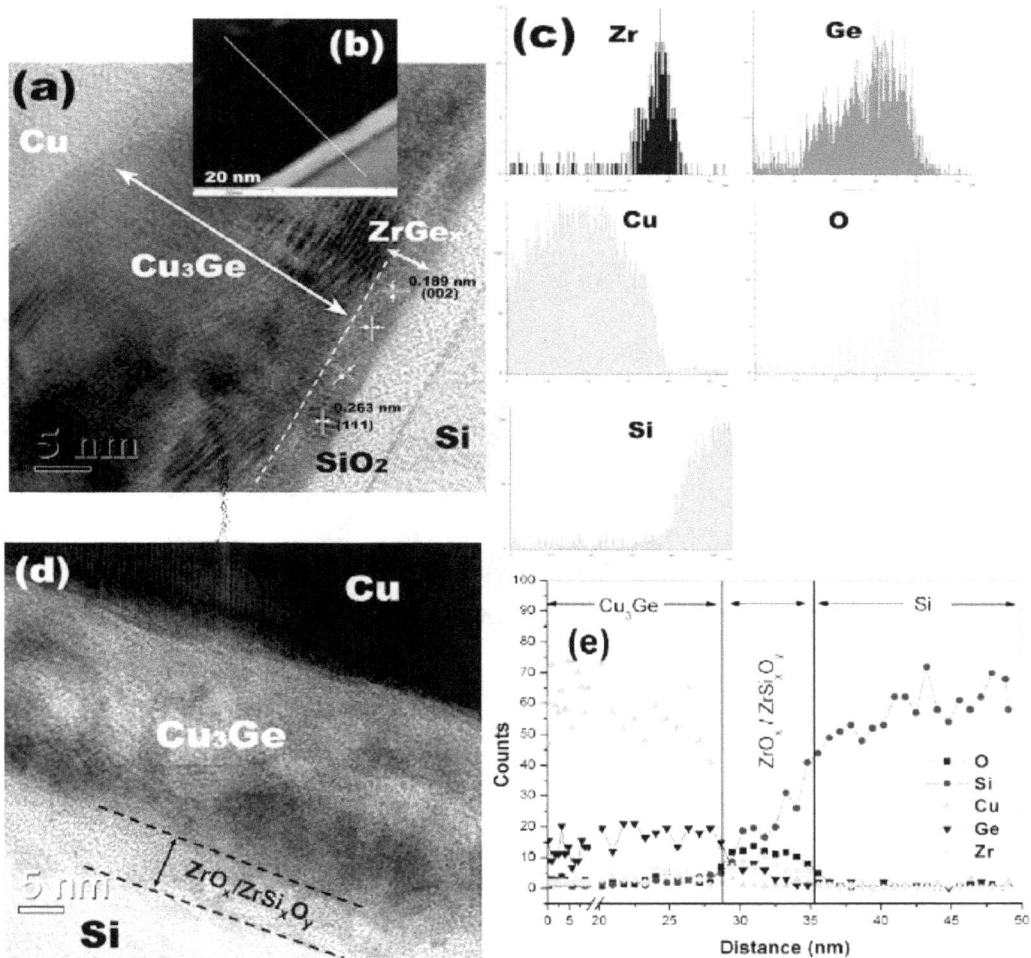

Fig. 7.8. (a) Cross-sectional HRTEM image of group A annealed at 550 °C for 1 h;
(b) the electron image of the zone of EDS line scan, and (c) the EDS line scans of Zr, Si, Cu, O
and Ge; (d) cross-sectional HRTEM image, and (e) EDS depth profile of the Cu/Cu(Ge, Zr)/Si
layer stack in the sample of group A annealed at 650 °C for 1 h [28].

The properties of Zr(Ge) alloy as an exhaustion interlayer between Cu(Zr) and p-SiOC:H dielectric layer were also studied. After annealed at 450 °C, a self-formed $CuGe_x/ZrO_x(ZrSi_yO_x)$ multilayer 5 nm in thickness appeared at the interface of CuZr film stacks (Fig. 7.9). The multilayer structures can keep integrity even when annealed at 500 °C [50]. Similar to Cu(Ge, Zr) alloy film, Zr atoms were precipitated at the Cu GBs and the film interface, which is beneficial to the formation of self-formed diffusion barrier [28]. In contrast, for the Cu(Zr) alloy film without Zr(Ge) interlayer, diffusion occurs after annealed at 450 °C for 2 h, which cannot satisfy the process requirement for Cu interconnect.

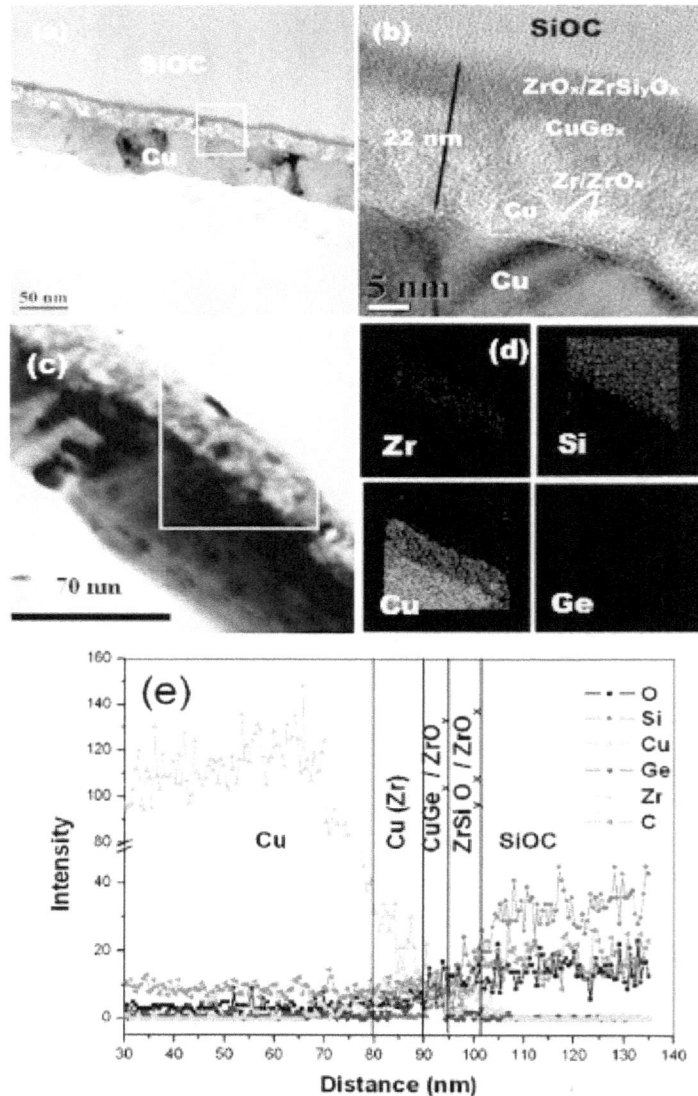

Fig. 7.9. Cross-section TEM image of the sample of group A annealed at 450 °C for 2 h: (a) bright-field; (b) HRTEM image of remarked zone in (a), (c) the electron image of the zone with EDS mapping scan; (d) EDS mapping images of Zr, Si, Cu, and Ge, and (e) EDS depth profile of Cu/Cu(Zr)/Zr(Ge)/*p*-SiOC:H film stack [50].

7.4. Metal Nitride Diffusion Barrier

Refractory metal nitrides are widely recognized as attractive diffusion barriers due to their high thermal stability and excellent conductivity [52-56]. Owing to the lower electrical resistivity (13.6 $\mu\Omega\cdot$cm) and larger negative heat of formation (-87.3 kal/mol) than that of TaN (-60.3 kal/mol) and TiN (-80.4 kal/mol) [57], Zr-N is a promising candidate diffusion barrier. Moreover, ternary Ti-Si-N [58], Ta-Si-N [59, 60], Zr-Si-N [61, 62] and Zr-Ge-N [63] exhibit excellent barrier performances.

7.4.1. Effect of Process Parameter on Zr-Si-N Diffusion Barrier

Deposition parameters like sputtering bias voltage and nitrogen partial pressure always affect the structure and properties of Zr-Si-N. Fig. 7.10 shows the XRD pattern of Zr-Si-N films sputtered at different bias voltages [61]. Only the diffraction peaks of ZrN appeared but no signals of Si_3N_4 or zirconium silicide, implying that Si might exist in amorphous state. As the bias voltage decreases, the XRD peaks of ZrN are reduced. As shown in Fig. 7.11, the XRD peaks of ZrN became weakened gradually, as the N_2 partial pressure was increased, and the peak of Si_3N_4 appeared at the N_2 partial pressure of 0.09 Pa. However, Nose et al. fabricated Zr-Si-N films with no bias voltage and substrate heating, and Si_3N_4 phase was not observed [64]. All in all, the crystallization of ZrN grains will be weakened as the concentration of Si is increased.

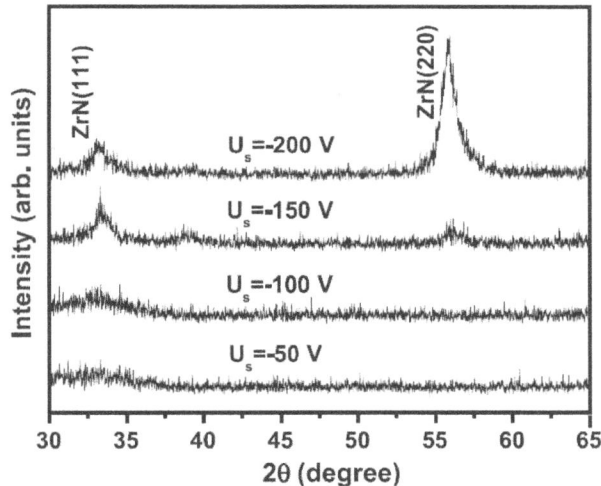

Fig. 7.10. XRD pattern of Zr-Si-N films sputtered with different sputtering bias voltage [61].

As shown in Fig. 7.12 and listed in Table 7.1, the thin films deposited at higher bias voltage and lower nitrogen partial pressure have lower resistivity. As the sputtering bias voltage is increased and the N_2 partial pressure is lowered, the Si content in Zr-Si-N film decreases. Since Si_3N_4 and S_3N_4-like phase is insulator, lower Si content results in low resistivity. Olowolafe et al. found similar behavior in Ta-Si-N diffusion barrier [66]. Additionally, crystallization of ZrN is improved at higher bias voltage and thus the resistivity of Zr-Si-N diffusion barrier decreases.

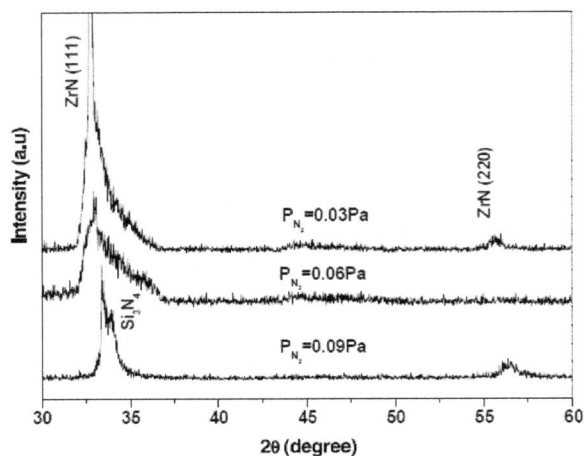

Fig. 7.11. XRD pattern of Zr-Si-N film sputtered with different nitrogen partial pressures [65].

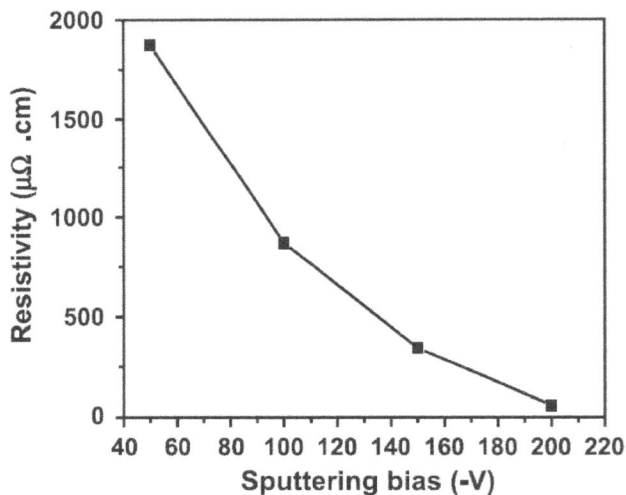

Fig. 7.12. The resistivity of Zr-Si-N films sputtered with different bias voltages [61].

Table 7.1. Composition and resistivity of the Zr-Si-N films sputtering with different nitrogen partial pressures.

Nitrogen partial pressure (Pa)	Composition (at.%)			Resistivity ($\mu\Omega$·cm)
	N	Si	Zr	
0.03	48.31	11.12	40.57	136
0.06	49.44	16.80	33.76	474
0.09	50.93	23.41	25.66	1325

7.4.2. Thermal Stability of Cu/Zr-Si-N/Si Systems

Barrier performance of Zr-Si-N films deposited under various N_2 partial pressures was compared. As shown in Fig. 7.13, no XRD peak of Cu_3Si appears even after annealed at 800 °C, suggesting excellent diffusion barrier performance of Zr-Si-N films. Indeed, ternary Zr-Si-N shows better thermal stability than that of Zr-N barrier, which was verified in Zr-Ge-N diffusion barrier [67]. The Cu/Zr-Ge-N/Si structure can be stably maintained even at 800 °C.

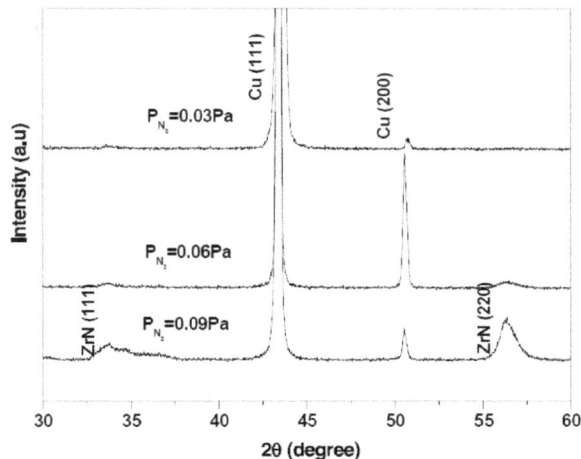

Fig. 7.13. XRD pattern of annealed Cu films on the Zr-Si-N diffusion barrier sputtered with different nitrogen partial pressure [65].

Barrier performance of Zr-Si-N film was further demonstrated by AES. Fig. 7.14 presents the AES depth analysis of Cu/(Zr-Si-N)/Si samples. No obvious diffusion of Cu in Zr-Si-N film after thermal annealing could be identified from the depth profiles, indicating excellent barrier performance. Detailed analysis of the AES profiles of Zr-Si-N barrier prepared under different bias voltage can be found in Ref. [68].

Fig. 7.15 displays the SEM images of the surface morphologies of Cu films on Zr-Si-N. There were cracks and voids on Cu thin films after thermal annealing, but the cracks and voids became smaller gradually as the N_2 partial pressure was increased from 0.03 Pa to 0.09 Pa. So the Zr-Si-N films deposited at higher N_2 partial pressure might exhibit better thermal stability as a diffusion barrier in Cu interconnection, which is consistent with the XRD and AES results.

Much effort has been done to study the diffusion barrier properties of ternary TM (TM=Ta, Ti, Mo, Zr, W)-Si-N alloys [60, 62, 69-72]. The failure temperature of amorphous Ti-Si-N is 650 °C, which is lower than the crystallization temperature (higher than 800 °C) [71]. In addition, Ta-Si-N diffusion barriers exhibited the sacrificial characteristics and the failure temperature varied from 650 °C~800 °C, dependent on the

N content [70]. All the Ta-Si-N barrier layers remain amorphous up to the failure temperature which is lower than the crystallization temperature.

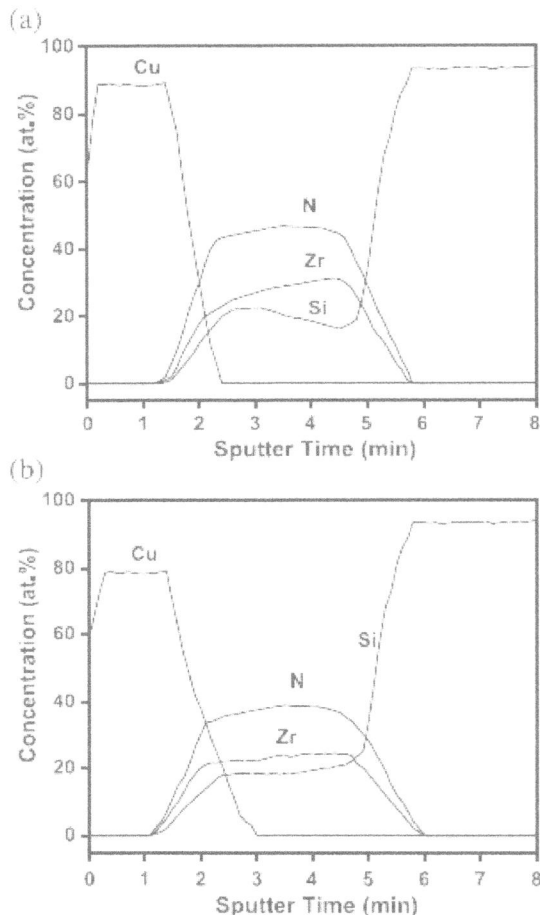

Fig. 7.14. AES depth profiles of Cu/(Zr-Si-N)/Si in (a) as-deposited, and (b) annealed state (-50 V bias for Zr-Si-N interlayer) [61].

The effect of Ge addition on the thermal stability of ZrN diffusion barrier was reported [63, 67, 73]. The Cu/ZrN(20 nm)/Si stacks and Cu/Zr-Ge-N(20 nm)/Si stacks were annealed in Ar atmosphere at 800 °C for 1 h [67]. According to the sheet resistance (Fig. 7.16) and the TEM images, Zr-Ge-N as a diffusion barrier is superior to ZrN. In essential, the as-deposited Zr-Ge-N film is mainly composed of amorphous ZrN and SiN_x phase, and has high crystallization temperature, while the pure ZrN film is composed of nano-grains which is less stable than the amorphous Zr-Ge-N film [67]. Moreover, Ge doping in ZrN thin films improves the oxidation resistance and thus improves the block function for Cu diffusion [73].

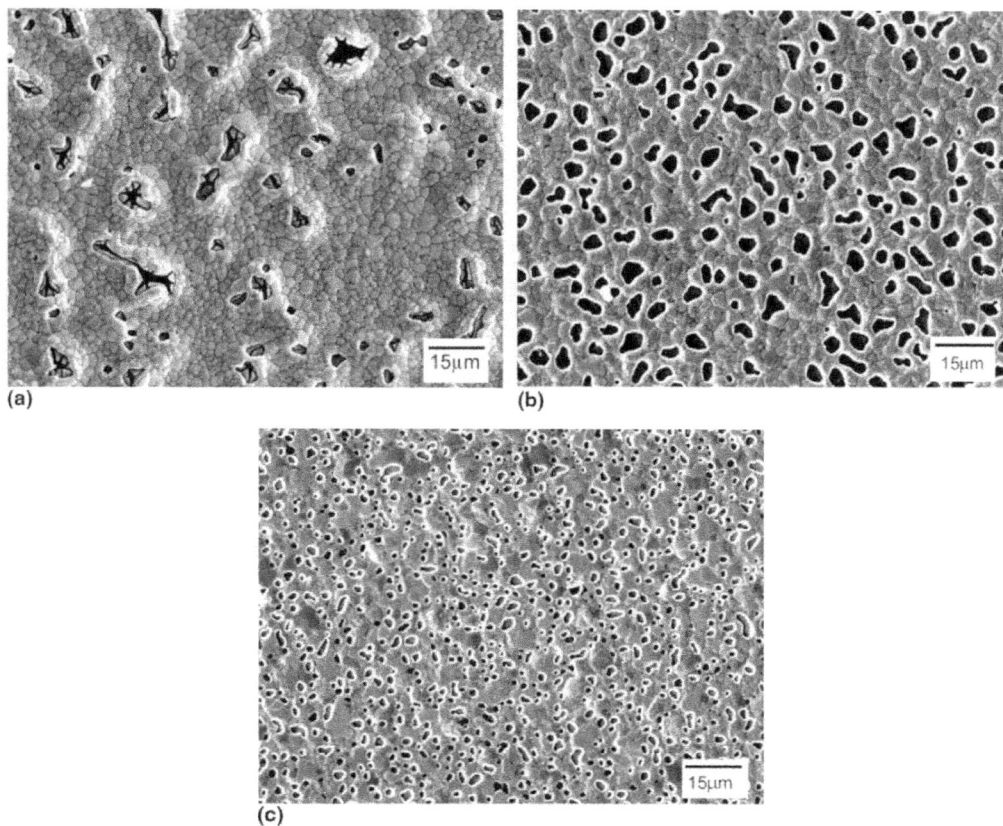

Fig. 7.15. The SEM images of the annealed Cu films on Zr-Si-N diffusion barrier sputtered with different nitrogen partial pressures: (a) P_{N2} =0.03 Pa; (b) P_{N2} =0.06 Pa; (c) P_{N2}=0.09 Pa [65].

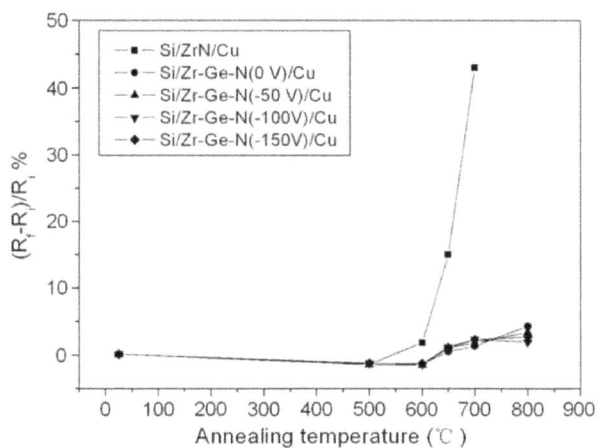

Fig. 7.16. Percentage of sheet resistance for the Cu/Zr-Ge-N/Si and Cu/ZrN/Si samples before and after several thermal treatments for 1 h [67].

7.5. Ru-Based Diffusion Barriers

Ruthenium (Ru), as a transition noble metal, has a low electrical resistivity (7.1 $\mu\Omega\cdot$cm), high melting point (2334 °C) and good chemical stability [20]. Since Cu thin films can be directly deposited on Ru barrier layer with good adhesion and negligible solubility [74], Ru can serve as a barrier layer of Cu interconnects. However, the GBs in polycrystalline Ru thin films might provide fast diffusion paths [75]. Specifically, the 5 nm thick Ru layer failed to block the diffusion of Cu atoms at 300 °C [76]. Many research works have been done to improve the barrier performance of Ru against Cu diffusion [76-78]. To this end, it is mandatory to eliminate GBs. So a small amount of Cr, Mo or Ta was introduced in Ru thin films in order to disturb the crystal lattice and suppress the crystallization [74, 79-81].

7.5.1. Element Doping in Ru-Based Films

He et.al prepared Ru-Ge films with different Ge at.% and studied the properties of Ru-Ge barrier. As shown in Fig. 7.17, the XRD peaks of Ru become broad with increasing Ge content and the thin films tend to be amorphous at 77.3 Ge at.% [82].

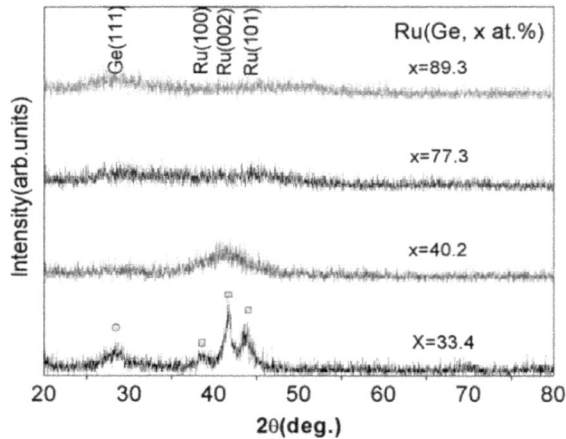

Fig. 7.17. XRD patterns of as-deposited Ru-Ge films with different Ge concentrations [82].

Mo and C were also incorporated to adjust the structure in Ru barrier layers [83]. For comparison, pure Ru and RuMoC films were deposited with different deposition power ratios (DPR, Ru:MoC). Fig. 7.18 a shows the diffraction patterns of the thin films prepared under the different DPR value from 0 to 2. Only Ru peaks were observed in the XRD patterns and pure Ru thin films exhibit polycrystalline structure. As the DPR was increased, the XRD peaks of Ru become weakened. The TEM image of RuMoC II indicates the amorphous characteristics of the diffusion barrier, as shown by the XRD pattern (Fig. 7.18 b). The doping atoms may exist in the form of partially substitution or the solid solution in Ru films [83].

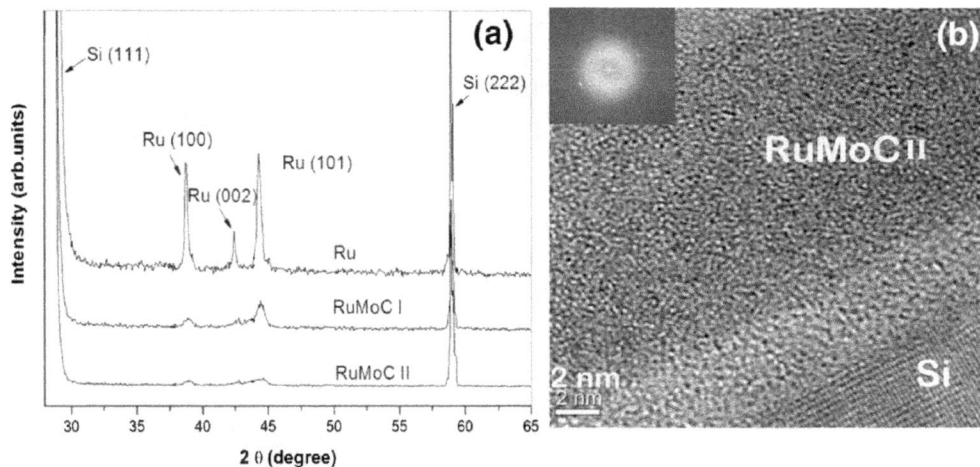

Fig. 7.18. (a) XRD patterns of the as-deposited Ru/Si, RuMoC I/Si and RuMoC II/Si samples. (b) Cross-sectional HRTEM image of the sample of RuMoC II/Si (as-deposited) [83].

Mun et al. examined the microstructure of Ru-N films deposited under different gas ratios, and Fig. 7.19 shows the TEM images [84]. When the gas ratio is 0.24, polycrystalline features with clear GBs can be observed. As the gas ratio is increased, the grain size is reduced and the GBs become ambiguous gradually, and the thin films become amorphous at the gas ratio of 0.88.

7.5.2. Thermal Stability of Ru-Based Diffusion Barriers

If the Ru thin films are changed from crystalline into amorphous, the diffusion barrier performance will be improved. Nano-structured and amorphous Ru-Ge thin films 15 nm in thickness were deposited and annealed to study the thermal stability. When the nano-structured Ru-Ge barrier layer was annealed at 773 K, Cu_3Si phase appeared (Fig. 7.20). However, the amorphous Ru-Ge thin films could effectively prevent Cu atom diffusion at 873 K (Fig. 7.21). This can be ascribed to the 2 nm thick self-formed amorphous $Ru(RuO_x)/RuGe_xCu_y$ multilayer (Fig. 7.22).

The thermal stability of amorphous RuZr diffusion barrier was also studied. Fig. 7.23 displays the cross-sectional TEM images and EDS profile of the Cu/RuZr thin films annealed at 450 °C [29]. The interfaces between RuZr barrier layer and Cu capping layer as well as Si substrate were distinct upon annealing at 450 °C for half an hour (Fig. 7.23a). The RuZr barrier maintained its amorphous nature even after thermal annealing, suggesting excellent barrier performance. But layer interface disappeared as a result of significant diffusion of Cu and Si atoms when the annealing temperature was elevated up to 500 °C (Fig. 7.24). The ratio of Cu and Si atoms is nearly 3:1, indicating the high resistance Cu_3Si phase, as displayed by the EDS spectra.

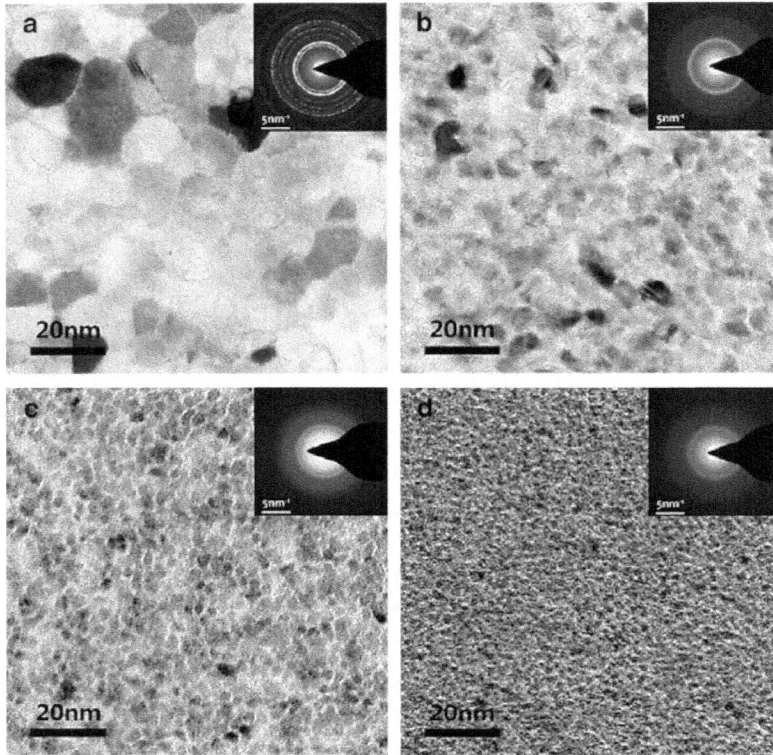

Fig. 7.19. Plan-view bright-field TEM images and corresponding selected area diffraction patterns of the films deposited at gas ratios of (a) 0.24; (b) 0.82; (c) 0.86, and (d) 0.88 [84].

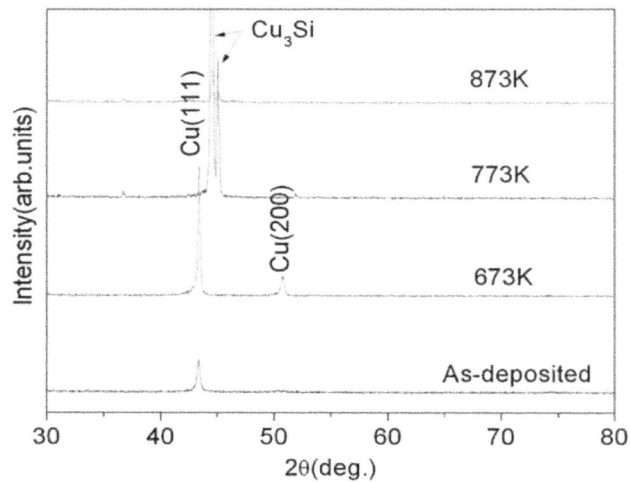

Fig. 7.20. XRD patterns of Cu/Cu(Ru)/Ru-Ge(40.2 %)/Si annealed at different temperatures [82].

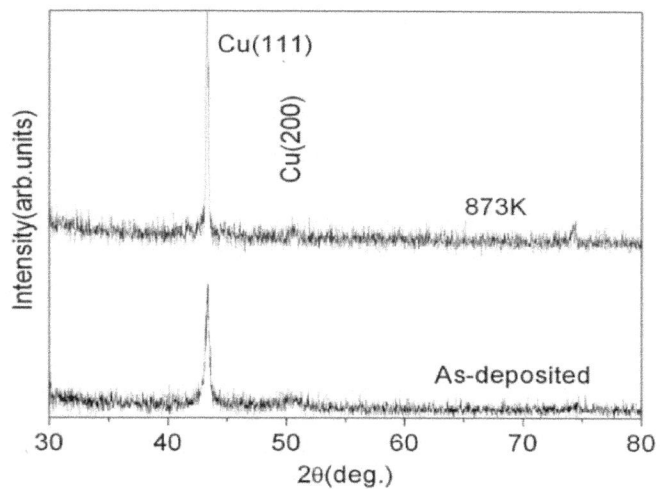

Fig. 7.21. XRD patterns of Cu/Cu(Ru)/Ru-Ge(77.3 %)/Si annealed at different temperatures [82].

Fig. 7.22. Cross-sectional TEM images of Cu/Cu(Ru)/Ru-Ge(77.3 %)/Si annealed at 473 K: (a) HRTEM image of remarked zone in (b); (b) bright field [82].

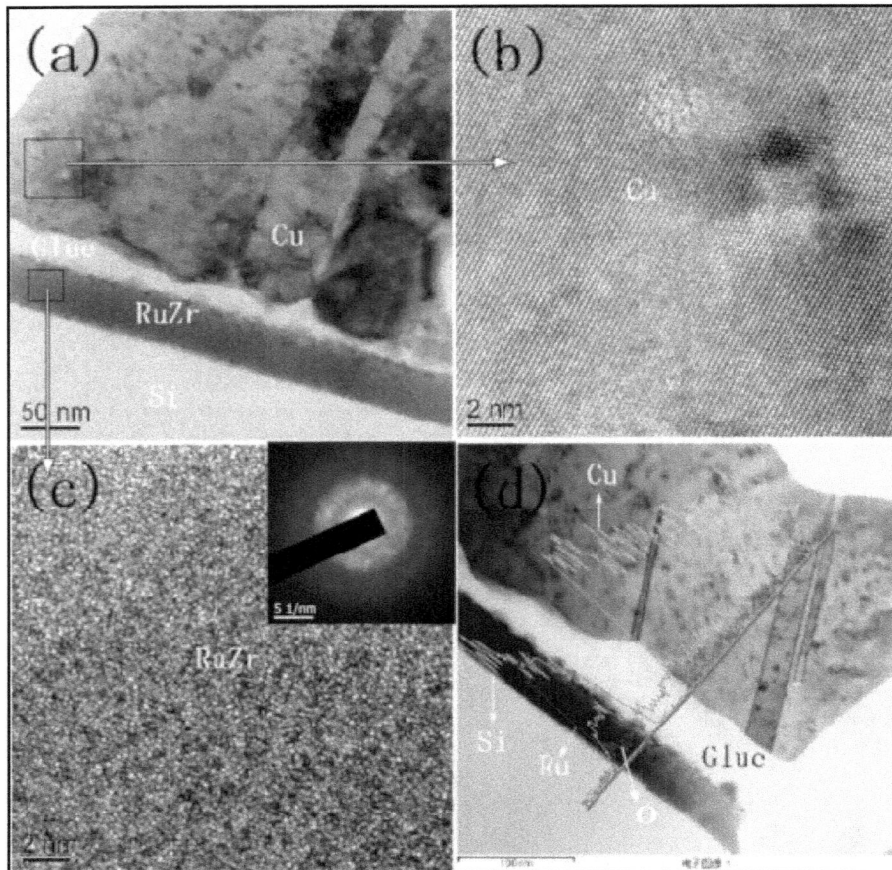

Fig. 7.23. Cross-sectional TEM images of Cu/RuZr thin films annealed at 450 °C for 30 min: (a) bright-field image; (b) HRTEM image of the local region in Cu thin film; (c) HRTEM image of the local region in RuZr barrier layer with the diffraction pattern in the inset; (d) EDS line scanning across the film interface [29].

7.6. ZrB$_2$-Based Diffusion Barriers

Metal borides have excellent physical properties, such as, low resistivity, high melting point, good chemical stability and excellent mechanical properties. Especially ZrB$_2$ has low resistivity (4.6 μΩ·cm) and high melting point (3245 °C) [85-87], moreover, the solid solubility of Zr and B in Cu is negligible, and ZrB$_2$ can be epitaxially grown on Si (111), 4H-SiC (0001), GaN (0001) and Si (001) substrates [88-91], highlighting good structural stability and barrier performance. Sung et al. prepared 20 nm thick ZrB$_2$ barrier layer by remote plasma chemical vapor deposition and the thermal stability could be maintained up to 750 °C [92]. Takeyama et al deposited 3-300 nm ZrB$_2$ film on glasses, SiO$_2$ and Cu substrates, and demonstrated the potential diffusion property in Cu metallization [85].

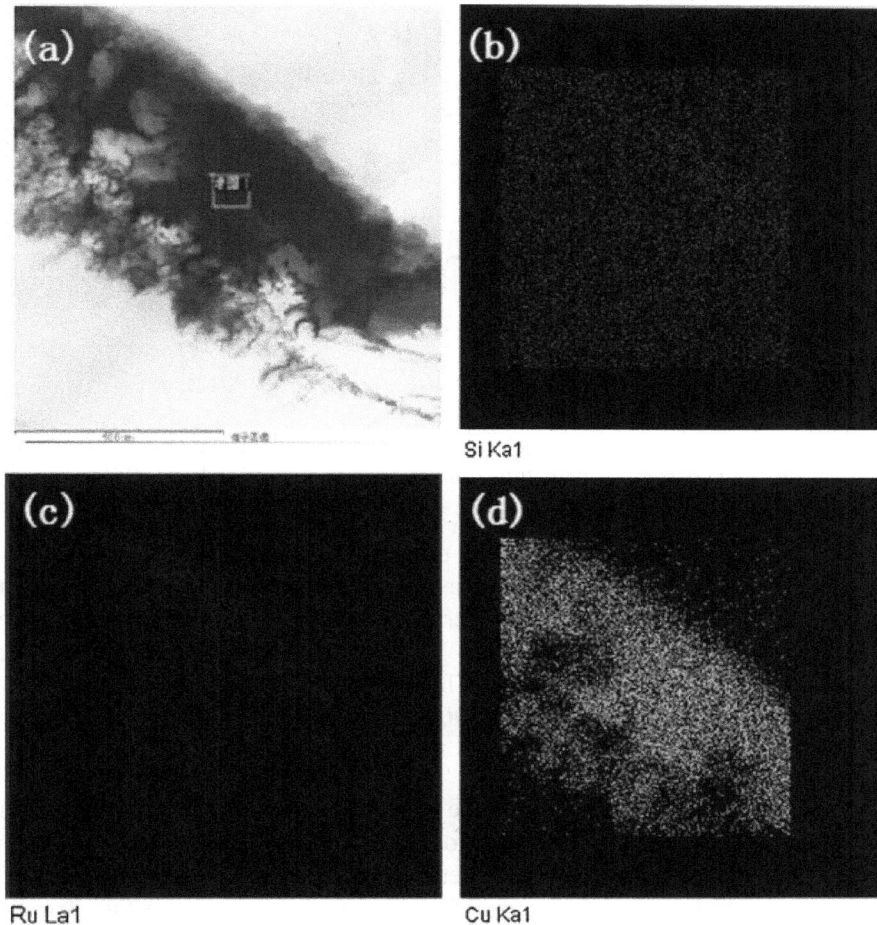

Fig. 7.24. Cross-sectional TEM images of Cu/RuZr thin films annealed at 500 for 30 min: (a) bright-field image and the EDS area scanning of (b) Si, (c) Ru, (d) Cu [29].

7.6.1. Nano-Grained ZrB$_2$ Diffusion Barrier

15 nm thick ZrB$_2$ diffusion barrier and 500 nm Cu capping layers were deposited on Si, respectively. The samples were then annealed in a N$_2$/H$_2$ (N$_2$:H$_2$=9:1) mixture atmosphere at 600-750 °C for 30 min. Fig. 7.25a shows the XPS spectra of Zr and B in the as-deposited ZrB$_2$ thin films. After Ar ion etching, Zr and B are in metallic bonding state of Zr-B, and the atomic ratio of Zr and B is 1:2.03. The single diffraction peak in Fig. 7.25b suggests the nano-grained characteristics in the as-deposited ZrB$_2$ films. The resistivity of the nano-grained ZrB$_2$ barrier layer is 234.8 $\mu\Omega\cdot$cm, and lower than the ternary alloys, Ta-Si-C (340$\mu\Omega\cdot$cm) [93], Ta-Si-N (>600 $\mu\Omega\cdot$cm) [72] and multi-element alloys (AlCrTaTiZr)N (580$\cdot\mu\Omega$cm) [94].

213

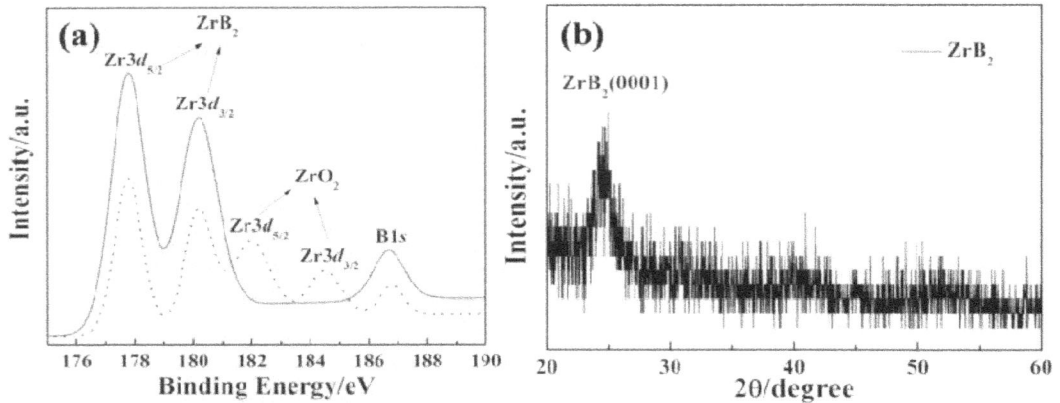

Fig. 7.25. (a) XPS spectra of as-deposited ZrB_2 films before and after Ar ion etching; (b) XRD pattern of ZrB_2 thin films without Cu capping layer [95].

The adhesion strength of the thin films was evaluated by using the nanoscratch test with a sphero-concial diamond tip indenter. A loading rate of 1 mN/s and a scratching rate of 5 μm/s were used. Five scratch tests were performed on each sample. The critical load for the delamination of thin films is 13 mN and 19 mN for the as-deposited and the annealed ZrB_2 thin films (400 °C), respectively. The adhesion strength is close to the value (15.55 mN) of Cu/Ta/DB/Si stacked layer [96]. Both are large enough for the applications as the barrier layer in Cu interconnection.

Fig. 7.26 shows the SEM images of the Cu films deposited on ZrB_2 films. The as-deposited thin films are smooth and contiguous but become rough upon thermal annealing at 650 °C. As the annealing temperature is elevated up to 700 °C, film aggregation occurs and several voids appear on the surface due to surface tension. The formation of $CuSi_x$ grains on the films at 725 °C indicates the failure of ZrB_2 layer.

According to the cross-sectional TEM images of $Cu/ZrB_2/Si$ thin films, sharp interface between layers is still clearly visible even annealed at 700 °C and no reaction occurred between ZrB_2 and Cu layer or Si substrate. It was reported that 15 nm thick amorphous ZrN and W_2N could only prevent the inter-diffusion between Cu and Si up to 600 °C [97, 98]. As the annealing temperature is elevated up to 725 °C, the nano-grains in ZrB_2 thin films grow into large ones and the GBs could provide the effective diffusion paths. When the annealing temperature is further elevated up to 750 °C, the barrier is seriously damaged and Cu thin films become discontinuous.

To better understand the failure mechanism of nano-grained ZrB_2 barrier, a schematic model is proposed, as shown in Fig. 7.27. In nano-grained barrier layers, the zigzag GBs will prolong the diffusion path and thus suppress the atom diffusion. The nano-grains in ZrB_2 thin films grow into large ones at 725 °C [95]. The more straight GBs provide fast diffusion path for Cu atoms (Fig. 7.27 c and 7.27 d). So it becomes easier for the failure of the nano-grained barrier.

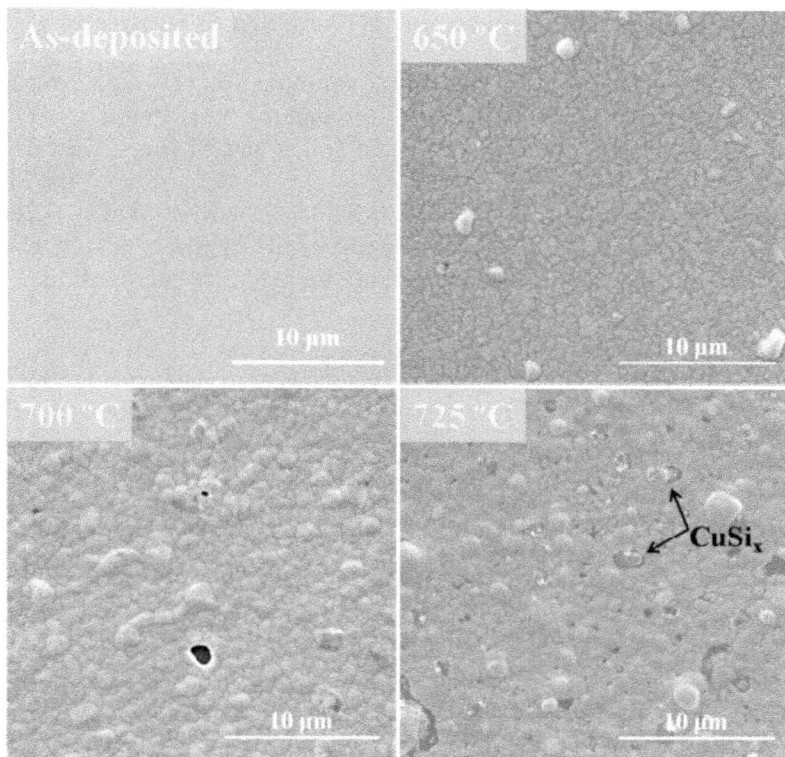

Fig. 7.26. SEM images of the surface morphologies of the Cu/ZrB$_2$ thin films [95].

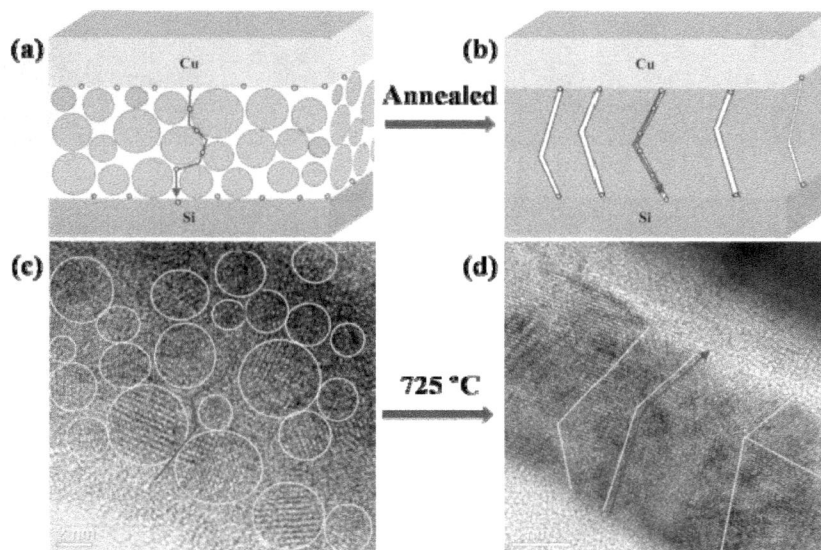

Fig. 7.27. HRTEM images of ZrB$_2$ films as-deposited (a) and annealed at 725 °C (b); schematic models of atomic diffusion in thin films of nano-grains (c) and large grains (d) [95].

7.6.2. Epitaxial Relationship Between Cu3Si Phase and Si Substrate

As shown in Fig. 7.28, the precipitated Cu_3Si compound is embedded in the Si matrix. The HR-TEM image of the interface (Fig. 7.28b) and the SAED patterns of Cu_3Si and Si demonstrate the epitaxial relationship between Cu_3Si and Si (Figs. 7.28 d-7.28 f). The detailed epitaxial relationship is Cu_3Si (000$\bar{1}$)//Si (1$\bar{1}$1) and Cu_3Si (1$\bar{1}$00)//Si ($\bar{2}\bar{1}$1) [95]. The edge dislocation density at the interface is 5.33×10^{12} cm^{-2} [95] (Fig. 7.28c). η" phase is an equilibrium phase at room temperature [99]. Thus the η" Cu_3Si phase and the diamond-cubic Si are mainly considered to study the epitaxial relationship.

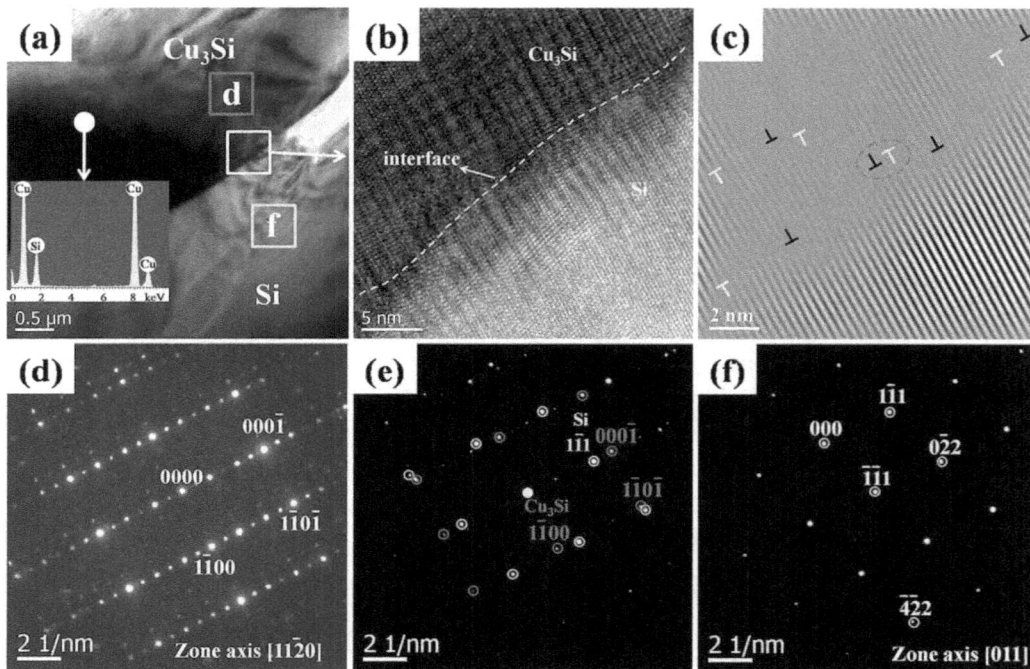

Fig. 7.28. Cross-sectional TEM images of Cu/ZrB$_2$/Si annealed at 750 °C: (a) the low magnification image with an EDX profile of Cu$_3$Si in the inset; (b) HRTEM image from the local region in (a); (c) the inverse FFT image from the (111) spot in (f); (d)-(f) the SAED patterns of Cu$_3$Si, the interface and Si respectively [95].

Fig. 7.29 presents the atomic arrangement model. As for $Cu_3Si(000\bar{1})$//Si(1$\bar{1}$1) configuration, Si atoms are arranged in a rhombohedral lattice in both Cu_3Si and Si, with two in-plane epitaxial directions along Cu_3Si[1$\bar{1}$00]//Si[$\bar{2}\bar{1}$1] and Cu_3Si[11$\bar{2}$0]//Si[011] with a mismatch of 6.25 % and 6.16 %, respectively. But for the Cu_3Si(1$\bar{1}$00)//Si($\bar{2}\bar{1}$1) configuration, Si atoms are arranged in rectangle structure with two in-plane epitaxial directions along Cu_3Si[000$\bar{1}$]//Si[1$\bar{1}$1] and Cu_3Si[11$\bar{2}$0]//Si[011] with a mismatch of 4.64 % and 6.16 %, respectively. The small mismatch indicates low dislocation density [95]. Similar epitaxial relationship was reported [100, 101].

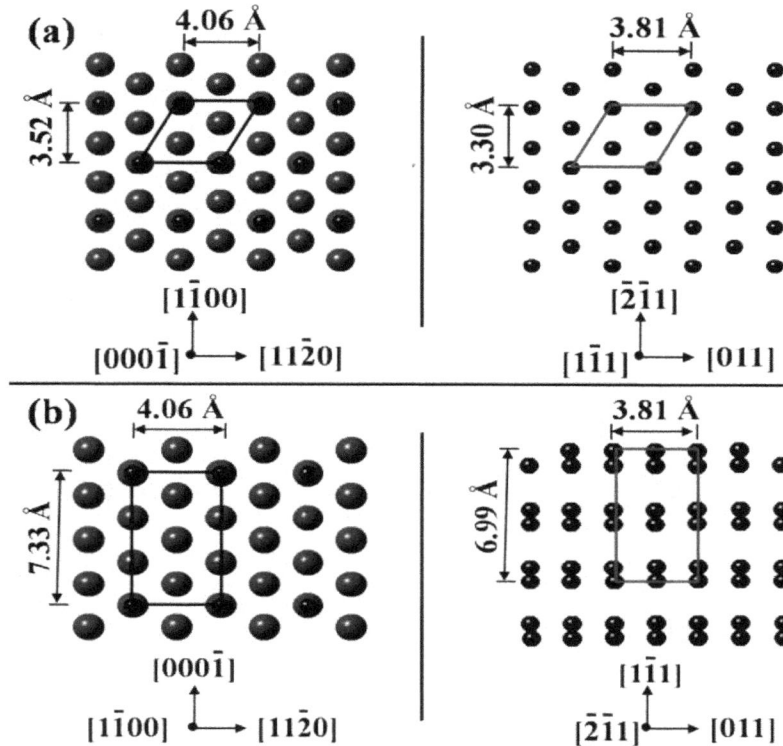

Fig. 7.29. Modeled atomic arrangements of Cu₃Si and Si in the epitaxial direction: (a) $Cu_3Si(000\bar{1})//Si(1\bar{1}1)$; (b) $Cu_3Si(1\bar{1}00)//Si(\overline{21}1)$ [95].

7.6.3. Amorphous ZrBₓOᵧ Diffusion Barrier

By doping a small amount of O element into ZrB_2 films, the nanocrystalline structure could be transformed into amorphous configuration [102, 103]. 5 nm ZrB_xO_y films were prepared by RF magnetron sputtering. The as-deposited ZrB_xO_y barrier layer shows an amorphous structure and is mainly composed of amorphous ZrO_2 and B_2O_3 in solid solution (Fig. 7.30) [104]. Oxygen atoms inhibit the crystallization of the barrier layer and improve the thermal stability.

Cu/5 nm ZrB_xO_y/Si samples were annealed in vacuum at temperature in the range of 200 °C~600 °C for 0.5 h. Fig. 7.31 shows the XRD patterns of the films. As the annealing temperature is elevated, the diffraction peaks of Cu(111) and Cu (200) become sharper and sharper as a result of grain growth. However, the XRD peaks of the barrier layers are still not evident, indicating amorphous nature. After annealed at 600 °C, Cu_3Si appears while the intensities of Cu peaks decrease. As shown in Fig. 7.32, the interface cannot be identified any longer after annealed at 600 °C for 30 min and, a large Cu_3Si grain appears on the surface (Fig. 7.32a). Hence, the inter-diffusion and reaction between Cu and Si atoms take place, and the ZrB_xO_y layer becomes invalid and unable to effectively block the atomic diffusion at 600 °C.

217

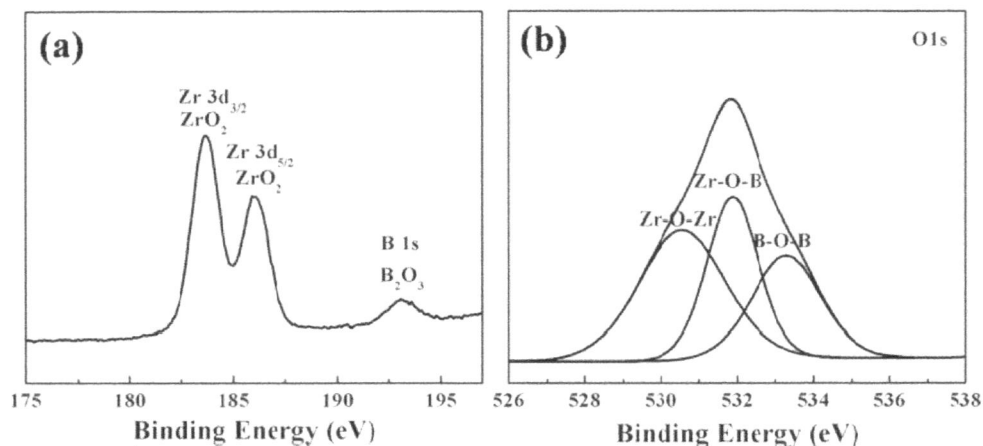

Fig. 7.30. XPS spectra of (a) Zr 3d and B 1s, and (b) O 1s in ZrB_xO_y [104].

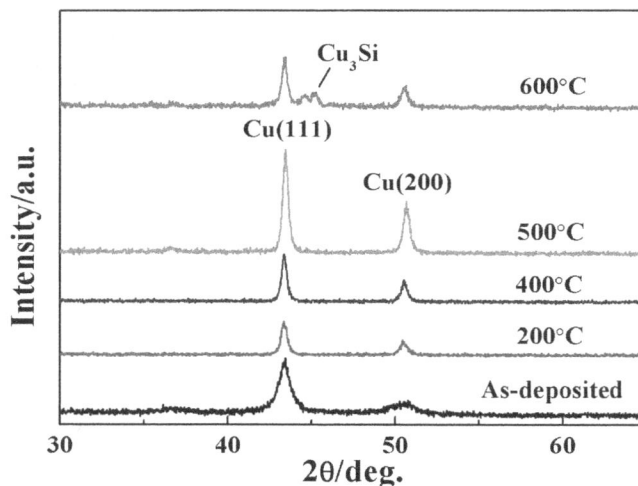

Fig. 7.31. XRD patterns of the Cu/ZrB_xO_y thin films as-deposited and annealed
at different temperatures [104].

Other ternary amorphous borides have also been reported. 30 nm amorphous Ta-B-N films became failure due to the formation of Cu-Si compounds after annealed at 800 °C for 60 min [105]. Leu et al found that 10 nm thick W-B-N is amorphous and has a low resistivity of about 240.4 $\mu\Omega \cdot cm$, moreover, it could prevent Cu diffusion even at 500 °C, suggesting W-B-N is a potential candidate as a diffusion barrier in Cu interconnection [106].

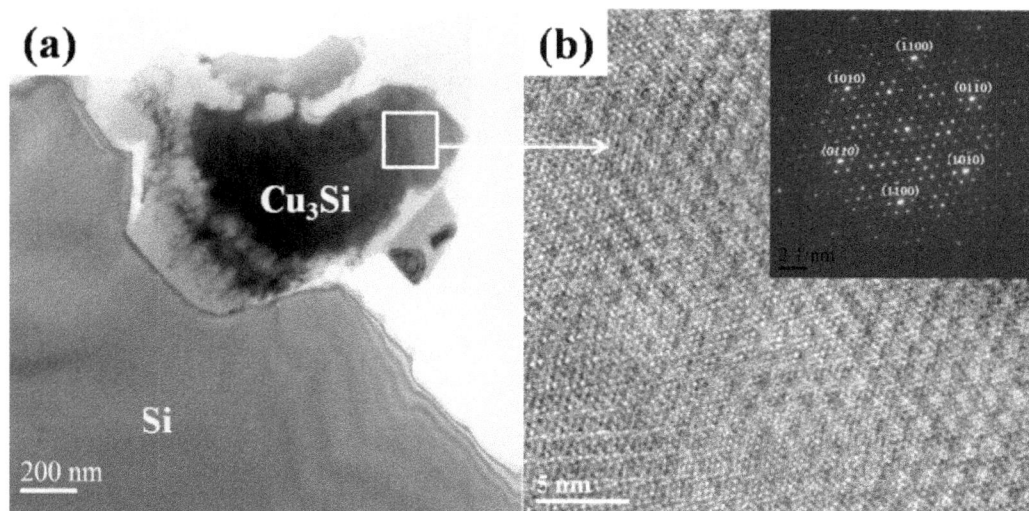

Fig. 7.32. HRTEM images of the cross-section morphologies of Cu/ZrB_xO_y thin films: (a) the sample annealed at 600 °C for 30 min; (b) HRTEM image of the newly formed phase denoted by while square in panel (a) and the corresponding electron diffraction pattern displayed in the inset of the figure [104].

7.7. Conclusion

Excellent diffusion barriers are required for Cu interconnects in smaller technology nodes of integrated circuit. To this end, higher thermal stability, ultra-thin thickness, lower resistivity, and better barrier performance must be satisfied. The fabrication approaches, the characterization method as well as four types of potential diffusion barriers are summarized. In particular, the process dependent diffusion barrier performances, the thermal stability and the mechanisms for the enhanced performance are well discussed in details. It can be concluded that Ru is the best diffusion barrier for Cu interconnects. For the next-generation technology node, the conformal diffusion barrier 1 nm in thickness can be deposited by state-of-art ALD. Moreover, graphene and nanotube interconnection have been proposed to replace Cu metallization. Therefore, in the near future, the appropriate diffusion barrier for the graphene or nanotube interconnection needs further researches.

Acknowledgements

This work was supported by the National Natural Science Foundation of China (Grant No. 51471130), the Natural Science Foundation of Shaanxi Province (No. 2017JZ015), the fund of the State Key Laboratory of Solidification Processing in NWPU (SKLSP201708), and Fundamental Research Funds for the Central Universities.

References

[1]. C. K. Hu, J. M. E. Harper, Copper interconnections and reliability, *Mater. Chem. Phys.,* Vol. 52, Issue 1, 1998, pp. 5-16.

[2]. Y. Wang, B.-H. Tang, F.-Y. Li, The properties of self-formed diffusion barrier layer in Cu(Cr) alloy, *Vacuum*, Vol. 126, 2016, pp. 51-54.

[3]. S. Sharma, M. Kumar, S. Rani, D. Kumar, Diffusion barrier characteristics of co monolayer prepared by Langmuir Blodgett technique, *Appl. Surf. Sci.*, Vol. 369, 2016, pp. 137-142.

[4]. M. P. Nguyen, Y. Sutou, J. Koike, Diffusion barrier property of $MnSi_xO_y$ layer formed by chemical vapor deposition for Cu advanced interconnect application, *Thin Solid Films,* Vol. 580, 2015, pp. 56-60.

[5]. B. Zhao, K. Sun, Z. Song, J. Yang, Ultrathin Mo/MoN bilayer nanostructure for diffusion barrier application of advanced Cu metallization, *Appl. Surf. Sci.,* Vol. 256, Issue 20, 2010, pp. 6003-6006.

[6]. R. P. Oleksak, A. Devaraj, G. S. Herman, Atomic-scale structural evolution of Ta-Ni-Si amorphous metal thin films, *Mater. Lett.,* Vol. 164, 2016, pp. 9-14.

[7]. W. K. Morrow, S. J. Pearton, F. Ren, Review of graphene as a solid state diffusion barrier, *Small*, Vol. 12, Issue 1, 2016, pp. 120-134.

[8]. H. Kim, T. Koseki, T. Ohba, T. Ohta, Y. Kojima, H. Sato, S. Hosaka, Y. Shimogaki, Effect of Ru crystal orientation on the adhesion characteristics of Cu for ultra-large scale integration interconnects, *Appl. Surf. Sci.*, Vol. 252, Issue 11, 2006, pp. 3938-3942.

[9]. O.-K. Kwon, S.-H. Kwon, H.-S. Park, S.-W. Kang, PEALD of a Ruthenium adhesion layer for copper interconnects, *J. Electrochem. Soc.*, Vol. 151, Issue 12, 2004, pp. C753-C756.

[10]. T. Hara, Y. Yoshida, H. Toida, Improved barrier and adhesion properties in sputtered TaSiN layer for copper interconnects, *Electrochem. Solid-State Lett.,* Vol. 5, Issue 5, 2002, pp. G36-G-39.

[11]. Q. Jiang, Y. F. Zhu, M. Zhao, Copper metallization for current very large scale integration, *Recent Patents on Nanotechnology*, Vol. 5, Issue 2, 2011, pp. 106-137.

[12]. Y. Sasajima, Y. Kimura, T. Tsumuraya, T. Nagano, J. Onuki, Search for barrier materials for Cu interconnects in integrated circuits, *ECS Journal of Solid State Science and Technology*, Vol. 2, Issue 9, 2013, pp. 351-356.

[13]. A. E. Kaloyeros, E. T. Eisenbraun, K. Dunn, O. van der Straten, Zero thickness diffusion barriers and metallization liners for nanoscale device applications, *Chem. Eng. Commun.*, Vol. 198, Issue 11, 2011, pp. 1453-1481.

[14]. C. Lee, Y.-L. Kuo, The evolution of diffusion barriers in copper metallization, *JOM*, Vol. 59, Issue 1, 2007, pp. 44-49.

[15]. H. Y. Wong, N. F. Mohd Shukor, N. Amin, Prospective development in diffusion barrier layers for copper metallization in LSI, *Microelectron. J.*, Vol. 38, Issue 6-7, 2007, pp. 777-782.

[16]. W. N. Gill, J. L. Plawsky, Design of diffusion barriers, *Thin Solid Films*, Vol. 515, Issue 11, 2007, pp. 4794-4800.

[17]. T. Osaka, H. Aramaki, M. Yoshino, K. Ueno, I. Matsuda, Y. Shacham-Diamand, Fabrication of electroless CoWP/NiB diffusion barrier layer on SiO_2 for ULSI devices, *J. Electrochem. Soc.*, Vol. 156, Issue 9, 2009, pp. H707-H710.

[18]. T. Osaka, N. Takano, T. Kurokawa, T. Kaneko, K. Ueno, Electroless nickel ternary alloy deposition on SiO_2 for application to diffusion barrier layer in copper interconnect technology, *J. Electrochem. Soc.*, Vol. 149, Issue 11, 2002, pp. C573-C578.

[19]. C. S. Hwang, C. Y. Yoo, Atomic Layer Deposition for Semiconductors, *Springer*, 2014.

[20]. M. Schaefer, R. Schlaf, Electronic structure investigation of atomic layer deposition ruthenium(oxide) thin films using photoemission spectroscopy, *J. Appl. Phys.*, Vol. 118, Issue 6, 2015, pp. 065306-1 – 065306-7.

[21]. B. H. Choi, Y. H. Lim, J. H. Lee, Y. B. Kim, H.-N. Lee, H. K. Lee, Preparation of Ru thin film layer on Si and TaN/Si as diffusion barrier by plasma enhanced atomic layer deposition, *Microelectron. Eng.*, Vol. 87, Issue 5-8, 2010, pp. 1391-1395.

[22]. K. Matsumoto, K. Maekawa, H. Nagai, J. Koike, Deposition behavior and substrate dependency of ALD MnO_x diffusion barrier layer, *IEEE 2013*, 2013, pp. 1-3.

[23]. J. S. Becker, R. G. Gordon, Diffusion barrier properties of tungsten nitride films grown by atomic layer deposition from bis(tert-butylimido)bis(dimethylamido) tungsten and ammonia, *Appl. Phys. Lett.*, Vol. 82, Issue 14, 2003, pp. 2239-2241.

[24]. P. Alén, M. Vehkamäki, M. Ritala, M. Leskelä, Diffusion Barrier Properties of Atomic Layer Deposited Ultrathin Ta_2O_5 and TiO_2 Films, J. Electrochem. Soc., Vol. 153, Issue 4, 2006, pp. G304-G308.

[25]. M. J. Frederick, R. Goswami, G. Ramanath, Sequence of Mg segregation, grain growth, and interfacial MgO formation in Cu-Mg alloy films on SiO_2 during vacuum annealing, *J. Appl. Phys.*, Vol. 93, Issue 10, 2003, pp. 5966-5972.

[26]. J. Koike, M. Haneda, J. Iijima, Y. Otsuka, H. Sako, K. Neishi, Growth kinetics and thermal stability of a self-formed barrier layer at Cu-Mn/SiO_2 interface, *J. Appl. Phys.*, Vol. 102, Issue 4, 2007, pp. 043527-1 - 043527-7.

[27]. C. Fei, W. Gao-hui, J. Long-tao, C. Guo-qin, Application of Cu-C and Cu-V alloys in barrier-less copper metallization, *Vacuum*, Vol. 122, 2015, pp. 122-126.

[28]. B. Liu, L. W. Lin, D. Ren, Y. P. Zhang, G. H. Jiao, K. W. Xu, Cu(Ge) alloy films with zirconium addition on barrierless Si for excellent property improvement, *J. Phys. D: Appl. Phys.*, Vol. 46, Issue 15, 2013, pp. 155305-1 – 155305-6.

[29]. Y. Meng, Z. X. Song, D. Qian, W. J. Dai, J. F. Wang, F. Ma, Y. H. Li, K. W. Xu, Thermal stability of RuZr alloy thin films as the diffusion barrier in Cu metallization, *J. Alloys Compd.*, Vol. 588, 2014, pp. 461-464.

[30]. C. H. Lee W, Cho B, Factors affecting passivation of Cu(Mg) alloy films, *J. Electrochem. Soc.*, Vol. 147, Issue 8, 2000, pp. 3066-3069.

[31]. J. J. H. Ko Y K, Lee S, Effects of molybdenum, silver dopants and a titanium substrate layer on copper film metallization, *J. Mater. Sc.*, Vol. 38, 2003, pp. 217-222.

[32]. S. Muranaka, M. Sueyoshi, K. Mori, K. Maekawa, M. Fujisawa, K. Asai, Effect of impurities and microstructure of Cu electroplated film on reliability of Cu interconnects using CuAl alloy seed, *Microelectron. Eng.*, Vol. 105, 2013, pp. 91-94.

[33]. X.-J. Wang, X.-P. Dong, C.-H. Jiang, Thermal performance of sputtered Cu films containing insoluble Zr and Cr for advanced barrierless Cu metallization, *Transactions of Nonferrous Metals Society of China*, Vol. 20, Issue 2, 2010, pp. 217-222.

[34]. S.-M. Yi, K.-H. Jang, J.-U. An, S.-S. Hwang, Y.-C. Joo, The self-formatting barrier characteristics of Cu-Mg/SiO_2 and Cu-Ru/SiO_2 films for Cu interconnects, *Microelectron. Reliab.*, Vol. 48, Issue 5, 2008, pp. 744-748.

[35]. B. Liu, Z. Song, K. Xu, The effect of zirconium dopant on the properties of copper films, *Surf. Coat. Technol.*, Vol. 201, Issue 9-11, 2007, pp. 5419-5421.

[36]. Y. Wang, X.-d. Yang, Z.-x. Song, Y.-t. Liu, Property improvement of copper films with zirconium additive for ULSI interconnects, *J. Alloys Compd.*, Vol. 486, Issue 1-2, 2009, pp. 418-422.

[37]. J. W. Lim, K. Mimura, K. Miyake, M. Yamashita, M. Isshiki, Effect of substrate bias voltage on the purity of Cu films deposited by non-mass separated ion beam deposition, *Thin Solid Films*, Vol. 434, Issue 1-2, 2003, pp. 30-33.

[38]. Y. Wang, F. Cao, M.-l. Zhang, Y.-t. Liu, Comparative study of Cu-Zr and Cu-Ru alloy films for barrier-free Cu metallization, *Thin Solid Films*, Vol. 519, Issue 10, 2011, pp. 3169-3172.

[39]. X. Y. Zhang, X. N. Li, L. F. Nie, J. P. Chu, Q. Wang, C. H. Lin, C. Dong, Highly stable carbon-doped Cu films on barrierless Si, *Appl. Surf. Sci.*, Vol. 257, Issue 8, 2011, pp. 3636-3640.

[40]. C. J. Liu, J. S. Chen, Y. K. Lin, Characterization of microstructure, interfacial reaction and diffusion of immiscible Cu(Ta) alloy thin film on SiO$_2$ at elevated temperature, *J. Electrochem. Soc.*, Vol. 151, Issue 1, 2004, pp. G18-G23.

[41]. F. Cao, G.-h. Wu, L.-t. Jiang, Evaluation of Cu(V) self-forming barrier for Cu metallization, *J. Alloys Compd.*, Vol. 657, 2016, pp. 483-486.

[42]. Y. Wang, F. Cao, M.-l. Zhang, T. Zhang, Property improvement of Cu-Zr alloy films with ruthenium addition for Cu metallization, *Acta Mater.*, Vol. 59, Issue 1, 2011, pp. 400-404.

[43]. C. J. Liu, J. S. Chen, High-temperature self-grown ZrO2 layer against Cu diffusion at Cu(2.5at.%Zr)/SiO$_2$ interface, *Journal of Vacuum Science & Technology B: Microelectronics and Nanometer Structures*, Vol. 231, 2005, p. 90.

[44]. M. Damayanti, T. Sritharan, Z. H. Gan, S. G. Mhaisalkar, N. Jiang, L. Chan, Ruthenium barrier/seed layer for Cu/Low-κ metallization, *J. Electrochem. Soc.*, Vol. 153, Issue 6, 2006, pp. J41-J45.

[45]. C. J. C. Barmak K., Rodbell K. P., On the use of alloying elements for Cu interconnect applications, *J. Vac. Sci. Technol.*, Vol. B24, Issue 6, 2006, pp. 2485-2498.

[46]. O. S. Borek M. A., Aboelfotoh M. O., Low resistivity Cu$_3$Ge on (100) Si for contacts and interconnections, *Appl. Phys. Lett.*, Vol. 69, Issue 23, 1996, pp. 3560-3562.

[47]. M. O. Aboelfotoh, M. A. Borek, J. Narayan, Microstructure and electrical resistivity of Cu and Cu$_3$Ge thin films on Si$_{1-x}$Ge$_x$ alloy layers, *J. Appl. Phys.*, Vol. 87, Issue 1, 2000, pp. 365-368.

[48]. M. O. Aboelfotoh, K. N. Tu, F. Nava, M. Michelini, Electrical transport properties of Cu$_3$Ge thin films, J. *Appl. Phys.*, Vol. 75, Issue 3, 1994, pp. 1616-1619.

[49]. C. P. Wang, T. Dai, Y. Lu, Z. Shi, J. J. Ruan, Y. H. Guo, X. J. Liu, Thermal stability of copper-aluminum alloy thin films for barrierless copper metallization on silicon substrate, *J. Electron. Mater.*, 2017, pp. 4891-4897.

[50]. B. Liu, Z. X. Song, Y. H. Li, K. W. Xu, An ultrathin Zr(Ge) alloy film as an exhaustion interlayer combined with Cu(Zr) seed layer for the Cu/porous SiOC:H dielectric integration, *Appl. Phys. Lett.*, Vol. 93, Issue 17, 2008, pp.174108-1 - 174108-3.

[51]. J. Koike, M. Wada, Self-forming diffusion barrier layer in Cu-Mn alloy metallization, *Appl. Phys. Lett.*, Vol. 87, Issue 4, 2005, pp. 041911-1 – 041911-3.

[52]. J. J. Araiza, O. Sánchez, Influence of aluminium incorporation on the structure of ZrN films deposited at low temperatures, *J. Phys. D: Appl. Phys.*, Vol. 4211, 2009, pp. 115422-1 - 115422.6.

[53]. Y. Wang, C.-H. Zhao, F. Cao, D.-W. Yang, Barrier capability of Zr-N films with different density and crystalline structure in Cu/Si contact systems, *Mater. Lett.*, Vol. 62, Issue 21-22, 2008, pp. 3761-3763.

[54]. J.-H. Huang, C.-H. Ho, G.-P. Yu, Effect of nitrogen flow rate on the structure and mechanical properties of ZrN thin films on Si(100) and stainless steel substrates, *Mater. Chem. Phys.*, Vol. 102, Issue 1, 2007, pp. 31-38.

[55]. K. Abe, Y. Harada, H. Onoda, Study of crystal orientation in Cu film on TiN layered structures, *Journal of Vacuum Science & Technology B: Microelectronics and Nanometer Structures,* Vol. 17, Issue 4, 1999, pp. 1464-1469.

[56]. G. S. Chen, S. C. Huang, S. T. Chen, T. J. Yang, P. Y. Lee, J. H. Jou, T. C. Lin, An optimal quasisuperlattice design to further improve thermal stability of tantalum nitride diffusion barriers, *Appl. Phys. Lett.*, Vol. 76, Issue 20, 2000, pp. 2895-2897.

[57]. C.-S. Chen, C.-P. Liu, H.-G. Yang, C. Y. A. Tsao, Influence of the preferred orientation and thickness of zirconium nitride films on the diffusion property in copper, *Journal of Vacuum Science & Technology B: Microelectronics and Nanometer Structures,* Vol. 22, Issue 3, 2004, pp. 1075-1083.

[58]. L. García González, J. Hernández Torres, M. G. Garnica Romo, L. Zamora Peredo, A. M. Courrech Arias, E. León Sarabia, F. J. Espinoza Beltrán, Ti/TiSiNO multilayers fabricated by Co-sputtering, *J. Mater. Eng. Perform.*, Vol. 22, 2013, pp. 2377-2381.

[59]. R. Hübner, M. Hecker, N. Mattern, V. Hoffmann, K. Wetzig, H. Heuer, C. Wenzel, H. J. Engelmann, D. Gehre, E. Zschech, Effect of nitrogen content on the degradation mechanisms of thin Ta-Si-N diffusion barriers for Cu metallization, *Thin Solid Films*, Vol. 500, Issue 1-2, 2006, pp. 259-267.

[60]. M. Ding, H. Zheng, L. Wang, H. Zhang, A study of Ta-Si-N/Ti bilayer diffusion barrier for copper/silicon contact systems, *Mater. Manuf. Processes*, Vol. 31, Issue 8, 2015, pp. 1009-1013.

[61]. Z. X. Song, K. W. Xu, H. Chen, The characterization of Zr-Si-N diffusion barrier films with different sputtering bias voltage, *Thin Solid Films*, Vol. 468, Issue 1-2, 2004, pp. 203-207.

[62]. Y. Wang, F. Cao, Y.-T. Liu, M.-H. Ding, Investigation of Zr-Si-N/Zr bilayered film as diffusion barrier for Cu ultralarge scale integration metallization, *Appl. Phys. Lett.*, Vol. 92, Issue 3, 2008, pp. 032108-1 – 032108-3.

[63]. L. C. Leu, P. Sadik, D. P. Norton, L. McElwee-White, T. J. Anderson, Comparative study of ZrN and Zr-Ge-N thin films as diffusion barriers for Cu metallization on Si, *Journal of Vacuum Science & Technology B: Microelectronics and Nanometer Structures*, Vol. 26, Issue 5, 2008, pp. 1723-1727.

[64]. M. Nose, M. Zhou, T. Nagae, T. Mae, M. Yokota, S. Saji, Properties of Zr-Si-N coatings prepared by RF reactive sputtering, *Surf. Coat. Technol.*, Vol. 132, Issue 2-3, 2000, pp. 163-168.

[65]. Z. X. Song, K. W. Xu, H. Chen, The Effect of nitrogen partial pressure on Zr-Si-N diffusion barrier, *Microelectron. Eng.*, Vol. 71, Issue 1, 2004, pp. 28-33.

[66]. J. O. Olowolafe, I. Rau, K. M. Unruh, C. P. Swann, Z. S. Jawad, T. Alford, Effect of composition on thermal stability and electrical resistivity of Ta-Si-N films, *Thin Solid Films*, Vol. 365, Issue 1, 2000, pp. 19-21.

[67]. L. W. Lin, B. Liu, D. Ren, C. Y. Zhan, G. H. Jiao, K. W. Xu, Effect of sputtering bias voltage on the structure and properties of Zr-Ge-N diffusion barrier films, *Surf. Coat. Technol.*, Vol. 228, 2013, pp. S237-S240.

[68]. Y. Wang, C. Zhu, Z. Song, Y. Li, High temperature stability of Zr-Si-N diffusion barrier in Cu/Si contact system, *Microelectron. Eng.*, Vol. 71, Issue 1, 2004, pp. 69-75.

[69]. T. E. Hong, J.-H. Jung, S. Yeo, T. Cheon, S. I. Bae, S.-H. Kim, S. J. Yeo, H.-S. Kim, T.-M. Chung, B. K. Park, C. G. Kim, D.-J. Lee, Highly conformal amorphous W-Si-n thin films by plasma-enhanced atomic layer deposition as a diffusion barrier for Cu Metallization, *The Journal of Physical Chemistry*, Vol. C119, Issue 3, 2015, pp. 1548-1556.

[70]. Y.-J. Lee, B.-S. Suh, M. S. Kwon, C.-O. Park, Barrier properties and failure mechanism of Ta-Si-N thin films for Cu interconnection, *J. Appl. Phys.*, Vol. 85, Issue 3, 1999, pp. 1927-1934.

[71]. J.-T. No, J.-H. O, C. Lee, Evaluation of Ti-Si-N as a diffusion barrier between copper and silicon, *Mater. Chem. Phys.*, Vol. 63, Issue 1, 2000, pp. 44-49.

[72]. J. S. Reid, E. Kolawa, R. P. Ruiz, M. A. Nicolet, Evaluation of amorphous (Mo,Ta,W)-Si-N diffusion barriers for <Si>|Cu| metallizations, *Thin Solid Films*, Vol. 236, Issue 1, 1993, pp. 319-324.

[73]. D. Pilloud, J. F. Pierson, A. Cavaleiro, M. C. Marco de Lucas, Effect of germanium addition on the properties of reactively sputtered ZrN films, *Thin Solid Films*, Vol. 492, Issue 1-2, 2005, pp. 180-186.

[74]. C. Y. Wu, Y. S. Wang, W. H. Lee, Copper electrodeposition on Ru-N barrier with various nitrogen content for 22 nm semiconductor manufacturing application, *J. Electrochem. Soc.*, Vol. 159, Issue 11, 2012, pp. D684-D689.

[75]. H. Wojcik, R. Kaltofen, U. Merkel, C. Krien, S. Strehle, J. Gluch, M. Knaut, C. Wenzel, A. Preusse, J. W. Bartha, M. Geidel, B. Adolphi, V. Neumann, R. Liske, F. Munnik, Electrical evaluation of Ru-W(-N), Ru-Ta(-N) and Ru-Mn films as Cu diffusion barriers, *Microelectron. Eng.*, Vol. 92, 2012, pp. 71-75.

[76]. T. N. Arunagiri, Y. Zhang, O. Chyan, M. El-Bouanani, M. J. Kim, K. H. Chen, C. T. Wu, L. C. Chen, 5 nm ruthenium thin film as a directly plateable copper diffusion barrier, *Appl. Phys. Lett.*, Vol. 86, Issue 8, 2005, pp. 083104-1 – 083104-3.

[77]. W. Wei, S. L. Parker, Y. M. Sun, J. M. White, G. Xiong, A. G. Joly, K. M. Beck, W. P. Hess, Study of copper diffusion through a ruthenium thin film by photoemission electron microscopy, *Appl. Phys. Lett.*, Vol. 90, Issue 11, 2007, pp. 111906-1 – 111906-3

[78]. D.-C. Perng, J.-B. Yeh, K.-C. Hsu, Y.-C. Wang, 5 nm amorphous boron and carbon added Ru film as a highly reliable cu diffusion barrier, *Electrochem. Solid-State Lett.*, Vol. 13, Issue 8, 2010, pp. H290-H293.

[79]. K.-C. Hsu, D.-C. Perng, Y.-C. Wang, Robust ultra-thin RuMo alloy film as a seedless Cu diffusion barrier, *J. Alloys Compd.*, Vol. 516, 2012, pp. 102-106.

[80]. C.-Y. Wu, W.-H. Lee, S.-C. Chang, Y.-L. Cheng, Y.-L. Wang, Effect of annealing on the microstructure and electrical property of RuN thin films, *J. Electrochem. Soc.*, Vol. 158, Issue 3, 2011, H338-H342.

[81]. H. Volders, L. Carbonell, N. Heylen, K. Kellens, C. Zhao, K. Marrant, G. Faelens, T. Conard, B. Parmentier, J. Steenbergen, M. V.d. Peer, C. J. Wilson, E. Sleeckx, G. P. Beyer, Z. Tőkei, V. Gravey, K. Shah, A. Cockburn, Barrier and seed repair performance of thin RuTa films for Cu interconnects, *Microelectron. Eng.*, Vol. 88, Issue 5, 2011, pp. 690-693.

[82]. G. He, L. Yao, Z. Song, Y. Li, K. Xu, Diffusion barrier performance of nano-structured and amorphous Ru-Ge diffusion barriers for copper metallization, *Vacuum*, Vol. 86, Issue 7, 2012, pp. 965-969.

[83]. G. Jiao, B. Liu, Q. Li, Investigation of amorphous RuMoC alloy films as a seedless diffusion barrier for Cu/p-SiOC:H ultralow-k dielectric integration, *Appl. Phys.*, Vol. A120, Issue 2, 2015, pp. 579-585.

[84]. K.-Y. Mun, T. E. Hong, T. Cheon, Y. Jang, B.-Y. Lim, S. Kim, S.-H. Kim, The effects of nitrogen incorporation on the properties of atomic layer deposited Ru thin films as a direct-plateable diffusion barrier for Cu interconnect, *Thin Solid Films*, Vol. 562, 2014, pp. 118-125.

[85]. M. B. Takeyama, A. Noya, Y. Nakadai, S. Kambara, M. Hatanaka, Y. Hayasaka, E. Aoyagi, H. Machida, K. Masu, Low temperature deposited Zr-B film applicable to extremely thin barrier for copper interconnect, *Appl. Surf. Sci.*, Vol. 256, Issue 4, 2009, pp. 1222-1226.

[86]. M. Samuelsson, J. Jensen, U. Helmersson, L. Hultman, H. Högberg, ZrB$_2$ thin films grown by high power impulse magnetron sputtering from a compound target, *Thin Solid Films*, Vol. 526, 2012, pp. 163-167.

[87]. D. M. Stewart, D. J. Frankel, R. J. Lad, Growth, structure, and high temperature stability of zirconium diboride thin films, *Journal of Vacuum Science & Technology A: Vacuum, Surfaces, and Films*, Vol. 33, Issue 3, 2015, pp. 031505-1 – 031505-6.

[88]. R. Roucka, J. Tolle, A. V. G. Chizmeshya, I. S. T. Tsong, J. Kouvetakis, Epitaxial film growth of zirconium diboride on Si(001), *J. Cryst. Growth*, Vol. 277, Issue 1-4, 2005, pp. 364-371.

[89]. L. Tengdelius, M. Samuelsson, J. Jensen, J. Lu, L. Hultman, U. Forsberg, E. Janzén, H. Högberg, Direct current magnetron sputtered ZrB$_2$ thin films on 4H-SiC(0001) and Si(100), *Thin Solid Films*, Vol. 550, 2014, pp. 285-290.

[90]. A. Fleurence, W. Zhang, C. Hubault, Y. Yamada-Takamura, Mechanisms of parasitic crystallites formation in ZrB$_2$(0001) buffer layer grown on Si(111), *Appl. Surf. Sci.*, Vol. 284, 2013, pp. 432-437.

[91]. J. Tolle, R. Roucka, I. S. T. Tsong, C. Ritter, P. A. Crozier, A. V. G. Chizmeshya, J. Kouvetakis, Epitaxial growth of group III nitrides on silicon substrates via a reflective lattice-matched zirconium diboride buffer layer, *Appl. Phys. Lett.*, Vol. 82, Issue 15, 2003, pp. 2398-2400.

[92]. J. Sung, D. M. Goedde, G. S. Girolami, J. R. Abelson, Remote-plasma chemical vapor deposition of conformal ZrB_2 films at low temperature: A promising diffusion barrier for ultralarge scale integrated electronics, *J. Appl. Phys.*, Vol. 91, Issue 6, 2002, pp. 3904-3911.

[93]. J.-S. Fang, W.-J. Su, M.-S. Huang, C.-F. Chiu, T.-S. Chin, Characteristics of plasma-treated amorphous Ta-Si-C film as a diffusion barrier for copper metallization, *J. Electron. Mater.*, Vol. 43, Issue 1, 2013, pp. 212-218.

[94]. S.-Y. Chang, M.-K. Chen, D.-S. Chen, Multiprincipal-element AlCrTaTiZr-Nitride nanocomposite film of extremely high thermal stability as diffusion barrier for Cu metallization, *J. Electrochem. Soc.*, Vol. 156, Issue 5, 2009, pp. G37-G42.

[95]. Y. Meng, F. Ma, Z. X. Song, Y. H. Li, K. W. Xu, Nano-grained ZrB_2 thin films as a high-performance diffusion barrier in Cu metallization, *RSC Adv.*, Vol. 6, Issue 2, 2016, pp. 844-850.

[96]. C. Liao, D. Guo, S. Wen, X. Lu, G. Pan, J. Luo, The assessment of interface adhesion of Cu/Ta/Black Diamond™/Si films stack structure by nanoindentation and nanoscratch tests, *Tribol. Lett.*, Vol. 53, Issue 2, 2013, pp. 401-410.

[97]. S. Cho, K. Lee, P. Song, H. Jeon, Y. Kim, Barrier characteristics of ZrN films deposited by remote plasma-enhanced atomic layer deposition using Tetrakis(diethylamino)zirconium Precursor, *Japanese Journal of Applied Physics*, Vol. 46, Issue 7A, 2007, pp. 4085-4088.

[98]. B. H. Lee, K. Yong, Diffusion barrier properties of metalorganic chemical vapor deposition -WN[sub x] compared with other barrier materials, *Journal of Vacuum Science & Technology B: Microelectronics and Nanometer Structures*, Vol. 22, Issue 5, 2004, pp. 2375-2379.

[99]. J. K. Solberg, The crystal structure of η-Cu_3Si precipitates in silicon, *Acta Crystallographica Section A*, Vol. 34, Issue 5, 1978, pp. 684-698.

[100]. F. W. Yuan, C. Y. Wang, G. A. Li, S. H. Chang, L. W. Chu, L. J. Chen, H. Y. Tuan, Solution-phase synthesis of single-crystal Cu_3Si nanowire arrays on diverse substrates with dual functions as high-performance field emitters and efficient anti-reflective layers, *Nanoscale*, Vol. 5, Issue 20, 2013, pp. 9875-9882.

[101]. L. T. Jung S J, Bell A P, Free-standing, single-crystal Cu_3Si nanowires, *Cryst. Growth & Des*, Vol. 12, Issue 6, 2012, pp. 3076-3081.

[102]. J. F. Pierson, A. Billard, T. Belmonte, H. Michel, C. Frantz, Influence of oxygen flow rate on the structural and mechanical properties of reactively magnetron sputter-deposited Zr-B-O coatings, *Thin Solid Films*, Vol. 347, Issue 1-2, 1999, pp. 78-84.

[103]. J. Ebisawa, Characterization of sputter-deposited ZrB_xO_y films, *Journal of Vacuum Science & Technology A: Vacuum, Surfaces, and Films*, Vol. 8, Issue 3, 1990, 1335.

[104]. Y. Meng, Z. X. Song, J. H. Chen, F. Ma, Y. H. Li, J. F. Wang, C. C. Wang, K. W. Xu, Ultrathin ZrBxOy films as diffusion barriers in Cu interconnects, *Vacuum*, Vol. 119, 2015, pp. 1-6.

[105]. S.-T. Lin, C. Lee, Characteristics of sputtered Ta-B-N thin films as diffusion barriers between copper and silicon, *Appl. Surf. Sci.*, Vol. 253, Issue 3, 2006, pp. 1215-1221.

[106]. L. C. Leu, D. P. Norton, L. McElwee-White, T. J. Anderson, Properties of reactively sputtered W-B-N thin film as a diffusion barrier for Cu metallization on Si, *Appl. Phys.*, Vol. A94, Issue 3, 2008, pp. 691-695.

Chapter 8

Metrology and Inspection Solutions for Fan-Out Wafer Level Packaging

Priya Mukundhan, Gurvinder Singh, Woo Young Han, Wayne Fitzgerald, Ben Meihack, Johnny Dai, Jian Ding, Robin Mair, Jay Chen, Fei Shen, Manjusha Mehendale, Mike Marshall, Timothy Kryman and Cheolkyu Kim

8.1. Introduction

The consumer electronics market is currently being driven by the demand for small form factor devices with a high level of integrated computing power, memory, and other embedded functions. Cost benefits derived from complementary metal oxide semiconductor (CMOS) scaling of transistors has ceased and in order to gain additional advantages, the industry has turned to advanced packaging technologies. In addition to serving as packaging support, advanced packaging is enabling more functionality to be integrated along with various types of devices in the same package and by extension driving cost reductions [1-4]. Additionally, they offer the following advantages: improved throughput between logic, memory and I/O as the interconnects are wider, improved performance and reduced RC delay. Also, by bridging two or more chips, the distance signals have to travel may be less than across a single device. This removes competition for resources, reduces congestion by using multiple chips in a package. Also, yields of smaller chips are higher than very large integrated system on chips (SoCs) [5].

Broadly referred to as wafer level packaging (WLP), several advanced packaging technology platforms have emerged to address various end application needs (Fig. 8.1). This encompasses fan-in wafer level chip scale packages (WLCSP), 3D WLP, fan-out wafer level packages (FO-WLP), 2.5D glass / silicon interposers and 3DIC with through silicon via (TSV) interconnects. Research and development efforts in WLP are underway at integrated device manufacturers (IDM), foundries and outsourced semiconductor assembly and test (OSAT) fabs around the world to find viable approaches that include optimizing various parts of the supply chain to bring competitive solutions to market. We

Priya Mukundhan
Rudolph Technologies, 550 Clark Drive, Budd Lake, NJ 07828

briefly describe the available packaging platforms but provide a more comprehensive overview of the FO-WLP.

Fan-in WLP (also WLCSP) has been used extensively in several consumer products for over a decade. Some of the advantages of WLCSP are low cost, low end, low profile, low pin count, small form factor, and high volume applications. Fan-in technology is being phased out in favor of the fan-out wafer/panel level packaging (FOW/PLP) or just FO-WLP technology. Fan-out works by embedding a die into a molding compound and redistributing (fanning out) the I/Os to outside of the die. It offers better electrical performance and facilitates integration of more advanced packaging architectures such as system-in-package (SiP), large system-on-chip (SoC) ICs such as application processors (AP) and 3D IC packaging [6].

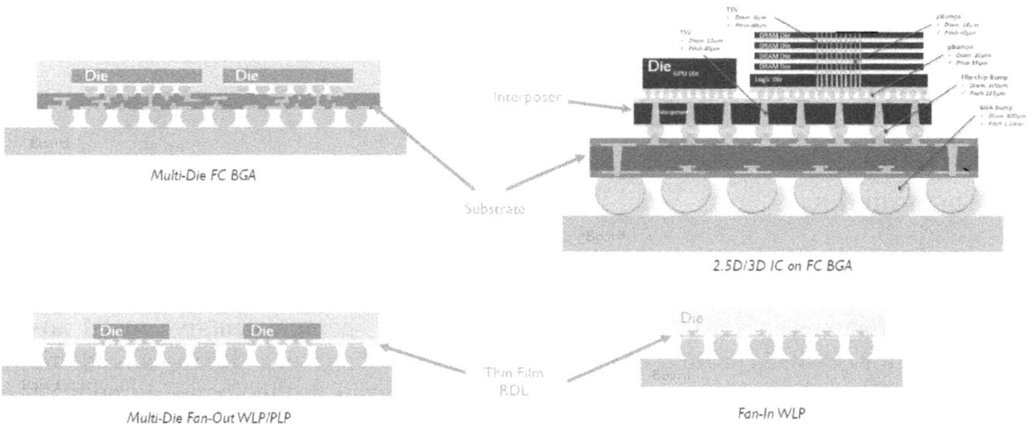

Fig. 8.1. Schematic representation of board level, substrate level and RDL level packages [7]. Courtesy: *Yole*.

SiP is one of the disruptive packaging platform that allows integration of heterogeneous technology (separating ICs with functional blocks), reduces cost and form and paves the way for Internet-of-Things (IoT). This architecture finds use in numerous applications within the mobile market, including smart phones and tablet applications, radio frequency (RF) applications and connectivity modules.

Another emerging advanced packaging platform is 3D integration. TSV technology has entered the mainstream for 3D ICs as they support heterogeneous integration of logic and memory devices resulting in significant improvements in performance. This solution is still cost prohibitive and is being driven primarily by applications, such as CMOS image sensors (CIS), high performance computers, networking, gaming and servers that demand the higher performance. TSVs allow devices to be stacked vertically while at the same time providing mechanical and electrical connections.

8.2. Fan-out Wafer Level Packaging

The FO-WLP market is expected to have the highest growth rate, exceeding $2 billion by 2020 including all platforms [1]. FO-WLP is showing a lot of versatility and thus lends itself attractive for a wide variety of applications. The fan-out process doesn't require a substrate and is relatively low cost as it eliminates the following process steps: wafer bumping, flip chip reflow, cleaning and underfill. Fig. 8.2 shows an example of the use of FO-WLP on an Apple A10 AP in iPhone 7/7+ [8]. This figure shows how the redistribution layer (RDL) has been integrated in the process resulting in lower cost, higher performance and lower profile package.

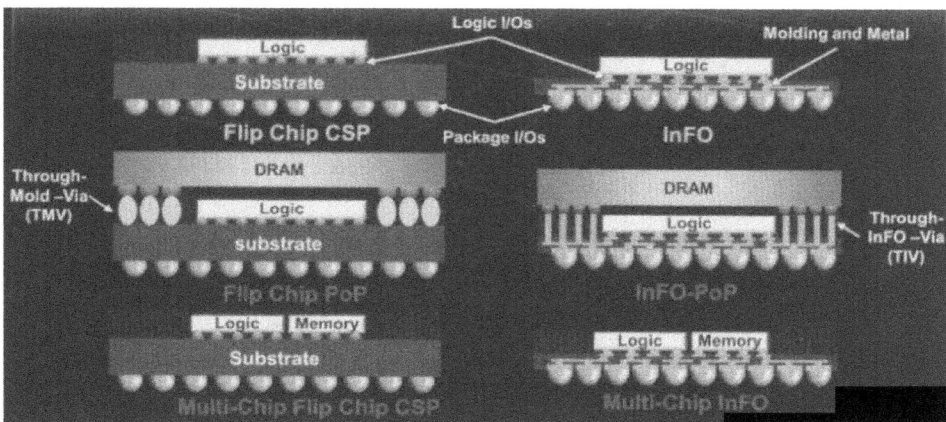

(a)

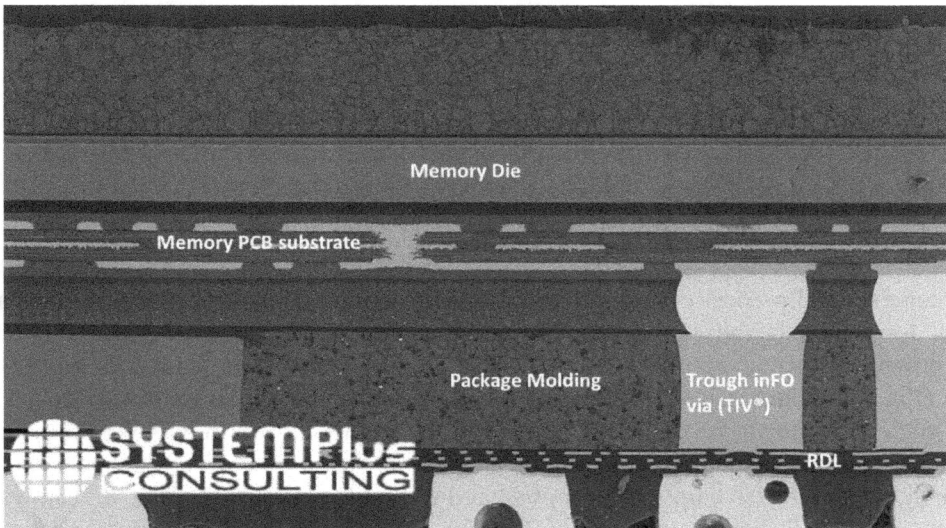

(b)

Fig. 8.2. (a) InFO process; (b) Cross-section image of the A10 processor in iPhone 7/7+ [4].

Advantages of FO-WLP over Fan-in WLP are the use of known good die (KGD), better wafer-level yield, using the best of silicon, multi-chip, embedded integrated passive devices, more than one RDL, higher pin counts, and better thermal performance. Additionally, fan-out enables handling multiple dies and discrete components in sharp contrast to fan-in that can handle only single die. Disadvantage of FO-WLP is that it cannot be used on large die size (> 12 mm × 12 mm) and large fan-out package size. This limitation comes from the large thermal expansion mismatch causing warpage. FO-WLP offers a compromise between performance and cost. In the following section we will focus on the details of the process flow, challenges encountered along with metrology and inspection insertion points that help in process monitoring and control.

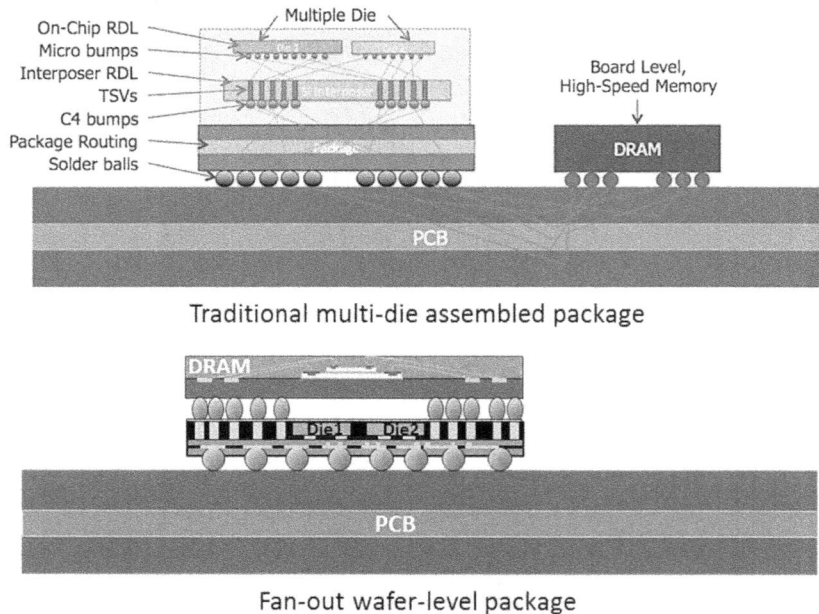

Fig. 8.3. Schematic representation comparing traditional multi-die assembled package and fan-out wafer level package [9].

There are several process options for the FO-WLP market. Fan-out wafer level packages can be formed using chip first or chip last process. Chip first can be processed using either face down or a face up approach. One can also process either a single layer or multi-layer RDL (Fig. 8.4). Chip last, also referred to as RDL-first, is only processed face down. Chip first with face down is the most conventional method to form FO-WLP. In some variations, passives are integrated into the package as they further shrink the package size and make it an attractive option for mobile and wearable applications. Fig. 8.5 shows the process flow for this method. Known Good Die (KGD) are picked and placed on a wafer or panel. This reconfigured carrier is over molded using epoxy mold compound (EMC). This is followed by building the RDLs for signals, power, and grounds from the Al or Cu pads. Finally, solder balls are mounted and the whole molding (with KGDs, RDLs, and solder balls) is diced into individual packages.

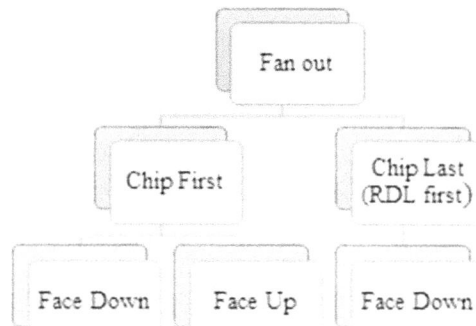

Fig. 8.4. FOWL processing options.

In the case of the chip first, die up approach, exemplified by TSMC's InFO process [6, 10], it can be seen that after the test of KGDs from the device wafer, a Ti/Cu under bump metallization (UBM) layer is sputtered using physical vapor deposition (PVD) on the Al or Cu pad, and a Cu contact pad (or stud) is electroplated on the UBM. This allows for building the RDLs later. This process is followed by spin coating a polymer, e.g., PI (polyimide), BCB (benzocyclobutene), or PBO (polybenzobisoxazole), on the whole wafer. The wafer is then diced into individual dies. The KGDs are then picked and placed face-up (die-up) on a double-sided thermal release tape, which is on a temporary carrier. These steps are followed by compression molding with EMC on the reconfigured (wafer or panel) carrier, and then back grinding the over-mold to expose the contact pad of the KGDs. The next step is to build up the RDLs from the contact pads and then mount the solder balls. Next comes the removal of the carrier and tape, and then the dicing of the whole molding into individual packages.

There are several technical challenges associated with FO-WLP that need to be addressed [11]. They are

a) Topography: when different device sizes are embedded, non-planarity becomes an issue for SiP;

b) Chip-to-mold non planarity: die protrusions at the interface can lead to RDL distortion;

c) Die shift: lack of precision during pick and place induces challenges for litho steps/alignments. Placement error is low at center and greater at the edge;

d) Warpage: thermal expansion of the carrier material during molding and shrinkage of mold compound can create non-uniform areas during process. This creates wafer flatness and stress on die;

e) Reliability: for large size die, creep fatigue and solder joint issues can occur;

f) Overall yield is dependent upon the quality of the RDL build-up technology. This is especially true for the die-first approach where the KGD is being committed to an RDL process and is at the mercy of the defect density of the wafer-level fan- out RDL build process.

(a)

(b)

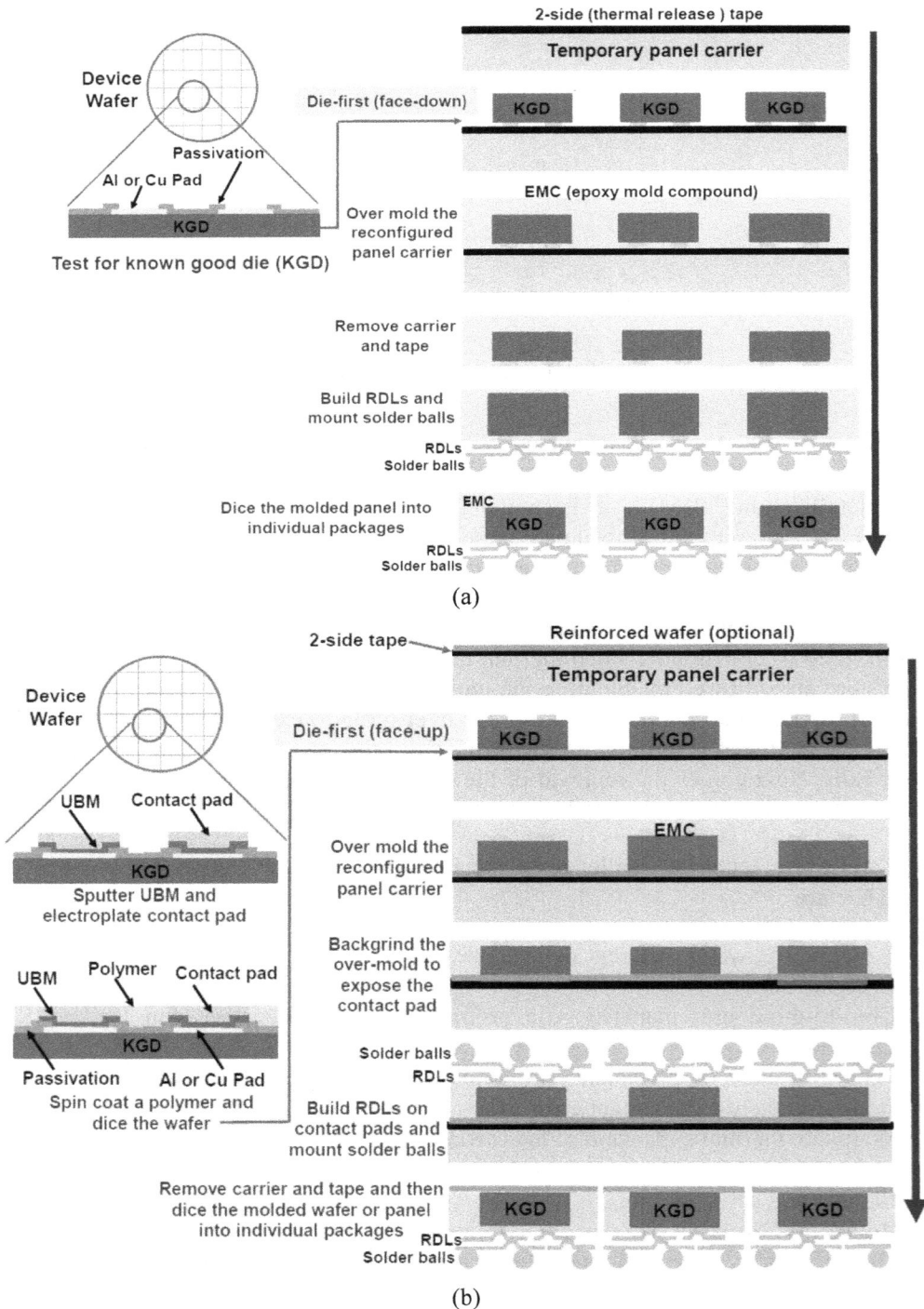

Fig. 8.5. (a) Schematic of Chip-First (Die-Down) FOW/PLP, and (b) Chip-first (Die-up).
Courtesy: John Lau, *ASM International*.

Some of the challenges involved in the fan-out process are being addressed by pursuing alternate approaches. Chip-last (RDL-first) overcomes the technical challenges associated with chip-first process. Readers are referred to [3] for details of this process. Suffice it to say that this is very different from chip-first and is afforded by very high density performance applications such as high end servers and components. An alternate approach is in Amkor's Silicon Wafer Integrated Fan-out Technology (SWIFT™) process [12]. The SWIFT process requires copper pillar processing at the wafer-level, it enables the die to be attached to good RDL structure with very fine pitch interconnect structures.

8.3. Metrology and Inspection in the FOWLP

In Fig. 8.6, key steps associated with the fan-out process along with challenges are identified. We discuss the metrology and inspection solutions available to address these challenges.

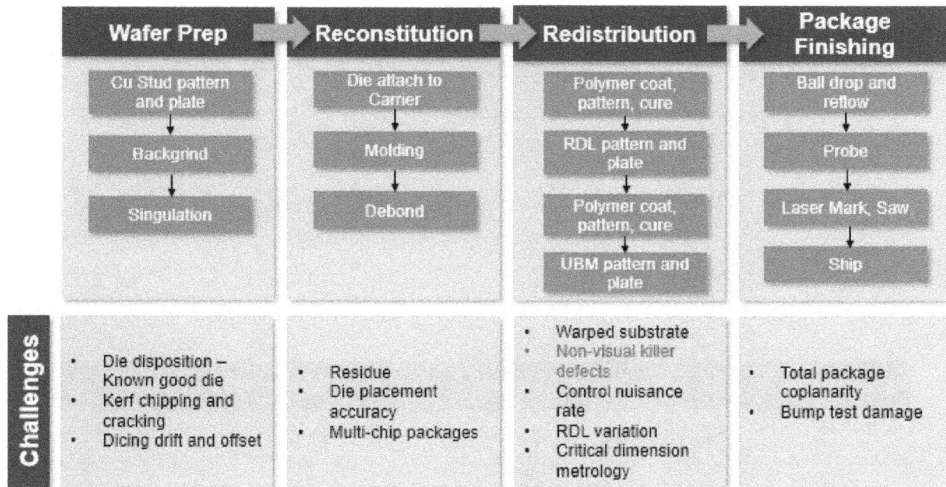

Fig. 8.6. FO-WLP process flow and challenges.

8.3.1. Singulation: Kerf Metrology

Rudolph Technologies' NSX® platform provides different approaches for detecting the defects generated during singulation: the first is an inspection approach that is focused on detecting defects after they have occurred and the second, a more sophisticated approach is to pre-empt occurrence of defects.

In the first approach, known as die seal ring inspection, algorithms have been developed around the die seal ring to ensure the die integrity. Typical process control parameters are identified in Fig. 8.8.

233

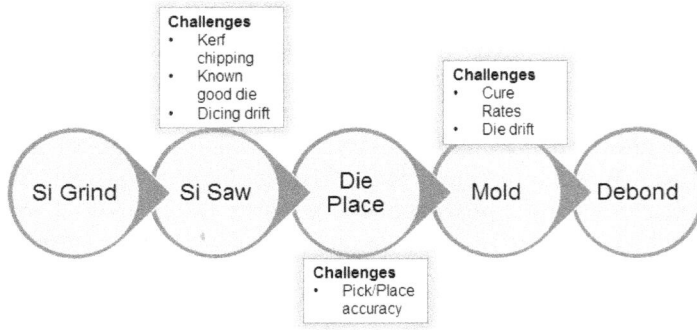

Fig. 8.7. Schematic representation of process challenges during wafer preparation.

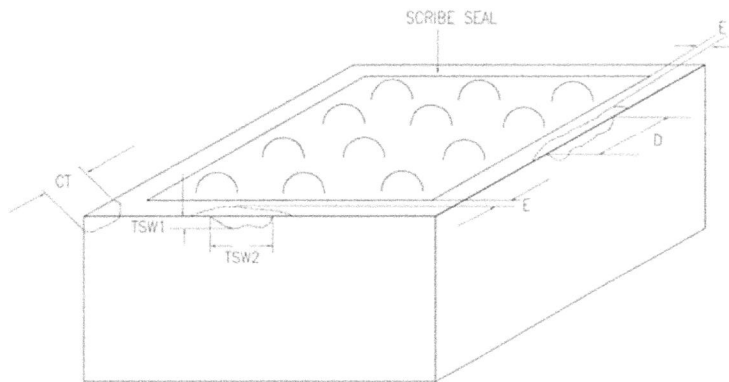

Symbol	Description	Customer criteria
D	Topside chipping (length)	≤ 50 um
E	Topside chipping (depth)	≥ 10 um away from scribe seal
CT	Topside corner chipping	≤ 20 um
TSW1	Topside sidewall chipping height (depth)	≤ 10 um
TSW2	Topside sidewall chipping length	≤ 50 um

Fig. 8.8. Typical process control parameters.

In Fig. 8.9 (a), we depict how regions of interest are set up per the inspection criteria. This approach uses a reticle-based reference model as opposed to a single die model. The advantage of this approach is that sensitivity is greatly enhanced and it minimizes false defect detection that is typically found due to street and reticle artifacts that are often falsely detected as chipping. In Fig. 8.9 (b), it is important to point out that with the die seal ring inspection enabled, one can see a dramatic decrease in nusiance defects, while detecting all of the critical defects such as seal ring contamination with more accurate binning. The algorithm also includes the capability to measure the defect size, defect position and frequency of defects. This allows one to perform advanced disposition and ensure an optimal balance between shipping good die and preventing the shipment of potential reliability fails. We have demonstrated the viability of the die seal ring inspection technique to help improve inspection yield, reduce review due to fall-out, reduce escapes and enhance capture rate of the critical saw defects.

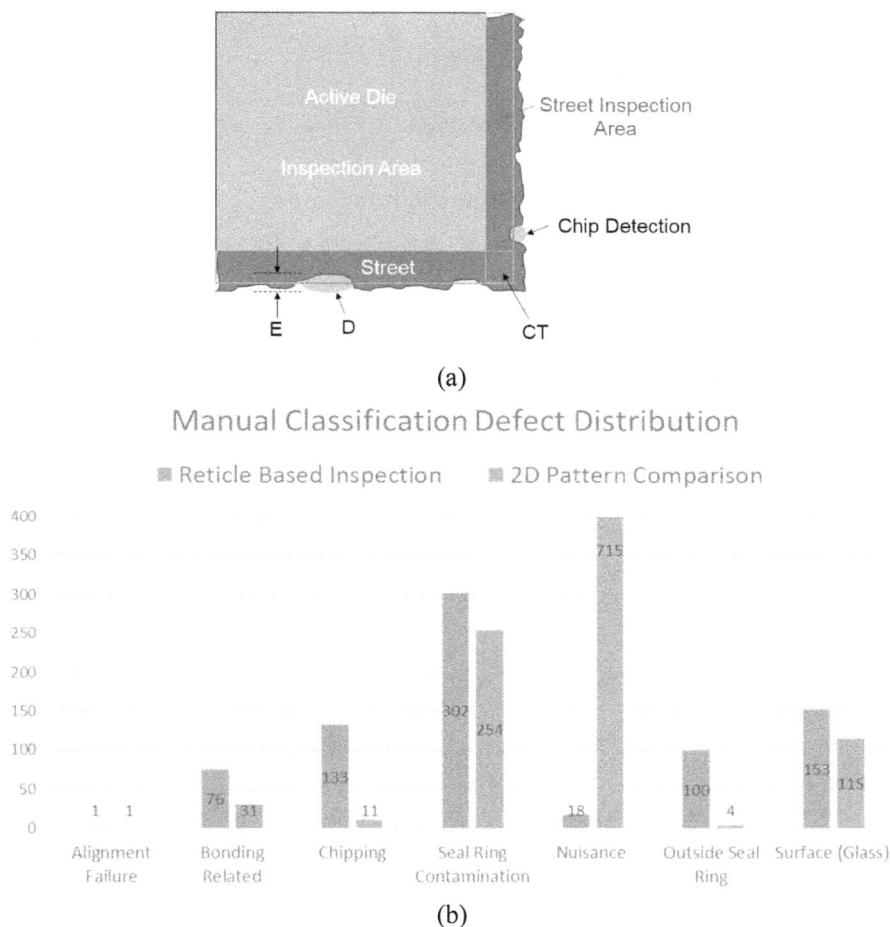

(a)

(b)

Fig. 8.9. (a) Typical inspection setup; (b) Nuisance reduction for reticle based layout.

In a second approach, we have developed and applied a sophisticated pre-emptive approach to delamintation and chipping control. Through a combination of kerf metrology and fault detection and classification (FDC) (Fig. 8.10), conditions of equipment that cause chipping and delamination may be detected and pre-empted resulting in significant savings in scrap.

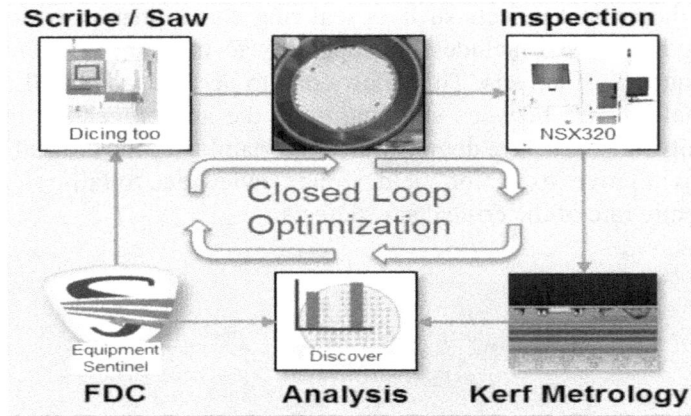

Fig. 8.10. Closed loop optimization solution for post-saw inspection.

Implementation of advanced process control (APC) techniques allowed for identification of problematic regions readily, improved recipe turnaround times, and adoption of flexible binning strategies to identify suspect dies. These translate not only to direct yield improvement but also help reduce maintenance costs by monitoring consumables and improve equipment up times. In the following example, we illustrate how kerf metrology works in tandem with the saw equipment parameters. The equipment parameters are monitored in real time and fed into Rudolph's Equipment Sentinel™ FDC software. The results of the FDC trace signals and the metrology can be overlaid to show the process changes based on saw recipe improvements (Fig. 8.11). This gives process engineers insight into which process signals they can change in order to reduce defects such as chipping [13].

8.3.2. Re-Constitution: Die Placement Accuracy

Die positioning control within the reconstituted wafer significantly affects downstream process requirements. The use of high productivity pick and place equipment with multiple gantries create challenges for the lithographic tool alignment when die placements from each gantry are not identical. The two operations that most affect die location error are the die placement operation and die migration during the compression molding process [14]. However the major source of error comes from shrinkage of the molding compound material during the compression molding process. This creates significant challenges in aligning subsequent metal layers to the device contacts. For back-end patterning, full wafer alignment tools (aligner) and step-and-repeat tools (stepper) are common optical patterning tools having different overlay performance characteristics. The full wafer aligner places the mask and wafer in close proximity and exposes the wafer in one shot, whereas the stepper exposes a smaller repeating pattern across the wafer. The aligner relies on the stability of the mask plate pattern to achieve repeatable results but does not have adjustments for dealing with common wafer distortions. Addressing or overcoming die offset and resulting overlay issues is one of the keys to making FO-WLP competitive with other package formats.

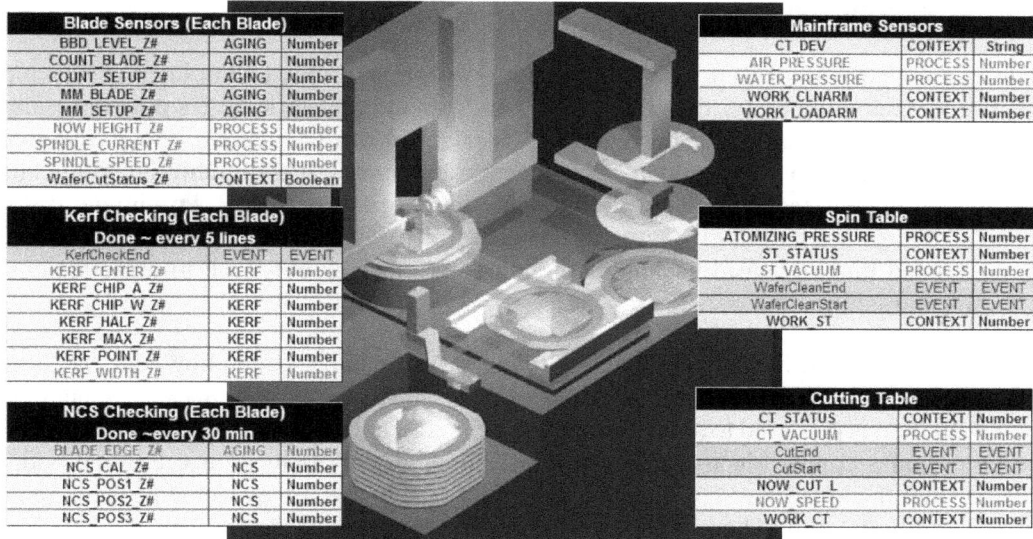

(a)

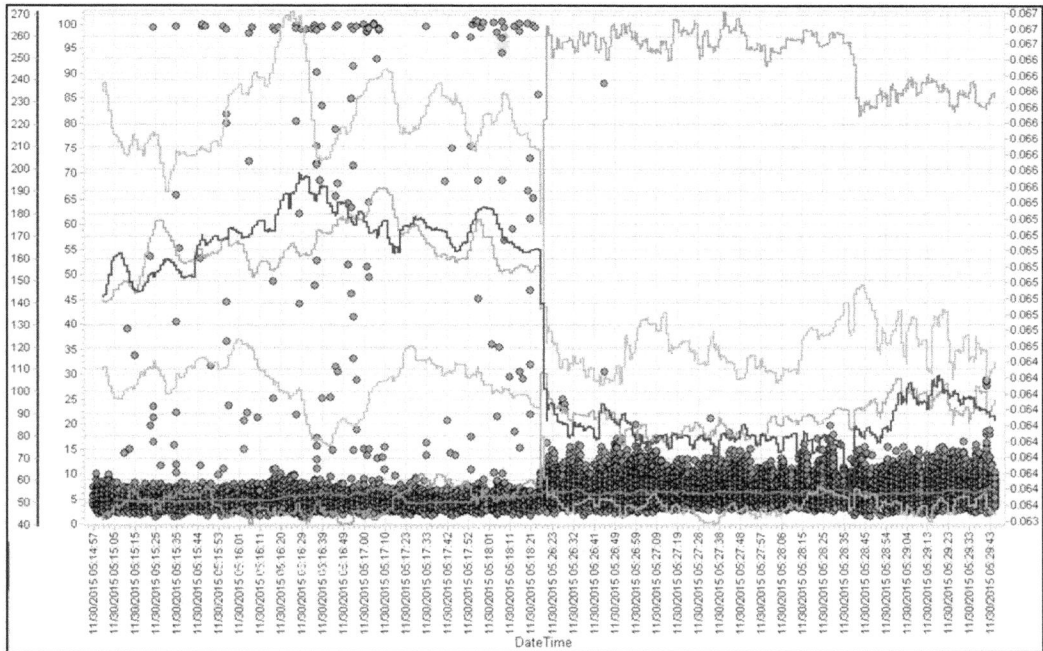

(b)

Fig. 8.11. (a) Multiple equipment sensors are monitored in real time;
(b) FDC-metrology correlation chart.

8.3.3. Redistribution Layer

The original purpose of RDL was to assist in the adaption of metal bumping and flip chip packaging technologies, by the addition of the metal and dielectric layers onto the wafer surface to re-route the legacy-designed irregular peripheral I/O layout, into a new area array bond pads layout to facilitate a balanced metal bumps and flip chip bonding. The redistribution layer technology required polymeric thin film (e.g. BCB, Polyimide, PBO) as insulator and a semi-additive metallization scheme (often Cu pattern plating). RDL use has extended into advanced packaging technologies, such as FO-WLP to drive the cost effective miniaturization of SiP.

There are three different methods for fabricating RDL depending on the line width/spacing of the conductor and the end application. In Fig. 8.12, a schematic representation is provided to show the process flow for an RDL, in a chip-up process using polymer and a copper electroplating process. For high end applications, the line width spacing and thickness are < 5 μm and 2 μm, respectively and are expected to scale down to < 2 μm and 1 μm. These RDLs are fabricated by the dual damascene method. For mid-performance applications, the Cu line width / spacing and thickness are 5-10 μm and 3 μm respectively and such RDL lines are fabricated by electrochemical deposition (ECD). For low performance applications, the Cu line width / spacing and thickness are > 10-2 μm and 5 μm respectively. They are fabricated by the resin-coated copper (RCC).

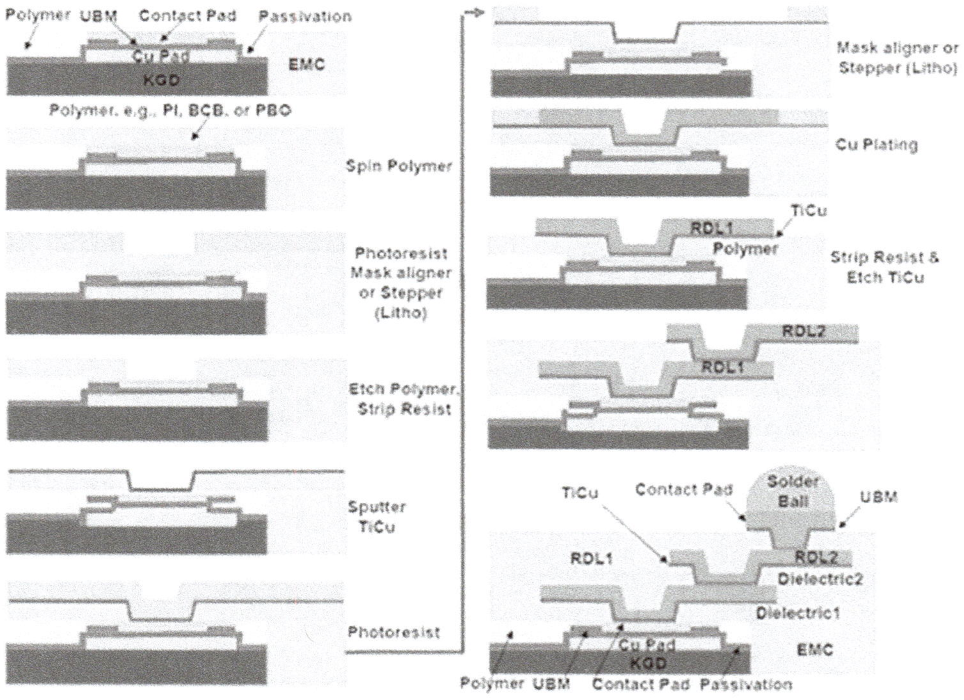

Fig. 8.12. Schematic of RDL process flow in a chip-up process.
Courtesy: John Lau, *ASM International*.

Table 8.1 summarizes the geometry material and process equipment needs for FO-WLP. First generation FO-WLP was geared towards mobile applications and RDL lines were typically 5/5 µm (line/space) and larger. Second generation growth is driven by the requirement and ability to integrate multiple chips on a single package with RDLs of tighter pitch 2/2 µm and in a smaller package [15, 16].

Table 8.1. Geometry, material, process and equipment for FOWLP.
Courtesy: John Lau, *ASM International.*

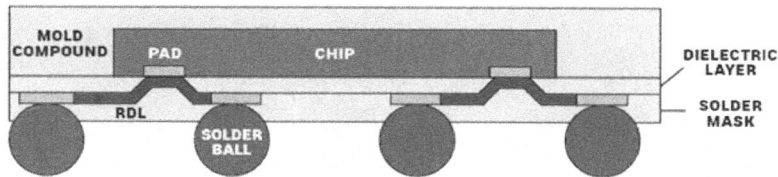

Temporary Carrier	Appl.	Line Width/ Spacing	Line Thick	Dielectric Mat. (Thick.)	Litho	Proc. / Equip
	High-end	≤ 2 - 5µm	1 - 2µm	SiO_2 (1µm)	Stepper	Cu Damascene Semi. Equip. High Precision P&P
	Middle-end	5 - 10µm	3µm	Polymers (4 - 8µm)	Mask aligner	Cu Plating Packaging Equip. Ordinary P&P
	Low-end	10 - 20µm	5µm	Resin (15 - 30µm)	Laser Direct Imaging	PCB Cu Plating PCB Equip. SMT P&P

Shown in Fig. 8.13 is the schematic representation of challenges encountered during the RDL process. Characterizing thickness of the resists, RDL and UBM are critical for process optimization and monitoring.

8.3.4. Characterization of Under Bump Metallization

UBM is typically a multi-layer metal stack that provides increased I/O between the device and the packaging substrate. UBM provides the critical interface between the metal pad of the integrated circuit (or the Cu or Al trace) and the solder (or gold) bump. Electroplating offers the advantage of depositing the UBM layer(s) only in the patterned area under the bump, simplifying the subsequent UBM etch processes. Different materials are used depending on the bumping process. To prevent direct reaction between solder or gold and the chip metallization, a diffusion barrier of Ti or TiW is deposited by sputtering on top of the IC metallization (Al or Cu). It also acts as an adhesion layer, for the wetting layer deposited on top of it in the case of solder bumping, and for the Au seed layer for electroplating of gold, in the case of gold bumping. TiW is used for gold bumping applications due to its superior barrier properties. In the case of solder bumping, Ti or TiW can be used, as the diffusion barrier is provided by the wetting layer (Ni or Cu) deposited on top of the adhesion/barrier layer (Ti or TiW) [17]. The solder wetting layer is a single layer or combination of Cu, Ni, (or NiV) and Au (Fig. 8.15).

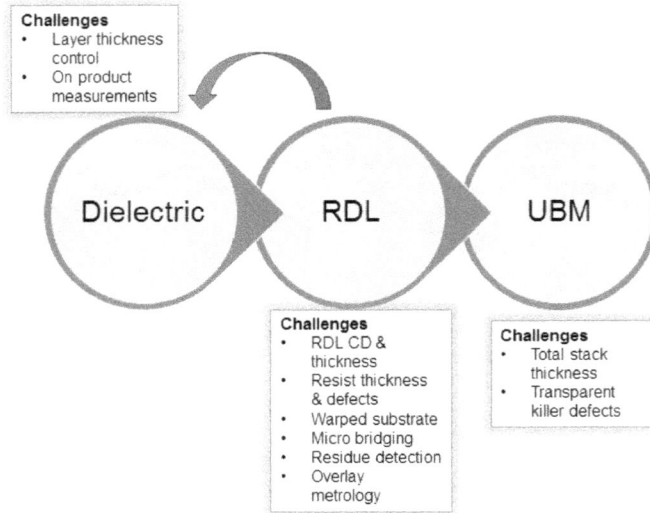

Fig. 8.13. Schematic representation of challenges encountered during RDL process.

Small spot X-ray fluorescence (XRF) techniques have found limited applications for UBM measurements. Measurements are challenging on smaller pads before the bump process and the technique is also prohibitively slow. However, during the bump process they are used for tracking the composition (Ag) in the solder. Thickness measurements of under bump metallization layers was performed using Picosecond Ultrasonic Laser (PULSE™) technique on different metal pads of varying sizes. PULSE™ technology is a proven workhorse for metal film metrology in leading-edge wafer fabs for front-end-of-the-line (FEOL) and back-end-of-the-line (BEOL) applications [18]. The non-contact, non-destructive, first principles technology allows measurement directly on-product wafers. In addition to thickness measurements, other parameters such as roughness, density, phase and modulus can be determined that provides users with additional information on their process. Thickness is calculated using the round trip transit time of the acoustics through the film using known speed of sound in the material. The technique provides accurate measurement of metal films, single or multi-layers ranging in thickness from 40 Å to 12 µm.

The optics schematic of this technology is shown in Fig. 8.14. The system uses pump-probe setup. A 0.1 ps laser PULSE™ (pump) is focused to about 5×7 µm^2 spot onto a wafer surface to create a sharp acoustic wave. The acoustic wave then travels away from the surface through the film at the speed of sound. At the interface with another material, a portion of the acoustic wave gets reflected and comes back to the surface while the rest is transmitted. When the reflected acoustic wave reaches the wafer surface, it is detected by another focused laser PULSE™ (probe) which was diverted from the pump PULSE™ by the beam splitter. There are two different methods of detecting the reflected acoustic wave at the surface. The first method is to detect the change of optical reflectivity that is caused by the strain of acoustic wave. The second method is to detect the deflection of reflected probe beam that is caused by the deformation of surface due to the acoustic wave. The second method requires a position sensitive detector (PSD), which is described in Fig. 8.14 (b).

The PULSE™ system was modified to accommodate a high resolution visible reflectometer to provide accurate transparent film thickness in addition to the RDL thickness on product wafers and a high resolution camera to provide critical dimension (CD) measurements in tandem with the thickness measurements thus providing users with complete information needed for process control.

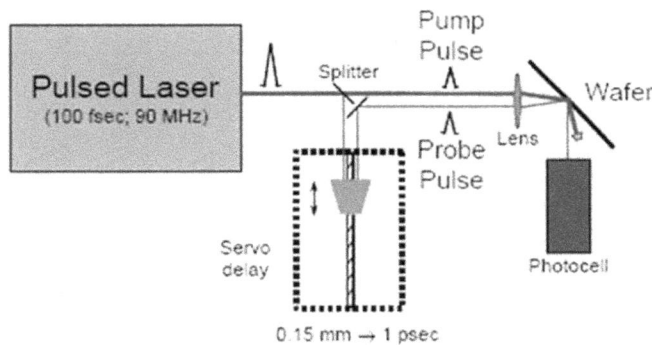

(a)

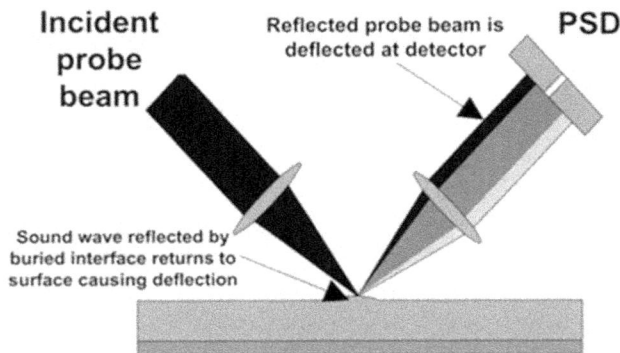

(b)

Fig. 8.14. (a) Schematic representation of picosecond ultrasonic technology;
(b) Position sensitive detection method.

241

(a)

Time

(b)

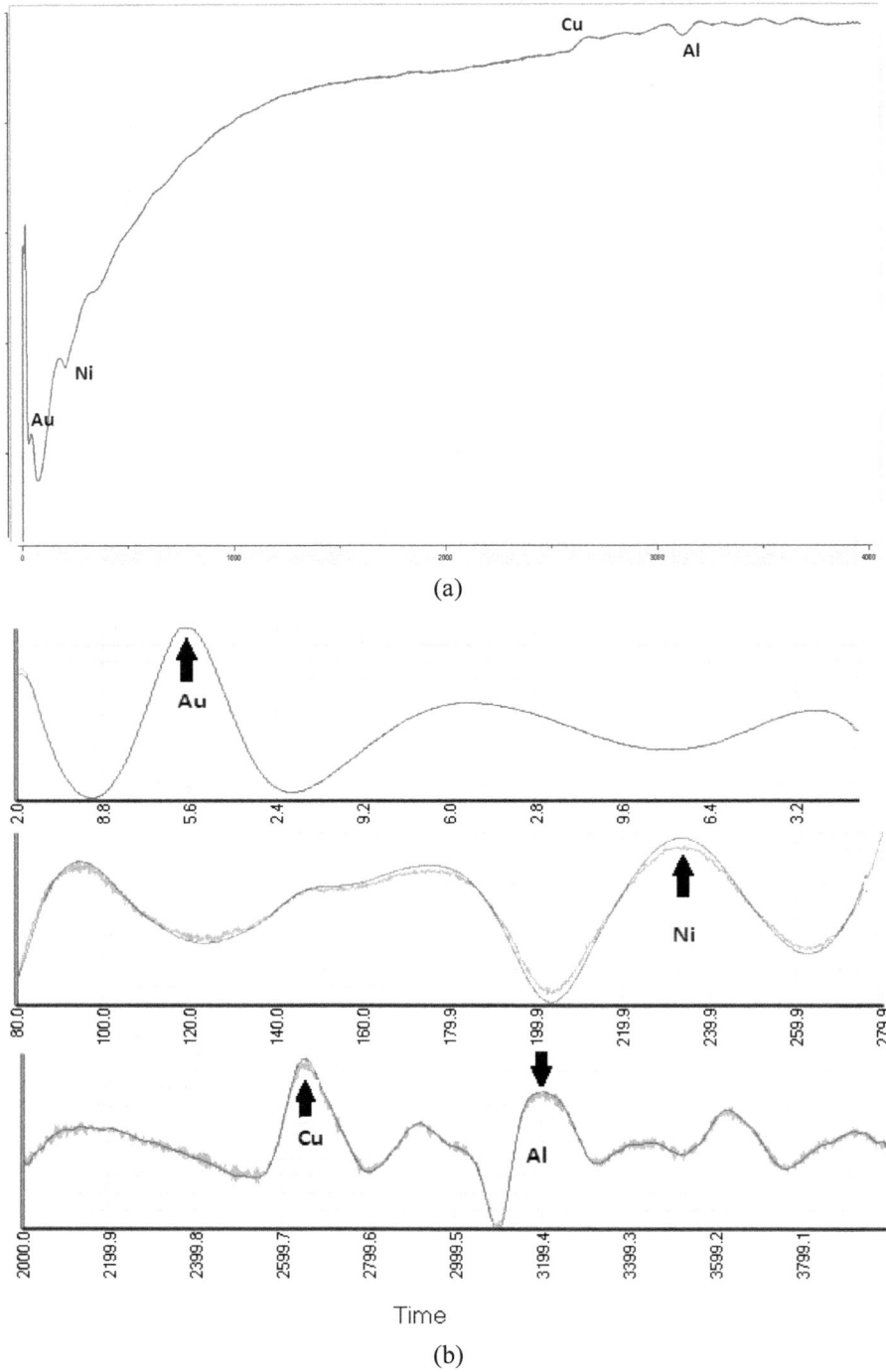

Fig. 8.15. (a) Raw data from the PULSE™ measurement with echoes labeled;
(b) Model fit to the measurements.

Multi-layer metal stack measurements of Au/Ni/Cu/Al from a 16μm bump pad is shown in Fig. 8.16. We can readily discern the echoes arriving from various interfaces as identified in (a) allowing simultaneous measurements of the multi-layers. In (b), modeled fit (in black) to measured data (green) is shown to illustrate how the data is fitted. Thickness of the layers is determined to be Au ~500 Å, Ni ~5800 Å, Cu ~6.3 μm and Al ~1.7 μm, respectively. Repeatability of each of the layers is 3 sigma < 0.5 %.

Copper seed layer uniformity is critical for good conductivity of the RDL layer. Shown in Fig. 8.16 is an example of measurement of bi-layer Cu seed/barrier. Measurements show excellent repeatability of the technique, typical 3 σ repeatability < 0.5 %.

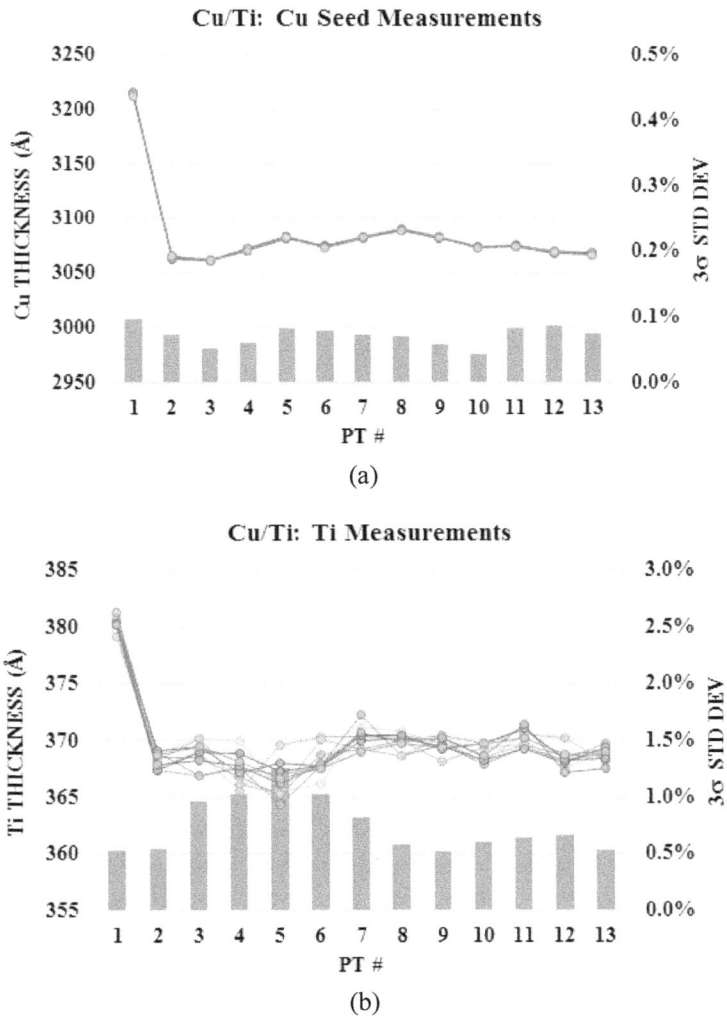

(a)

(b)

Fig. 8.16. Simultaneous measurements of Cu seed (a), and Ti barrier (b), respectively.

8.3.5. Characterization of Photoresists and Polymeric Materials

Photoresists and polymeric materials find applications as thin film dielectrics in WLP and as thick films to define structures. When the thin films are used as permanent dielectrics, they remain in the component and serve as insulators. Thick photoresists are required during electroplating to realize high density metal structures for applications such as copper pillars and bump plating and at deep dry etching in TSV when polymer resists are used as masks.

Critical inspection and metrology steps during the polymer/resist deposition and patterning are

1) Thickness and uniformity characterization;
2) Polyimide (PI) uniformity issues leading to nuisance defects;
3) Residue detection;
4) CD overlay metrology.

8.3.6. Thickness and Uniformity Characterization

Optical techniques such as reflectometers and interferometers are commonly used to characterize the dielectric films as well as the resists. Here, we describe the use of a patented fully automated focused beam ellipsometry (FBE) system to meet the growing characterization needs of the advanced packaging process. Measurements of dielectric layers of oxide/nitride stacks, thick photoresists, and polymers on glass substrates are now routinely made at post-deposition, post-etch, pre-and post-CMP process during RDL, UBM, and micro bump formation. Also, feed forward metrology, a concept that has gained widespread attention and acceptance in the front-end semiconductor manufacturing for measuring complex multi-layer, multi-parameter film stacks has been explored and we demonstrate the application of this concept for multi-layer dielectric stack measurements and its potential advantage during both R&D and in-line monitoring.

During spin-on deposition of the resists, the thickness uniformity across the wafers can vary significantly and it is important to perform full wafer map thickness characterization. Within wafer uniformity profiles up to 3 mm edge exclusion are shown in Fig. 8.17 (a) and (b) for 3 μm and 60 μm resist films, respectively. Typical non- uniformity for the resists < 10 μm are about ~6-7 % and are ~20-25 % for films that are > 50 μm. Also, shown in Fig. 8.17 (c) is the excellent linear correlation between the FBE and the reference tool across the process range. Measurements are very repeatable and the 3 σ standard deviation is better than 0.3 %. In addition to measurements of polymers on silicon substrates, FBE was used to characterize polymeric films on glass substrates. This was a critical fab process step and there was no robust technique available at that time for characterization. During the temporary bonding process, the polymer acts as a release layer to bond with the adhesive of the silicon substrate. If the process is not monitored, the polymeric film with poor uniformity will lead to issues during subsequent processing. In Fig. 8.18, high resolution line scans performed on polymeric films (2000 Å-1 μm) is shown. The thicker films show more non-uniformity at the edge than the thinner films.

(a)   (b)

(c)

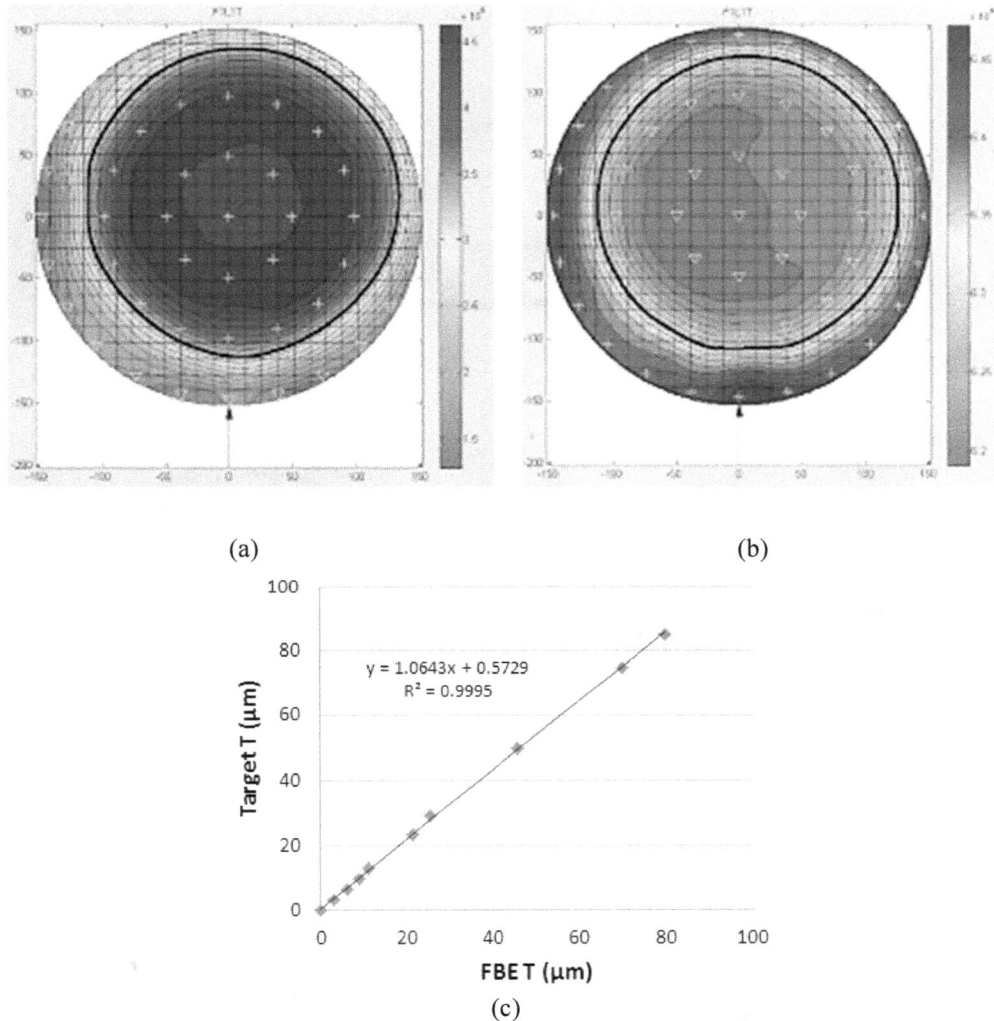

Fig. 8.17. 49 pt within wafer uniformity profiles (a) 45 μm, and (b) 6 μm resist wafer; (c) Correlation between reference metrology and FBE.

8.3.7. Detection of Residue and Non-Visual Killer Defects

Examples of non-visible defects range from voids in TSVs to faint organic photoresist or cleaning chemical residues and incomplete etch on the bump pad. Organic residue-based defects have been difficult to detect using conventional inspection techniques using bright field or dark field illumination. They escape detection with impact to yield and device performance. Residue can lead to higher contact resistance. While an open/short can be identified by electrical testing, a reduction in space between conductor lines can be tricky. It can cause a leakage issue or potential reliability issue with electromigration when used in the end application [5].

(a)

(b)   (c)

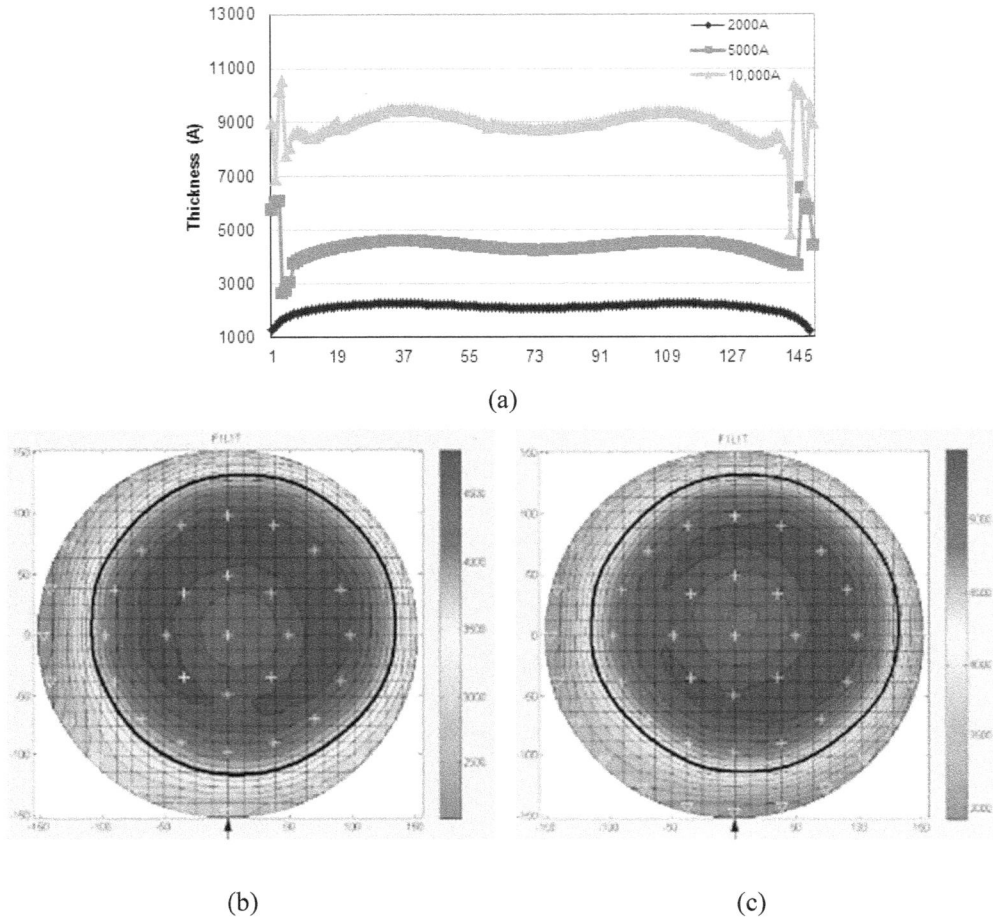

Fig. 8.18. (a) High resolution line scans across the polymer films (2000 Å-1 μm) on glass substrates (b) and (c) within wafer uniformity profiles of a 5000 Å film on glass and silicon, substrate, respectively.

Manual microscopes find limited use as they are slow and often rely on operator judgment resulting in high variability in the results. Results from introducing a patented automated high speed inspection technique (Clearfind™ Technology) is presented here. Briefly, the technique incorporates high sensitivity camera sensor technology, a dual inline focus system, and a patented illumination technology useful in distinguishing low contrast organic defects from the high contrast grain structure of underlying metals, or conversely, in finding shorts or opens in metal lines that overlay highlighted organic layers.

The technique has been successfully used in detecting defects at or below 1 μm. In Fig. 8.19 (a), using bright field illumination, no defects are detected. However, using Clearfind technology, one of the bond pads appears brighter while the rest of them are dark. Failing to identify such residues can result in subsequent failure during electrical test or an even costlier field failure.

(a)

(b)

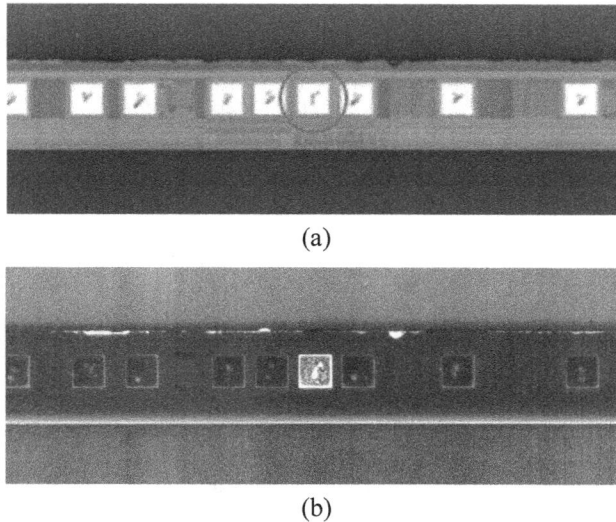

Fig. 8.19. (a) Bright field image shows no defects on bond pad; (b) Clearfind™ Technology shows presence of organic residue on bond pad.

Another area where the technology provides unique capability is in inspecting advanced RDL. In packaging applications with 2 μm RDL L/S, RDL defects of interest are half the size of the RDL width i.e. 1 μm for 2 μm RDL. However, in many cases, the acceptable metal graininess can be larger than the detection size. This leads to a high nuisance rate impacting the total throughput of the wafer which includes manual review. Using the patented Clearfind™ technology, we have been able to identify defects of interest such as finding 1 μm open or short in the RDL lines. Nuisance defects due to graininess of the metals that would have normally obscured the detection are not an issue anymore. Fig. 8.20 (a) and (b) show the bright field and Clearfind™ images respectively, for a dense RDL application. In the white light image, it is difficult to detect 1 μm defects as the metal graininess is large [19].

8.3.8. Characterization of RDL

Depending on the process step, development state and the level of accuracy needed, there are multiple options available for RDL thickness and CD measurements. Scanning white light interferometer (SWLI) systems rely on interference patterns for providing thickness information and are sufficient for monitoring RDL thickness > 5 μm. They are all equipped with a camera and microscope that additionally provide CD measurements, although not simultaneously in most cases. However, as the RDL films get thinner, interference signals are complicated resulting in poor accuracy. Manual optical CD tools are tedious to use. Wafer fabs also rely on CD-scanning electron microscope (SEM) and cross-sectional SEM for more accurate CD measurements but it is not viable for high volume measurements. As pointed out earlier, the FO-WLP wafers exhibit significant warpage and the CD-SEM tools to date do not meet the automation requirements for measurements on product wafers. Cross-section SEM is destructive.

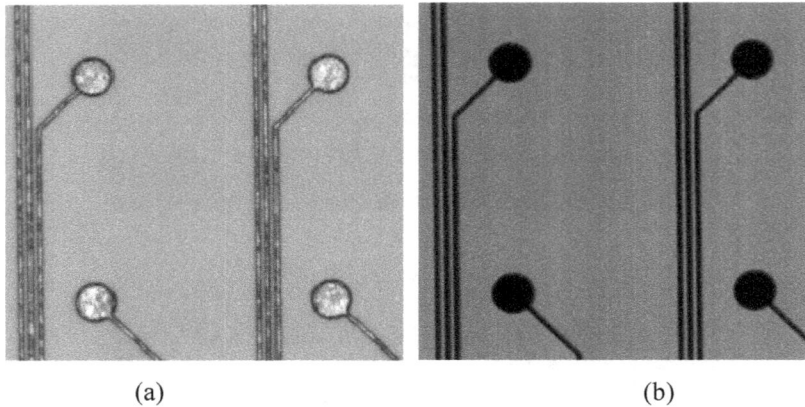

(a)   (b)

Fig. 8.20. (a) Bright field inspection of the RDL lines reveal the metal graininess; (b) Clearfind™ technology not impacted by metal graininess and not susceptible to nuisance defects.

At Rudolph, we have integrated reflectometer and visible light interferometer systems to our inspection platform. This allows us to directly measure the transparent film thickness while the interferometry captures topography (distance from the sensor), thus enabling the system to measure the thickness of the opaque metals by scanning over the edge of the feature. This technique is called the visible thickness and shape sensor (VT-SS).

In Fig. 8.21, we show how the step height is calculated using the VT-SS. The VT-SS allows capturing both the transparent polyimide thickness and opaque Cu feature height with a single scan from polyimide layer to Cu feature (*up*). From the part of the scan covering the polyimide, signals representing the direct measure of the polyimide thickness, the distance to the first surface of the polyimide, and the distance to a metal surface under the passivation stack are measured. The direct measure of the polyimide thickness is a standard reflectometer measurement. In the part of the scan where the sensor spot illuminates the Cu step height, the direct thickness peak and one of the distance peaks disappear. Only a distance peak to the surface of the Cu feature is present since the copper is opaque. The Cu step height above the first polyimide layer is then determined from the appropriate distance measures from each part of the scan. Thus, all the desired thickness and Cu thickness measurements are reported. The system is capable of making CD measurements and meet the P/T < 10 % needed for process control.

At the 2 μm and below RDL line widths, accuracy of the thickness measurements becomes even more critical for process monitoring and control. CD measurements of the RDL lines and via CD are also important. Using the PULSE™ technology, RDL thickness capability (2 μm-10 μm) were evaluated on various structures- pads, isolated and dense line arrays at various levels. Fig. 8.22 shows modeled fit to measured data from a 2 μm pad (a) and 4 μm pad (b).

The raw data is characterized by a high signal-to-noise ratio that contributes to the excellent repeatability. The arrival times of the echo are obtained from the measurement, and with the known speed of sound in the material, we can readily calculate the thickness of the films. The acoustic echo, in Fig. 8.22 occurs at ~800 ps corresponding to an RDL thickness of 2 µm. In (b), the arrival time of the echo is ~1600 ps and the thickness is determined to be ~4 µm. We have observed that the thicker RDL films are also characterized by higher surface roughness and local non-uniformity.

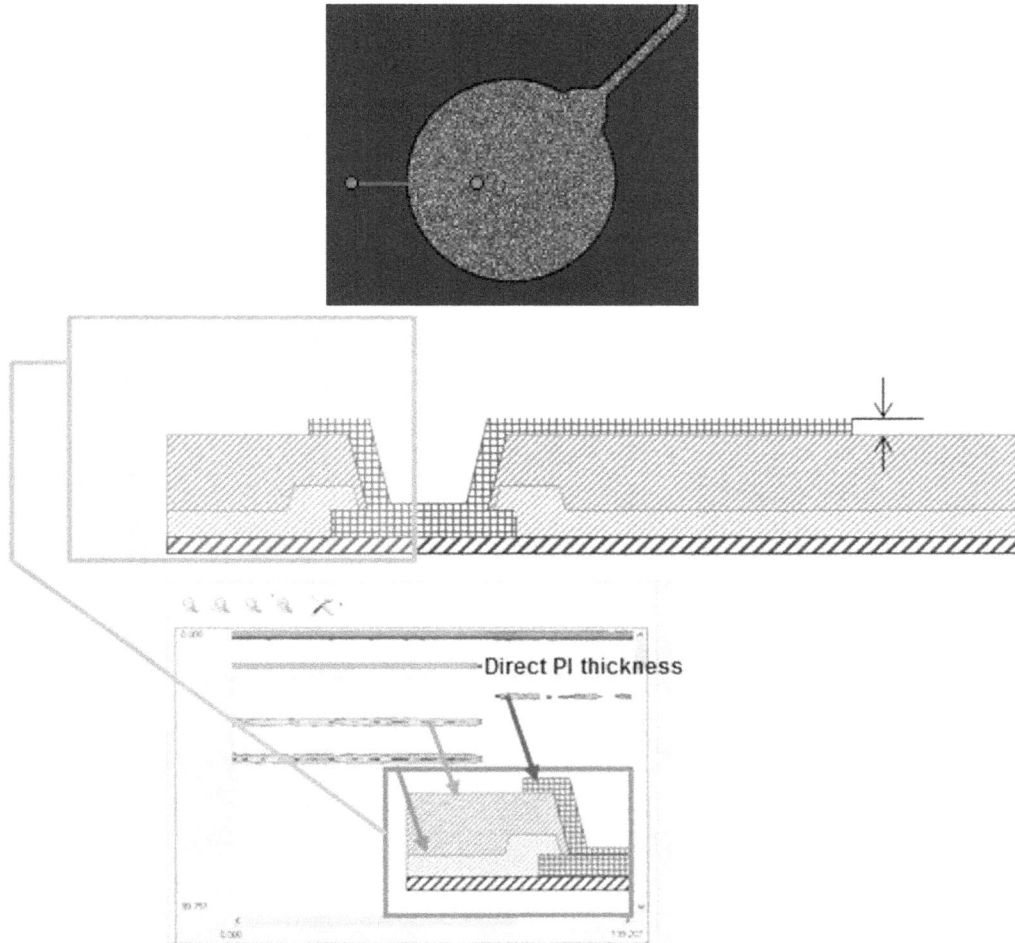

Fig. 8.21. RDL step height measurement using reflectometer and interferometer combination on the VT-SS.

Shown in Fig 8.23 (a) and (b) are isolated and dense RDL lines for CD measurements. As pointed out earlier, the ultra-thin IPD enabled shrinking of the InFO package size. The cost of IPD by itself is insignificant but the cost penalty of failing IPD in an InFO package is significant [20]. Hence, understanding the RDL line/spacing simultaneously with thickness in a fine-pitch process provides information that was not previously available. Early test results from a 2 µm RDL process show that P/T requirements of < 0.1 µm are met.

MetaPULSE (522.0nm)

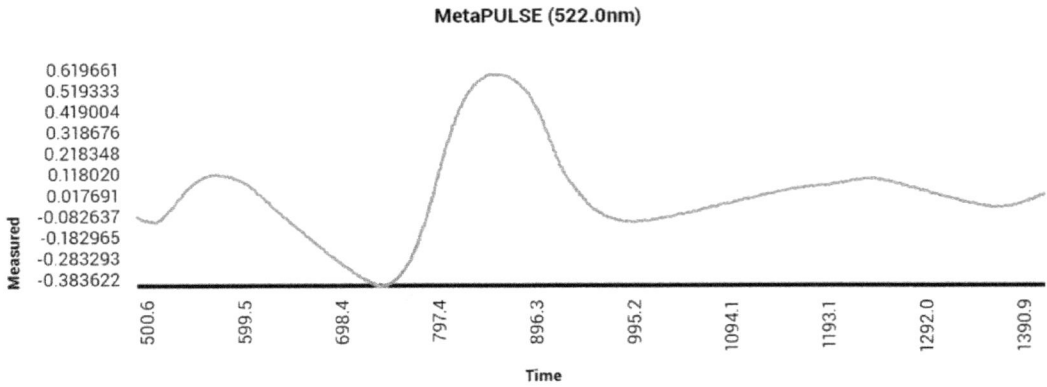

0.619661
0.519333
0.419004
0.318676
0.218348
0.118020
0.017691
-0.082637
-0.182965
-0.283293
-0.383622

Measured

500.6 599.5 698.4 797.4 896.3 995.2 1094.1 1193.1 1292.0 1390.9

Time

(a)

MetaPULSE (522.0nm)

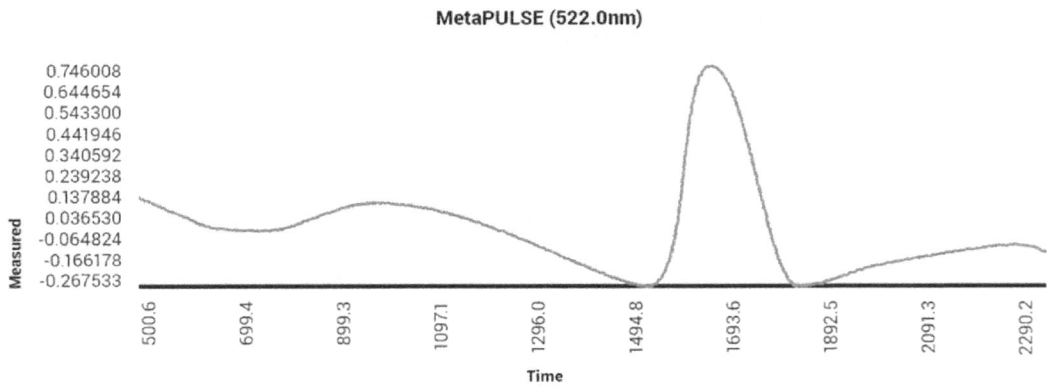

0.746008
0.644654
0.543300
0.441946
0.340592
0.239238
0.137884
0.036530
-0.064824
-0.166178
-0.267533

Measured

500.6 699.4 899.3 1097.1 1296.0 1494.8 1693.6 1892.5 2091.3 2290.2

Time

(b)

RDL Measurement

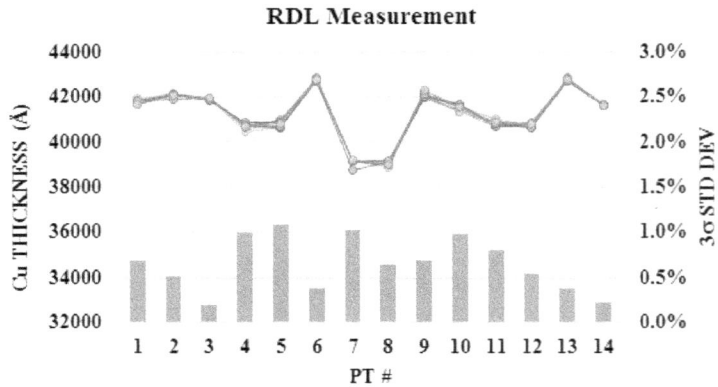

(c)

Fig. 8.22. (a) Modeled fit to measured data from a 2 μm RDL film. Echo arrives at ~800 ps corresponding to a 2 μm film; (b) Echo arrival time ~1600 ps corresponding to a 4 μm RDL film; (c) Cross-wafer thickness variation and repeatability performance.

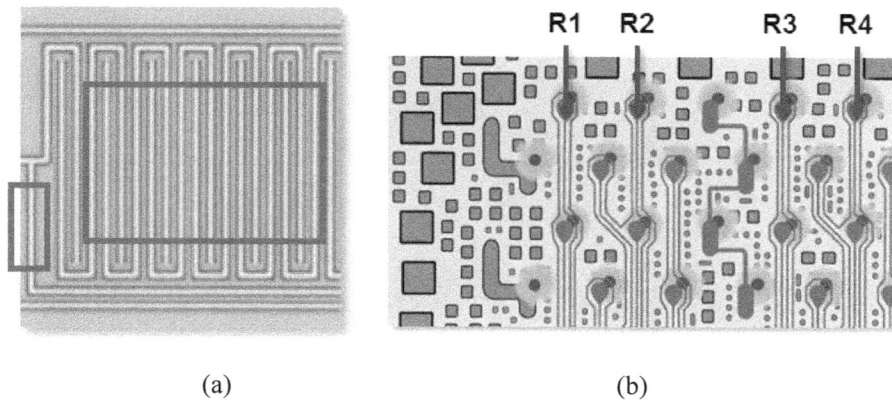

(a)   (b)

Fig. 8.23. (a) Isolated and dense RDL regions identified for measurement;
(b) RDL measurements in several regions of interest.

Any inaccuracy in wafer placement manifests itself as observable site level variations in repeatability performance. The repeatability demonstrated here has been confirmed to be more than adequate for process monitoring. Site level performance can be further improved, if desired, by averaging over the local non-uniformity with some trade-off to throughput.

8.3.9. CD Overlay Metrology

Adaptive patterning has been created to account for the inherent die shifts in fan out processing. Alternate methods such as adaptive alignment have been developed in chips face-up panelization process where the whole RDL pattern and via 1 layers are shifted to account for the actual die position and rotation [21].

During the PI via exposure process, ensuring that the PI via is correctly exposed and via is well centered is important for connectivity. Alignment process window tolerances are extremely tight for both via diameter as well as the pad. This has a direct impact on the yield. CD diameter is determined using tools available on the system software (Fig. 8.24) and using data analytic options available on the NSX® platform. Information such as via position with respect to the pad position along with any relevant offsets is provided. For a package with a single fan-out routing layer, the overlay problem is moved to the next via layer underlying the UBM. It is not desirable to allow the position of the UBM pattern to shift with respect to the edge of the package – it must be held constant. Therefore, if the entire RDL pattern shifts with respect to the fixed UBM pattern, the shift must be accommodated either by increasing the size of the underlying capture pad on the RDL layer or by reducing the diameter of the via connecting the UBM to the RDL layer. Typical 3 σ repeatability for CD and offset measurements < 0.1 μm.

251

(a)

(b)

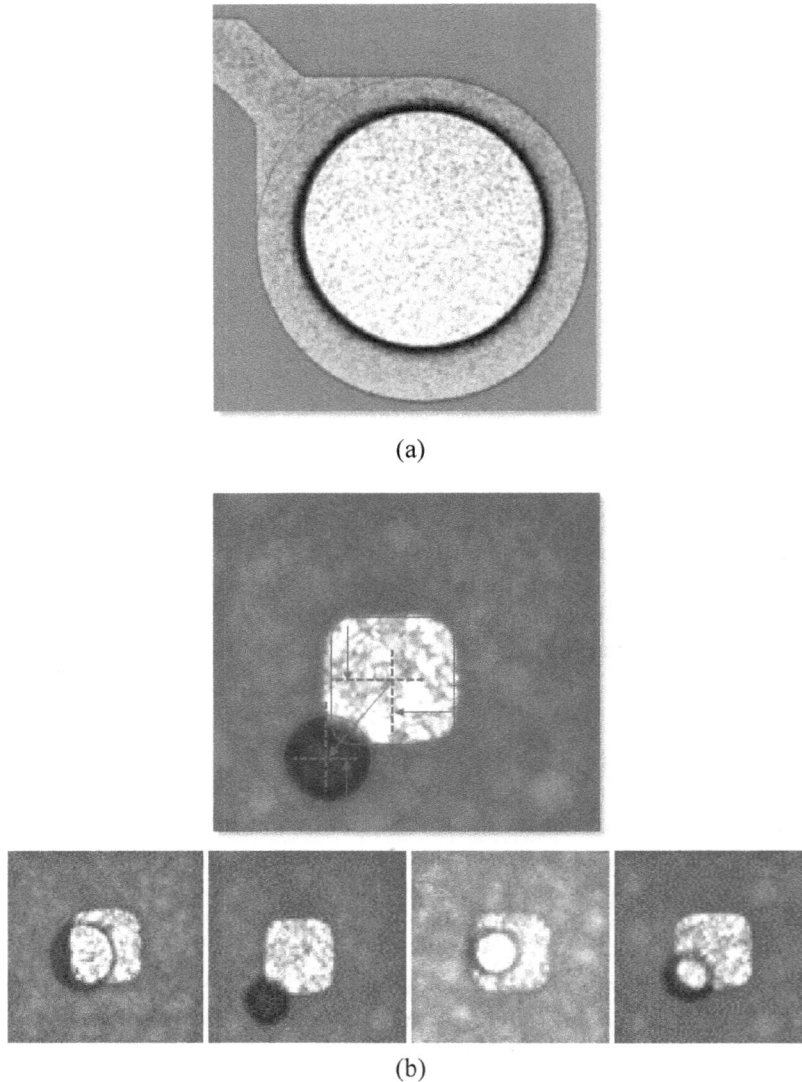

Fig. 8.24. (a) Via well centered on pad; (b) example of via not well centered on pad. Offsets (x, y, distance) can be determined and provided for metrology fine tuning.

8.3.10. Characterization of Wafer Bow

Interlayer dielectric film stress measurement results help OSATs monitor the mechanical integrity of the process during development and to ensure yield at volume manufacturing. The ability to perform fully automated measurements and characterize within wafer profiles via line scans and polar maps is desirable. Briefly, the focusing system of our ellipsometer is used in the determination of radius of curvature. Wafers are measured before and after film deposition to obtain the true radius of curvature and to minimize any artifacts that could interfere with the surface profile measurement. Using the radius of curvature information thus obtained along with the measured thickness of the film and the substrate, we calculate wafer stress using Stoney's equation [22].

Oxide and nitride film thickness and bow were determined by performing line scans in both x and y directions (Fig. 8.25) that were subsequently used in the stress calculation.

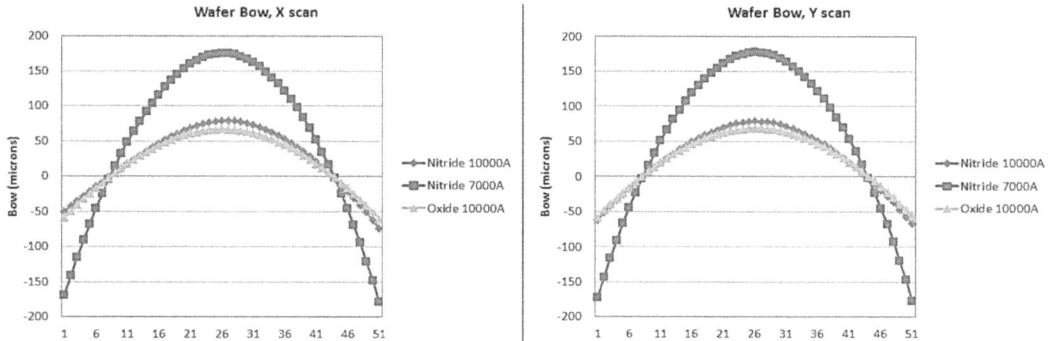

Fig. 8.25. X scan and Y scan profiles showing the wafer bow on nitride and oxide single layer blanket wafers.

8.4. Bump Metrology

Shown in Fig. 8.26 is a schematic representation of the bump process. In this section, we will focus on the challenges in bump metrology and inspection.

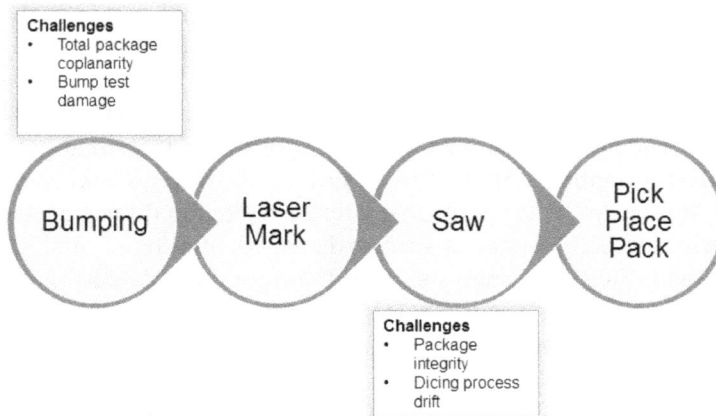

Fig. 8.26. Schematic representation of the bump process.

Bump interconnect technology plays a critical role in 3D integration of semiconductor devices and is a key enabler of cost effective middle-end supply chain processes. Copper

253

pillar bumps are not a packaging type. Instead, they serve as interconnect between two dies or a die to a board (Fig. 8.27). Copper pillars enable more I/Os, smaller pitches and better thermal conductivity, compared to traditional solutions. In addition, they offer excellent electromigration performance for high current carrying applications resulting in superior reliability of the end products. Fine pitch copper pillar bumps have replaced conventional solder bumps in high end processors, graphics, FPGAs, power amplifiers, handheld consumer electronic devices as well as in servers and network applications [23].

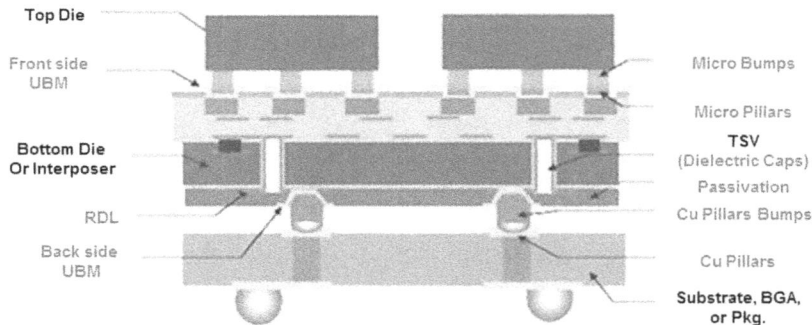

Fig. 8.27. Schematic representation of copper pillars and microbumps used in advanced packaging process.

An integral part of the transition to 3D packaging is microbump scaling. The aggressive scaling in microbump dimensions (height and CD) is entering the 10 μm territory [24] enabling higher interconnect density which is a defining feature of 3D stacking. In order for this technology to be viable for stacking, microbumps at both die-level and wafer-level must meet certain key criteria. From a measurement perspective, the individual height and CD as well as the die-level coplanarity have to be measured with a high accuracy and precision such that the desired P/T ≤ 10 % can be achieved. During process development, as many bumps as possible need to be characterized while for process monitoring, a meaningful subset of bumps per die and dies per wafer can be qualified [25]. The variation can be significant enough to have serious implications for succeeding process steps like wafer-level underfill applications, die-to-die and die-to-wafer stacking. Inspection and metrology data are needed to properly characterize the bump dimensions, coplanarity of each die, and detect defects of interest such as damaged, missing or mislocated bumps. A comprehensive and thorough analysis would ensure that defects are flagged. The consequence of not flagging these issues will impact yield during stacking, namely open and short circuits, die cracking and thermal sinks.

Copper pillar bumps are manufactured with the same materials and processing techniques as conventional solder bumps. The bump base is defined by lithographic patterning of a layer of photoresist, and the bump metal stacks are electroplated in the open features on top of an UBM. The resist is subsequently stripped and any UBM not protected by the bump is etched away. The wafer then goes through reflow and cleaning processes. Typical bump metal stacks are bi-layer Cu/SnAg and tri-layer Cu/Ni/SnAg. Laser-based acoustic system for determining the thickness of individual layers in copper pillar stacks have shown some promising results but additional work and development is needed to have a viable high volume solution [23].

Careful monitoring and control of the copper pillar's critical features and physical properties are essential to ensuring a high quality, reliable bump structure. Metrology measurements of height, coplanarity, position, diameter and volume are critical for monitoring the integrity of the pillars. In addition, inspection capabilities must include the ability to detect bump defects, including missing bumps, bridged bumps, and misshapen bumps. Contact-based profilometers are commonly used in wafer bumping for measurement of metal feature due to their ease of use and their low cost of ownership. However, the method of measurement is largely semi-automatic, and hence only a sampling approach can be used and also the identification of exact features and measurement locations becomes challenging. An automated system that performs 100 % die inspection and having an ability to adapt to different features and product types in a high volume manufacturing system becomes essential [26]. There are currently devices with as many as 50,000 bumps at a die level and over 20 million bumps at the wafer-level on a 300 mm wafer. With the introduction of micro bumps, it is predicted that there may be as many as 50 million bumps on a single 300 mm wafer in the near future. The task of analyzing metrology data (bump height, coplanarity, position, and diameter) in real time on each of the bumps (x wafer, x wafer lots) becomes monumental. Using our next generation high throughput bump inspection and metrology system, combined with Discover® data analytics software we have demonstrated the readiness of the technology for advanced process monitoring and control [26].

Shown in Fig. 8.27 is a laser triangulation system for 3D measurements of bump height and coplanarity integrated to the NSX® inspection platform. A laser, incident on the wafer surface at an angle of 45 degrees and focused to a spot size of 5 μm, scans a line over 1 mm in length at a rate of 8 MHz on the wafer surface while the wafer is transported in a direction perpendicular to the scanned line. A lens collects the reflected/scattered laser light and focuses it on a position sensitive detector. Changes in the location of the collected light on the detector provide height dimensions. A range of resolution settings permits the user to select the optimal balance of throughput and resolution.

In Fig. 8.29, bump height measurements from a whole wafer is presented. As can be seen from the figure, the bumps are shorter in the center and gradually increases and are taller at the edges. Using Discover® software, additional analysis of the scan is performed in both X and Y. Fig. 8.29 (b) and (c) show the average bump height per die and average bump area per die. One can readily observe the trend that as the bump height increases, the bump area decreases.

Additional correlation analysis of this data was completed and revealed a unique correlation between the height and bump area. In this particular study, ~20,000 bumps were sampled. Typical bump height repeatability for the 9 μm bumps was 3 sigma ~0.27 μm. Co-planarity of the bumps was found to be 3 sigma ~0.26 μm.

(a)

(b)

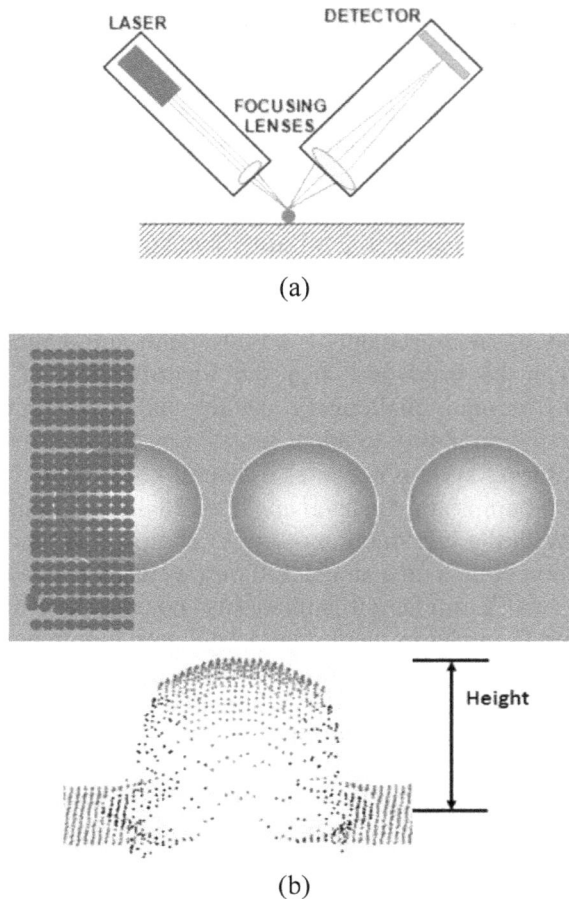

Fig. 8.28. (a) Laser triangulation uses a finely focused laser and a position sensitive detector to measure bump height; (b) Height determined from 3D image generated.

8.5. Data Analytics

We would be remiss if we didn't highlight the significance of data analytics that ties together the data generated from metrology and inspection steps. Such process control software at the tool level (tool-centric) and at fab-level are important, especially in a cost-sensitive market. They help to improve yield and lower the manufacturing costs. Throughout this Chapter, we have identified several areas in the FO-WLP process where yield improvements and better process control can be realized via pre-emptive analysis, feed forward controls, and analyzing correlations between large volumes of data (> 100 GB per day). This daunting task can be simplified by using advanced analytics in Discover® software. There is no better example to demonstrate this than bump measurements. In Fig. 8.30, we show how data can be used to analyze process variations at wafer level, individual bump level, die level and wafer lot level. Binning strategies can be set up to flag different types of failure.

(a)

(b)

(c)

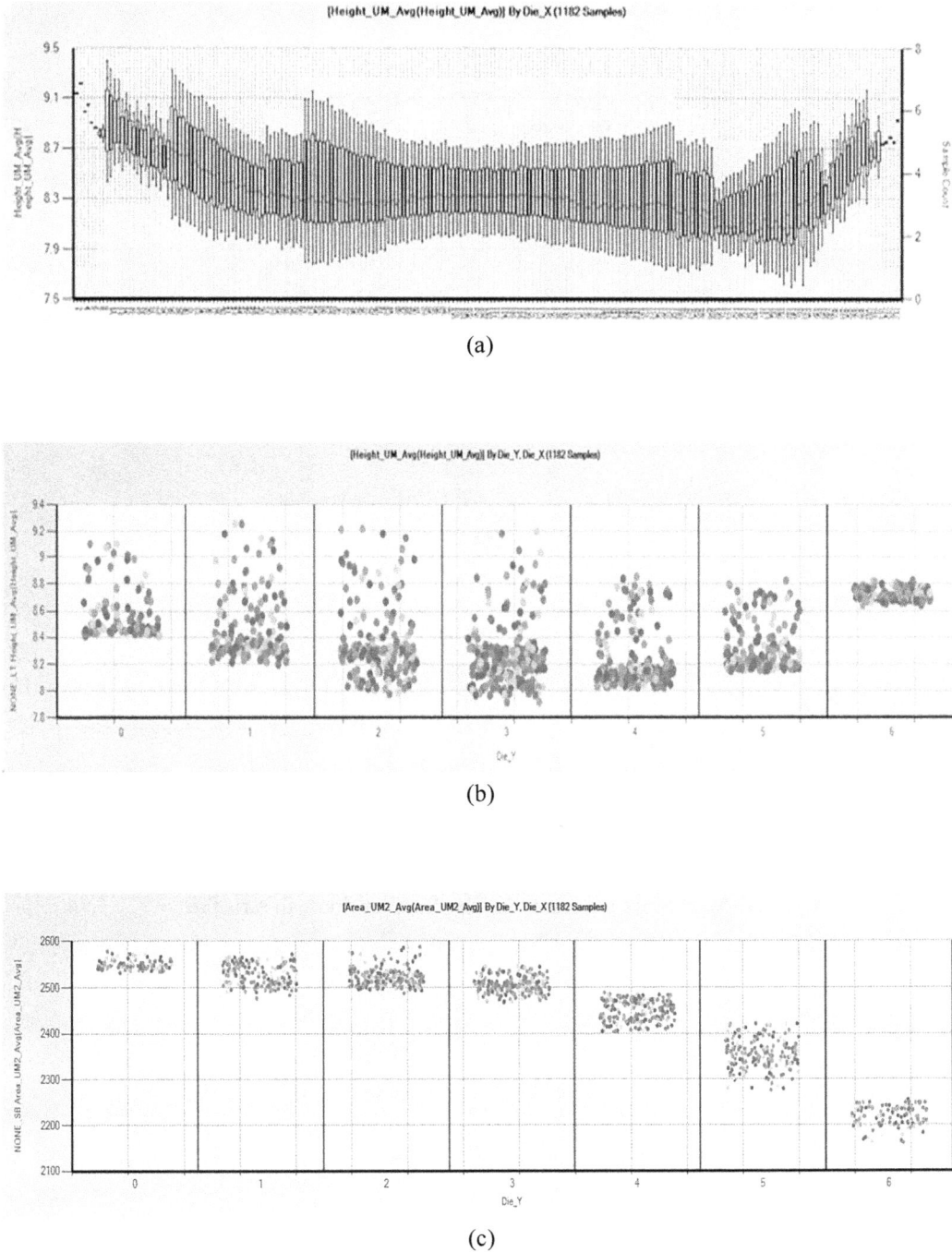

Fig. 8.29. (a) 3D bump height summary across the whole wafer; (b) average bump height per die from Y-scan, and (c) average bump area per die from Y scan. Note that as the bump height increases in (b), the area decreases (c).

Parameters of interest

Wafer-level Analysis

Bump diameter changes from left to right of the wafer

Bump position changes from left to right of the wafer

Process Window Analysis

Lot-Level Analysis; Good process (green), Bad process (Red)

Die-level maps generated to turn bump attribute to pass/fail

Understanding Yield Loss

Bin die by failure type

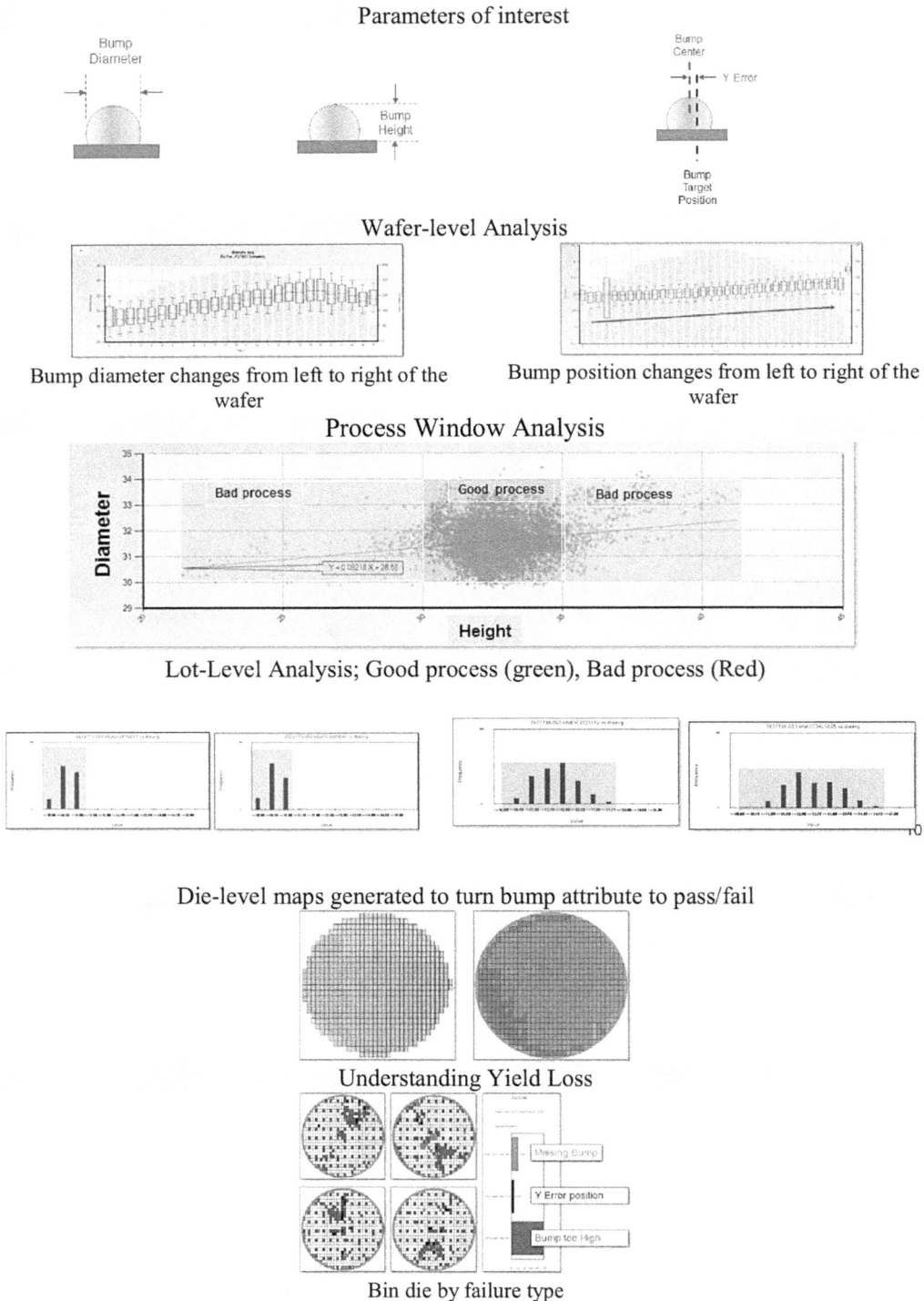

Fig. 8.30. Data analytics on Discover® software provides several options for analyzing large amount of data.

8.6. Conclusions

Fan-out wafer level packaging platform offers tremendous opportunities for advancing growth in mobile and networking applications. As more complex chips are integrated in fan-out, the processing is more complicated and in order to increase and maintain yield, the production process should be continually monitored. Metrology and inspection solutions are critical not only for meeting current challenges but should be extendible to address next generation needs. Automated metrology and inspection solutions discussed meet the repeatability, reproducibility and throughput requirements of the fan-out process. When combined with the advanced data analytics software. They provide users with feed forward information and pre-emptive solutions for process control thus influencing process yields and helping lower costs.

Acknowledgments

The authors would like to thank the teams in New Jersey, Minnesota, Wilmington and Taiwan for the extensive work that went into data collection and analysis.

References

[1]. T. Buisson, S. Kumar, What is driving advanced packaging platforms development ?, *Chip Scale Review*, May-June 2016, pp. 32-36.

[2]. R. Tummala, V. Sundaram, P. Raj, V. Smet, T. Shi, Future of embedding and fan-out technologies, *Chip Scale Review*, March-April 2017, pp. 20-28.

[3]. J. Lau, N. Fan, L. Ming, Design, material, process, equipment of embedded fan-out wafer/panel-level packaging, *Chip Scale Review*, May-June 2016, pp. 38-44.

[4]. J. Lau, Recent advances and trends in advanced packaging, *Chip Scale Review*, May-June 2017, pp. 46-54.

[5]. E. Sperling, 2.5D, Fan-out inspection grows, *Semiconductor Engineering*, May 2017.

[6]. J. Lau, Patent issues of fan-out wafer/panel-level packaging, *Chip Scale Review*, Nov.-Dec. 2015, pp. 42-46.

[7]. Status of Advanced Packaging Report, *Yole Dev.*, June 2017.

[8]. C. Tseng, C. Liu, C. Wu, D. Yu, InFO (wafer-level integrated fan out) technology, in *Proceeding of the IEEE Electronic Components and Technology Conference (IEEE/ECTC'16)*, Las Vegas, May 2016, pp. 1-6.

[9]. Mentor Graphics Corporation, https://blogs.mentor.com/calibre/blog/2016/07/12/not-yet-a-fan-of-fan-out-why-you-should-be/

[10]. J. Lin, J. Hung, N. Liu, Y. Mao, W. Shih, T. Tung, Packaged semiconductor device with a molding compound and a method of forming the same, *US Patent 9*, 000, 584, December 2011.

[11]. Fan-out: Technologies & Market Trends 2016, *Yole Dev.*, 2016.

[12]. R. Huemoeller, C. Zwenger, Silicon wafer integrated fan-out technology, *Chip Scale Review*, March 2015, pp. 34-37.

[13]. W. Fitzgerald, R. Roy, Advanced low-k die singulation defect inspection and pre-emptive singulation defect detection, in *Proceeding of the 37th International Electronic Manufacturing Technology Conference (IEMT'16)*, Penang, 2016.

[14]. W. Flack, R. Hsieh, G. Kenyon, K. Nguyen, M. Ranjan, N. Silvo, P. Cardoso, E. O'Toole, R. Leuschner, W. Robl, T. Meyer, Lithography technique to reduce the alignment errors from die placement in FO-WLP applications, in *Proceeding of the Electronic Components and Technology Conference (ECTC'11)*, Lake Buena Vista, 2011.

[15]. Y. Jin, X. Baraton, S. Yoon, Y. Lin, P. Marimuthu, V. Ganesh, Next generation eWLB packaging, in *Proceedings of the 12th Electronics Packaging Technology Conference (EPTC'10)*, Singapore, 2010, pp. 1388-1393.

[16]. P. Garrou, Solid State Technology, http://semimd.com/insights-from-leading-edge/page/5/.

[17]. K. O'Donnell, Solid State Technology, http://electroiq.com/blog/2008/04/ubm-creating-the-critical-interface/

[18]. P. Huang, B. Chiu, J. Chao, C. Hung Lu, S. Chen, J. Chen, F. Shen, J. Ding, J. Dai, P. Mukundhan, T. Kryman, Optical and acoustic metrology techniques for 2.5D and 3D advanced packaging, in *Proceeding of the IMAPS Device Packaging*, San Diego, 2014.

[19]. G. Singh, C. Suresh, J. Thornell, W. Young Han, High speed fluorescent inspection of non-visible defects, in *Proceeding of the IMAPS Device Packaging*, Pasadena, 2016.

[20]. T. Chiu, Y. Lin, Y. Wang, C. Wu, H. Lin, M. Wang, IPD robustness test methodology for InFO, in *Proceeding of the e-Manufacturing & Design Collaboration Symposium (eMDC'16)*, Hsinchu, 2016.

[21]. B. Rogers, D. Sanchez, C. Bishop, C. Sandstrom, C. Scanlan, T. Olson, Chips "face-up" panelization approach for fan-out packaging, *Chip Scale Review*, May-June 2016, pp. 24-30.

[22]. P. Huang, Y. Liu, J. Chao, C. Lu, S. Chen, J. Chen, F. Shen, J. Ding, P. Mukundhan, T. Kryman, In-line process monitoring of advanced packaging process using Focused Beam Ellipsometry, *Microelectronic Engineering*, Vol. 137, April, 2015, pp. 111-116.

[23]. T. Murray, A. Bakir, D. Stobbe, M. Kotelyanskii, R. Mair, M. Mehendale, X. Ru, J. Cohen, M. Schulberg, P. Mukundhan, T. Kryman, A new in-line laser-based acoustic technique for pillar bump metrology, *Journal of Microelectronics and Electronic Packaging,* April 2016, pp. 58-63.

[24]. Y. Ohara, A. Noriki, K. Sakuma, K.-W. Lee, M. Murugesan, J. Bea, F. Yamada, T. Fukushima, T. Tanaka, M. Koyanagi, 10 μm fine pitch Cu/Sn micro-bumps for 3-D super-chip stack, in *Proceeding of the IEEE International Conference on 3D System Integration (3DIC'09)*, San Francisco, 2009.

[25]. L. Haensel, M. Liebens, T. Vandeweyer, A. Miller, E. Beyne, M. Wiesiollek, H. Eisenbach, M. Filzen, Y. Wen, S. Sood, A Study of microbump metrology and defectivity at 20 μm pitch and below for 3D TSV stacking, in *Proceeding of the International Wafer-Level Packaging Conference (IWLPC'15)*, San Jose, USA, 2015.

[26]. J. Zao, J. Thornell, Controlling Measurements of WLP in High Mix, *High Volume Manufacturing*, 2015.

[27]. W. Y. Han, M. Marshall, Next generation bump inspection and metrology: Big data analysis, in *Proceeding of the International Wafer-Level Packaging Conference (IWLPC'17)*, San Jose, USA, 2017.

Chapter 9

Electrothermal Modeling of GaN HEMTs

Anwar Jarndal

9.1. Introduction

The future wireless communication systems are planned to offer broadband and high data rate services based on smaller size base and satellite stations. The transmitting power amplifiers (PAs) for such wideband applications must meet stringent specifications. These specifications are aimed mainly to reduce the signal distortion, which is mainly attributed to the power amplifier nonlinearity. The power efficiency as another specification, which is responsible for significant portion of the station size and cost, should also be considered. Research activities in the last decade have proven that the emerging Gallium nitride (GaN) high-electron mobility transistor (HEMT) is the best candidate for designing linear and high efficient PAs for the future communication systems. This requires a comprehensive understanding of the GaN HEMT physics (parasitic and thermal effects) and an accurate modeling of its nonlinear behavior.

Drain current is the key nonlinear element of the transistor model. The current nonlinearity is related to intrinsic nonlinearities (trans-conductance and output conductance bias dependency) and strong nonlinearities in Ohms, breakdown, pinch-off and forward regions [1]. For GaN HEMT as a high-power transistor, self-heating represents another regenerative process that strongly impacts the drain current [2]. The self-heating due to high power dissipation degrades the electrons saturation velocity and thus reduces the current. Further significant effects of the GaN transistor with respect to other technologies are surface and buffer trapping [3]. The surface trapping, which is related to polarization-induced surface states, can be reduced by proper surface passivation [4]. However, the buffer trapping cannot be ignored, especially for GaN on Si substrate [5]. The stronger lattice mismatch between GaN and Si results in free ions that behave as electron traps [6]. The kink effect in the DC characteristic (see Fig. 4 (a)) can be assumed as a signature of the buffer trapping. This effect is attributed to hot electrons being injected into the buffer traps under the influence of high drain voltage [7]. These trapped electrons deplete the

Anwar Jarndal
Department of Electrical and Computer Engineering, University of Sharjah, 27272 Sharjah, UAE

2DEG (Two Dimensional Electron Gas) channel and result in a reduction of the drain current for subsequent V_{DS} traces [8].

In the past few years, many modeling techniques for GaN HEMT devices have been developed [9-16]. Most of these models are based on table-based, analytical or artificial neural networks (ANN) modeling techniques. The first technique is based upon lookup tables developed from measured data of drain and gate currents. During simulation, the current, at any drain and gate voltages, is calculated using the interpolation or approximation technique [10]. This modeling approach is easy to implement; however, it cannot be used to predict out-of-range measurements and has limited nonlinearities simulation (lower convergence rate). ANN modeling has higher accuracy even under strong nonlinear operating conditions. This modeling technique, however, has limited prediction capability. The analytical modeling is more efficient in terms of rate of convergence and prediction capability; however, higher effort is required to formulate the model and optimize its fitting parameters. In this chapter, two modeling techniques will be presented: table-based and analytical. In the first technique, B-spline based smoothing procedure is developed and applied to the measured data to improve the model nonlinear simulation and its rate-of-convergence. For the analytical modeling a part of the problem of higher effort has been resolved solved by developing an automatic genetic algorithm optimization procedure to find the fitting parameters of a modified version of the reported model in [17]. The model formula is enhanced by adding extra fitting parameters to consider the thermal [18] and trapping-induced kink effects. The genetic-algorithm-based global optimization procedure is developed and programmed in Matlab$^{©}$ to find optimal values for the model fitting parameters. The adopted method improves the automaticity of the process and reduces the well-known local minima problem. The whole large-signal equivalent circuit model using both table-based and analytical techniques is implemented in ADS and validated by DC and RF (small- and large-signal) measurements.

The Chapter is organized as follows. The physics behind the kink and thermal effects is explained in Section 9.2. The large-signal equivalent circuit model is also presented in the same section. The table-based modeling technique is presented in Section 9.3. Section 9.4 presents the adopted analytical model and the developed genetic-algorithm-based extraction procedure for the model fitting parameters. In Section 9.5 the model is demonstrated by designing an inverse class-F power amplifier based on the same considered packaged GaN HEMT on Si substrate. Finally, a conclusion is drawn in Section 9.6.

9.2. Thermal and Trapping Induced Kink Effect

To reduce the cost and improve the circuit integration capability, GaN HEMT is currently fabricated on Si substrate. With respect to SiC substrate, there are some challenges including the lattice mismatch between GaN and Si (≈ 17 %) [6], which results in higher density of dislocations in the GaN-substrate interface. These dislocations manifest themselves as electrons or holes traps [19-22]. These traps are responsible for the last mentioned IV kink and DC-RF dispersion. The latter effect is attributed to the longer trapping time, which prevents electrons from following higher frequency stimulus and participating in the channel conduction. Furthermore, the negative charge due to the trapped electrons depletes the channel (backgating) and reduces the drain current under RF operation [8]. In the circuit level, this has been accounted for by adding series C_{rf} and R_{rf} in the drain side (see Fig. 9.1) to simulate the channel conductance dispersion. The low thermal conductivity of silicon (1.3 W/cm/°C) and low thermal expansion coefficient (2.6x10^{-6} °C) present another challenge for GaN HEMT [18]. This could be observed clearly from the self-heating induced current collapse in the DC IV characteristics of the considered packaged GaN HEMT from Nitronex corporation at high-power dissipation region (see Fig. 9.4 (a)).

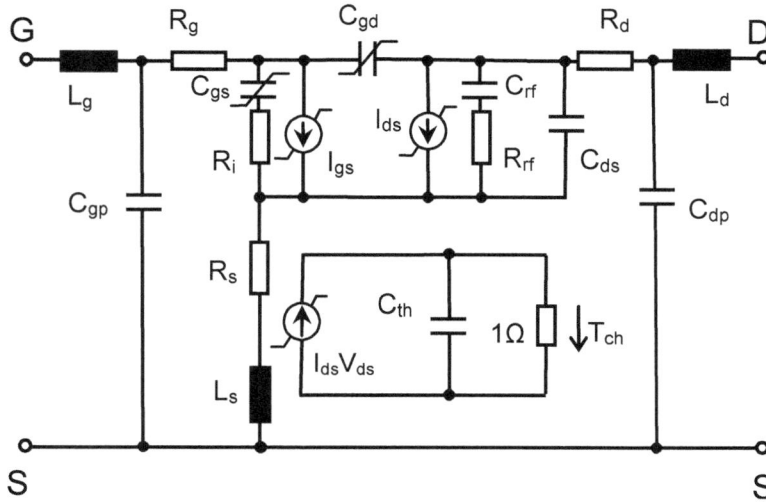

Fig. 9.1. Large-signal equivalent circuit model for GaN HEMTs including self-heating and output conductance dispersion effects.

For the device under investigation, the extracted unity gain frequency (f_T) under active bias condition is around 8 GHz. This parameter defines the maximum achievable performance of the device, which is determined mainly by the parasitic effects. At frequencies smaller than f_T, the device can be simulated by the equivalent circuit model shown in Fig. 9.1. Therefore, all the extrinsic and intrinsic model elements are extracted from S-parameter measurements taken from 30 MHz up to 3 GHz. This range was selected to respect the low frequency requirement for capacitances extraction and the operating frequency of the power amplifier to be designed (2.35 GHz). The pad capacitances C_{gp} and C_{dp} are determined from cold pinch-off S-parameters following the same procedure presented in [23]; where C_{ds} is assumed bias-independent and its value is absorbed in C_{dp}. For this reason C_{ds} is assign zero value in Table 9.1. The extrinsic resistances and inductances are extracted from cold forward S-parameters at 2V gate voltage. These S-parameters are converted first to Z-parameters. Then, the inductances are extracted from the curves of the imaginary parts of ωZ_{ij} vs. ω^2 by linear data fitting. The extrinsic resistances are then extracted from the real parts of ωZ_{ij} vs. ω^2 also by linear data fitting after de-embedding the extrinsic capacitances C_{gp} and C_{dp} [24]. Table 9.1 presents the extracted extrinsic elements of the employed 4-W GaN HEMT. The intrinsic elements under certain bias condition are extracted from the intrinsic Y-parameters by linear data fitting after de-embedding the extracted extrinsic elements from the measured S-parameters. R_i and C_{ds} are taken as bias-independent elements and kept at their values, which are obtained under pinch-off bias voltage. The extracted values of C_{gd} show significant variation with V_{ds} and it is noted that this variation has strong impact on the simulated gain and power added efficiency. For that reason, its extracted values at different gate-drain voltages are fitted by a simple polynomial function, as presented in Fig. 9.2, to describe the nonlinear behavior of this capacitance. Part of the device nonlinearity is related to the bias dependency of C_{gs}. Therefore, and in order to improve the device simulation under different classes of operation, C_{gs} is modeled by simple 1-D tangent function as shown in Fig. 9.2.

Table 9.1. Extrinsic parameters of 4-W packaged GAN HEMT.

C_{gp} [pF]	C_{dp} [pF]	L_g [nH]	L_d [nH]	L_s [nH]	R_g [Ω]	R_d [Ω]	R_s [Ω]
1.39	1.65	1.58	1.30	0.14	0.77	0.91	0.38

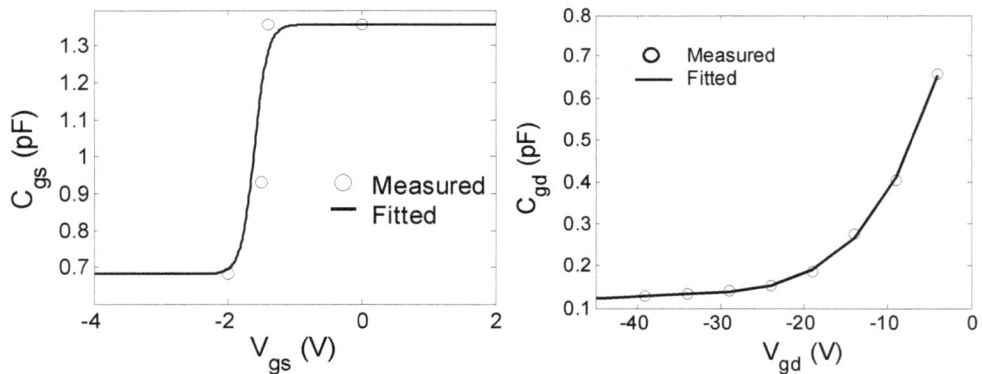

Fig. 9.2. Fitting of C_{gs} by the function $0.68+0.34\ (1+tanh(4(V_{gs}+1.4)))$ and C_{gd} by the function $0.89+0.06V_{gd}+0.15*10^{-2}V_{gd}^2+0.17*10^{-4}V_{gd}^3+0.7*10^{-7}V_{gd}^4$ for a 4-W GaN HEMT [28].

The passive elements C_{rf} and R_{rf} have no physical meaning and are only used to simulate the output conductance dispersion. Therefore, their values can be estimated by fitting S-parameter measurements under high drain and high-negative gate bias voltages. The high stress voltages under this condition stimulate more free carriers in the device channel to be injected in trapping states. The injected charge results in an electrical field, which modulates the channel and enhances the frequency dependence of the output conductance [25].

9.3. Table-Based DC IV Model

Because of its higher accuracy, a table-based model is used to represent the drain and gate currents [26, 27]. To build the model table, the measured data of these currents are referenced to the intrinsic gate and drain voltages, which can be calculated as

$$V_{gs} = V_{GS} - (R_g + R_s)I_{gs} - R_sI_{ds} ,$$ (9.1)

$$V_{ds} = V_{DS} - (R_d + R_s)I_{ds} - R_sI_{gs} .$$ (9.2)

However, constructing the model in this way will provide a non-equidistant grid of currents with respect to V_{gs} and V_{ds} voltages, which is not convenient to handle in Advanced Design System (ADS) simulator environment. B-spline approximation technique is used to provide a uniform data grid for the currents [28]. This technique preserves the continuity of the data and its higher derivatives; therefore, it smoothes the progress of the model's simulation and enhance its accuracy in predicting the device nonlinear behavior in soft turn-on and turn-off conditions as illustrated in Fig. 9.3.

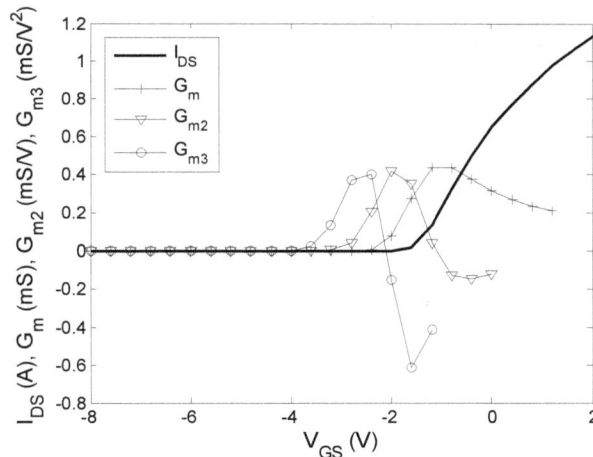

Fig. 9.3. Simulated drain current and its derivatives at $V_{DS} = 10$ V for a 4-W GaN HEMT. G_m, G_{m2} and G_{m3} are the first, second and third derivatives of the drain current with respect to the gate-source voltage.

Using this technique, a model for a packaged 4-W GaN HEMT device from Nitronex corporation was constructed using standard DC IV measurements. As it can be seen in Fig. 9.3, the model accurately reproduces the gate and drain IV characteristics in the switching (pinch-off and forward) as well as in the transition region. The model also accurately simulates the typically observed self-heating induced current collapse in the DC IV characteristics in the high power dissipation region. In general, this current reduction is significant under static and quasi-static operation [29]. However, it is reduced by increasing the frequency of operation since the input signal is not slow enough to heat up the device.

To simulate this effect the drain current is formulated as

$$I_{ds}(V_{ds},V_{gs}) = I_{ds,DC}(V_{ds},V_{gs}) \left[1 + K_T H(\omega) \begin{pmatrix} I_{ds,DC}(V_{ds},V_{gs})V_{ds} \\ -I_{ds,DC}(V_{dso},V_{gso})V_{dso} \end{pmatrix} \right] \quad (9.3)$$

$I_{ds,DC}$ is the measured DC drain current and K_T is a thermal constant describing the dependency of the drain current on the device self-heating. V_{ds} and V_{gs} are the instantaneous intrinsic drain and gate voltages, which are dynamically changed around their average voltages or DC values V_{dso} and V_{gso}. $H(\omega)$ is a thermal frequency response function, which can be defined as

$$H(\omega) = j\omega\tau / (1 + j\omega\tau), \quad (9.4)$$

where ω is the operating frequency and τ is a thermal time constant. This function describes the smooth transition of the drain current from static or quasi-static to RF and it is implemented using the high-pass thermal sub-circuit in the model (see Fig. 9.1). In this sub-circuit, the value of the thermal capacitance C_{th} is selected so that a value in the order of 1 ms [30] can be obtained for τ. The thermal resistance R_{th} is normalized to one because its value is incorporated in the thermal constant K_T. With respect to the original model presented in [28], this presentation for the drain current considers the self-heating due to the average dissipated power under RF operation. Thus, it improves the model accuracy for simulating lower efficiency classes of operation as will be verified in the next section. Under RF small-signal condition the current in (3) can be linearized around the generic quiescent voltages V_{gso} and V_{dso} to its differential transconductance and output conductance as

$$g_m^{RF}(V_{dso},V_{gso}) = g_m^{DC}(V_{dso},V_{gso}) \left[1 + K_T I_{dso} V_{dso} \right], \quad (9.5)$$

$$g_{ds}^{RF}(V_{dso},V_{gso}) = g_{ds}^{DC}(V_{dso},V_{gso}) \left[1 + K_T I_{dso} V_{dso} \right] + K_T I_{dso}^2, \quad (9.6)$$

where $I_{dso} = I_{ds,DC}(V_{dso},V_{gso})$. g_m^{RF} and g_{ds}^{RF} can be extracted from measured intrinsic Y-parameters under suitable bias condition. The intrinsic Y-parameters can be determined from the measured S-parameters after de-embedding the effect of the extrinsic elements as will be discussed in Section 9.4. The bias condition can be selected in the saturation region, for low V_{dso} just above the knee voltage, and high $I_{ds,DC}$. Here, the device will not be under high stress voltages to stimulate additional surface or buffer trapping effects [7] and the current dispersion can be attributed mainly to the self-heating effect. g_m^{DC} and g_{ds}^{DC} can be obtained through numerical differentiation of the measured $I_{ds,DC}$ at the same bias condition. Linear least square method can then be used to solve (5) and (6) and to find the optimal value of K_T. The calculated value of K_T for the analyzed 4-W device is 0.09. For the analyzed device the passivation process is optimized to reduce the surface traps [31]. However, the kink in the IV characteristics (see Fig. 9.4 (a)) can be considered as a signature of buffer trapping effect [7]. This effect causes variation of the output conductance under static and RF operation. To account for this effect, the series elements R_{rf} and C_{rf} (see Fig. 9.1) are added in the intrinsic drain side of the model [32].

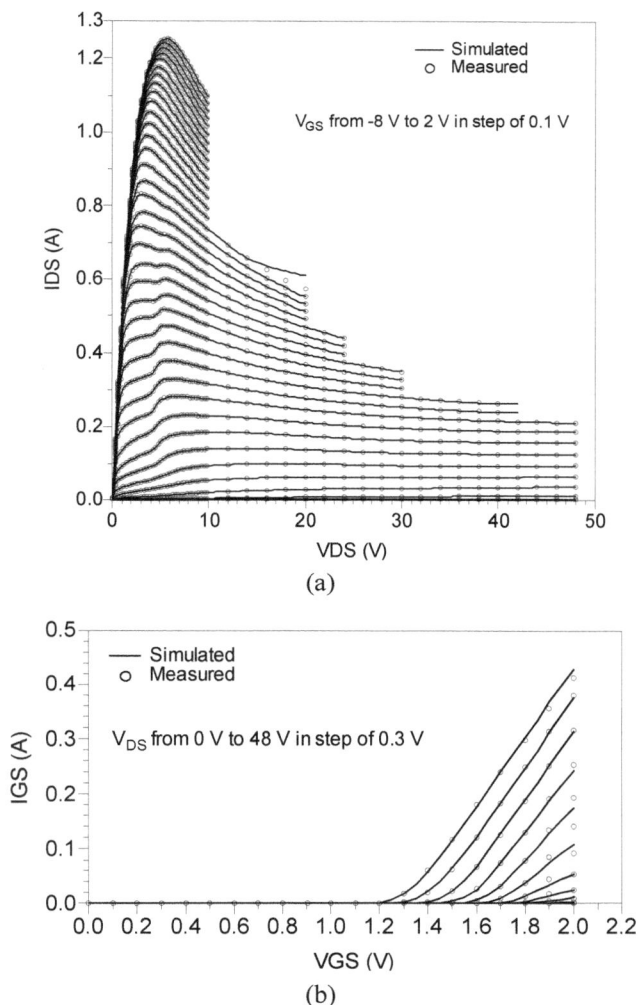

Fig. 9.4. Simulated: (a) DC drain current and (b) DC gate current of a 4-W GaN HEMT in comparison with measurements.

The ADS large-signal model has been also validated by RF small- and large-signal measurements for the same considered 4-W device. Fig. 9.5 shows simulated and measured S-parameters under cold pinch-off and forward bias conditions. Except from the small frequency-lag (delay) in the curves at higher (lower) frequency due to the fact of the neglected inductive (capacitive) effects in the cold pinch-off (forward) device model a relatively good agreement between the simulations and the measurements has been obtained.

267

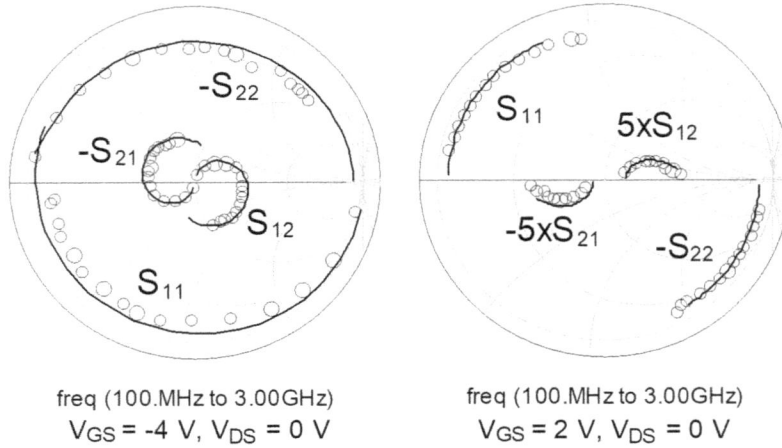

freq (100.MHz to 3.00GHz)   freq (100.MHz to 3.00GHz)

V_{GS} = -4 V, V_{DS} = 0 V V_{GS} = 2 V, V_{DS} = 0 V

Fig. 9.5. Measured (symbols) and simulated (lines) S-parameters for a 4-W GaN HEMT at pinch-off (left chart) and at forward (right chart) bias condition.

Single-tone large-signal measurements are performed for the same 4-W GaN HEMT in 50 Ω source and load terminations under different bias conditions (classes of operation) and different input drive levels. The corresponding simulations have been performed using the recent model and compared with the measurements. The simulations are also repeated using the model presented in [28]. The results of this comparison are presented in Fig. 9.6. As it can be seen, distinguishing of the current model becomes clearer with increasing the gate bias voltage toward lower efficiency class of operation where the effect of the average dissipated power and C_{gs} nonlinearity are more significant.

9.4. Analytical DC IV Model

As it has been mentioned analytical model has better prediction and convergence capability. In this part the model in Fig. 9.1 will be presented with improved analytical modeling of the drain current. The drain current I_{ds} is modeled by the following formulas [33]:

$$I_{dso} = \beta \frac{V_{gs3}^2}{1+\frac{V_{gs3}^{plin}}{VL}} * \tanh\left(\frac{\alpha*ln(1+V_{ds})}{V_{gs3}^{psat}}\right), \tag{9.7}$$

where

$$V_{gs3} = V_{ST} * Ln\left(1 + e^{\frac{V_{gs2}}{V_{ST}}}\right), \tag{9.8}$$

$$V_{gs2} = V_{gs1-}\frac{1}{2}\left(V_{gs1} + \sqrt{\left(V_{gs1} - V_K\right)^2 + \Delta^2} - \sqrt{V_K^2 + \Delta^2}\right), \tag{9.9}$$

$$V_{gs1} = V_{gs} - V_{to}, \tag{9.10}$$

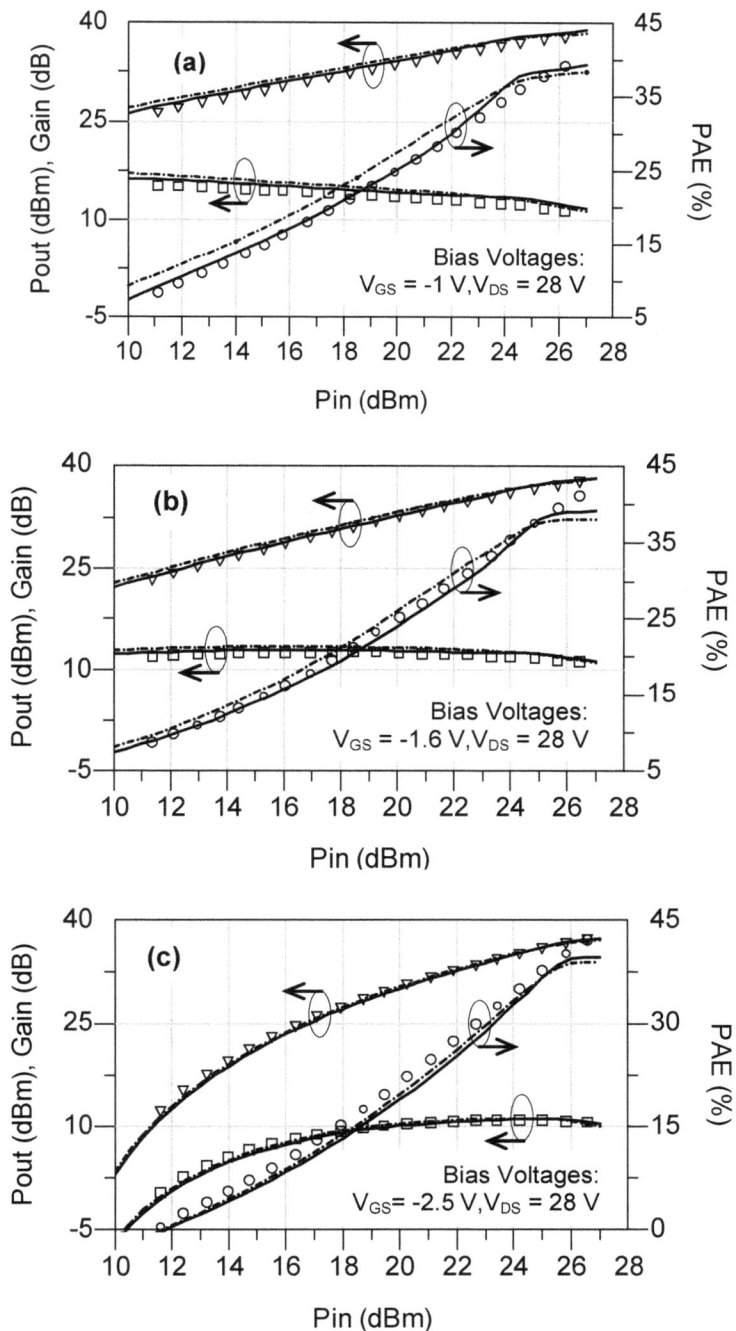

Fig. 9.6. Measured (symbols) and simulated output power, gain and efficiency using the proposed model (lines) and the last model in [28] (discrete lines) at different classes of operation: (a) class-AB, (b) class-B and (c) class-C for a 4-W GaN HEMT in a 50 Ω source and load environment at 2.35 GHz.

$$\beta = \beta_0 - \beta_1 * T_{ch}, \qquad (9.11)$$

$$T_{ch} = R_{th}(V_{ds}I_{ds}) + (T_A - T_{ref}), \qquad (9.12)$$

$$R_{th} = R_{th0} * \left(1 - K_c tanh\left(K_x(V_{gs} - V_c)\right)\right), \qquad (9.13)$$

$$\alpha = -\alpha_o * tanh\left(p1(V_{gs} - V_{c1})\right). \qquad (9.14)$$

In comparison with the model in [17], the term V_{ds} in (9.1) is replaced by $ln(1+V_{ds})$ and α is represented as a *tanh* function of V_{gs} instead of the linear relation. It has been found that this description improves the model accuracy especially in the forward region. The fitting parameters VL and β are used to control the transition from the quadratic to the linear regions of the I_{ds}-V_{gs} characteristics. The slope of the trans-conductance in the linear region is adjusted by *plin*, and the turn-on process is controlled by V_{ST} and V_{to} (the pinch-off voltage). To accurately simulate the typical smooth pinch-off of the GaN device, a second-order polynomial with V_K and Δ fitting parameters is used. The transition from the triode region to the saturation region of the I_{ds}-V_{ds} characteristics is modeled by two hyperbolic functions with fitting parameters of *psat*, α_o, *p1* and V_{c1} (the knee voltage at V_{gs}=0 V). The first *tanh* function is to simulate the transition from triode to saturation, while the second *tanh* is to account for the dependency of the knee voltage on V_{gs}. The fitting parameter β has been modified to consider the complicated thermal behavior of such a high-power device. The dependence of thermal conductivity of the GaN HEMT on self-heating [34] is modeled indirectly by formulating the thermal resistance R_{th} in terms of the gate voltage. The trapping-induced kink effect has been accounted for by the additional term I_k in (9.9) to simulate the gate- and drain-pumping-induced kink effects [35]. Thus the complete formula of I_{ds} is

$$I_{ds} = I_{dso} + I_k\left(1 + tanh(k_r(V_{ds} - V_{c3}))\right), \qquad (9.15)$$

where

$$I_k = r * \left(1 - abs\left(tanh\left(s * (V_{gs} - V_{c2})\right)\right)\right). \qquad (9.16)$$

As illustrated in Fig. 9.7, the optimization process is carried out through three phases. Multi-region optimization is used to find optimal values for the influencing parameters at each region. As can be seen in Fig. 9.8, the entire IV characteristics are divided in to three zones: zone 1, zone 2 and zone 3. The first zone ($V_{DS} < 4$ V) is used to characterize the general IV without considering the kink or thermal effects. The second zone is an extension of the first one at higher V_{DS} (< 6 V) to characterize the inherent kink of the drain current. The last zone represents the whole IVs subjected to the kink and thermal effects. The IV measurements at zone 1 are used to optimize the related fitting parameters of α_o, *p1*, V_{c1}, β_o, V_{to}, Δ, V_K, V_{ST}, *psat*, *plin*, and V_L. As can be seen, the kink effect is obvious in the IVs of zone 2 and thus optimized values for *r*, *s*, V_{c2}, K_r and V_{c3} parameters are determined based on these measurements. The thermal fitting parameters of β_1, R_{tho}, K_c, K_x and V_c are determined by optimizing the measurements at zone 3, which includes the high-power dissipation area.

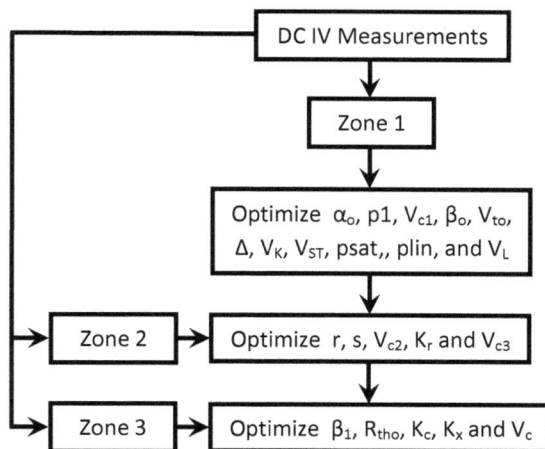

Fig. 9.7. Flow chart of the drain current fitting parameters optimization process.

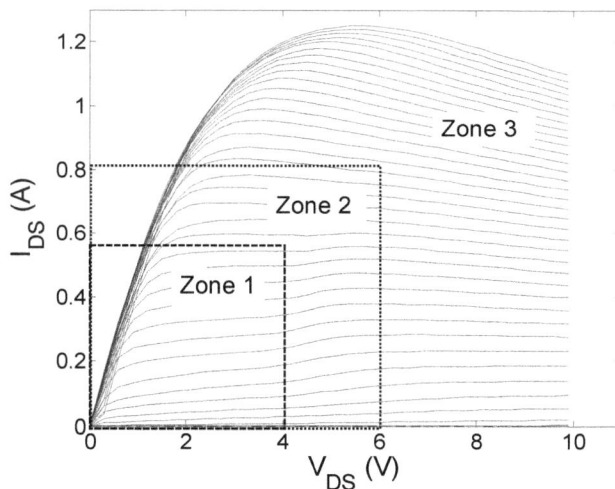

Fig. 9.8. Measured drain current of 4-W GaN HEMT over three zones for fitting parameters optimization. V_{GS} is from -8V to 2V in step of 0.2 V. V_{GS} and V_{DS} are the gate and drain extrinsic voltages.

The gate current I_{gs} is modeled as

$$I_{gs} = I_{go} * e^{\frac{\alpha_g(V_{gs}-V_{go})}{I_{go}}}, \tag{9.17}$$

where

$$V_{go} = a_1 V_{ds} + a_2, \tag{9.18}$$

$$\alpha_g = a_3 V_{ds} + a_4. \tag{9.19}$$

271

As can be seen in Fig. 9.9, increasing V_{ds} will increase the turn-on voltage V_{go} and reduce the decaying rate of α_g. This effect has been considered simply by linear formulas for these parameters in terms of V_{ds}. The same genetic algorithm optimization technique is used to determine optimal values for I_{go}, a_1, a_2, a_3 and a_4 fitting parameters of the gate current model.

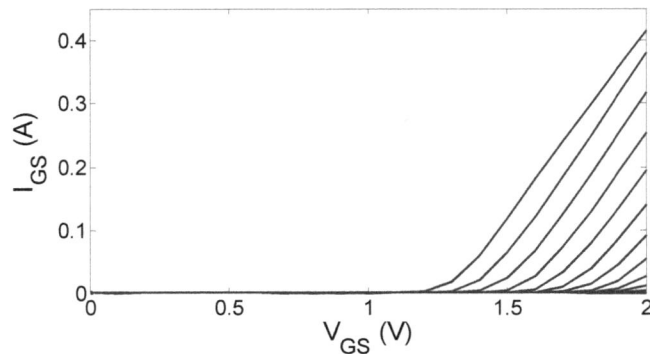

Fig. 9.9. Measured gate current of 4-W GaN HEMT for V_{DS} is from 0V to 10 V. V_{GS} and V_{DS} are the gate and drain extrinsic voltages.

9.4.1. Model Parameters Optimization Using Genetic-Algorithm

In principle, a local optimization technique such as the simplex method can be used to find optimal values for the fitting parameters of the I_{ds} and I_{gs} models. However, with such a large number of variables, the optimization problem is nonlinear with multiple local minima and thus the quality of final optimized values depends on the initial guess. In this case, a local optimization technique is not efficient and in order to overcome this problem and find the global minimum, the genetic algorithm (GA) as a global optimization technique has been adopted. The GA is a widely used technique and well-matched to our problem because of its instinctiveness, ease of implementation, and the ability to efficiently solve highly nonlinear problems [36]. The optimization process has been applied to the fitting parameters of I_{ds} and I_{gs}. The steps of the implemented genetic optimization are illustrated in Fig. 9.10 and the whole procedure can be summarized as follows:

1. Generating of initial population of vectors. Each vector includes randomly assigned values for the considered model fitting parameters. These initial generation of vectors are the parents of the next-generation vectors that will undergo the optimization process until the maximum number of generation N_{max} is reached.

2. Computing the error between simulated and measured DC IVs for each vector. I_{ds} or I_{gs} are calculated using the assigned model fitting parameters over the entire measured grid of voltages V_{gs} and V_{ds} for the considered zone. The total error between the simulated I_{ds} or I_{gs} and the corresponding measured ones is determined by

$$Error = \frac{1}{N}\sum_{m=1}^{N}\left(I_{ds,gs}^{meas} - I_{ds,gs}^{sim}\right)^{2}, \qquad (9.20)$$

where N is the total number of measurements, $I_{ds,gs}^{meas}$ is the measured DC drain or gate current and $I_{ds,gs}^{sim}$ is the corresponding simulated currents.

3. Sorting all vectors and their errors to reject some of the maximum error vectors in the population of the current generation.

4. Crossover reproduction of the selected vectors by using double-point crossover routine. The vectors are ordered such that vectors in odd numbered positions are crossed with the vectors in the adjacent even numbered positions.

5. Mutation reproduction by altering randomly the values of each vector to generate offspring from the previous crossover process.

6. Calculating the corresponding errors using (9.20) for the reproduced offspring vectors.

7. Reinsertion by replacing the vectors with most errors in the old population (parents) with vectors in the new reproduced population (offspring).

8. The generational counter is incremented, and the steps from 4 to 8 are repeated until generation Number N= N_{max} or when the error of one of the vectors is less than or equal to a fixed threshold value ε. Under this condition the process stops and the values included in the minimum error vector are chosen as optimal values for the fitting parameters.

Matlab program has been developed to implement the mentioned GA optimization steps. The proposed modeling approach has been applied to the same DC drain and gate currents of the considered 4-W packaged GaN HEMT device over a wide range of bias conditions (V_{DS}: 0 to 48 V and V_{GS}: -6 to 2 V). The optimization process is started by generating an initial population of 1000 vectors. Each vector includes randomly assigned values for the model fitting parameters. Two stop conditions are used concurrently for the procedure: the maximum number of generation (N_{max} = 100) and the minimum error (ε = 0.001). Thus the optimization process stops if the error reaches ε before attaining the N_{max}'s iteration, otherwise the process will continue up to N_{max}. Initially, the optimization procedure has been applied to the zone 1 measurements to find optimal values for α_o, $p1$, V_{c1}, β_o, V_{to}, Δ, V_K, V_{ST}, $psat$, $plin$, and V_L fitting parameters. The obtained optimal values for these fitting parameters are listed in Table 9.2. Fig. 9.11 (a) shows variation of the minimum error versus the number of generation during optimization. Fig. 9.11 (b) shows an excellent fitting for the drain current in zone 1 using these optimized parameters. This process is re-applied on the measurements of zones 2 and 3. Model fitting in zone 2 is presented in Figs. 9.12 and the optimized values of the corresponding fitting parameters are listed in Table 9.2. The same optimization program has been implemented to find the fitting parameters of the gate current model. The complete optimized values of the fitting parameters are listed in Table 9.2.

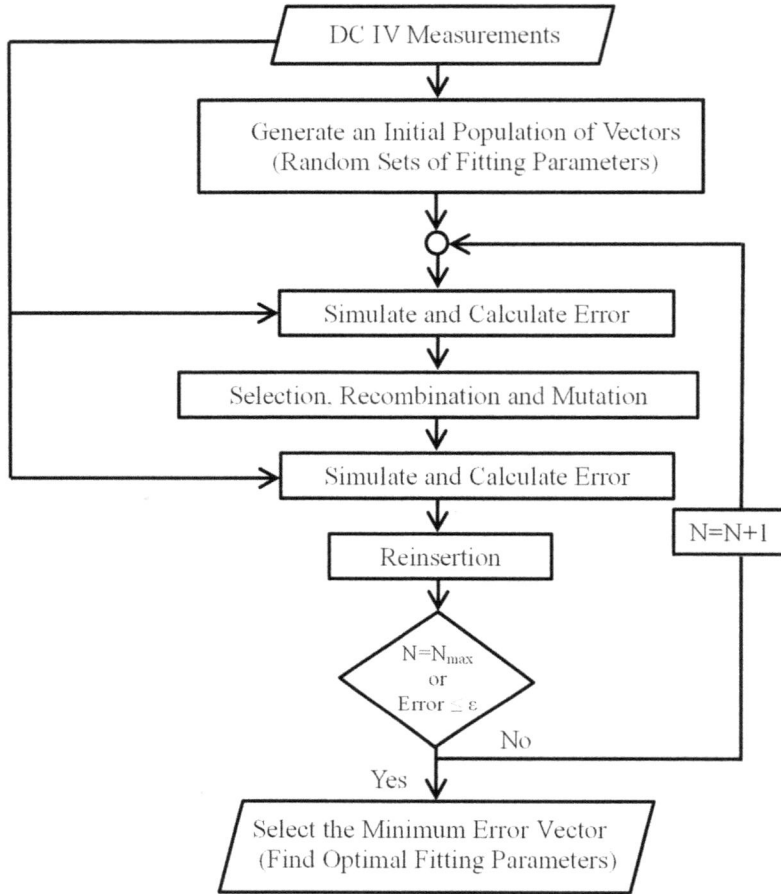

Fig. 9.10. Flow chart of the fitting parameters optimization using genetic algorithm.

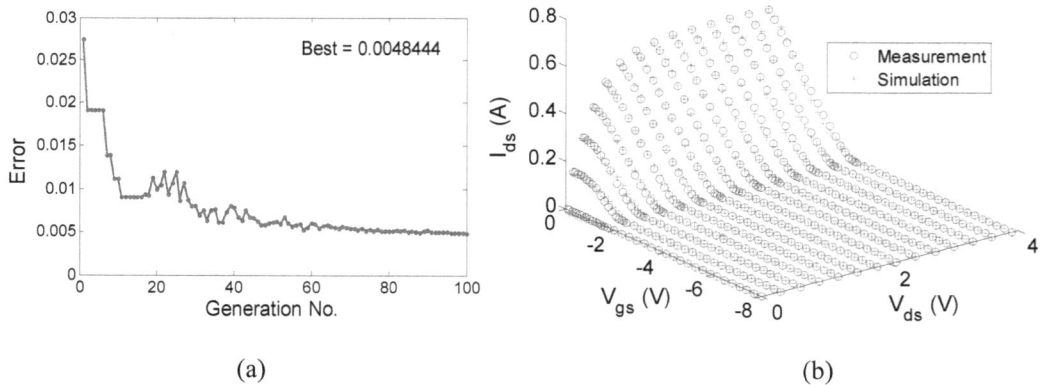

(a)   (b)

Fig. 9.11. (a) Variation of the error with the number of generation during fitting parameters optimization of the drain current model in zone 1. (b) Simulated and measured DC IV of a 4-W GaN HEMT in zone 1.

(a)   (b)

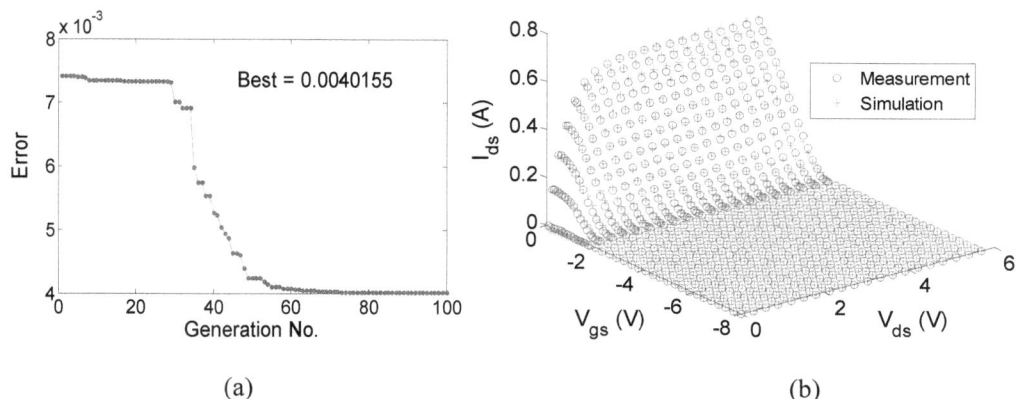

Fig. 9.12. (a) Variation of the error with the number of generation during fitting parameters optimization of the drain current model in zone 2; (b) Simulated and measured DC IV of a 4-W GaN.

Table 9.2. Optimized fitting parameters of the models of the drain and gate currents.

Model	FITTING PARAMETERS			
I_{ds}	plin = 2.36 VL = 2.28 [V] α_o = 1.845 [1/V] p1 = 2.69 [1/V] V_{c1} = 2.8 [V]	r = 0.041 [A] s = 0.903 [1/V] psat = 1.23 VST = 0.03315 [V] VK = 1.8277 [V]	Δ = 4.3 [V] V_{to} = -1.476 [V] Kr = 1.69 [1/V] β_o = 1.464 [A/V²] $\beta1$ = 0.485 [A/kV²]	R_{tho} = 0.0737 [k/W] Kc = 2.1685 Kx = 0.2734 [1/V] Vc = 2.2 [V] Vc2 = -0.7537 [V] Vc3 = 3.806 [V]
I_{gs}	Igo = 0.04 [A] a1 = 0.1504	a2 = 0.51 [V]	a3 = 0.224 [A/V²] a4 = 1.34 [A/V]	

The developed currents model has been embedded in the large-signal equivalent circuits and implemented in ADS. The ADS-implemented model has been used to simulate DC IV and S-parameters measurements. As can be seen in Fig. 9.13, the model can efficiently reproduce the nonlinear behavior of the drain and gate currents. The drain current model also simulates the kink effect in a very good manner. The model also accurately predicts the typical self-heating induced collapse of the DC drain current in the high power dissipation area. The results of S-parameters simulations are shown in Fig. 9.14 with a very good agreement between the simulations and measured data. Single-tone large-signal RF measurements under different bias conditions (classes of operation) are carried out for the same device and compared with the model simulations. The results of this comparison are presented in Figs. 9.15, 9.16 and 9.17, which show output power, gain and power-added-efficiency versus input power. As can be observed, the model can efficiently predict the AM-AM/AM-PM characteristics of the device under different operating conditions up to its 4 W (36 dBm) rated power.

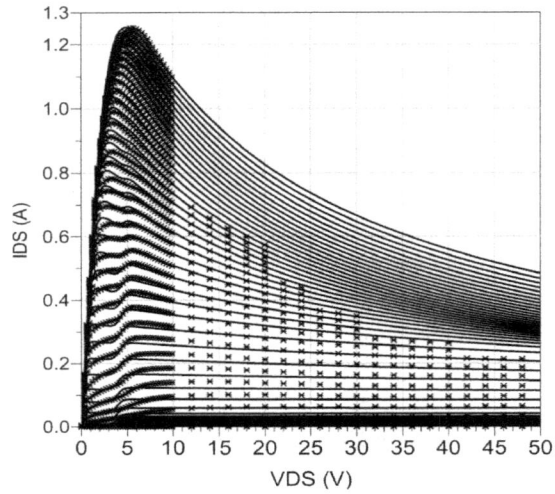

(a)

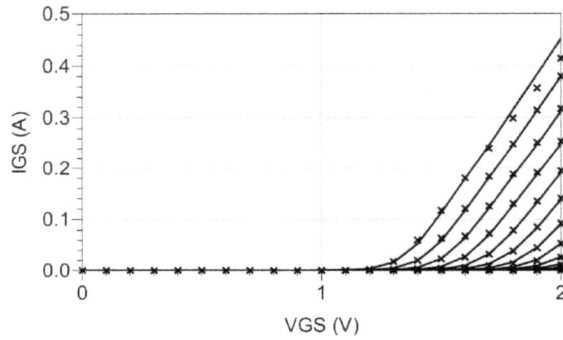

(b)

Fig. 9.13. (a) Simulated (lines) and measured (cross) DC IVs of a 4-W GaN HEMT. The drain current (a), and gate current (b) versus the extrinsic voltages.

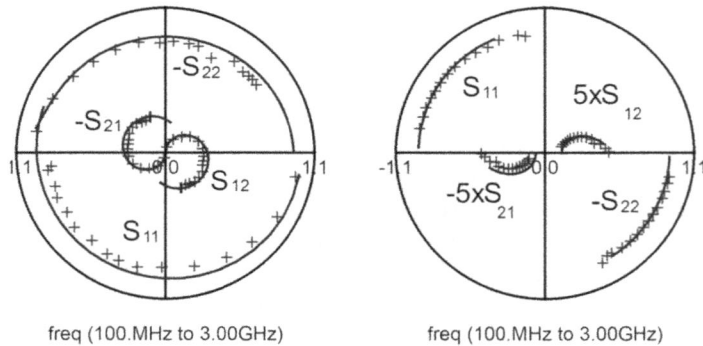

freq (100.MHz to 3.00GHz)   freq (100.MHz to 3.00GHz)

Fig. 9.14. (a) Measured (cross) and simulated (lines) S-parameters of a 4-W GaN HEMT at: (a) V_{GS} = -4 V, V_{DS} = 0 V and (b) V_{GS} = 2 V, V_{DS} = 0 V.

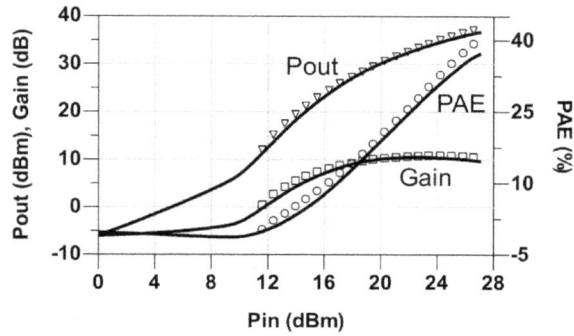

Fig. 9.15. (a) Comparison between measured (symbols) and simulated (lines) output power, gain and efficiency for a class-C (V_{GS} = -2.5 V and V_{DS} = 28 V) operated packaged 4-W GaN HEMT in a 50 Ω source and load environment at 2.35 GHz.

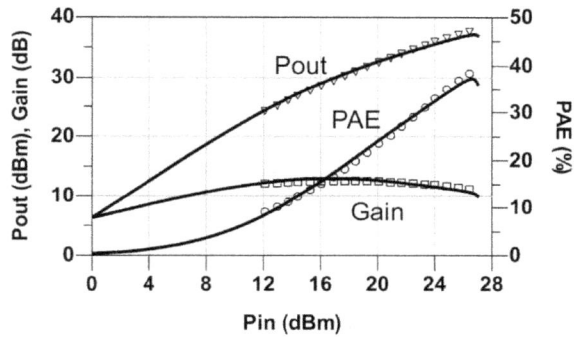

Fig. 9.16. Comparison between measured (symbols) and simulated (lines) output power, gain and efficiency for a class-B (V_{GS} = -1.6 V and V_{DS} = 28 V) operated packaged 4-W GaN HEMT in a 50 Ω source and load environment at 2.35 GHz.

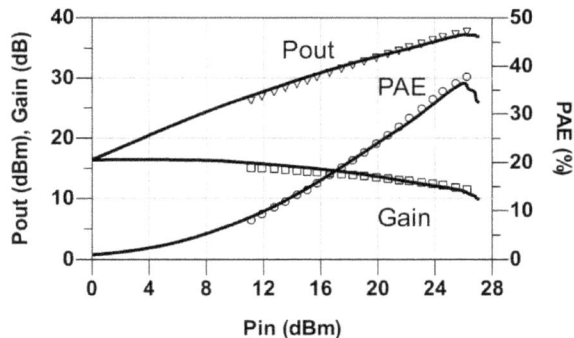

Fig. 9.17. Measured (symbols) and simulated (lines) output power, gain and efficiency for a class-AB (V_{GS} = -1 V and V_{DS} = 28 V) operated packaged 4-W GaN HEMT in a 50 Ω source and load environment at 2.35 GHz.

9.5. Inverse Class-F Power Amplifier

The developed model has been demonstrated by designing and simulating an inverse class-F switching-mode power amplifier based on the same considered 4-W packaged device [28, 33]. The amplifier has been designed and fabricated at 2.35 GHz operating frequency (see Fig. 9.18 (a)). Input 50-Ω matching circuit was used to enhance the output power and the gain. Output 50-Ω matching network was designed and optimized to provide the fundamental and harmonic impedances for inverse class-F and maximize the power-added-efficiency. As illustrated in Fig. 9.18 (b), the amplifier was implement in ADS using the developed model and simulated under 2.35-GHz single-tone stimulus with 50-Ω terminations (see Fig. 9.19). The simulated power-sweep was compared with real measurements for the fabricated inverse class-F amplifier. Fig. 9.19 shows simulated and measured output power, gain and PAE for the amplifier at 2.35 GHz and under -2.5 and 28 V gate and drain bias voltages, respectively. As can be seen, the model can accurately simulate the device and this accordingly validates its applicability for designing nonlinear circuits.

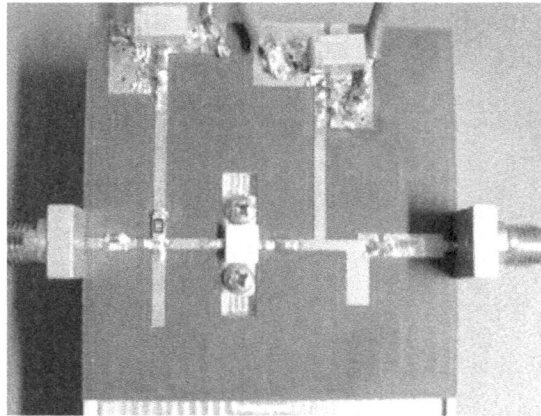

(a)

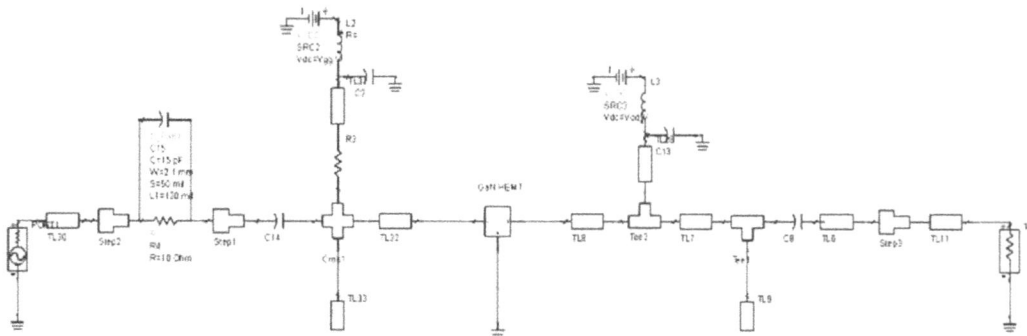

(b)

Fig. 9.18. Realized (a) and simulated (b) inverse class-F switching mode power amplifier in 50 Ω source and load environment at 2.35 operating frequency.

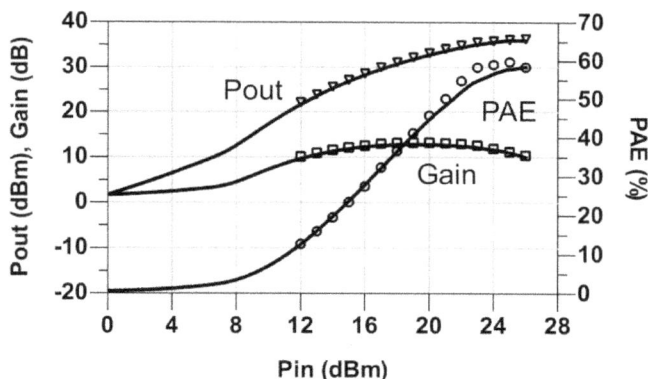

Fig. 9.19. Measured (symbols) and simulated (lines) output power, gain, and efficiency at 2.35 GHz frequency for the inverse class-F switching-mode PA with V_{DS} = 28 V and V_{GS} = - 2.5 V.

9.6. Conclusion

In this chapter two techniques of electrototothermal modeling have been presented and applied to GaN transistors. The first approach is table-based with adopted B-spline approximation procedure to build look-up tables of uniform smooth data for the drain and gate currents. This technique preserves the continuity of the data and its higher derivatives; therefore, it smoothes the progress of the model's simulation and enhance its accuracy in predicting the device nonlinear behavior. In the second approach, an enhanced analytical current model that predicts self-heating and trapping-induced kink effects was developed along with its improved fitting parameters GA-based optimization procedure. Instead of using the whole IV measurement, different zones are used for fitting parameters optimization. The first zone of self-heating and kink-free measurements are used to optimize the fitting parameters of the general nonlinear behavior of the drain current. The second zone of kink-affected measurements are used to optimize the related fitting parameters of the model. The whole measurements, including the high-power dissipation area, are then used to optimize the remaining fitting parameters. Both modeling approaches have been applied to 4-W packaged GaN HEMT and validated by DC and RF measurements. The developed models show very good results for simulating the transistor. The models have been used to design an inverse class-F switching mode power amplifier. The fabricated amplifier has been tested by RF measurements at 2.35 GHz and a very good agreement is obtained between these measurements and the corresponding simulated ones.

Acknowledgements

The author acknowledges the support from the University of Sharjah, Sharjah, United Arab Emiratis.

References

[1]. N. B. Carvalho, J. C. Pedro, Large-and small-signal IMD behavior of microwave power amplifier, *IEEE Transactions on Microwave Theory & Techniques*, Vol. 47, 1999, pp. 2364-2374.

[2]. A. E. Parker, J. G. Rathmell, Broad-band characterization of FET self-heating, *IEEE Transactions on Microwave Theory & Techniques*, Vol. 53, 2005, pp. 2424-2429.

[3]. G. Meneghesso, G. Verzellesi, R. Pierobon, F. Rampazzo, A. Chini, U. K. Mishra, C. Canali, E. Zanoni, Surface-related drain current dispersion effects in AlGaN-GaN HEMTs, *IEEE Transactions on Microwave Theory & Techniques*, Vol. 51, 2004, pp. 1554-1561.

[4]. R. Vetury, N. Q. Zhang, S. Keller, U. K. Mishra, The impact of surface states on the DC and RF characteristics of AlGaN/GaN HFETs, *IEEE Transactions on Electron Devices*, Vol. 1, Issue 48, 2001, pp. 560-566.

[5]. S. Yang, C. W. Bay, C. Zhou, O. Jiang, J. Lu, B. Huang, K. J. Chen, Investigation of buffer traps in AlGaN/GaN-on-Si devices by thermally stimulated current spectroscopy and back-gating measurement, *Applied Physics Letters*, Vol. 104, 2014, pp. 013504-1 - 013504-4..

[6]. L. Liu, J. H. Edgar, Substrates for gallium nitride epitaxy, *Materials Science and Engineering*, Vol. R37, 2002, pp. 61-127.

[7]. S. C. Binari, K. Ikossi, J. A. Roussos, W. Kruppa, D. Park, H.B. Dietrich, D. D. Koleske, A. E. Wickenden, R. L. Henry, Trapping effect and microwave power performance in AlGaN/GaN HEMTs, *IEEE Transactions on Electron Devices*, Vol. 48, 2001, pp. 465-471.

[8]. G. Kompa, Basic Properties of III-V Devices - Understanding Mysterious Trapping Phenomena, *Kassel University Press GmbH*, Kassel, 2014.

[9]. J.-W. Lee, K. Webb, A temperature-dependent nonlinear analytic model for AlGaN-GaN HEMTs on SiC, *IEEE Transaction on Microwave Theory and Techniques*, Vol. 52, 2004, pp. 2-9.

[10]. A. Jarndal, G. Kompa, Large-signal model for AlGaN/GaN HEMT accurately predicts trapping and self-heating induced dispersion and intermodulation distortion, *IEEE Transaction on Electron Devices*, Vol. 54, 2007, pp. 2830-2836.

[11]. A. Jarndal, P. Aflaki, R. Negra, A. Kouki, F. M. Ghannouchi, Large-signal modeling methodology for GaN HEMTs for RF switching-mode power amplifiers design, *International Journal of RF and Microwave Computer-Aided Engineering*, Vol. 21, 2010, pp. 45-50.

[12]. A. Jarndal, A. Z. Markos, G. Kompa, Improved modeling of GaN HEMT on Si substrate for design of RF power amplifiers, *IEEE Transactions on Microwave Theory and Techniques*, Vol. 59, 2011, pp. 644-651.

[13]. D. Root, J. Xu, F. Sischka, M. Marcu, J. Horn, R. Biernacki, M. Iwamoto, Compact and behavioral modeling of transistors from NVNA measurements: New flows and future trends, in *Proceeding of the IEEE Custom Integrated Circuits Conference (CICC'12)*, San Jose, CA, USA, 2012, pp. 1–6.

[14]. O. Jardel, F. De Groote, T. Reveyrand, J.-C. Jacquet, C. Charbonniaud, J.-P. Teyssier, D. Floriot, R. Quéré, An electrothermal model for AlGaN/GaN power HEMTs including trapping effects to improve large-signal simulation results on high VSWR, *IEEE Transactions on Microwave Theory and Techniques*, Vol. 55, 2007, pp. 2660-2669.

[15]. I. Angelov, V. Desmaris, K. Dynefors, P. A. Nilsson, N. Rorsman, H. Zirath, On the large-signal modelling of AlGaN/GaN HEMTs and SiC MESFETs, in *Proceeding of the Gallium Arsenide and Other Semiconductor Application Symposium (EGAAS'05)*, Paris, October 2005, pp. 309-312.

[16]. A. Jarndal, Genetic-algorithm based neural-network modeling approach applied to AlGaN/GaN devices, *International Journal of RF and Microwave Computer Aided Engineering*, Vol. 23, 2013, pp. 149-156.

[17]. C. Fager, J. C. Pedro, N. B. Carvalho, H. Zirath, Prediction of IMD in LDMOS transistor amplifiers using a new large-signal model, *IEEE Trans Microwave Theory Technol*, Vol. 50, pp. 2834-2842.

[18]. I. Saidia, Y. Cordierb, M. Chmielowskab, H. Mejric, H. Maarefa, Thermal effects in AlGaN/GaN/Si high electron mobility transistors, *Solid-State Electronics*, Vol. 61, 2011, pp. 1-6.

[19]. J. K. Kaushik, V. R. Balakrishnan, B. S. Panwar, R. Muralidharan, On the origin of kink effect in current-voltage characteristics of AlGaN/GaN high electron mobility transistors, *IEEE Transactions on Electro Devices*, Vol. 60, 2013, pp. 3351-3357.

[20]. J. Haruyama, H. Negishi, Y. Nishimura, Y. Nashimoto, Substrate-related kink effects with a strong light-sensitivity in AlGaN/InGaAs PHEMT, *IEEE Transactions on Electron Devices*, Vol. 44, 1997, pp. 25-33.

[21]. D. DiSanto, H. Sun, C. Bolognesi, Ozone passivation of slow transient current collapse in AlGaN/ GaN field effect transistors: The role of threading dislocations and the passivation mechanism, *Applied Physics Letters*, Vol. 88, 2006, pp. 013504-1 - 013504-3.

[22]. S. Ghosh, S. Dinara, P. Mukhopadhyay, S. Jana, A. Bag, A. Chakraborty, Y. E. Chang, S. Kabi, D. Biswas, Effects of threading dislocations on drain current dispersion and slow transients in unpassivated AlGaN/GaN/Si heterostructure field-effect transistors, *Applied Physics Letter*, Vol. 105, 2014, pp. 073502-1 - 073502-5.

[23]. G. Dambrine, A. Cappy, F. Heliodore, E. Playez, A new method for determining the FET small-signal equivalent circuit, *IEEE Trans Microwave Theory & Tech*, Vol. 36, 1988, pp. 1151-1159.

[24]. A. Jarndal, AlGaN/GaN HEMTs on SiC and Si substrates: a review from the small-signal-modeling's perspective, *International Journal of RF and Microwave Computer Aided Engineering*, Vol. 24, 2014, pp. 389-400.

[25]. S. S. H. Hsu, P. Nguyen-Tan, D. Pavlidi, E. Alekseev, Frequency dependent output resistance and transconductance in AlGaN/GaN MODFETs, in *Proceeding of the International Semiconductor Devices Symposium (ISDRS'99)*, Charlottesville, VA, USA, December 1999, pp. 315-317.

[26]. D. E. Root, S. Fan, J. Meyer, Technology-independent large-signal FET models: a measurement-based approach to active device modeling, in *Proceeding of the 15th ARMMS Conf.*, Bath, U.K., September 1991, pp. 1-21.

[27]. A. Werthof, G. Kompa, A unified consistent DC to RF large-signal FET model covering the strong dispersion effects of HEMT devices, in *Proceeding of the European Microwave Conference (EuMC'92)*, Helsinki, Finland, October 1992, pp. 1091-1096.

[28]. A. Jarndal, P. Aflaki, L. Degachi, A. Birafane, A. Kouki, R. Negra, F. M. Ghannouchi, Large-signal model for AlGaN/GaN HEMTs suitable for RF switching-mode power amplifiers design, *International Journal of Solid State Electronics*, Vol. 54, 2010, pp. 696-700.

[29]. I. Melczarsky, J. A. Lonac, F. Filicori, A. Santarelli, Compact empirical modeling of nonlinear dynamic thermal effects in electron devices, *IEEE Trans. Microwave Theory Tech.*, Vol. 56, No. 9, 2008, pp. 2017-2024.

[30]. E. Kohn, I. Daumiller, M. Kunze, J. Van Nostrand, J. Sewell, T. Jenkins, Switching behavior of GaN-based HFETs: thermal and electronics transients, *Electron. Letters*, Vol. 38, 2002, pp. 603-605.

[31]. P. Rajagopal, J. Roberts, J. Cook, J. Brown, E. Piner, K. Linthicum, MOCVD AlGaN–GaN HFETs on Si: challenges and issues, in *Proceeding of the Materials Research Society Fall Meeting (MRS'03)*, Boston, MA, USA, December 2003, p. Y7.2.

[32]. A. Camacho-Penasola, C. Aitchison, Modeling frequency dependence of output impedance of a microwave MESFET at low frequencies, *Electron. Letters*, Vol. 21, 1985, pp. 528-529.

[33]. A. Jarndal, F. M. Ghannouchi, Improved modeling of GaN HEMTs for predicting thermal and trapping-induced-kink effects, *Journal of Solid State Electronics*, Vol. 123, 2016, pp. 19-25.

[34]. A. Darwish, A. J. Bayba, H. A. Hung, Channel temperature analysis of GaN HEMTs with nonlinear thermal conductivity, *IEEE Transactions on Electron Devices*, Vol. 62, 2015, pp. 840-846.

[35]. M. Wang M, K. J. Chen, Kink effect in AlGaN/GaN HEMTs induced by drain and gate pumping, *IEEE Electron Devices Letters*, Vol. 32, 2011, pp. 482-484.

[36]. R. Hassan, B. Cohanim, O. deWeck, G. Venter, Comparison of particle swarm optimization and the genetic algorithm, in *Proceeding of the 46th AIAA/ASME/ASCE/AHS/ASC Structures, Structural Dynamics and Materials Conference*, Austin, TX, USA, April 2005, pp. 1-13.

Chapter 10

Hot Carrier Injection (HCI) of High-k/Metal Gate MOSFET with Gate-Last Process

Hong Yang, Wenwu Wang and Weichun Luo

10.1. Reliability of HKMG MOSFET with Gate-Last Process

High-k/Metal Gate (HKMG) has been applied as gate stack in manufactory since 45 nm node technology, including planar device and 3D FinFET device [1-4]. According to CMOS integration technology, there are two kinds of HKMG process, one is gate-first, the other is gate-last, including high-k first and high-k last. Recently, HKMG gate-last process is widely accepted due to the effectively reducing the EOT, however, the low-temperature annealing in gate-last process induced more defects and traps in high-k material and interface between high-k and metal gate, which raise more heavier and complex reliability issues in front-end line of manufactory, such as Bias Temperature Instability (BTI), Time-Dependent Dielectrics Breakdown (TDDB), Stress Induced Leakage Current (SILC) and Hot Carrier Injection (HCI) [2-5].

10.2. Hot Carrier Injection Characteristics of HKMG MOSFET

Among the reliability issues of MOSFET with HKMG stack, HCI is becoming more and more important due to thin effective equivalent oxide (EOT) and short channel length, especially for n-type device [4, 5]. The "hot" in HCI means the carrier (usually electron) with high energy from impact ionization due to the accelerated carrier in the pinch-off region at saturation mode. The hot carrier could inject into the gate stack, then induce the electric degradation, such as threshold voltage (V_{th}), transconductance (g_m) and drain current in saturation (I_{dsat}) et.al. Among them, the Idsat is the most direct parameter to show the HCI degradation in manufactory. There are two HCI stress cases for MOSFET with HKMG stack, one is the peak-Isub case, the other one is Vg = Vd case.

Hong Yang
Key Laboratory of Microelectronics Devices & Integrated Technology, Institute of Microelectronics, Chinese Academy of Sciences, Beijing 100029, China

10.2.1. Peak-Isub Stress

For MOSFET with Poly/SiO$_2$ gate stack, the worst hot carrier degradation occurs at maximum of the substrate current (peak-Isub), when the impact ionization is the heaviest. There is still peak-Isub for MOSFET with high-k/Metal gate stack [1-3, 7]. This probably explained by aggressive scaling of EOT, especially for sub-1 nm case, for example, the Intel's HCI stress of HKMG device for 45 nm, 32 nm node technology still is peak-Isub case, as shown in Fig. 10.1 [3].

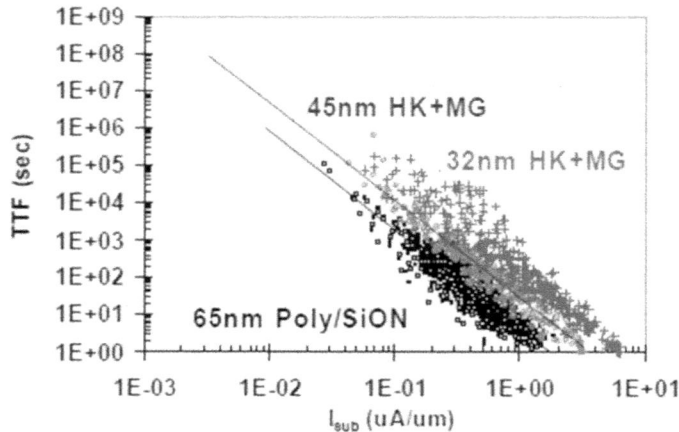

Fig. 10.1. Intel's HCI characteristics of core device in 65/45/32 nm node under peak-Isub stress [3].

10.2.2. Vg=Vd Stress

In traditional MOSFET with poly/SiO$_2$ gate stack, the peak-Isub stress is the worst case, however, with the shrinking of EOT, the Vg=Vd is becoming the worst HCI case due to the large BTI degradation [6, 7]. In our work, the relation between HCI lifetime and gate stress (Vg) under different drain stress is shown in Fig. 10.2 [8]. Here, the device with 0.8 nm EOT, Vgs ranges from 0.5 V to 2.4 V and Vds are 2.0/2.2/2.4 V. There is obvious two transitions in the curve of HCI lifetime and Vg. The 1st transition is related to the peak-Isub mode, which is the maximum electron impact ionization to induce worst lifetime. The 2nd transition is the shift from HCI to PBTI. So, the worst HCI stress isn't peak-Isub mode but Vg = Vd case.

Actually, it's difficult to accurately separate the influence of HCI and BTI, especially for device with short channel length, so in order to evaluate the HCI lifetime, the Vg = Vd stress is chosen in manufactory. Fig. 10.3 shows the Intel's HCI is measured under Vg = Vd case for core device in 22 nm and 14 nm node technology [4]. However, the HCI of I/O device in 22/14 nm node technology is still measured under peak-Isub stress.

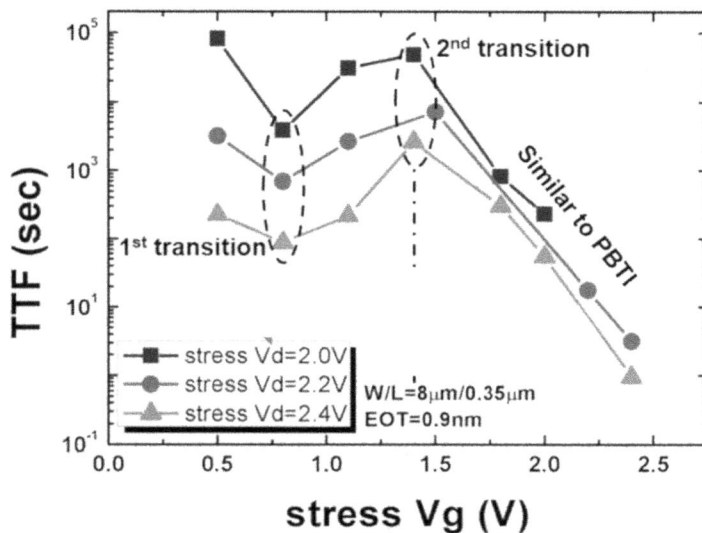

Fig. 10.2. The curve of HCI lifetime and Vg stress. There are two transitions, one is due to peak-Isub, the other is shift from HCI to PBTI [8].

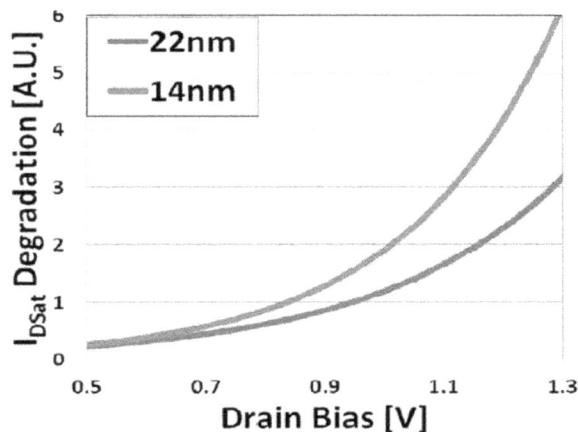

Fig. 10.3. Intel's HCI characteristics of core device in 22/14 nm node under Vg = Vd case [4].

10.3. Coupling between HCI and BTI

With the thin EOT, there is the coupling between HCI and PBTI [9, 10]. Physically, the "cold carrier" from PBTI is easy to inject to gate dielectrics under HCI stress due to thin EOT as shown in Fig. 10.4. There are "hot" carrier from HCI stress near to the drain part at pinch-off region and "cold" carrier from PBTI-like stress near to middle and source part. Usually, the "hot" carrier induces both interface trap and bulk trap degradation near to drain region, while the "cold" carrier does bulk trap degradation in gate stack.

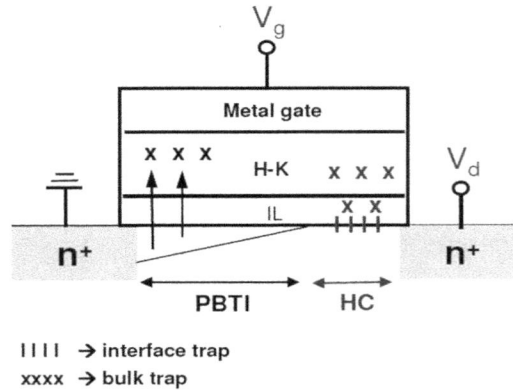

Fig. 10.4. The physical mechanism of coupling between HCI and PBTI [9].

10.3.1. A Method to Separate "Hot" Carrier and "Cold" Carrier Using DIBL

In order to separate "hot" carrier and "cold" carrier under HCI stress, in our work, the DIBL is focused and the Normal HCI stress and Reverse HCI stress are applied. Here, Reverse HCI stress means the drain bias (Vds) is applied in "source" terminal, in other word, the bulk trap is near source region and drain region under Normal HCI mode and Reverse HCI mode, respectively. The DIBL is simulated for both bulk trap in oxide near to source and drain region based on Synopsys tool. Simulation results show the obvious DIBL shift for both bulk trap near source and drain region. Fig. 10.5 shows the change of conduction band along channel length direction with and w/o bulk trap [11]. However, there isn't DIBL shift due to the interface trap under HCI stress. So, the DIBL shift is the key to difference "hot" carrier from HCI and "cold" carrier from PBTI. This method works for both device with poly/SiO$_2$ and HKMG gate stack.

10.3.2. Separation of "Hot Carrier" and "Cold Carrier"

Based on the method in 10.3.1, the DIBL is measured under Normal HCI stress and Reverse HCI stress with high and low Vg stress. Here, Vd is 2.4 V, Vg is 2.0 V and 0.8 V, stress time is 1000 seconds at room temperature. Our results show there is DIBL shift under high Vg stress (Vg=2.0 V), while there isn't DIBL shift under low Vg stress (Vg=0.8 V) in Fig. 10.6. It means that the "hot" carrier is dominant under low Vg stress case, while "cold" carrier does under high Vg stress case [11]. This is consistent to our results in Fig. 10.2 [8].

10.4. Physical Mechanisms of HCI

In HCI stress, the carrier accelerates through pinch-off region at saturation mode, and get high energy called "hot" carrier, then induces impact ionization and breaks the bond in the interface between silicon and oxide to generated interface trap. Next, the interface trap generation by hydrogen will be introduced.

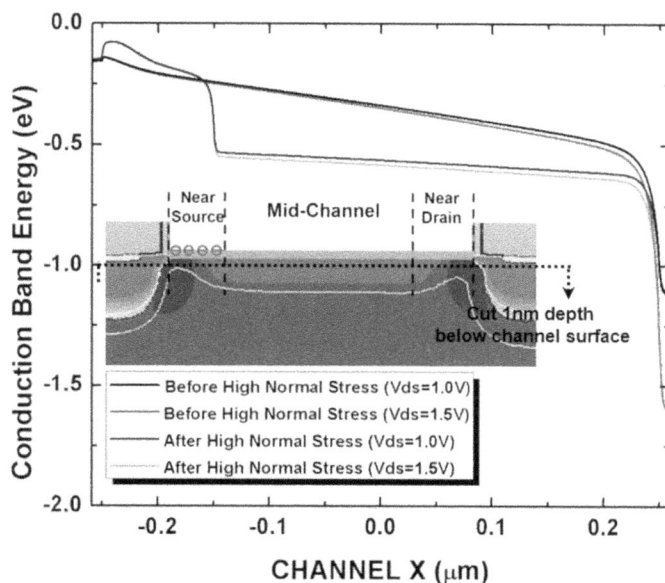

(a)

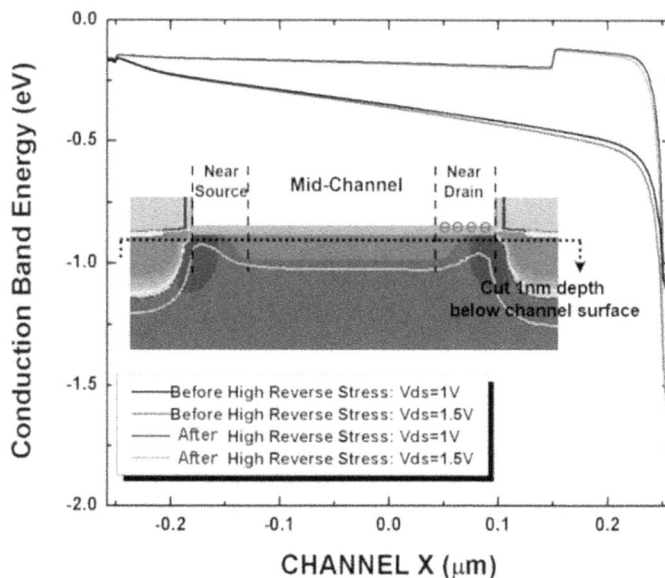

(b)

Fig. 10.5. The conduction band shift along channel length direction with bulk trap: (a) Near to source region (Normal HCI stress); (b) Near to drain region (Reverse HCI stress). The conduction band shift directly induces DIBL change [11].

(a)

(b)

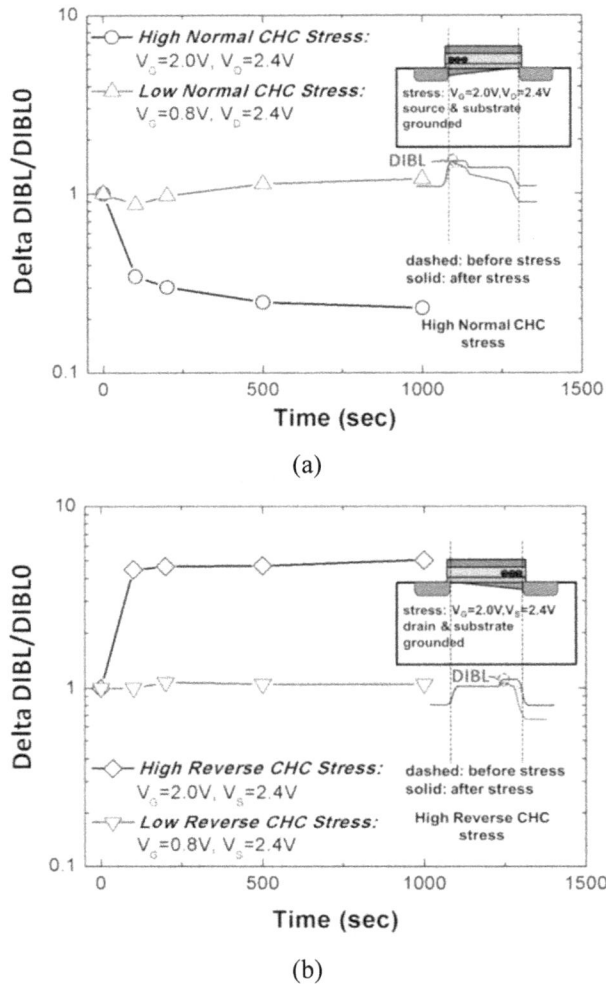

Fig. 10.6. DIBL shift of MOSFET (a) Under Normal HCI stress, and (b) Under Reverse HCI stress [11].

Regards to the interface trap generation during HCI, the R-D model by hydrogen is widely accepted [12]. Usually, the hot carrier with high energy to break the bond of silicon and hydrogen in the interface of silicon and gate dielectrics, then the hydrogen diffuses to the gate stack by hydrogen atom, hydrogen ion or hydrogen molecule, as shown in Fig. 10.7.

10.5. HCI Lifetime Extrapolation

The HCI lifetime extrapolation is to extract the drain voltage (or substrate leakage) in which bias the device could work well at least 10 years, based on the results under high voltage stress case. Usually, the 10 years lifetime is expected to larger than 1.1 times of Vcc.

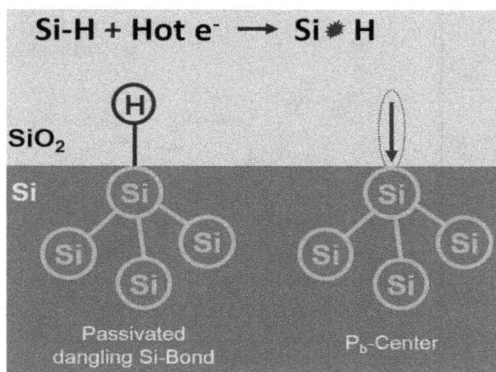

Fig. 10.7. Diagram of the Si-H bond broken and interface trap generation.

10.5.1. Power-law of TTF and Isub

In peak-Isub HCI stress mode, the curve of Time-to-Failure (TTF) and Isub is the key to extrapolation HCI lifetime, as shown in Fig. 10.1. There is the power-law of TTF and Isub, which is consistent to JEDEC's Eq. 10.1 [13]. Here, TTF is Time to Fail; B is random constant, which is dependent on doping distribution, the spacer size et.al.; Isub is the peak-Isub; N is from 2 to 4; Eaa is action energy, and from -0.2~0.4 eV; k is Boltzmann constant; T is kelvin's temperature. Usually, in the curve of TTF and Isub to extract Isub then to get the drain voltage.

$$TTF = B * x + b\left(I_{sub}\right)^{-N} * \exp\left(E_{aa} / kT\right).$$ (10.1)

10.5.2. Two-Step in Curve of TTF and Isub

In our work, different from Fig. 10.1, the two-step curve of TTF and Isub is shown in Fig. 10.8 [14]. If extrapolate the HCI lifetime for whole curve, the lifetime is underestimated using power-law of TTF and Isub in blue points. Physically, the blue case shows two-step, one is called lower Vd case, the other one is called higher Vd case. Based on our method to separate "hot" and "cold" carrier, under higher Vd the bulk trap from PBTI is dominant, while under lower Vd the interface trap from HCI is dominant. So, it's not always accurate to extract HCI lifetime for full curve of TTF and Isub.

10.6. Conclusions

In this Chapter, the HCI of HKMG MOSFET with gate-last process is summarized, including the measurement, coupling of HCI and BTI, physical mechanism and lifetime extrapolation. With the shrinking of channel length, the coupling of HCI and BTI is becoming more and more serious, so the worst HCI case isn't peak-Isub but Vg = Vd, however, peak-Isub stress is the case to show the intrinsic HCI.

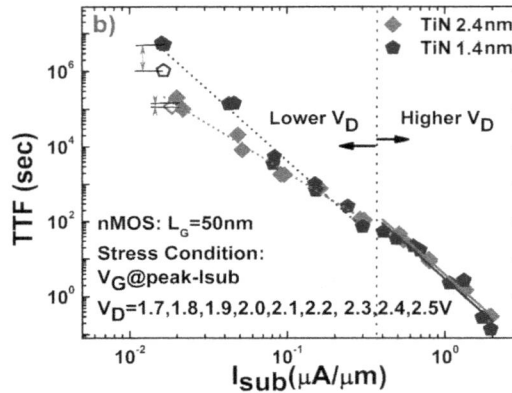

Fig. 10.8. Two-step in curve of TTF and Isub as shown in blue points [14].

References

[1]. C. Auth, A. Cappellani, J. S. Chun, A. Dalis, A. Davis, T. Ghani, G. Glass, T. Glassman, M. Harper, M. Hattendrof, P. Hentges, S. Jaloviar, S. Joshi, J. Klaus, K. Kuhn, D. Lavric, M. Lu, H. Mariappan, K. Mistry, B. Norris, N. Rahhal-orabi, P. Ranade, J. Sandford, L. Shifren, V. Souw, K. Tone, F. Tambwe, A. Thomspon, D. Towner, T. Troeger, P. Vandervoorn, C. Wallace, J. Wiedemer, C. Wiegand, 45 nm high-k + metal gate strain-enhanced transistors, in *Proceedings of the Symposium on VLSI Technology (VLSIT'08)*, June 2008, pp. 128-129.

[2]. P. Packan, S. Akbar, M. Armstrong, D. Bergstrom, M. Brazier, H. Deshpande, K. Dev, G. Ding, T. Ghani, O. Golonzka, W. Han, J. He, R. Heussner, R. James, J. Jopling, C. Kenyon, S-H. Lee, M. Liu, S. Lodha, B. Mattis, A. Murthy, L. Neiberg, J. Neirynck, S. Pae, C. Parker, L. Pipes, J. Sebastian, J. Seiple, B. Sell, A. Sharma, S. Sivakumar, B. Song, A. St. Amour, K. Tone, T. Troeger, C. Weber, K. Zhang, Y. Luo, S. Natarajan, High performance 32 nm logic technology featuring 2nd generation high-k + metal gate transistors, in *Proceedings of the IEEE International Electron Devices Meeting (IEDM'09)*, December 2009, pp.1-4.

[3]. S. Pae, A. Ashok, J. Choi, T. Ghani, J. He, S. H. Lee, K. Lemay, M. Liu, R. Lu, P. Packan, C. Parker, R. Purser, A.S. Amour, B. Woolery, Reliability characterization of 32 nm high-k and metal-gate logic transistor technology, in *Proceedings of the International Reliability Physics Symposium (IRPS'10)*, 2010, pp. 287-292.

[4]. S. Novak, C. Parker, D. Becher, M. Liu, M. Agostinelli, M. Chahal, P. Packan, P. Nayak, S. Ramey, S. Natarajan, Transistor Aging, Reliability in 14 nm Tri-Gate technology, in *Proceedings of the IEEE International Reliability Physics Symposium (IRPS'15)*, 2015, pp. 2.F.2.1-2.F.2.5.

[5]. C. Prasad, K. W. Park, M. Chahal, I. Meric, S. R. Novak, S. Ramey, Transistor reliability characterization and comparisons for a 14 nm Tri-gate technology optimized for system-on-chip and foundry platforms, in *Proceedings of the IEEE International Reliability Physics Symposium (IRPS'16)*, 2016, pp. 4B-5-1 - 4B-5-8.

[6]. J. H. Sim, B. H. Lee, C. Rino, S. C. Song, G. Bersuker, Hot carrier degradation of HfSiON gate dielectrics with TiN electrode, *IEEE Transactions on Device and Materials Reliability*, Vol. 5, No. 2, June 2005, pp. 177-182.

[7]. E. Amat, T. Kauerauf, R. Degraeve, R. Rodrıguez, M. Nafrıa, X. Aymerich, G. Groeseneken, Channel hot-carrier degradation in short-channel transistors with high-k/metal gate stacks, *IEEE Transactions on Device and Materials Reliability*, Vol. 9, Jun. 2009, pp. 425-430.

[8]. W. Luo, H. Yang, W. Wang, H. Xu, S. Ren, B. Tang, Z. Tang, Y. Xu, J. Xu, J. Yan, C. Zhao, D. Chen, T. Ye, Channel hot-carrier degradation characteristics and trap activities of high-k/metal gate nMOSFETs, in *Proceedings of the 20th IEEE International Symposium on the Physical and Failure Analysis of Integrated Circuits (IPFA'13)*, 2013, pp. 666-669.

[9]. N. H. H. Hsu, J. W. You, H. C. Ma, S. C. Lee, E. Chen, L. S. Huang, Y.C. Cheng, O. Cheng, I. C. Chen, Intrinsic hot-carrier degradation of nMOSFETs by decoupling PBTI component in 28 nm high-k/metal gate stacks, in *Proceedings of the IEEE International Reliability Physics Symposium (IRPS'12)*, 2012, pp. XT.13.1-XT.13.4.

[10]. E. Amat, R. Rodriguez, M. Nafria, X. Aymerich, T. Kauerauf, R. Degraeve, G. Groeseneken, New insights into the wide Id range channel hot-carrier degradation in high-k based devices, *in Proceedings of the IEEE International Reliability Physics Symposium (IRPS'09)*, 2009, pp. 1028-1032.

[11]. W. Luo, H. Yang, W. Wang, L. Zhao, H. Xu, S. Ren, B. Tang, Z. Tang, Y. Xu, J. Xu, J. Yan, C. Zhao, D. Chen, T. Ye, Physical understanding of different drain-induced-barrier-lowering variations in high-k/metal gate n-channel metal oxide semiconductor field effect transistors induced by charge trapping under normal and reverse channel hot carrier stresses, *Applied Physics Letters*, Vol. 103, 2013, pp. 183502-1 – 183502-4.

[12]. H. Chenming, C. T. Simon, H. Fu-Chieh, K. Ping-Keung, C. Tung-Yi, K. W. Terrill, Hot-electron-induced MOSFET degradation - model, monitor, and improvement, *IEEE Journal of Solid-State Circuits*, Vol. 20, 1985, pp. 295-305.

[13]. Failure Mechanisms and Models for Semiconductor Devices, *Jedec Solid State Technology Association*, JEP122G, 2010.

[14]. W. Luo, H. Yang, W. Wang, Y. Xu, B. Tang, S. Ren, H. Xu, Y. Wang, L. Qi, J. Yan, H. Zhu, C. Zhao, D. Chen, T. Ye, Accurate lifetime prediction for channel hot carrier stress on sub-1 nm equivalent oxide thickness HK/MG nMOSFET with thin titanium nitride capping layer, *Microelectronics Reliability*, Vol. 62, 2016, pp. 70-73.

Chapter 11

Sol-Gel Oxides Spin-Coating
and Dye-Sensitized Solar Cell Performance

Amar Merazga and Ateyyah Al-Baradi

11.1. Introduction

Dye-Sensitized Solar Cells (DSSCs) based on nanocrystalline TiO_2 have attracted extensive attention in academic research and industrial application since O'Regan et al. reported their breakthrough DSSC discovery in 1991 [1]. DSSCs may offer an alternative to conventional semiconductor p-n junction solar cells due to their low cost and relatively high efficiency.

In the literature, applications of sol-gel oxides are mostly related to the study of film properties [2-9]. Apart from TiO_2 and, less frequently ZnO, none is found to be directly related to DSSC as the main semiconductor component. Applications of sol-gel oxides in DSSCs [10-12] are rather concerned with the active issue related to surface modification of the TiO_2 electrode by thin oxide layers to improve the associated DSSC performance [13-17]. In particular, TiO_2 surface modification by the TiO_2 material itself has been considered as a standard post-treatment. It consists of coating the nanoporous TiO_2 film of the DSSC photo-electrode using a $TiCl_4$ chemical bath [13, 14]; The photo-electrode with screen-printed TiO_2 film is sintered and freshly immersed into diluted aqueous solution of $TiCl_4$ before being sintered again. This post-treatment has been the best surface cure to apply towards a homogeneously thick nanoporous TiO_2 film in DSSC applications. The effects of this treatment, which manifest in DSSC performance as an increase in the short-circuit photocurrent J_{SC}, are a downward shift in the TiO_2 conduction band (CB) potential and a reduction in the rate of electron back-recombination to the electrolyte [14]. The consequent increase in the efficiency of interfacial charge separation is the reason for the J_{sc} increase, whereas the expected shift of the open-circuit V_{OC} due to TiO_2 CB shift is offset by charge accumulation at V_{OC} due to retarded electron recombination [14].

Spin-coating is a useful deposition technique to produce nanostructured materials in the form of thin films. It is often applied in connection with the sol-gel preparation technique

Amar Merazga
Physics department, Taif University, Taif, Saudi Arabia

[18-21] to form the so-called sol-gel spin-coating method to prepare homogeneous thin nanostructured films. The present work is concerned with the study of the effect on DSSC performance of an oxide layer as deposited onto the TiO_2 photo-electrode of the DSSC by means of the sol-gel spin-coating method. The effect of two sol-gel oxides, taken as typical examples, will be studied, namely the ZnO with comparable band-gap to that of the post-treated TiO_2 film (3.4 eV) and the MgO with much wider band-gap (7.8 eV).

In general, the effects of an oxide post-treatment of the TiO_2 surface, causing DSSC efficiency (η) enhancement, are mainly changes in electron recombination or injection rates and changes in light harvesting efficiency resulting from changes in the amount of adsorbed dye and the proportion of dye-absorbed light [17]. Different coating techniques of ZnO and MgO oxides onto the nanoporous TiO_2 film were reported and various interpretations of the enhancement of η were proposed [10, 11, 22-24].

Liu et al. [25] have deposited a thin ZnO blocking layer by spin-coating as a buffer at the interface between the F-doped SnO_2 (FTO) layer and the TiO_2 film. They have been able to monitor the level of the DSSC short circuit current J_{sc} by adjusting the ZnO precursor concentration. As the latter decreases, J_{sc} increases to approach the bare TiO_2 case. The accumulation of electrons in the TiO_2 conduction band, due to the energy barrier created by the ZnO layer between the TiO_2 and the FTO films, shifts up the Fermi-level and increases V_{oc} as a result. An optimum of the precursor concentration at 0.2 M was found to promote η from 5.85 % to 6.7 %. Preceding this study, Kao et al. [26] employed the sol-gel spin-coating technique to fabricate uncoated and ZnO spin-coated TiO_2 electrodes for DSSCs. Multilayer coating was applied to TiO_2 to reach the required thickness of nanoporous film and a monolayer of ZnO was applied to coat the TiO_2 film. All the photovoltaic (PV) parameters were increased upon ZnO spin-coating, resulting in an increase of η from 2.5 % to 3.25 % (30 % enhancement). This was attributed to recombination rate reduction following creation of energy barrier by the coating ZnO layer.

In early work on solid state DSSC by Taguchi et al. [22], TiO_2 nanoporous film was soaked in 0.05 M magnesium methoxide solution at 40 °C for 10 min before being sintered at 450 °C for 20 min. The 36 % enhancement in DSSC efficiency was explained to be essentially due to V_{OC} increase from 0.4 V to 0.5 V as a result of reduced recombination rate and consequent upward shift of the Fermi-level. The slight increase of J_{SC} was attributed to about 20 % increase in the adsorbed dye amount. Sol-gel MgO post-treatment of the TiO_2 electrode in solid state DSSC was further studied by Kumara et al. [23] as a function of MgO layer thickness using magnesium acetate solution for chemical bath preparation. They observed a simultaneous decrease in J_{SC} and increase in V_{OC} with increasing MgO thickness and determined an optimum at 0.5 nm for maximum efficiency. They interpreted this effect of MgO post-treatment in terms of a balance between reduction of electron injection rate and reduction of electron recombination rate, both caused by the MgO layer of increasing thickness. Li et al. [11] used a similar chemical bath method to optimize the performance of electrolyte-based DSSC by adjusting the dipping time of the TiO_2 electrode. Here, V_{OC} decreased monotonically with increasing dipping time and η followed the photocurrent J_{SC} in having a maximum at an optimal dipping time of 20 min. 21 % improvement over the uncoated TiO_2 case was achieved. A similar behavior of J_{SC} and η was measured by Wu et al. [24] as a function of sputtering time when using the dc-magnetron sputtering for MgO coating. They observed maxima in J_{SC} and η at the sputtering time of 3 min. The study results revealed that the MgO layer corresponding to 3 min time has the optimal thickness which can enhance J_{SC} by passivation of the TiO_2 trap states and reduction of electron recombination rate without seriously affecting the electron injection rate. Jang et al. chose a different sol-gel MgO technique [10], where magnesium methoxide solution in methanol was mixed to TiO_2 nano-powder at an optimized proportion of 0.6 % wt before deposition and sintering on a transparent conducting substrate. Both J_{SC} and V_{OC} shifted considerably upward in this MgO post-treatment resulting in 45 % increase in electrolyte-based DSSC efficiency. In addition to recombination suppression effect causing V_{OC} shift, the topotactic thermal transfer from hydroxide $Mg(OH)_2$ to lower density oxide MgO in the deposition process was believed to be the cause of excess porosity in the TiO_2/MgO film. The excess porosity in the film implies an excess amount of adsorbed dye which is the reason behind the J_{SC} and η increase.

We further examine the variation of η as a function of the deposited amount of oxide. It was realized that the oxide amount can be expressed in terms of the oxide precursor concentration and/or in terms of the number of oxide sol drops during the spin-coating operation [16]. Optimisation of the oxide amount for the highest possible DSSC η can then be achieved by combined adjustment of these two oxide parameters. We prepared ZnO and MgO colloidal sols at different precursor concentrations (from 0.1 M to 0.6 M) using a sol-gel technique. We then deposited a ZnO (or MgO) layer by spin-coating an amount of one or more sol drops onto the TiO_2 electrode and subsequently drying and sintering it. The so-prepared ZnO (or MgO)-coated TiO_2 electrodes were dye-loaded and used as working electrodes in DSSCs. The variation of the DSSC η showed maximum around an optimum of a single drop at 0.2 M concentration for ZnO and a single drop at 0.1 M concentration for MgO. DSSC η enhancement of over 100 % was recorded for the ZnO post-treatment and over 50 % η enhancement was recorded for the MgO treatment.

It is shown that this optimal amount of oxide forms the thinnest oxide layer that can create the required energy barrier to reduce considerably the electron back-recombination, thus enhancing the photocurrent without seriously affecting the dye-adsorption efficiency of the TiO_2 film. Further increase of oxide amount beyond the optimum results in oxide layers thick enough to decrease the J_{SC} and η through decreasing the dye-adsorption efficiency of the TiO_2 and the electron injection rate.

11.2. Experimental Methods

TiO_2 was deposited on FTO-coated glass using the screen printing technique [27]. Two layers of 4 μm were successively printed and dried at 120 °C. The 8 μm film was then fired for 1 h at 475 °C. The purchased 10 cm × 10 cm TiO_2-coated FTO-glass (Solaronix) was cut into similar 2 cm × 2 cm TiO_2 electrodes to be used for preparation of our DSSCs. XRD patterns from the bare FTO-coated substrate and from the FTO-coated substrate with deposited TiO_2 film are shown in Fig. 11.1. The TiO_2 peaks indicated by asterisks at the angles 26.08°, 48.76° and 55.22° correspond to anatase phase [28, 29] with the predominant crystal orientation (101) [29].

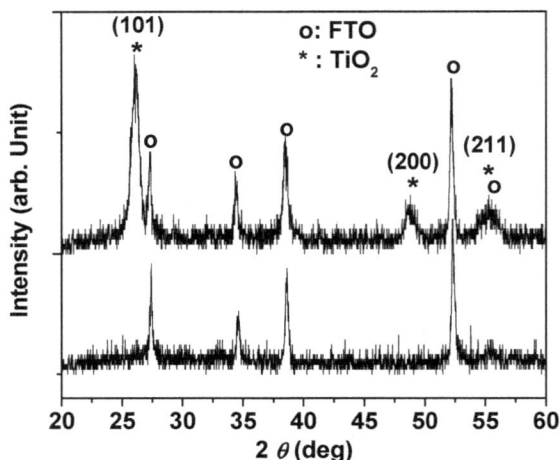

Fig.11.1. XRD patterns from the FTO-coated substrate (bottom curve) and the TiO_2 film on FTO-coated substrate (top curve). The TiO_2 peaks are indicated by asterisks.

The sol-gel spin-coating method was employed to deposit thin layers of ZnO (or MgO) on the surface of the TiO_2 electrode. In the case of ZnO, the precursor, zinc acetate dihydrate $[Zn(CH_3COO)_2.2H_2O]$, was mixed with the solvent, 2-methoxyethanol, at different concentrations. Mono-ethanolamine, as a stabilizer, was added dropwise until satisfying the molar ratio 1:1 with the precursor while the solution was being heated at 60 °C under continuous magnetic stirring [25]. In the case of MgO, magnesium acetate tetrahydrate $[Mg(CH_3COO)_2.4H_2O]$ was used as MgO precursor at different concentrations in ethanol. Few drops of nitric acid were added to serve as stabilizer, while the solution under continuous magnetic stirring was being heated at 60 °C [8]. The prepared solution containing suspended ZnO (or MgO) nanoparticles was then kept in a clean and firmly closed flask before spin-coating. The spin-coater (SPS Spin-150) was adjusted to run at a speed of 3000 rpm for 30 s to produce reasonable thin films.

The oxide amount at a fixed precursor concentration was monitored by the number of oxide sol drops injected through the needle of a 'smart' dispenser by carefully advancing the plunger into the barrel. In Fig. 11.2, we plot the weight (mg) of released ZnO sol amount versus the contained number of drops in this amount. Such a calibration of the sol dispenser clearly indicates that the different drops are identical with a homogeneous material of about the same weight. It also indicates that the number of drops in the released sol content is proportional to its weight, and therefore the number of drops is reliable as a parameter to monitor the coating ZnO amount.

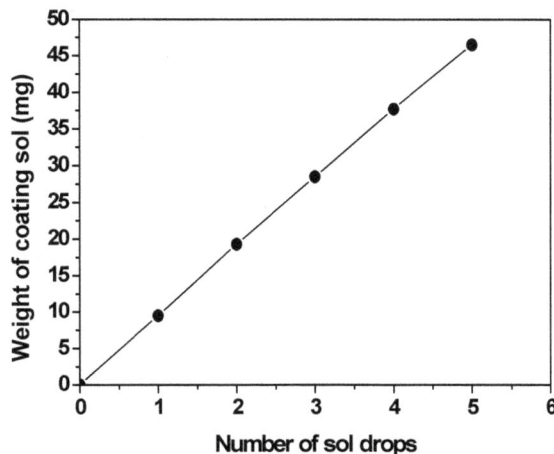

Fig. 11.2. Calibration of the ZnO sol dispenser: Weight of ZnO sol amount versus the contained number of drops.

The ZnO (MgO)-coated TiO_2 electrodes were then dried and sintered at 400 K (500 K) for one hour. Ruthenium-based dye ($C_{26}H_{20}O_{10}N_6S_2Ru$) known as N3 (Solaronix) was used to sensitize both uncoated and ZnO (or MgO)-coated TiO_2 electrodes by immersing in 3×10^{-4} M solution in ethanol. After dye-loading, half cm of the material was peeled off from the side to leave uncovered the FTO for electrical contact. Another FTO-glass substrate coated with platinum catalyst (Solaronix) was assembled as a counter-electrode against the dye-loaded electrode by means of two paper clips. Iodide-based electrolyte (Solaronix) containing Iodide/ Triodide redox couple with 100 mM tri-iodide concentration was injected through a capillary channel originally drilled across the counter-electrode. The illuminated area of 0.25 cm^2 was determined using a black tape mask on the glass side of the photo-electrode. The current/voltage (*I-V*) characteristics of the DSSCs were measured using the photovoltaic system consisting of a solar simulator (Solar-Light) and an electrometer (Keithley 2400). The latter is computer-controlled to acquire and plot the *I-V* data, while AM 1.5-filtered light from the 300 W-Xenon lamp of the solar simulator shines the DSSC at a power density $P_i = 100$ mW/cm^2. The absorbance spectra of the different films and dye solutions were measured by means of a UV/VIS/NIR spectrophotometer (Jasco V-570).A Tektronix 500 MHz oscilloscope was used to measure the DSSC photovoltage V_{oc}-decay over a time range of 10s for recombination test. Field emission scanning electron-microscope (FESEM, quanta FEG 250, FEI) was used in SEM imaging for surface morphology.

11.3. Results and Discussions

11.3.1. ZnO Results

A TEM image of the prepared 0.4 M-ZnO colloidal solution is shown in Fig. 11.3 to identify the nature and size of the coating ZnO particles prior to optical and photovoltaic characterization of the ZnO-coated electrodes and DSSCs. Well observed crystallites of an average size of 100 nm are illustrated. Song et al. [30] have reported similar TEM-observed crystallites of ZnO prepared by the same sol-gel method described above.

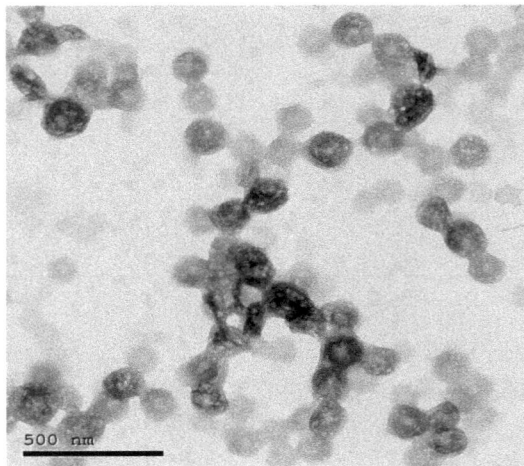

Fig.11.3. Suspended 0.4M-ZnO particles as observed by TEM.

Fig. 11.4 a shows the *I-V* characteristics of the DSSC with bare TiO_2 electrode as a reference to those of the DSSCs with ZnO-coated TiO_2 electrodes prepared at 5 different ZnO amounts (1-5 drops). In Table 11.1, we present the PV parameters extracted from the *I-V* curves of Fig. 11.4a. At the smallest ZnO amount (1 drop), there is a clear enhancement of the short circuit photocurrent J_{SC} from 5.73 to 7.01 mA/cm^2, while a smaller upward shift from 0.58 to 0.65 V is observed for the open-circuit voltage V_{OC} and a practically constant fill factor $FF\sim62.5$ % is found at this level of ZnO amount. Further increase of the ZnO amount causes J_{SC} to fall drastically to zero. V_{OC} shows a small decrease to 0.54 V in the range between 1 drop and 2 drops, whereas FF remains practically unchanged in this range, beyond which both V_{OC} and FF decrease remarkably.

The resulting conversion efficiency $\eta = [(FF \times J_{SC} \times V_{OC})/P_i]$ is plotted in Fig. 11.4 b, where a maximum $\eta \sim 3$ % is observed at the smallest ZnO amount (1 drop) followed by a sharp decrease to 0 %.

It is known in the literature that the presence of a metal oxide layer deposited on the TiO_2 electrode with a lower electronic affinity (higher conduction band energy minimum), such as the ZnO case, creates an energy barrier that can prevent the injected electrons from the dye to recombine back to dye molecules or to electrolyte species [25, 26, 31]. However, there must be an optimum of the amount of ZnO in the deposited layer which can be considered as a critical transition between two regions: Region of small amounts below the optimum (<1 drop), where the energy barrier vanishes and the back-recombination rate regains the case of bare TiO_2 electrode and the region above the optimum (>1 drop), where the dye-adsorption efficiency of the TiO_2 film is reduced by the coating ZnO material, which reduces the light harvesting rate and the electron injection rate as a consequence.

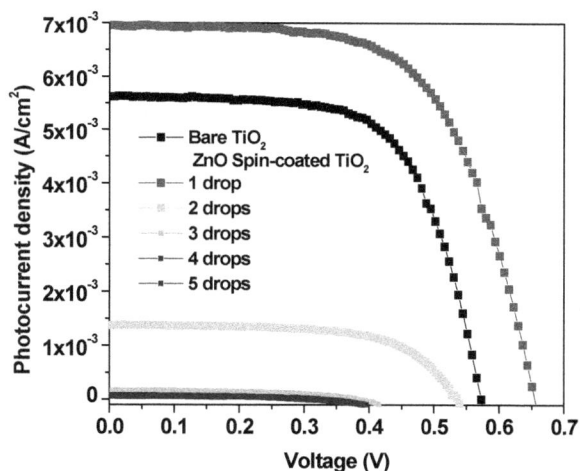

Fig. 11.4a. I-V-characteristics of ZnO spin-coated TiO₂ DSSCs at different 0.4M-ZnO amount (1, 2, 3, 4 and 5 drops).

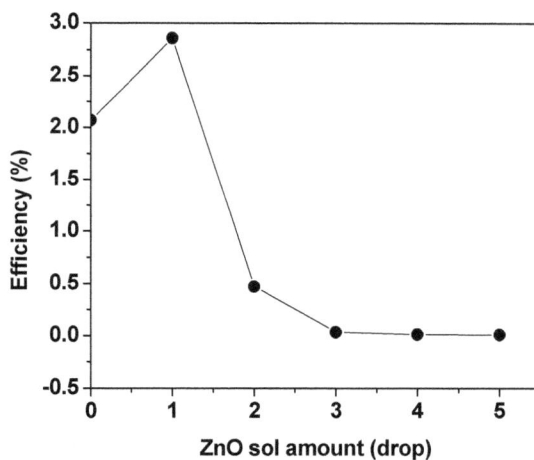

Fig. 11.4b. Variation of η with ZnO amount (number of sol drops).

Table 11.1. PV parameters of 5 DSSCs fabricated with ZnO-coated TiO₂ electrodes for 5 different amounts of 0.4M-ZnO (1, 2, 3, 4 and 5 drops).

ZnO amount:(Nº of sol drops)	PV parameters		
	J_{SC} (mA/cm²)	V_{OC}(V)	FF (%)
0	5.73	0.576	62.6
1	7.01	0.654	62.3
2	1.40	0.536	63.1
3	0.179	0.386	46.7
4	0.084	0.337	40.1
5	0.174	0.209	25.1

Fig. 11.5 shows the dark *I-V* curves of the DSSCs for the five different amounts (1-5 drops) of the 0.4 M-ZnO sol, as compared to the dark *I-V* curve associated with the uncoated TiO₂ DSSC. In dark, the electron injection from the dye is totally absent and the current is limited to electron diffusion from the TiO₂ semiconductor to the electrolyte [32].

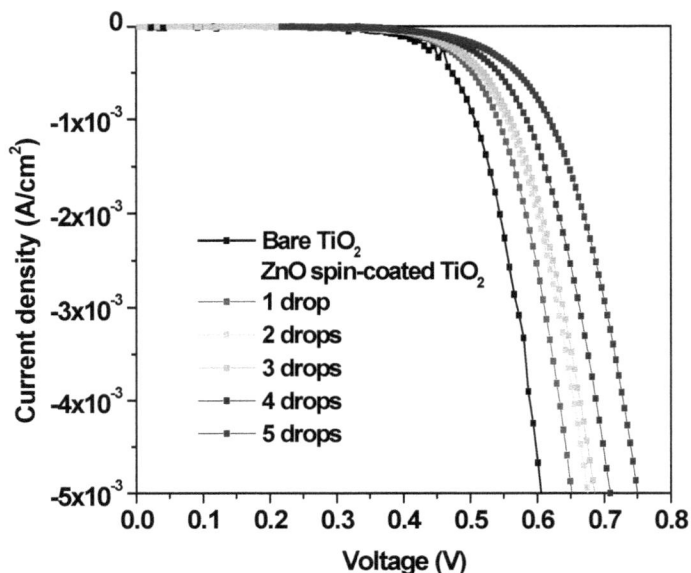

Fig. 11.5. Dark *I–V* characteristics of the DSSCs with 0.4 M-ZnO spin-coated TiO₂ electrodes for 5 different ZnO amounts (1, 2, 3, 4 and 5 drops), as compared to that of the DSSC with bare TiO₂ electrode.

As the ZnO amount increases, the current decreases and requires the assistance of a higher voltage to regain the value of the uncoated TiO₂ case. In other words, the electrons require higher voltages to overcome energy barriers created by larger amounts of spin-deposited ZnO. The dark *I–V* characteristics of the DSSCs have also been reported by Liu [25] to explain the reduction of the back-recombination rate under illumination with increasing the ZnO precursor concentration. Despite the existence of the required energy barrier to reduce the rate of back-recombination more effectively in the region above the optimum, the photocurrent and the DSSC efficiency show sharp decreases with increasing ZnO sol amount. The reasonable cause of this must be a sharp decrease of the injection rate as a result of the decrease of dye-adsorption efficiency of the TiO₂ film. The factors at the origin of this decrease can be revealed by the measurement and comparison of the absorbance spectra of the different ZnO-coated TiO₂ films with loaded dye. In Fig. 11.6, we present the absorbance spectra of 4 different dye-loaded TiO₂ films that were pre-coated by 0.4 M–ZnO using 4 different amounts (1-5 drops). The absorbance spectra of the dye-loaded uncoated TiO₂ film and the bare TiO₂ film are included for comparison.

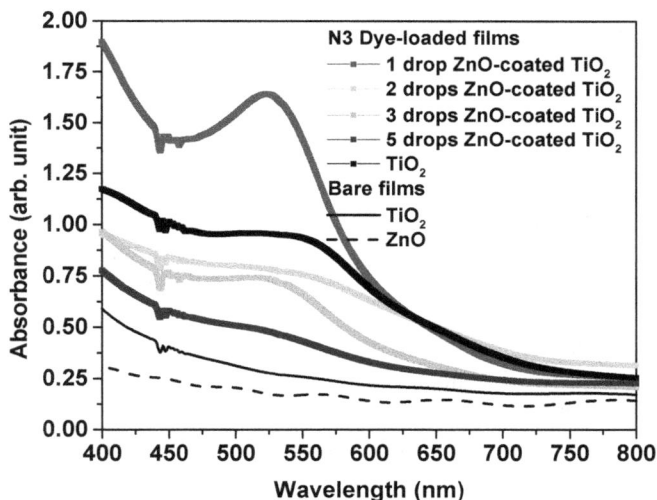

Fig. 11.6. Absorbance spectra of 4 different dye-loaded TiO_2 films pre-coated with 0.4 M-ZnO layers using 4 different ZnO amounts (1, 2, 3 and 5 drops), as compared to that of the dye-loaded uncoated TiO_2 film (curve with symbol ■). The spectra of bare TiO_2 (solid line) and ZnO (dashed line) films are included for comparison.

We can first compare the spectrum of the dye-loaded TiO_2 with that of the bare TiO_2. It is noticeable that the spectrum of the bare TiO_2 film (solid line) is much lower than that of the dye-loaded uncoated TiO_2 film (curve with symbol ■), which means that the dye molecules form the main light absorber in the visible range. Fig. 11.6 contains also the absorbance spectrum of the bare 0.4 M-ZnO film (dashed line) as spin-coated on FTO-glass substrate using 10 sol drops, which is even lower than that of the bare TiO_2 film. Therefore, one would expect that the dye molecules will still form the main absorber in the presence of the coating ZnO layer and that the spectra of the dye-loaded ZnO-coated TiO_2 films will be similar to that of the dye-loaded uncoated TiO_2 film. In contrast, the coating of the TiO_2 film using only a single drop of 0.4 M–ZnO sol results in a much higher spectrum (red curve) than that of the uncoated TiO_2 film. In fact, the chemical instability of ZnO against acidic dye molecules results in partial dissociation of Zn^{2+} ions from ZnO particles during the dye-loading process [33, 34]. The Zn^{2+}-dye aggregates thus formed on the TiO_2 film should be responsible of the observed high excess absorption in this case of a single drop of ZnO sol. This light loss by Zn^{2+}-dye absorption does not affect the photocurrent because the one-drop ZnO layer is very thin and the surface density of the ZnO particles on the TiO_2 film is so low that the dye adsorption efficiency of the TiO_2 film is still high and not seriously affected by the ZnO coating. Yet, the necessary energy barrier to minimize the rate of electron back-recombination is established, causing the enhancement of the photocurrent and DSSC efficiency as detailed above (Fig. 11.4 a and 11.4 b). With increasing the amount of coating ZnO, this will partially screen the dye molecules from being efficiently adsorbed into the pores of the TiO_2 film. The effective dye-covered area reduces and the electron light harvesting efficiency reduces as a consequence. The ZnO precursor concentration of 0.4 M used here is high enough, so that only a second drop of ZnO sol can cause a remarkable reduction in the dye adsorption and

301

light harvest efficiency. Thereby, the DSSC photocurrent (Fig. 11.4 a, green curve) shifts well below that of the bare TiO$_2$ case (Fig. 11.4 a, black curve), and the DSSC efficiency reduces to about 0.5 % (Fig. 11.4 b). Further increase of the ZnO amount to 3 and 5 drops will vanishingly reduce the dye-adsorption efficiency, resulting in absorbance spectra even further lower than that of the bare TiO$_2$ film (Fig. 11.6, purple and brown curves). The ZnO layer will entirely screen the dye and one should rather consider dye adsorption by the ZnO which is known to be of much lower efficiency than that of the TiO$_2$ [35].

An estimation of the amount of adsorbed dye in the ZnO-coated TiO$_2$ electrodes can be achieved by proceeding with the dye desorption technique [36, 37]. Fig. 11.7 (a) shows the absorbance spectra for solutions of desorbed dye from these ZnO-coated TiO$_2$ electrodes. The solution for dye-desorption consists of 4 ml of 0.1 M sodium hydroxide (NaOH) and ethanol, mixed at equal volumes, in which the dye-loaded electrode is immersed. As can be observed, the lower the number of ZnO sol drops, the higher is the absorbance of the desorbed dye, which means that the amount of adsorbed dye is higher in TiO$_2$ films of lower coated ZnO sol amount. The absorbance spectrum plotted with black line in Fig. 11.7 (a) is measured for a prepared N$_3$-dye solution at 0.1 mg/ml. This is to serve as a reference to estimate the adsorbed dye amount (mg) in each ZnO-coated TiO$_2$ film from the corresponding absorbance spectrum. In Fig. 11.7 (b) we plot the desorbed dye amount (mg) against the ZnO sol amount (number of drops) as estimated from Fig. 11.6 (a) at the peak wavelength around 500 nm with assumed proportionality between the absorbance and concentration of the dye.

It is worth noting that the trend of the curve related to adsorbed dye amount with increasing ZnO amount (Fig. 11.7 (b)) is not as sharp as that of the DSSC efficiency in Fig. 11.4 (b). Therefore, the decrease of TiO$_2$ dye-adsorption efficiency with increasing ZnO amount is not the only factor to account for the sharp fall of the DSSC efficiency, but also the increase of light loss due to absorption by ZnO particles and Zn-dye aggregates (Fig. 11.6).

The above study has led to an optimal ZnO amount in terms of the number of drops of the coating ZnO sol. However, it is fairly reasonable that one should maintain the amount of ZnO sol at this optimum of 1 drop and vary the ZnO precursor concentration [25] to achieve a complete optimization of the ZnO content for a maximum DSSC efficiency. In Fig. 11.8 (a) are shown 5 *I-V* characteristics of DSSCs with ZnO-coated TiO$_2$ at 5 different precursor concentrations (0.1, 0.2, 0.3, 0.4 and 0.5 M) using a single drop of ZnO sol. The *I-V* characteristic of the bare TiO$_2$ DSSC is also plotted for comparison. Note that the set of *I-V* curves in Fig. 11.8 (a) are results of independent measurements that can differ from the set of *I–V* curves in Fig. 11.4 (a) due to different measurement conditions.

Table 11.2 presents the photovoltaic parameters deduced by data fitting to the curves of Fig. 11.8 (a).

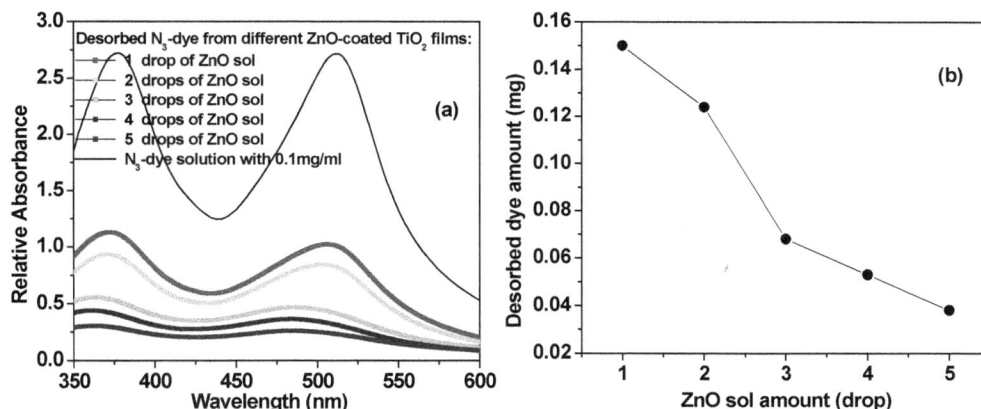

Fig. 11.7. Absorbance spectra of desorbed N_3-dye from different ZnO-coated TiO_2 electrodes (a), and the estimated amount (mg) of the desorbed dye from the ZnO-coated TiO_2 electrodes at different ZnO sol amount (b).

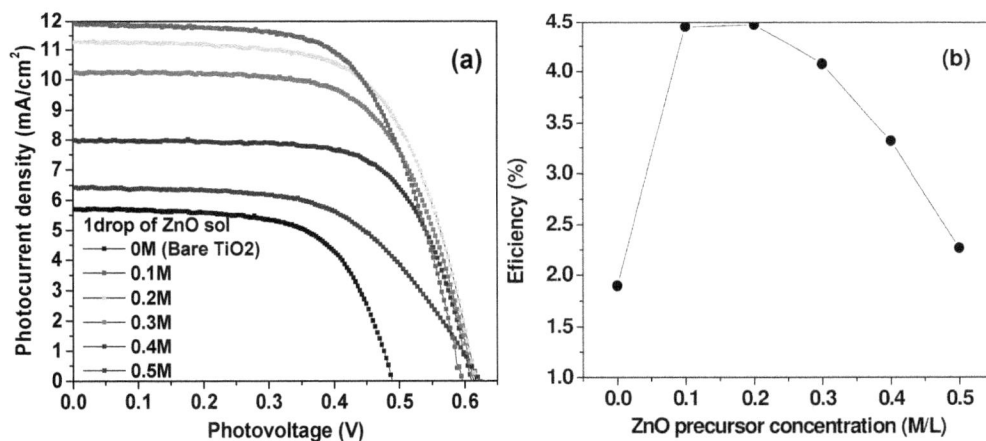

Fig. 11.8. (a) *I–V* characteristics of ZnO spin-coated TiO_2 DSSCs using a single sol drop different ZnO precursor concentrations. The *I–V* characteristic of the bare TiO_2 DSSC is also shown for comparison and (b) variation of the DSSC efficiency with ZnO precursor concentration.

Table 11.2. Photovoltaic parameters of DSSCs with ZnO-coated TiO_2 at different ZnO precursor concentrations.

ZnO precursor concentration	J_{SC}(mA/cm^2)	V_{OC}(V)	FF (%)
0M(Bare TiO_2)	5.616	0.490	63.7
0.1M	11.908	0.596	62.7
0.2M	11.372	0.615	64.0
0.3M	10.298	0.612	64.7
0.4M	8.032	0.609	67.7
0.5M	6.441	0.622	56.6

It can be seen that the DSSC short circuit current J_{SC} increases with decreasing ZnO precursor concentration to reach the maximum limit at 0.1 M. The open circuit voltage V_{OC} also increases but it shows, on the overall, a shift from about 0.5 V, in the case of bare TiO$_2$ DSSC, to about 0.6 V, for the DSSCs with ZnO-coated TiO$_2$. The resulting variation of the DSSC efficiency as a function of ZnO precursor concentration is plotted in Fig. 11.8 (b). The curve shows a clear optimum at the ZnO precursor concentration between 0.1 M and 0.2 M, for which the DSSC efficiency is at maximum with a remarkable enhancement of over 100 %.

An important remark to draw from this variation is that the efficiency of the DSSC with ZnO-coated TiO$_2$ remains at all precursor concentrations above that of the bare TiO$_2$ DSSC. This means that the single drop amount of ZnO sol is too low that even at 0.5M the deposited ZnO amount in the ZnO layer could not lower the dye-adsorption efficiency of the TiO$_2$ film below the required level to counterbalance the opposite effect of the energy barrier created by this layer. The observed 0.1 V shift (from ~ 0.5 V to ~ 0.6 V) of the open-circuit voltage V_{OC} (Fig. 11.8 (a), Table 11.2), apparently independent of ZnO concentration, can be readily attributed to an upward shift of the Fermi-level in the semiconductor side by about 0.1 eV upon ZnO coating. This is due to the increase of free electron density in the TiO$_2$ CB as a result of the decrease of the rate of electron recombination [25].

11.3.2. MgO Results

In the following, the results in the MgO case will be presented in a comparative approach with the above results of the ZnO case. The oxide sol parameters determining the sol amount were found to be 0.1 M precursor concentration and a single sol drop. Therefore, only MgO results of uncoated and 0.1 M-coated TiO$_2$ will be presented.

A TEM image of the prepared 0.1 M-MgO solution is shown in Fig. 11.9 (a) to identify the shape and size of the coating MgO nanoparticles prior to photovoltaic characterization of the MgO-coated TiO$_2$ electrode and associated DSSC. Well observed spherical MgO aggregates of different sizes ranging from 40 nm to 100 nm are illustrated.

It can be observed that these MgO aggregates are made of small clustered nanoparticles with sizes of about 10 nm to 20 nm. Upon spin-coating onto a substrate surface, the clustered particles (aggregates) are expected to separate and disperse under the effect of the centrifugal force to form a thin layer of nanoparticles. Indeed, Fig. 11.9 (b) shows the SEM image of the MgO nanoparticles, as spin-coated onto a micro-structured FTO film where they form a thin layer of nanoparticles covering the FTO grains. Similar spherical aggregates were observed for sol-gel ZnO where, unlike spin-coating, the drop-casting deposition conserved the spherical aggregate form of the deposited particles [17].

Fig. 11.10 shows the *I-V* characteristics of the DSSC based on the uncoated TiO$_2$ electrode and the DSSC based on the 0.1 M MgO-coated TiO$_2$ electrode. The photovoltaic parameters deduced from each characteristic are indicated in Table 11.3.

Fig. 11.9. (a) TEM image of 0.1 M MgO aggregates as suspended in sol, and (b) SEM image of MgO nanoparticles as spin-coated on FTO-glass.

Precautions were taken to maintain the same measurement conditions for the TiO_2 film in both uncoated and MgO-coated electrodes. The uncoated electrode was heated simultaneously with the MgO-coated electrode while sintering the deposited MgO layer at 500 K. Thus, the TiO_2 film in both electrodes experienced the same heat effect. Both electrodes were also dye-loaded simultaneously in the same dye solution to avoid any systematic difference before *I-V* characterization of the associated DSSCs.

There is a clear enhancement in the short-circuit photocurrent density J_{SC} from 7.1 mA/cm^2 to 8.7 mA/cm^2 accompanied with a considerable upward shift of the open-circuit voltage V_{OC} from 0.60 V to 0.66 V and an increase of the fill factor *FF* from 0.64

to 0.70. Thus, the resulting power conversion efficiency $\eta = (FF \times J_{SC} \times V_{OC})/P_i$ shows an enhancement from 2.68 % to 4.06 %, meaning an improvement of about 51.5 %.

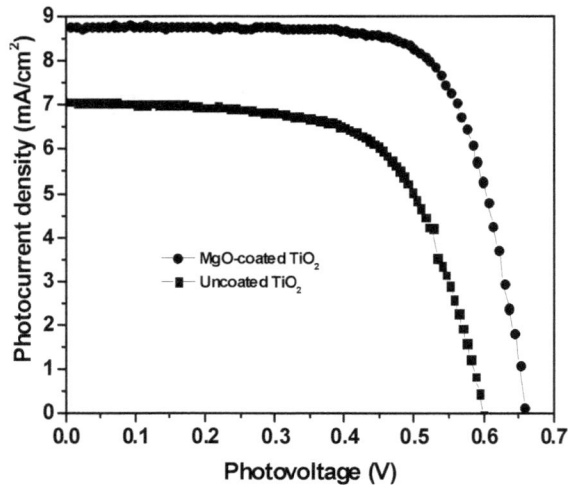

Fig. 11.10. *I-V* Characteristics of the DSSC with MgO-coated TiO₂ (●) and the DSSC with uncoated TiO₂ (■).

Table 11.3. PV-parameters of the DSSC with uncoated TiO₂ and the DSSC with MgO-coated TiO₂.

DSSC with	PV parameters			
	J_{SC}(mA/cm²)	V_{OC}(V)	*FF*	η(%)
Uncoated TiO₂	7.08	0.60	0.64	2.68
MgO-coated TiO₂	8.71	0.66	0.70	4.06

The dark *I-V* characteristics have been reported to support the interpretation of enhancement of the DSSC efficiency in terms of a blocking energy barrier caused by a suitable coating metal oxide layer onto the TiO₂ [25, 26, 31]. Fig. 11.11 shows the dark I-V characteristics of the DSSC with 0.1M MgO-coated TiO₂ (●) and the DSSC with uncoated TiO₂ (■).

Similarly to the ZnO case (Fig. 11.5), upon MgO-coating the current drops and requires the assistance of a higher voltage to regain the value of the uncoated TiO₂ case. In other words, the electrons require a higher voltage to overcome the energy barrier created by the spin-coated MgO layer.

The transient V_{OC}-decay following suppression of light is a useful technique to demonstrate the reduction of the rate of electron back-transport and recombination in DSSCs [16, 38, 39]. Fig. 11.12 shows the V_{OC}-decay for the uncoated TiO₂ DSSC (■) and the MgO-coated TiO_2 DSSC (●). The curves are raw data from a 500 MHz oscilloscope (Tektronix). It is clear that the V_{OC}-decay is much slower for the DSSC with MgO-coated TiO_2. Since the rate of V_{OC}-decay is proportional to the rate of recombination, this gives direct evidence that the rate of electron recombination reduces considerably with MgO-coating of the TiO_2 electrode, which is due to the energy barrier created by the MgO layer.

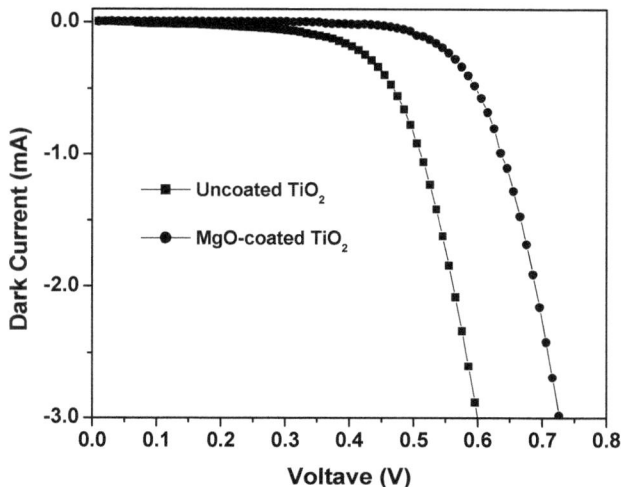

Fig. 11.11. Dark *I-V* characteristics of the DSSC with MgO-coated TiO_2 (●) and the DSSC with uncoated TiO_2 (■).

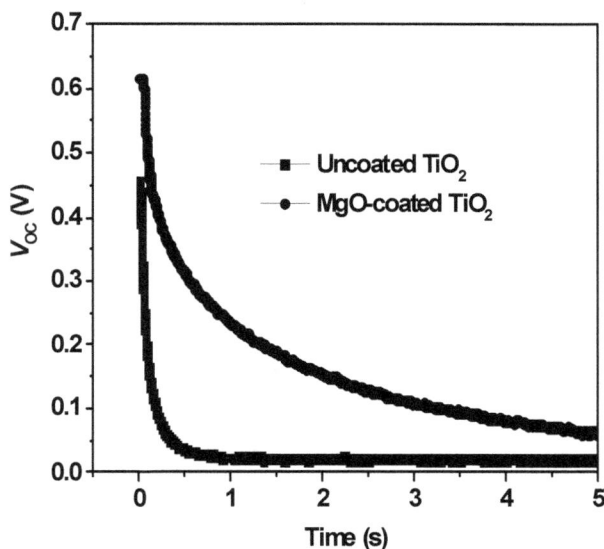

Fig. 11.12. V_{oc}-decay following suppression of illumination for the uncoated TiO_2(■) and the MgO-coated TiO_2 (●).

Fig. 11.13 shows the absorbance spectra of the uncoated (black curve) and MgO-coated (red curve) dye-loaded TiO_2 electrodes. The spectra before dye-loading are also included for comparison (curves with open symbols). Below about 370 nm, in the UV region, the two spectra tend to rejoin those of the pure TiO_2 and MgO-coated TiO_2 to show the predominance of their intrinsic absorption in this region and reflect the band to band electron excitation in the TiO_2 or the MgO-coated TiO_2 semiconductors. Above 370 nm, in the visible region, the two spectra are much higher than those of the uncoated TiO_2 and MgO-coated TiO_2 films. They further present the two characteristic peaks of the N3-dye spectrum (black line), related to the metal-to-ligand charge transfer (MLCT) around 520 nm and the ligand-centered charge transfer (LCCT) around 380nm. This implies that the dye molecules form the main light absorber in the visible range (from 370 nm to 700 nm) for both cases of dye-loaded TiO_2 (uncoated and MgO-coated). However, a downward shift of the spectrum associated with the MgO-coated TiO_2, with respect to that associated with the uncoated TiO_2, is observed. Since the dye-adsorption efficiency of the TiO_2 film depends essentially on the proportion of absorbed light by the dye molecules, the result of Fig. 11.13 indicates a decrease in dye-adsorption efficiency upon MgO spin-coating. This result is to be compared with the corresponding ZnO result in Fig. 11.6. Unlike the MgO case here, the spectrum of the ZnO-coated TiO_2 there is much higher than that of the uncoated TiO_2. As argued above, the excess light absorption by the ZnO-coated film may be due to Zn^{2+}-dye aggregates participation [5, 17].

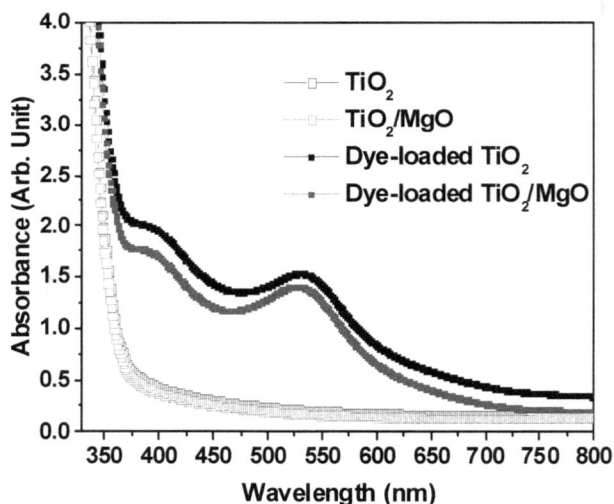

Fig. 11.13. Absorbance spectra of the dye-loaded TiO_2 electrode, for the uncoated (black curve) and MgO-coated (red curve) cases. The spectra before dye-loading are included (open symbols) together with the spectrum of the employed dye solution (black line).

The small decrease in the dye-adsorption efficiency of the TiO_2 film caused by MgO-coating (Fig. 11.13) has a direct consequence to decrease the electron injection rate. However, this can not affect the photocurrent increase induced by the reduction of the electron recombination rate caused by the energy barrier created by the MgO layer. It is also argued that a very thin insulating coating layer does not hinder the transfer of energetic electrons from the dye molecules to the TiO_2 conduction band that is controlled by coupling between the excited states of the dye molecules and the extended states of the TiO_2 CB [29]. Thus, electron injection occurs by tunneling through the thin MgO layer. The latter forms at the same time the energy barrier that reduces the rate of electron back-recombination [40], thus increasing the photocurrent and consequently the DSSC efficiency. This is not the case for the ZnO where the band-gap energy (3.4 eV) is comparable to that of the TiO_2 (3.2 eV). In this case the excited energy level of the dye molecules is higher than the conduction band minimum of both ZnO and TiO_2 semiconductors and electron injection occurs more favorably. Moreover, the high electron mobility of the ZnO nanoparticles contributes to additional increase of the photocurrent. These advantages of the ZnO can readily explain the much more enhancement of the DSSC efficiency by ZnO spin-coating (100 %) than by MgO spin-coating (50 %). Note, however, that the enhancement of TiO_2 DSSC efficiency by coating metal oxides is relative depending on the quality of the original uncoated TiO_2 DSSC. One would expect less efficiency enhancement with thicker and pre-treated TiO_2 film where the original efficiency before post-treatment is much higher than 2 %.

11.4. Conclusion

The role of a spin-coated thin layer of sol-gel metal oxide on the TiO_2 electrode to enhance the power conversion efficiency of the corresponding DSSC was investigated. Two oxide cases were examined and compared, the ZnO of comparable band-gap energy to that of the TiO_2 and the MgO of much larger band-gap energy. The coating oxide amount is monitored by the number of sol drops during spin-coating and the oxide precursor concentration. An optimum of a single sol drop at the precursor concentration between 0.1 M and 0.2 M was determined in the ZnO case, for which the DSSC efficiency is at maximum with an enhancement of over 100 %. The optimum in the MgO case was found at the MgO precursor concentration 0.1 M for a single sol drop with an efficiency enhancement of about 50 %. In both cases, the efficiency enhancement was mainly attributed to reduction of the rate of electron back-recombination as a result of an energy barrier created by the spin-coated oxide layer. Beyond the optimum, the oxide amount in the coating layer will decrease the dye-adsorption efficiency of the TiO_2 film. The electron injection rate will consequently reduce, causing the fall of the DSSC efficiency.

Acknowledgements

Taif University is acknowledged for financial support for the project No. 1434-2702. The MSc students Fahd Al-juaid and Fahhad Al-Subai are acknowledged for their contributions.

References

[1]. B. O'Regan, M. Gratzel, A low-cost, high-efficiency solar cell based on dye-sensitized colloidal TiO_2 films, *Nature*, Vol. 353, Issue 6346, 1991, pp. 737-740.

[2]. X. Fu, Z. Song, G. Wu, J. Huang, X. Duo, C. Lin, Preparation and characterization of MgO thin films by a novel sol-gel method, *Journal Sol-Gel Science and Technology*, Vol. 16, Issue 3, 1999, pp.277-281.

[3]. J. W. Bullard, Z. Xu, M. Menon, A novel thin film phase of oriented MgO grown from a liquid solution, *Journal Crystal Growth*, Vol. 233, Issue 1-2, 2001, pp. 389-398.

[4]. S. Chakrabarti, D. Ganguli, S. Chaudhuri, A. K. Pal, Crystalline magnesium oxide films on soda lime glass by sol-gel processing, *Materials Letters*, Vol. 54, Issue2-3, 2002, pp. 120-123.

[5]. S. Chakrabarti, D. Ganguli, S. Chaudhuri, A. K. Pal, Preparation of hydroxide-free magnesium oxide films by an alkoxide-free sol–gel technique, *Materials Letters*, Vol. 57, Issue29, 2003, pp. 4483-4492.

[6]. C. Bondoux, P. Prene, P. Belleville, F. Guillet, S. Lambert, B. Minot, R. Jerisian, MgO insulating films prepared by sol-gel route for SiC substrate, *J. European Ceramic Society*, Vol. 25, Issue12, 2005, pp. 2795-2798.

[7]. M. S. Mastuli, N. S. Ansari, M. A. Nawawi, A. M. Mahat, Effects of cationic surfactant insol-gel synthesis of nano-sized magnesium oxide, *APCBEE Procedia*, Vol. 3, 2012, pp. 93-98.

[8]. Z. Habibah, A. N. Arshad, L. N. Ismail, R. A. Bakar, M. Rusop, Chemical solution deposited magnesium oxide films: Influence of deposition time on electrical and structural properties, *Procedia Engineering*, Vol. 56, 2013, pp. 737-742.

[9]. S. Valanarasu, V. Dhanasekaran, M. Karunakaran, T. A. Vijayan, R. Chandramohan, T. Mahalingam, Microstructural, optical and electrical properties of various time annealed spin coated MgO thin films, *Journal Mater. Science, Mater. Electronics*, Vol. 25, Issue 9, 2014, pp. 3846-3853.

[10]. H. S. Jung, J. K. Lee, M. Nastasi, S. W. Lee, J. Y. Kim, J. S. Park, K. S. Hong, H. Shin, Preparation of nanoporous MgO-coated TiO_2 nanoparticles and their application to the electrode of dye-sensitized solar cells, *Langmuir*, Vol. 21, Issue23, 2005, pp. 10332-10335.

[11]. B. Li, G. Lu, L. Luo, Y. Tang, TiO_2/MgO core-shell film: fabrication and application to dye-sensitized solar cells, *Wuhan University Journal of Natural Sciences*, Vol. 15, Issue4, 2010, pp. 325-329.

[12]. C. Photiphitak, P. Rakkawamsuk, P. Muthitamongkol, C. Thanachayanont, Performance enhancement of dye-sensitized solar cells by MgO coating on TiO_2 electrodes, *International Journal of Chemical, Molecular, Nuclear, Materials and Metallurgical Engineering*, Vol. 6, Issue 5, 2012, pp. 465-469.

[13]. P. M. Sommeling, B. C. O'regan, R. R. Haswell, H. J. P. Smit, N. J. Bakker, J. J. T. Smits, J. M. Kroon, J. A. M. van Roosmalen, Influence of a $TiCl_4$ Post-treatment on nanocrystalline TiO_2 films in dye-sensitized solar cells, *Journal of Phys. Chem. B*, Vol. 110, Issue 39, 2006, pp. 19191-19197.

[14]. B. C. O'regan, J. R. Durrant, P. M. Sommeling, N. J. Bakker, Influence of the $TiCl_4$ treatment on nanocrystalline TiO_2 films in dye-sensitized solar cells. 2. Charge density, band edge shifts, and quantification of recombination losses at short circuit, *Journal Phys. Chem. C*, Vol. 111, Issue 37, 2007, pp. 14001-14010.

[15]. S. Xuhui, C. Xinglan, T. Wanquan, W. Dong, L. Kefei, Performance comparison of dye-sensitized solar cells by using different metal oxide coated TiO_2 as the photoanode, *AIP Advances 4*, 2014, pp. 031304-1 – 031304-7.

[16]. F. Al-juaid, A. Merazga, A. M. Al-Baradi, F. Abdel-wahab, Effect of sol–gel ZnO spin-coating on the performance of TiO$_2$-based dye-sensitized solar cells, *Solid-State Electronics*, Vol. 87, 2013, pp. 98-103.

[17]. Q. Zhang, K. Park, G. Cao, Synthesis of ZnO aggregates and their application in dye-sensitized solar cells, *Material Matters*, Vol. 5, Issue 2, 2010, pp. 32-39.

[18]. S. Ilican, Y. Caglar, M. Caglar, Preparation and characterization of ZnO thin films deposited by sol-gel spin coating method, *Journal Optoelectronics and Advanced Materials*, Vol. 10, Issue 10, 2008, pp. 2578-2583.

[19]. D. Raoufi, T. Raoufi, The effect of heat treatment on the physical properties of sol-gelderived ZnO thin films, *Applied Surface Science*, Vol. 255, Issue 11, 2009, pp. 5812-5817.

[20]. M. Smirnov, C. Baban, G. I. Rusu, Structural and optical characteristics of spin-coated ZnO thin films, *Applied Surface Science*, Vol. 256, Issue 8, 2010, pp. 2405-2408.

[21]. K. Balachandra, P. Raji, Synthesis and characterization of nano zinc oxide by sol-gel spin-coating, *Recent Research in Sci. and Tech.*, Vol. 3, Issue 3, 2011, pp. 48-52.

[22]. T. Taguchi, Improving the performance of solid-state dye-sensitized solar cells using MgO-coated TiO$_2$ nanoporous film, *Chem. Commun.*, Vol. 9, 2003, pp. 2480-2481.

[23]. G. R. A. Kumara, M. Okuya, K. Murakami, S. Kaneko, V. V. Jayaweera, K. Tennakone, Dye-sensitized solid-state solar cells made from magnesiumoxide-coatednanocrystalline titanium dioxide films: enhancement of the efficiency, *Journal of Photochemistry and Photobiology A: Chemistry,* Vol. 164, Issue 1-2, 2004, pp. 183-185.

[24]. S. Wu, Enhancement in dye-sensitized solar cells based on MgO-coated TiO$_2$ electrodes by reactive DC magnetron sputtering, *Nanotechnology*, Vol. 19, Issue 21, 2008.

[25]. Y. Liu, Efficiency enhancement in dye-sensitized solar cells by interfacial modification of conducting glass/mesoporous TiO$_2$ using a novel ZnO compact blocking film, *Journal Power Sources*, Vol. 196, Issue 1, 2011, pp. 475-481.

[26]. M. C. Kao, H. Z. Chen, S. L. Young, Effects of ZnO coating on the Performance of TiO$_2$ nanostructured thin films for dye-sensitized solar cells, *Appl. Phys. A*, Vol. 97, Issue 2, 2009, pp. 469-474.

[27]. Frederic Oswald, *Solaronix Company*, private communication.

[28]. Y. Zou, Z.-A. Wang, X.-H. Lan, N.-K. Huang, Anatase phase TiO$_2$ anode of a dye-sensitized solar cell prepared by using RMFMS, *Journal Korean Phys. Society*, Vol. 55, Issue 6, 2009, pp. 2650-2653.

[29]. H. Wang, M. Liu, C. Yan, J. Bell, Reduced electron recombination of dye-sensitized solar cells based on TiO$_2$ spheres consisting of ultrathin nanosheets with [001] facet exposed, *Beilstein Journal of Nanotechnology*, Vol. 3, 2012, pp. 378-387.

[30]. Y. Song, M. Zheng, L. Ma, W. Shen, Anisotropic growth and formation mechanism investigation of 1D ZnO nanorods in spin-coating sol-gel process, *Journal of Nanoscience and Nanotechnology*, Vol. 10, Issue 1, 2010, pp. 426-432.

[31]. S.S. Kim, J. H. Yum, Y. E. Sung, Improved performance of a dye-sensitized solar cellusing a TiO2/ZnO/Eosin electrode, *Solar Energy Mater. and Solar Cells*, Vol. 79, Issue 4, 2003, pp. 495-505.

[32]. M. Penny, T. Farrell, G. Will, J. Bell, Modeling interfacial charge transfer in dye-sensitized solar cells, *Journal of Photochemistry and Photobiology A: Chemistry*, Vol. 164, 2004, pp. 41-46.

[33]. Y. J. Shin, Enhancement of photovoltaic properties of Ti-modified nanocrystalline ZnO electrode for dye-sensitized solar cells, *Bull. Korean Chem. Soc.*, Vol. 26, Issue 12, 2005, pp.1929-1930.

[34]. I. Bedja, P. V. Kama, X. Hua, P. G. Lappin, S. Hotchandani, Photosensitization of nano-crystalline ZnO films by bis (2,2'-bipyridine) (2,2'-bipyridine-4,4'-dicarboxylic acid) Ruthenium (II), *Langmuir*, Vol. 13, Issue 8, 1997, pp. 2398-2403.

[35]. T. Soga, Nanostructured Materials for Solar Energy Conversion, *Elsevier B.V.*, 2006.

311

[36]. Q. Zhang, Effects of Lithium ions on dye-sensitized ZnO aggregate solarcells, *Chem. Mater.*, Vol. 22, Issue 8, 2010, pp. 24-27.

[37]. A. Umar, A. A. Alharbi, P. Singh, S. A. Sayari, Growth of aligned hexagonal ZnO nanorods on FTO substrate for dye-sensitized solar cells application, *Journal of Nanoscience and Nanotechnology*, Vol. 11, Issue 4, 2011, pp. 3560-3564.

[38]. P. V. Kamat, I. Bedja, S. Hotchandani, L. K. Patterson, Photosensitization of nano-crystalline semiconductor films: Modulation of electron transfer between excited ruthenium complex and SnO_2 nanocrystallites with an externally applied bias, *Journal Phys. Chem.*, Vol. 100, Issue 12, 1996, pp. 4900-4908.

[39]. H. Yu, S. Zhang, H. Zhao, G. Will, P. Liu, An efficient and low-cost TiO_2 compact layer for performance improvement of dye-sensitized solar cells, *Electrochimica Acta*, Vol. 54, Issue 4, 2009, pp. 1319-1324.

[40]. E. Palomares, J. N. Clifford, S. A. Haque, T. Lutz, J. R. Durrant, Control of charge recombination dynamics in dye-sensitized solar cells by the use of conformally deposited metal oxide blocking layers, *Journal Am. Chem. Soc.*, Vol. 125, Issue 2, 2003, 475-482.

Chapter 12

Review of the Reliability of Flexible Packaging of Thin and Ultra-Thin ICs

I. R. Bose, N. Palavesam, C. Landesberger and C. Kutter

12.1. Introduction and Background

Sub-millimetre package heights are becoming prevalent due to the market-pull that applications in the mobile and wearable segments. Lower height for chip packages is a dominant requirement for electronic components that are targeting this segment. In the past decade, the use of thinned silicon dies for the fabrication of Integrated Circuits (IC) has become commonplace in applications such as mobile devices, health monitoring, consumer electronics, sensors, robotics, wearables, aerospace, automotive, displays and large area electronics, flexible lighting and the "Internet of Things" (IoT) [1-4]. However, the challenge has been to package these thinned ICs effectively in a sub-millimetre package, which is not rigid yet reliable, thus allowing them to be bent and mounted onto a curved surface or used in applications where the housings of the ICs themselves are an active part of the sensing mechanism.

Packaging of thin and ultra-thin silicon ICs in flexible foil based packages present a paradigm shift to conventional IC packaging. This would allow thinned ICs to be integrated into any object of daily life, thus acting as a true enabler for the IoT market. As shown in Fig. 12.1, till date the existing market for IoT has been limited to automotive, personal computing and handheld devices. Nevertheless, the flexible chip foil packaging has the ability to integrate IoT nodes i.e. microcontroller ICs, sensing elements and power management without compromising on performance, thereby opening up an entire new market segment, which is referred to as the "Internet of Everything" (IoE).

Fig. 12.2 shows the three general methods in which ICs can be integrated with flexible substrates. The most common method and also currently the flex-printed circuit board (PCB) industry standard is the use of packaged surface mount devices (SMD) directly soldered to the PCB board. The standard substrate material is glass-reinforced epoxy laminate sheets such as G10, G11, FR4, FR5 and FR6. Of these, FR4 is the grade most

Dr.-Ing. Indranil Ronnie Bose
Fraunhofer Research Institution for Microsystems and Solid State Technologies EMFT, Germany

313

widely used due to its flame retardant properties. The flexibility of the overall system is rather limited and although the PCB system can be bent to a certain extent. Although, this method is not truly flex, neither is it a technique that uses thinned ICs nor is it really a packaging technique, nevertheless, this method has been included here as a lot of applications utilize this strategy to integrate ICs onto (semi) flexible substrates. This method has been in production since a while now and is the defacto standard for applications such as LED lighting stripes, hearing aids, fitness trackers, smart watches etc.

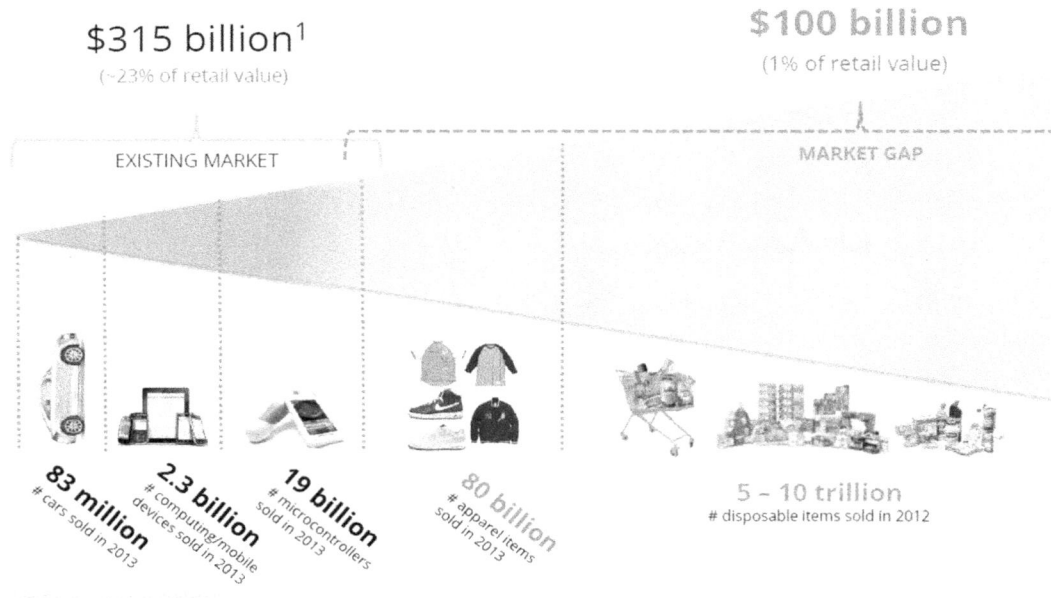

Fig. 12.1. Overview of the existing IoT market and the future IoE market [5].

The second integration strategy, the chip on foil (COF) approach, involves the flipchip assembly of thinned ICs directly onto the flexible substrates. Here, the IC is thinned down to 70-150 μm and then flipchip positioned onto the backplane board, where the contact pads have been structured already, and the interconnections are enabled either by low temperature soldering or anisotropic conductive adhesives (ACA) or Non Conductive Adhesive (NCA) or Isotropic Conductive Adhesive (ICA). The resulting package height is about 0.5 mm and is partly bendable. The stability provided by the flexible backplane substrate, which usually comprises of poly carbonate (PC), polyethylene naphthalate (PEN), polyethylene terephthalate (PET), polyimide (PI) etc., is sufficient to address the needs of applications such as bank chip cards, RFID label tags, flexible display interconnects etc.

The third integration approach is the chip foil package. Here, the ultra-thin IC is fully embedded into the flexible substrate, resulting in a setup wherein the ultra-thin IC and substrate form fully flexible system. The term ultra-thin has been used here to signify that the silicon dies have been thinned down to well below 40 μm. Under these thicknesses, the silicon wafer is no longer rigid but is completely flexible. The resulting package heights are about 150 μm. At present, the technology readiness level (TRL) of this technology is at the nascent stage, thus it is not yet in production but exists as prototypes and demonstrators in leading R&D laboratories. The intended applications for this fully flexible packaging are in the field of robotics, such as e-skin, IoT sensor nodes, micro System in Package (SIP) applications.

	SMD Mounting onto Flexible Substrates	Flip-Chip assemblies of of thinned bare dies	Embedded IC: Chip-Foil-Package
	 IC	 IC	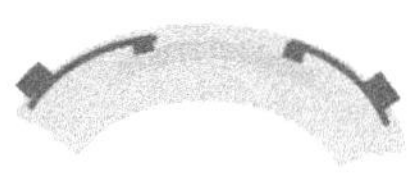
IC thickness	< 1 mm	70-150 μm	12-40 μm
Package Height	~ 1 mm	~ 0.5 mm	<< 0.2 mm
Flexibility	partly rigid	limited bendability	fully flexible
TRL/Maturity	In production	In production	Prototypes and Demonstrators at R&D Institutes
Applications	Wearables, Smart Watches, LED-Stripes, Hearing Aids etc.	Smart Cards, RFID-Labels, Flex-Interconnects to Displays, etc.	"electronic-skin" (Robotics), IoT, System in Packages (micro-SiP) etc.

Fig. 12.2. Integration strategies to bring ICs on/into flexible substrates.

In this Chapter, the focus will be to examine the reliability and stability of the chip on foil (COF) and chip in foil i.e. embedded package, wherein the effect of the flexible substrates to offer mechanical stability fragile thin ICs will be studied in detail. The other focus of this Chapter is the manufacturing and processing of the fully flexible embedded approach. The authors have purposefully excluded the standard SMD assembly on flex and the fabrication of flip chip assembly on flex substrates, as these technologies have been well documented and are mature technologies that are already in production since a while. Rather, the focus lies on the packaging aspect of thin and ultra-thin ICs with respect to flexible foil substrates and their reliability. The review encompasses and summarizes the research and previous work done at the Fraunhofer EMFT [6-12].

12.2. Measurement of Fracture Strength of Thin and Ultra-Thin ICs

The advantage of COF assembly is that the handling and processing problems can be addressed while achieving good mechanical flexibility [13]. Several research groups besides the Fraunhofer EMFT have published about the mechanical strength and flexibility of thin and ultra-thin silicon dies [14, 15]. However, the shortcomings of most studies has been the use of the standard testing methods whereby sample chip sizes of ~15 mm^2 have been used to measure the breaking strength, albeit for current applications, the thin and ultra-thin IC chips range about ~ 4 mm^2.

The fracture strength of silicon ICs can be measured either by a uniaxial or by a biaxial bending test. Uniaxial bending tests are more sensitive to the damages occurring during dicing, on the sidewalls and the edges of the chips, while biaxial bending tests are more sensitive to the damages occurring on the surface of the chips during back grinding [9]. Biaxial point load test methods such as the ball-breaker test have been manifested to determine the fracture strength of very small sized samples [16]. However, the ball-breaker test measures the fracture strength of Si chips by analyzing mainly the surface quality of the Si chips and neglects the impact of the sidewall quality on the fracture strength of the Si chips. Sidewall quality is more dominant than surface quality in determining the fracture strength of the chips under uniaxial loading conditions and when the chips are prepared using different dicing techniques [17].

Due to the above reasoning, there is a preference to opt for uniaxial load conditions to test the fracture strength of small thin ICs that are used in real world applications. The Line-Load test is an example of such a test methodology, which is based on uniaxial loading and can be used to measure the fracture strength of ICs having a thickness of ~150 μm [18]. However, the with chip thickness < 25 μm, most of the commonly used load sensors used in the Line-Load setup fail to read the drop in the force feedback due to the fracture of the thin and ultra-thin ICs and is unable to reliably detect the breaking point in the load-time curve as shown in Fig. 12.3 (a).

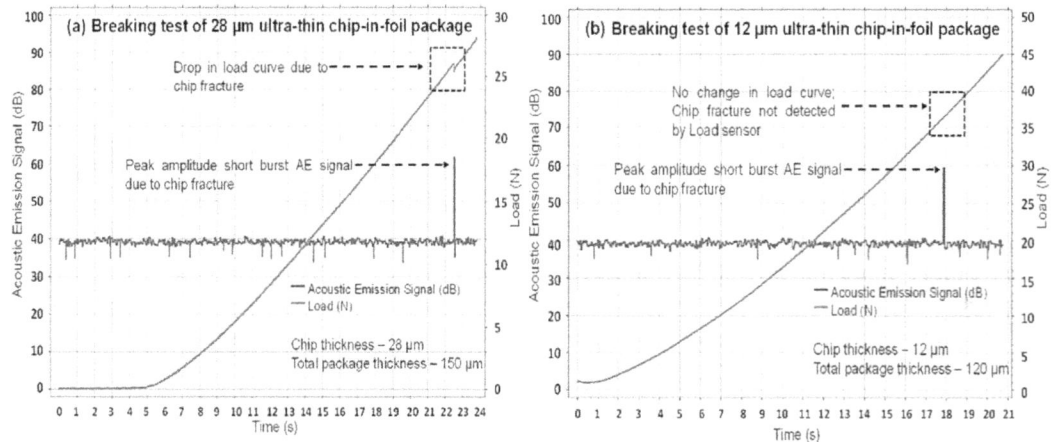

Fig. 12.3. Line-Load test plot and the augmented acoustic emission signal to determine the fracture strength of (a) a 28 μm, and (b) 12 μm ultra-thin chip in foil package [9].

To circumvent this problem, an acoustic emission (AE) augmented Line-Load test is implemented. An auxiliary system consisting of an AE sensor is added to the line-load test setup and by correlating the peak amplitude short burst AE signal with the corresponding load value on the load-time curve, the exact fracture strength can be ascertained. This setup works reliably for ultra-thin chips as low as 12 μm, as shown in the Fig. 12.3 (b).

12.3. Mechanical Stress Analysis of Thin and Ultra-Thin Silicon Dies in Flex Packaging

In order to establish the mechanical properties of any material, a small tensile force needs to be applied and the parameters measured. This tensile force approach is rather cumbersome for such small chip dimensions, and instead bending tests are implemented. As previously mentioned in the Section 12.2 that among uniaxial or biaxial bending tests, uniaxial is preferred, which may be implemented as a 4-point or 3-point bending test. For small die sizes, the 3-point uniaxial bending test is preferred due to the easy setup and ease of carrying out the test. The equation for calculating the bending stress of silicon dies using 3-point-bending test is defined by the SEMI G86-0303 test standard [19]. However, the limitation of this test standard is that when the bending deflection is several times higher than the chip thickness itself without the chip breaking, which is normally the case with dies under 40 μm and with lower bending radii, this standard by itself can no longer be used to calculate the bending stresses in the chip packages.

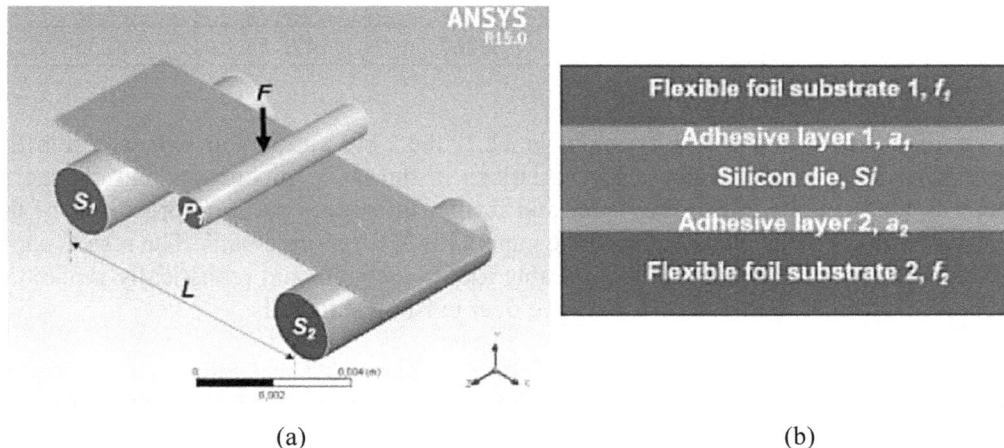

(a)   (b)

Fig. 12.4. (a) The 3-point bending test finite element model, and (b) the thin chip stack cross section [7].

A finite element model and its consequent experimental validation is the defacto norm to calculate and model the mechanical stresses in thin and ultra-thin silicon dies. Fig. 12.4 (a) shows the schematic of the Finite Element Model (FEM) normally used to analyze the bending stress of the chip-in-foil package, whereas Fig. 12.4 (b) shows the chip-in-foil stack itself.

Table 12.1 shows the comparative results of the FEA simulation with the actual experimental results. Here, three different chip thicknesses (h_{Si}) of 30 μm, 65 μm and 130 μm were used, thickness of the adhesive layers were 5 μm and PI foil thickness was 50 μm. The pressing rod (P1) for load application had a diameter of 1 mm and the supporting rods (S1, S2) each had a diameter of 2 mm. The load span (L) was 10 mm for h_{Si} =130 μm and 8 mm for h_{Si}=30 μm and 65 μm respectively [7]. The simulations were performed with displacement control using the mean fracture displacement values obtained from the experiments reported in [6].

Table 12.1. Comparative results of the FEA simulation to the experimentally measured values [7].

Chip Thickness	Sample Type	Fracture Strength (N)		FEA Calculated Bending Stress (MPa)	Experimental mean fracture displacement (mm)
		FEA Simulated	Experimentally Measured		
30	bare	0.19	0.202	653.43	1.2
	embedded	0.64	0.583	1191.1	2.1
65	bare	0.74	0.780	511.77	0.46
	embedded	1.12	1.021	605.37	0.53
130	bare	1.96	2.312	425.58	0.31
	embedded	2.40	2.487	475.53	0.35

As can be seen from the results in Table 12.1, the FEA simulation values match the experimentally measured values. For variations in the layer thicknesses and change in materials for the chip embedding layer and further inferences and interpretations of the FEA simulation results please refer to the papers [6, 7] for more details. The results show that the FEA simulation method is a reliable tool to estimate and realistically model the mechanical stresses in the flexible package over existing standards.

Additionally, the experimental results of Table 12.1 show that when the forces needed to fracture bare silicon versus silicon packaged in foil are compared to one another, the flex packaged silicon chips show greater stability, and the stability increases with the reduction in chip thickness. The results of Table 12.1 are plotted in Fig. 12.5. Here, it can be seen that the thinner silicon samples break at lower forces and embedding always increases the sturdiness against external load. In fact, in the case of ultra-thin samples (30 μm), the fracture force was increased by a factor of nearly three when the chips were packaged in the flexible foils [10]. It can also be seen that the mean displacement at the instance of breaking of the silicon samples was increased by a factor of two [7]. These results again confirm the improvement of mechanical reliability of thin silicon chips when embedded in laminates, this had been the original motivation for the development of thin and bendable IC packages at Fraunhofer EMFT [10, 13].

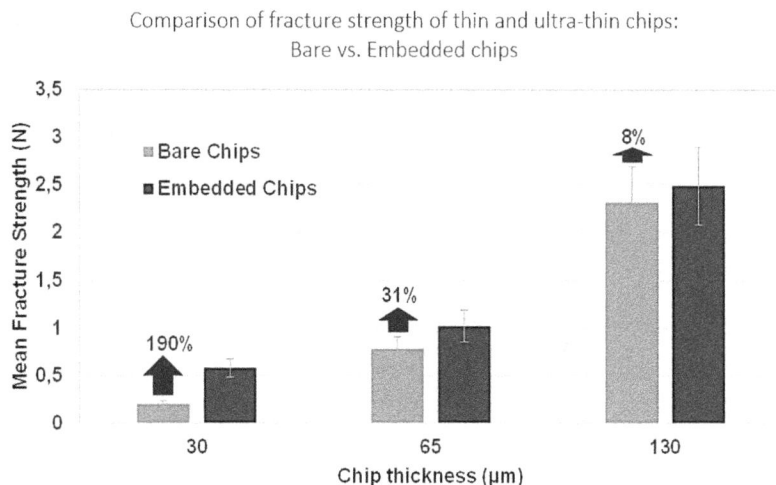

Fig. 12.5. Comparison of the mean fracture strengths for bare versus flexible foil embedded chips [10].

12.4. Recurrent Bending Tests of the Flexible Packages

To simulate the real life bending and flexing of the packages during their intended product life time in a specific application, the flexible packages need to be subjected to recurrent bending and the electrical and mechanical changes characterized as a result thereof. There are two major methods used for performing such tests, namely the Fixed-Radius bending test and the Free form bending tests. Both testing methods should be performed on the flexible packages because they highlight the effect of recurrent bending on slightly different parts of the package. Fig. 12.6 shows the difference in the major bending axis for the two bending test types, here a thin IC is flip chip assembled via an ACA to a base PI flexible foil, on which the metal layer is already patterned. In this setup, it can be seen that the Fixed-radius bending test's axis runs along the centre line of the chip itself, i.e. here the chip itself plays a dominant role. Whereas in the case of the Free form bending test the weakest point, i.e. the interface from the (semi) rigid IC to the flexible foil and the wiring layers themselves, play a dominant role.

12.4.1. Fixed-Radius Bending Test

This testing methodology uses a cylindrical fixture of a defined diameter, with Polytetrafluoroethylene (PTFE) being the most common material to construct the cylinder. The cylinder diameter corresponds to the bending radius of the flexible package being tested. The flexible package is partially wrapped around cylinder face and is placed under constant tension as shown in Fig. 12.7. One measurement cycle involves the displacement of the flexible package over the face of the cylinder. This procedure is repeated thousands of time to simulate real life usage scenarios. Most Fixed-radius bending test setups are offline, i.e. after a set number of repetitions an intermediate electrical and/or mechanical measurement is carried out and then the process is repeated till the entire number of cycles are completed. Although, in certain cases, such as a daisy chain setup, the clamps can be actively connected, i.e. an online electrical test may be performed during the entire bending cycle.

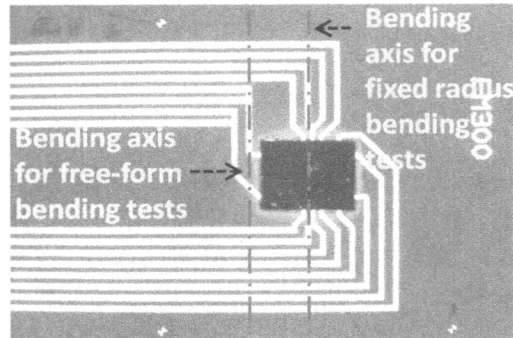

Fig. 12.6. The bending axes for the Fixed-radius bending test and the Free-form bending test [11].

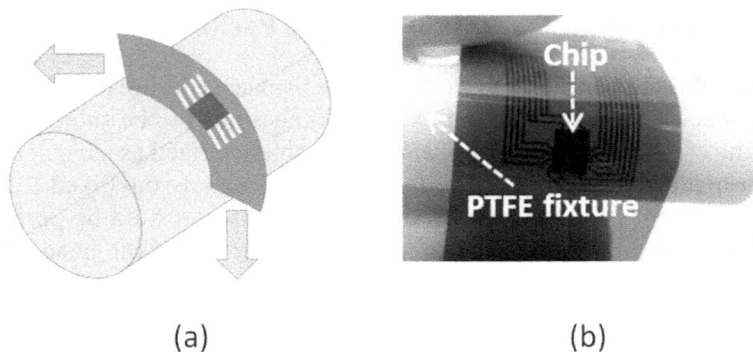

(a)   (b)

Fig. 12.7. (a) The schematic of the Fixed-radius bending test setup, (b) the actual test setup using a 10 mm diameter cylinder to test a COF assembly [11].

Typical results from such a Fixed-radius bending test look like Fig. 12.8. Here, two different chip thickness are compared i.e. a 28 μm and a 20 μm COF assembly. Fig. 12.8 clearly shows after how many cycles the deterioration starts. Normally, as can be seen also, the thinner chips perform better than their thicker counter parts. In this setup, the chip was connected in the daisy chain configuration to enable online testing during the testing procedure, and shows that the 20 μm ultra-thin package was able to resist three and half thousand bending cycles before the failure set in, whereas the thicker 28 μm chip failed just under two thousand cycles. However, what cannot be inferred from the results of these tests is the cause of the deterioration, which could be due to a number of factors ranging from metal interconnects delamination, ACA bonds breaking, chip fracture, etc. For this, a detailed scanning electron microscope (SEM) image would be necessary to determine the cause of the failure mechanism.

Usually, this method is more sensitive to the surface condition of the chip itself, i.e. the wafer processing, the thinning process, stress relief and the die singulation by sawing or by plasma dicing. This is due to the fact that the major bending axis lies on the centerline of the chip itself. Publications [6, 10, 11] have shown the differences that chip surface roughness and edge roughness play in the reliability of these flexible packages under recurrent bending cycles, wherein plasma dicing seems to offer the best results for ultra-thin dies.

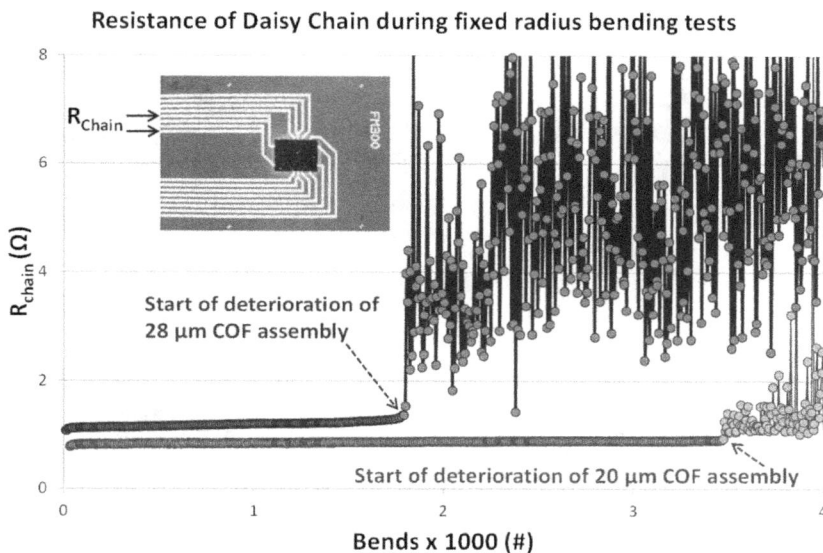

Fig. 12.8. Fixed-radius recurrent bending cycles exhibiting the deterioration of the daisy chain resistance of 12 μm and 20 μm COF assemblies [11].

12.4.2. Free-Form Bending Test

In this case, the flexible samples bend freely without a clearly defined bending radius. The samples are mounted in such a way that during the bending cycle, the compression on the flexible package has an axis that lies at the interface between the chip and the foil. This allows predesignating the maximum stress zone on the most critical parts i.e. the wiring lines of the COF assembly.

As there is no standardized general Free-form bending tester available in the market, with most equipment being custom specific to the applications they are intended, the Fraunhofer EMFT designed and constructed its own bending machine to ideally test flexible packages for the small form factors of real life ICs. The testing machine, as shown in Fig. 12.9 (a), has a fixed base plate and another movable arm. The movable arm can trace an arc ranging from 0° to 180°. The test samples are mounted in between the fixed base plate and the movable plate and are electrically connected using coaxial cables. The tester is capable of handling multiple small samples placed side by side and measured simultaneously.

One complete bending test cycle constitutes of four intermediate steps, wherein the arm moves to the following angular positions and multiple measurements are made at each position. The sequence is the following: start position of 0°, arm movement from 0° to 90°, rest position at 90° and lastly the movement of the arm from 90° back to 0°. The testing procedure and the respective swing arm positions are shown in Fig. 12.9 (b). Electrical measurements are made at each of these four steps. The frequency of each measurement cycle can be altered by simply adjusting the swing arm velocity to suit and simulate the application for which the flexible package is intended.

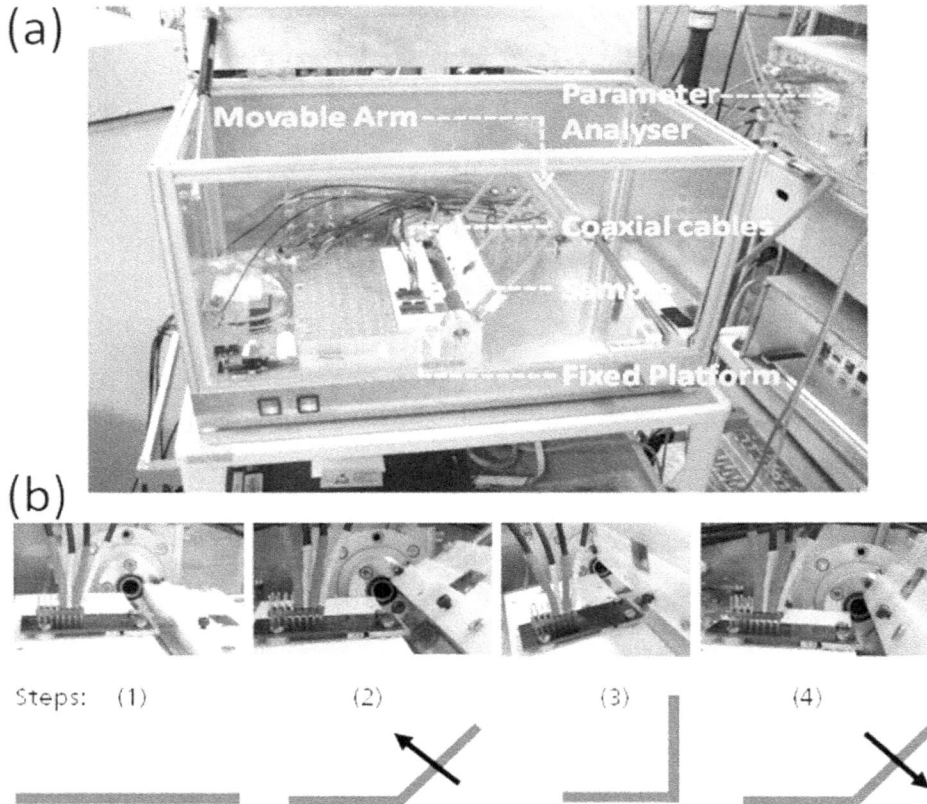

Fig. 12.9. (a) Custom built Free-form bending tester, (b) test procedure at the specific arm positions [6].

A similar experiment to the one used for the Fixed-form bending test was setup to exhibit the effects of the Free-form bending test. Here, two COF assemblies one with 28 µm and the other with 250 µm chip thickness was subjected to recurrent bending using this machine, such that the bending axis lay just at the edge of the chip i.e. along the metallic contacts. The results are shown in Fig. 12.10. It can be seen that the 28 µm ultra-thin chip could be subjected to 6000 bending cycles before the deterioration could be observed, whereas in the case of the thicker 250 µm chip, the deterioration was after 3500 bending cycles.

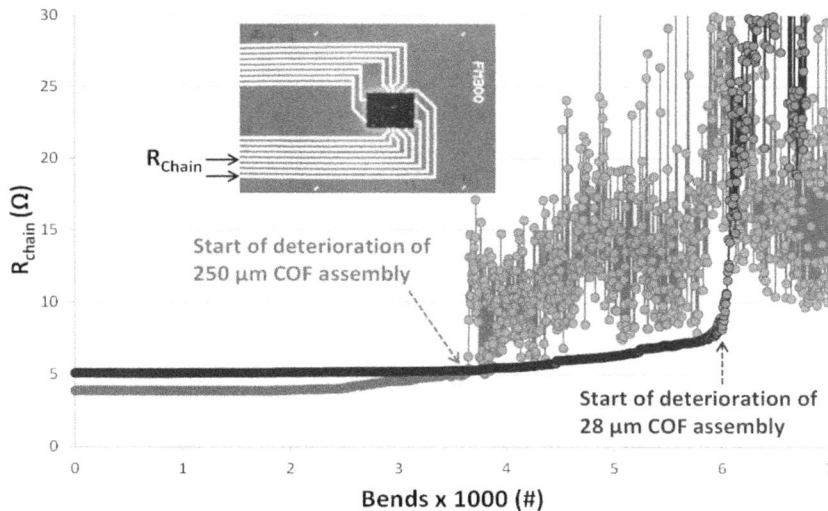

Fig. 12.10. Free-form bending test results on a 28 μm and 250 μm thin chips in a COF assembly [11].

To enable a better understanding of the defect mechanism resulting in the deterioration, the 28 μm COF was used to perform a Computed tomography (CT) image. The CT image is shown in Fig. 12.11. It can be seen here that there are several cracks running through the entire depth of the wiring lines, these breakages also explain the noisy behavior of the Free-form bending test results. During the inward stroke, i.e. position 2 (refer to Fig. 12.9b), and the intermediate position 3, the compression resulted in the lines still conducting the measurement signal, albeit with slightly increased resistance. Whereas, during the return stroke, i.e. position 4, till the final resting position, i.e. position 1, the resulting elongation and tension on the sample caused the cracks to open and act as open circuits, leading to a high resistance value being measured. From Fig. 12.11 it is also evident that the chip itself did not break nor where there cracks observed at the chip-foil interface. Additionally, nor did the ACA fail, which would have resulted in delamination on the chip. Hence, it could be concluded that the rupture of the foil wiring layer was the major cause of failure in the COF samples [11].

12.5. Embedded IC – Chip-in-Foil Approach

As can be seen from the results in Fig. 12.5, bare ultra-thin monocrystalline silicon chips are relatively fragile requiring low mechanical forces to break. This is a cause of concern due to the high risk of breakage during the assembly process. An ideal packaging technology should be able to use these ultra-thin chips to utilize the flexibility aspect, while still being mechanically stable during the assembly and manufacturing process, and lastly post-packaging reliability of the interconnects and flexible package itself when subjected to recurrent bending tests. The chip-in-foil packaging technology developed at the Fraunhofer EMFT, as shown in Fig. 12.12, addresses these needs [10, 12].

323

Additionally, it has the added advantage that no additional wafer processing steps such as redistribution layers (RDL) or under bump metallization (UBM) of the contact pads are necessary. The entire process is Roll-to-Roll (R2R) compatible and relies solely on high-resolution photolithography processes.

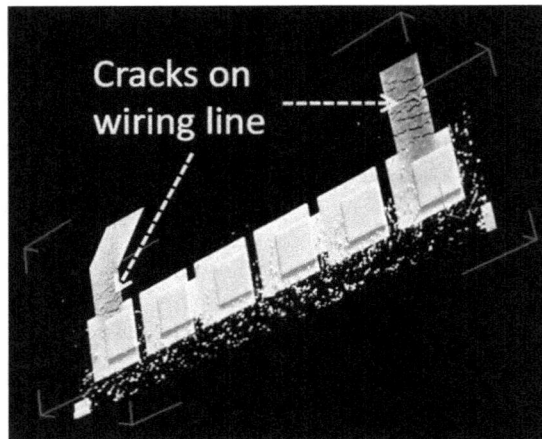

Fig. 12.11. Computed Tomography image of the COF sample showing the failure mechanism [11].

Fig. 12.12. (a) Schematic cross-section of a chip embedded in foil, (b) SEM image of a cross section of an actual embedded chip in foil package [10]

A detailed process flow of the packaging process can be found in [10]. As a proof of concept, a PIC16 series microcontroller IC from the supplier Microchip was packaged using this technology. The start point of this process was a whole standard wafer from the supplier from which ultra-thin chips (25-30 μm) were enabled using the "dicing by thinning" process. This process involves drawing trenches using a standard sawing process on the front side of the wafer. The trench depth is slightly deeper than the intended final ultra-thin chip thickness. Then backside thinning was done using the grinding equipment from the supplier DISCO, followed by a short chemical mechanical polishing (CMP) step, leading to the singulation of the ultra-thin dies. These dies were then embedded in a flexible chip-in-foil package using a 50 μm thick PI foil as the base mounted onto a 150 mm wafer backing. The fan-out structures of the chip were photo-lithographically structured and metallized using a sputtered copper process. Finally, the individual chip-in-foil packages were singulated from the backplane foil via a laser cutting process. The results of the process are shown in Fig. 12.13.

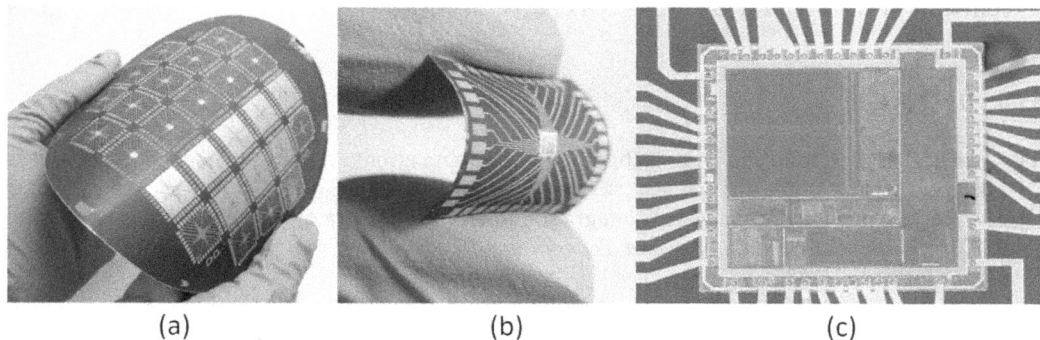

(a)　　　　　　　　(b)　　　　　　　　(c)

Fig. 12.13. (a) Batch of chip-in-foil packages of the PIC16 micro controller on a 50 μm PI film laminate after wafer level processing, (b) singulated individual chip-in-foil packages and (c) microscopic view of the embedded and interconnected microcontroller chip, with minimum distance between two interconnects being 30 μm [10].

The flexible chip-in-foils were tested electrically for functionality. A low Ohmic resistance at the common VDD was measured at all the pads of the microcontroller IC. This also proves proper alignment of all chips and interconnects on the PI foil. Optical inspection revealed just a few interrupts at fan-out interconnects. Consequently, the micro controller was programmed and its functional output was verified. A further important observation being that no chip cracking was detected after removal of the PI foil from the carrier and various handling steps. This reinforces the claim of having successfully developed a process technology for a robust yet flexible multi-thin-chip-module-packaging [10].

12.6. Conclusions

Reliable and robust flexible packaging of thin and ultra-thin silicon chips will enable a lot of new applications and market segments, especially in the IoT field. The ability to cheaply produce flexible packages via the use of high throughput technologies like Roll-to-Roll manufacturing and photolithography processes will further aid hybrid integration and system-in-package technologies to offer even more innovative yet high-performance silicon based solutions. Ultimately, this will be the packaging technology of choice, which should facilitate the "Trillion-Sensor" concept in becoming reality.

Acknowledgements

The authors would like to thank the colleagues at the department of Flexible Systems at the Fraunhofer EMFT along with the partners who supported and enabled this research. Additionally, parts of the research work was funded by the Bavarian Government under contract no. VI/3-3622/452/3, as well as the European Commission under the grant agreement PITN-GA-2012-317488-CONTEST.

References

[1]. J. Burghartz, Ultra-thin chip technology and applications, *Springer Science & Business Media*, 2010.

[2]. K. Bock, Multifunctional system integration in flexible substrates, in *Proceedings of the 64th IEEE Electronic Components and Technology Conference (ECTC'14)*, Orlando, FL, USA, 2014, pp. 1482-1487.

[3]. S. Priyabadini, 3-D stacking of ultrathin chip packages: an innovative packaging and interconnection technology, *IEEE Transactions on Components, Packaging and Manufacturing Technology*, Vol. 3, Issue 7, 2013, pp. 1114-1122.

[4]. J. van den Brand, Flexible and stretchable electronics for wearable health devices, *Solid-State Electronics*, Vol. 113, 2015, pp. 116-120.

[5]. S. Davor, The Smarter Way to Build the Internet of Everything, *Tsensors Summit*, San Diego, November 2014.

[6]. N. Palavesam, C. Landesberger, K. Bock, Investigations of the fracture strength of thin silicon dies embedded in flexible foil substrates, in *Proceedings of the IEEE 20th International Symposium for Design and Technology in Electronic Packaging (SIITME'14)*, 2014, pp. 267-271.

[7]. N. Palavesam, C. Landesberger, C. Kutter, K. Bock, Finite element analysis of uniaxial bending of ultra-thin silicon dies embedded in flexible foil substrates, in *Proceedings of the 11th Conference on Research in Microelectronics and Electronics (PRIME'15)*, 2015, pp. 137-140.

[8]. N. Palavesam, D. Bonfert, W. Hell, C. Landesberger, H. Gieser, C. Kutter, K. Bock, Electrical behaviour of flip-chip bonded thin silicon chip-on-foil assembly during bending, in *Proceedings of the 21st International Symposium for Design and Technology in Electronic Packaging (SIITME'15)*, 2015, pp. 367-372.

[9]. N. Palavesam, C Landesberger, C. Kutter, K. Bock, A novel test method for robustness assessment of very small, functional ultra-thin chips embedded in flexible foils, in *Proceedings of the Smart Systems Integration Conference (SSI'16)*, Munich, 2016, pp. 122-129.

[10]. C. Landesberger, N. Palavesam, W. Hell, A. Drost, R. Faul, H. Gieser, D. Bonfert, K. Bock, C. Kutter, Novel processing scheme for embedding and interconnection of ultra-thin IC devices in flexible chip foil packages and recurrent bending reliability analysis, in *Proceedings of the International Conference on Electronics Packaging (ICEP'16)*, 2016, pp. 473-478.

[11]. N. Palavesam, D. Bonfert, W. Hell, C. Landesberger, H. Gieser, C. Kutter, K. Bock, Mechanical Reliability Analysis of Ultra-thin Chip-on-Foil Assemblies under different types of recurrent bending, in *Proceedings of the IEEE 66th Electronic Components and Technology Conference (ECTC'16)*, 2016, pp. 1664-1670.

[12]. C. Landesberger, N. Palavesam, A. Drost, W. Hell, R. Faul, C. Kutter, Thin chip foil packaging: an enabling technology for ultra-thin packages, *Chip Scale Review*, Vol. 20, Issue 6, 2016, pp. 13-14.

[13]. G. Klink, Innovative packaging concepts for ultra thin integrated circuits, in *Proceedings of the 51st IEEE Electronic Components and Technology Conference (ECTC'01)*, 2001.

[14]. S. Schönfelder, Investigations of strength properties of ultra-thin silicon, in *Proceedings of the 6th International Conference on Thermal, Mechanical and Multi-Physics Simulation and Experiments in Micro-Electronics and Micro-Systems (EuroSimE'05)*, 2005, pp. 105-111.

[15]. D. A. Van Den Ende, Mechanical and electrical properties of ultra-thin chips and flexible electronics assemblies during bending, *Microelectronics Reliability*, Vol. 54, Issue 12, 2014, pp. 2860-2870.

[16]. H. Guojun, Characterization of silicon die strength with application to die crack analysis, in *Proceedings of the 33rd IEEE/CPMT International Electronic Manufacturing Technology Symposium (IEMT'08)*, 2008, pp. 1-7.

[17]. S. Schoenfelder, Investigations of the influence of dicing techniques on the strength properties of thin silicon, *Microelectronics Reliability*, Vol. 47, Issue 2, 2007, pp. 168-178.

[18]. M. Y. Tsai, C. H. Chen, Evaluation of test methods for silicon die strength, *Microelectronics Reliability*, Vol. 48, Issue 6, 2008, pp. 933-941.

[19]. Test Method for Measurement of Chip (Die) Strength by Mean of 3-Point Bending, *SEMI G86-0303 Standard*.

Chapter 13

Thermal Contact Resistance within Press Pack IGBTs

Erping Deng, Zhibin Zhao and Yongzhang Huang

13.1. Background and Challenges

13.1.1. Opportunities for Press Pack IGBTs

Nowadays, the challenges such as globally increasing demand of electrical energy, the stringency of conventional energy resources (such as oil, gas and coal) and so on, are arising in the field of electrical energy supply [1]. The High Voltage Direct Current (HVDC) transmission system, especially the flexible HVDC transmission system with voltage source converters (VSC), is an innovative solution because of its advantages of the ability to supply the power to the passive power grid (i.e. islet), the independent control of the active and reactive power, the flexible operation modes and so on [2]. The most important part of the flexible HVDC transmission system is the converter valve and HVDC breaker, which are based on Insulated Gate Bipolar Transistors (IGBTs), briefly shown in Fig. 13.1 below.

The flexible HVDC transmission system has been successfully applied in the developed countries for many years and it is prosperous in China recent years. More and more projects with higher voltage and higher capacity ratings are developing to meet the requirements and the reliability is most important issue. This high voltage and high reliability application therefore has greatly promoted the development of IGBTs. There are two packaging styles for high power IGBT devices: typical wire-bonded IGBT module and press pack IGBTs. The high power IGBT module of 3300 V/1500 A had been widely used in the flexible HVDC transmission system. While with the growing demand of capacity, the IGBT module can not meet the increasing voltage and capacity requirements and press pack IGBTs is gradually applied with its advantages of higher power density, easy to connect in series and short-circuit failure mode and so on [3].

Erping Deng
North China Electric Power University, Beijing, China

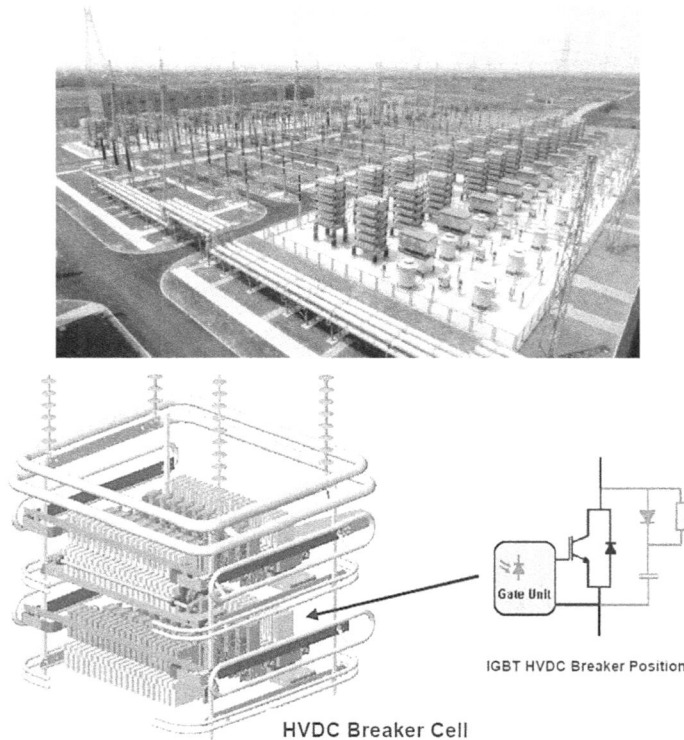

Fig. 13.1. Two most important components in the flexible HVDC transmission system: converter valver (*top*), HVDC breaker (*bottom*).

The first press pack IGBTs used in the converter valve of the flexible HVDC transmission system in China is 4500 V/1500 A. And then the press pack IGBTs of 3300 V/2000 A, 3300 V/3000 A, 4500 V/2000 A, and 4500 V/3000 A are required in the future flexible project because of the higher capacity demand, for example the 4500 V/3000 A is needed in the 500 kV/3000 MW or 800 kV/3000 MW flexible project. This gives a good opportunity for the development of press pack IGBTs.

13.1.2. Challenges for Press Pack IGBTs

The press pack packaging style for high voltage and high power density IGBT can be divided into StakPak (Fig. 13.2) and press pack (Fig. 13.3). The original motivation in most cases was the poor power cycling capability of early versions of wire-bonded modules and their explosion behaviour [4]. The StakPak packaging style is patent protected by ABB and the research on the StakPak is very limited. The press pack is widely used by Poseico, Fuji, Westcode and Toshiba because of the experience with the packaging of high power devices, such as gate turn off thyristors, diodes, IGCT, etc. [5], and many researches are based on this packaging style. Therefore, the press pack IGBTs discussed in this Chapter is the press pack style as shown in Fig. 13.3.

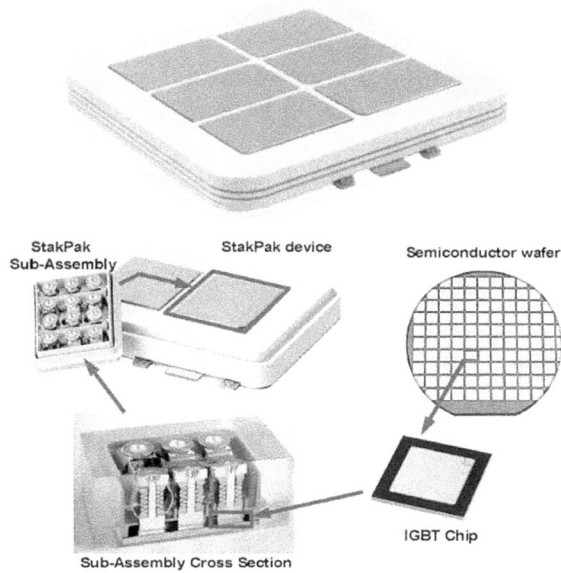

Fig. 13.2. StakPak packaging style from ABB.

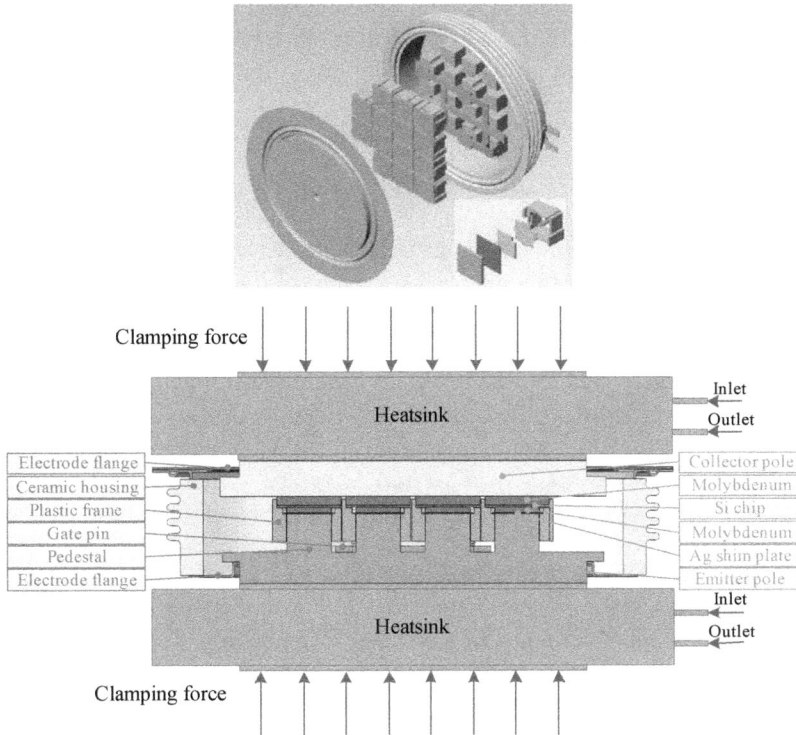

Fig. 13.3. Press pack style for press pack IGBTs: exploded views (*top*), cross-section view (*bottom*).

Fig. 13.3 shows that the press pack IGBTs has a multi-layered structure. The electrical and thermal paths for the silicon chips are supplied by the collector and emitter copper electrodes. Furthermore, the needed clamping force is also applied on the two electrodes. Two molybdenum plates surrounding the silicon chips are to uniform the clamping force distribution and reduce the thermal expansion/contraction between the molybdenum plates and silicon chips when the press pack IGBT undergoes high temperature variations. A silicon chip subassembly is consisted of a silver shim plate, together with a silicon chip and two molybdenum plates. Many silicon chip subassemblies connected in parallel to form a press pack IGBT and the current rating is determined by the paralleled number.

With the increasing demand of higher voltage and current ratings, more and more silicon chip subassemblies are needed to connect in parallel in a press pack IGBT. Therefore, there are many challenges in the packaging technology, especially the long life time reliability when applied in the flexible HVDC transmission system.

- Current distribution among silicon chips [6];
- Clamping force distribution among silicon chips [7-9];
- Junction temperature distribution among silicon chips [10];
- The internal insulation problem [11];
- Long time reliability [12].

Meanwhile, the current distribution and junction temperature distribution also influenced by the clamping force distribution a lot [13] through the electrical and thermal contact resistance [14]. The issue of both the electrical and thermal contact resistance therefore is also very important in press pack IGBTs.

13.1.3. Thermal Contact Resistance within Press Pack IGBTs

As press pack IGBTs has a multi-layers structure shown in Fig. 13.3 all the components are stacked together through the external clamping force. There exist many contact surfaces and the contact surfaces are rough as the Fig. 13.4 shows the micro cross-section diagram of an IGBT chip [15]. An aluminium thickness of several μm is needed to cover the chip surface to realize the connection with the external electrodes. When two rough surfaces are bought into contact, actual contact only occurs at certain discrete spots or micro-contacts, while the non-contacting areas form vacuums or are filled with some medium (such as air, water or oil). Thermal interface material is therefore the most commonly used material to fill the gap to increase the contact area as shown in Fig. 13.5.

The actual contact area accounts for approximately 0.01 % - 0.1% of the nominal contact area and the proportion only increases to 1 % - 2 % under a contact pressure of 10 MPa [16]. Because the actual contact surface area is relatively small, as is the thermal conductivity of the interfacial gases, heat flow across the interface experiences a relatively large thermal resistance, commonly referred to as thermal contact resistance [17, 18]. The interface specific thermal contact resistance can be determined experimentally by measuring the temperature drop across the contact interface at a given clamping force. Another possibility is to use interface specific constants and based on the clamping force calculate the thermal contact resistance. Busca et al. [10] shows that the thermal contact resistance within press pack IGBTs at nominal clamping force accounts for about 51 % of the totally thermal resistance and the result is shown in Fig. 13.6. Furthermore, as the thermal contact resistance is mainly affected by the clamping force, it also affects the transient thermal impedance a lot as shown in Fig. 13.7.

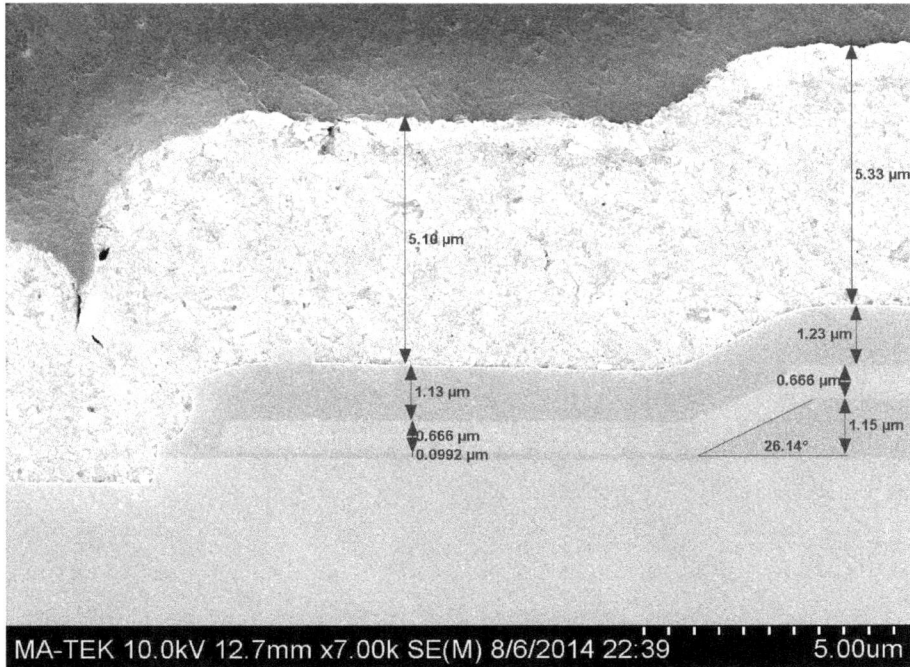

Fig. 13.4. Micro cross-section diagram of an IGBT chip: micro cross-section diagram (*top*), amplified diagram of specific area (*bottom*).

333

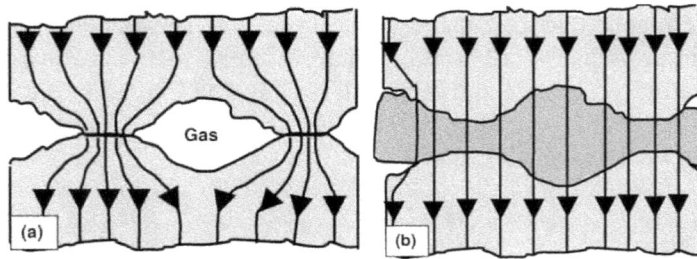

Fig. 13.5. Micro-contact interface schematic diagram: contact interface (*left*), contact interface with thermal interface material (*right*).

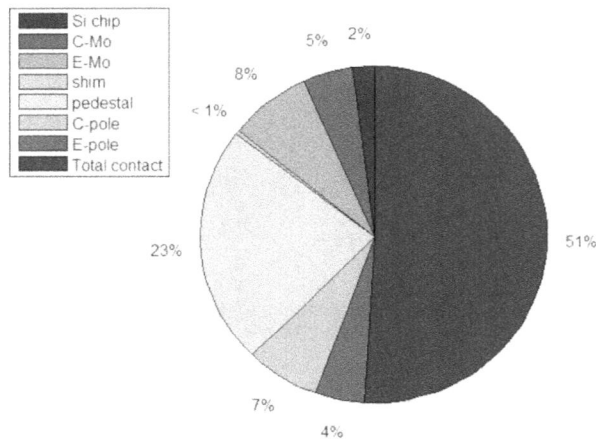

Fig. 13.6. Share of the thermal resistances among the layers in press pack IGBTs.

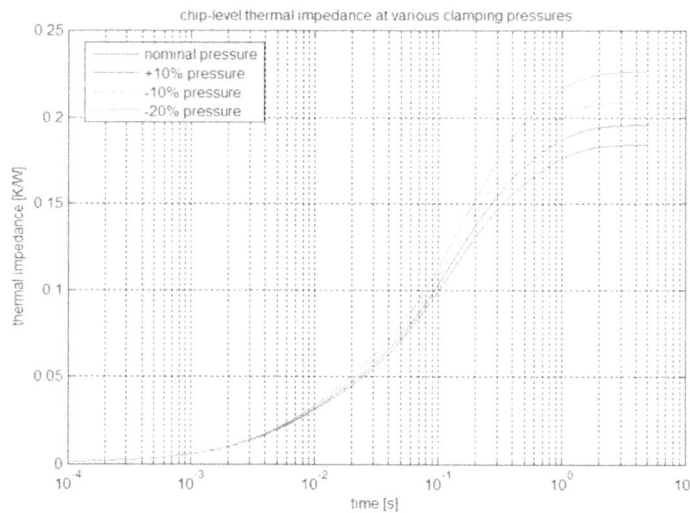

Fig. 13.7. Chip-level thermal impedance at various clamping pressures (at 1.77 kN nominal pressure).

Fig. 13.7 shows that the clamping force has a significance influence on transient thermal impedance because of the thermal contact resistance. If the clamping force of one silicon chip among those paralleled chips is lower than others, thermal contact resistance will increase a lot and eventually leads this silicon chip to thermal damage.

13.2. Finite Element Simulation

13.2.1. Theoretical Model

As shown in Fig. 13.5 the contact spot is always named micro-contact, and the non-contact area is filled with air. Because the thermal conductivity of the interfacial gases is relatively low (about 0.023 W/(m·K)), heat flow across the interface experiences a relatively large thermal resistance, commonly referred to as thermal contact resistance. Micro-contacts, which have a relatively long distance among them, are general distributed randomly when two rough surfaces came into contact. Only thermal conduction is considered between the micro-contact and the interfacial gaps, which are usually filled with air in this situation [19]. In most practical situations concerning thermal contact resistance, the gap thickness between two contacting bodies is quite small (< 10 μm), and thus the Grashof number based on the gap thickness is less than 2000. Consequently, in most instances, the heat transfer of convection through the interstitial gas in the gap is neglected [20]. Thermal radiation across the interfacial gaps is generally considered as insignificant as the surface temperature is less than approximately 700 K [21]. In conclusion, the thermal contact conductance h_c of the contact interfaces, which can be calculated by formula (13.1), consists of the micro-contact thermal contact conductance h_s and interfacial gap thermal contact conductance h_g. Furthermore, thermal contact resistance R_c can be obtained through formula (13.2), as the thermal contact conductance h_c and nominal contact area A is given.

$$h_c = h_g + h_s. \tag{13.1}$$

$$R_c = A \cdot h_c. \tag{13.2}$$

13.2.1.1. Micro-Contact Conductance h_s

Williamson et al. [22] have shown experimentally that many of the techniques used to produce engineering surfaces give a Gaussian distribution of surface heights. The contact between Gaussian rough surfaces can be considered as the contact between a single Gaussian surface, having the effective surface characteristics, placed in contact with a perfectly smooth flat surface, as shown in Fig. 13.8 [23].

The equivalent roughness and surface slope can be calculated from formula (13.3).

$$\sigma = \sqrt{\sigma_1^2 + \sigma_2^2} \text{ and } m = \sqrt{m_1^2 + m_2^2}, \tag{13.3}$$

335

where subscripts 1 and 2 represent the two contact surfaces, respectively. The surface roughness of the rough contact interface is largely influenced by the machining precision and can be measured via a surface roughness tester. The surface slope *m* affected by the manufacture craft and surface roughness, etc., can be approximately predicted through a function of surface roughness with theoretical calculation or experimental data. The reason is that the surface topography is determined by the manufacture craft or different process [24] and four typical surfaces are shown in Fig. 13.9. Meanwhile, the surface roughness and slope is depended on the surface topography to a large extent as shown in Fig. 13.8.

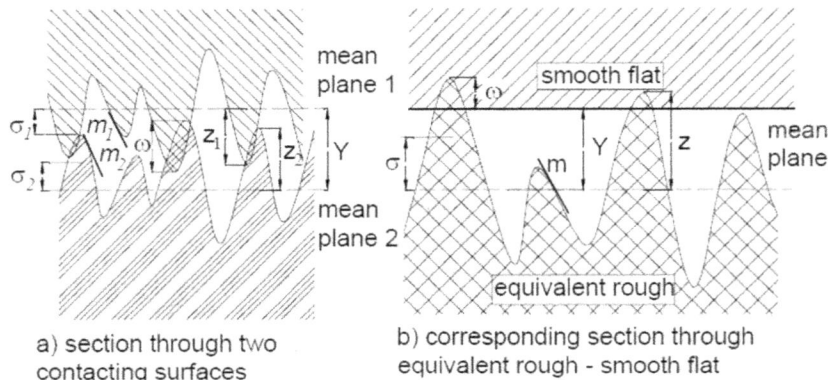

Fig. 13.8. The roughness surfaces equivalent schematic diagram.

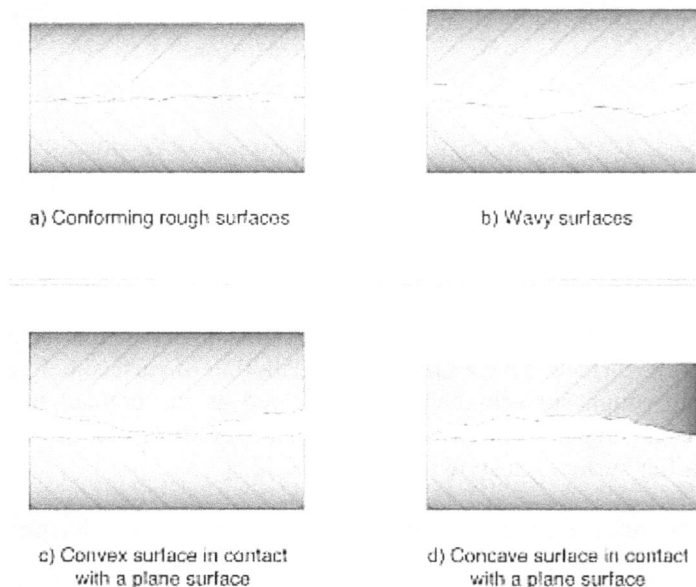

Fig. 13.9. Contact configurations.

Correlations for surface slope are summarized in Table 13.1. As the measured surface roughness of those interfaces in press pack IGBTs is much lower than 1.6 μm [15], the surface slope is identified by [25] shown in Table 13.1.

Table 13.1. Correlations for surface slope.

Reference	Correlation
Tanner [26]	$m = 0.152 \cdot \sigma^{0.4}$
Antonetti et al. [25]	$m = 0.124 \cdot \sigma^{0.743}, \quad \sigma \leq 1.6 \mu m$
Lambert [27]	$m = 0.076 \cdot \sigma^{0.52}$

Thermal resistance associated with micro-contact is generally modelled by applying the so-called flux cube solution to each micro-contact with the assumption that the asperities undergo plastic deformation. According to the theory of Bahrami et al., thermal contact conductance of the contact interface can be calculated as long as a single micro-contact thermal model is given [28, 29].

$$h_s = 1.25 \cdot k_s \cdot \frac{m}{\sigma} \cdot \left(\frac{p}{H_c} \right)^{0.95}, \tag{13.4}$$

where p is the contact pressure and k_s is the harmonic mean of solid thermal conductivity of the two bodies, which is shown in formula (13.5).

$$k_s = \frac{2 \cdot k_1 \cdot k_2}{k_1 + k_2}. \tag{13.5}$$

H_c is the surface microhardness of the softer of the two contacting solids. The microhardness is generally complex because it depends on several geometric and physical parameters, such as the Vickers microhardness correlation coefficients, surface roughness and contact pressure. As shown in formula (13.6), Song and Yovanovich related H_{mic} to the surface parameters and nominal pressure [30].

$$\frac{p}{H_{mic}} = \left(\frac{p}{c_1 \cdot \left(1.62 \cdot \frac{\sigma}{\sigma_0} \cdot m \right)^{c_2}} \right)^{\frac{1}{1+0.071 \cdot c_2}}. \tag{13.6}$$

13.2.1.2. Micro-Gap Conductance h_g

Many theoretical models of the gap conductance h_g are proposed by scholars, and the most commonly used is given by the approximation of Yovanovich [31], which is shown in (13.7).

$$h_g = \frac{k_g}{Y+m}, \tag{13.7}$$

where k_g is the thermal conductivity of the gap substance and the gas parameter M accounts for rarefaction effects at high temperatures and low gas pressures. The effective gap thickness Y, shown in Fig. 13.8, can be calculated accurately by means of the simple power-law correlation equation proposed by Antonetti and Yovanovich [32].

$$Y = 1.53 \cdot \sigma \cdot \left(\frac{p}{H_c}\right)^{-0.097}. \tag{13.8}$$

13.2.2. Finite Element Model

In the high-voltage and high-power-density application, multiple silicon chips, including the IGBT chips and Fast Recovery Diode (FRD) chips anti-paralleled with IGBT chips, are connected in parallel in press pack IGBTs to increase the current rating to meet the application requirements. Therefore some difference exists inevitably among these chips, especially the clamping force and temperature distribution. The thermal contact resistance between multi-chips is also connected in parallel because multi-chips in press pack IGBTs are connected in parallel. As the thermal contact resistance is significantly influenced by the clamping force and temperature within press pack IGBTs [14, 33], and the difference in the clamping force and temperature among those chips also leads to some distinction in the thermal contact resistance. Thus, the thermal contact resistance of each contact layer cannot be clearly revealed through the thermal contact resistance of press pack IGBTs.

Furthermore, different from the IGBT chip used in most wire bond packaging, the gate of the IGBT chip used in press pack packaging is located in the corner of the chip. Thus, grooves are shaped into the pedestals, as well as other components on the emitter side, for the gate pin of the housing. In order to exclude interference factors and reduce the complexity of the experiment, a single FRD chip submodule can use to predict the behaviour of thermal contact resistance within press pack IGBTs because the heat path of IGBT chip and FRD chip are the same. The three-dimensional structure diagram and equivalent thermal network are shown in Fig. 13.10 and Fig. 13.11.

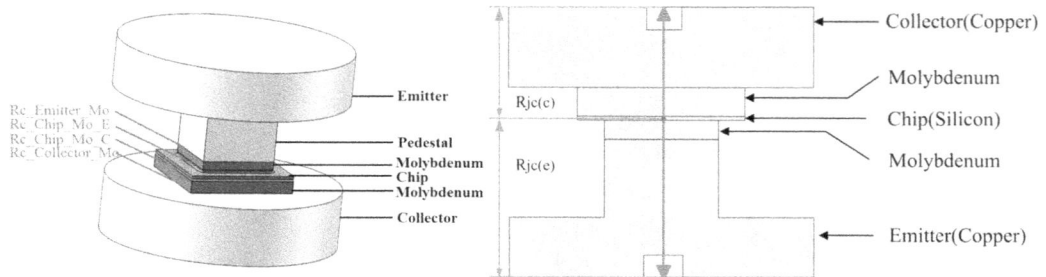

Fig. 13.10. A single FRD chip submodule: three-dimensional structure diagram (left), section view and the relationship of thermal resistance (right).

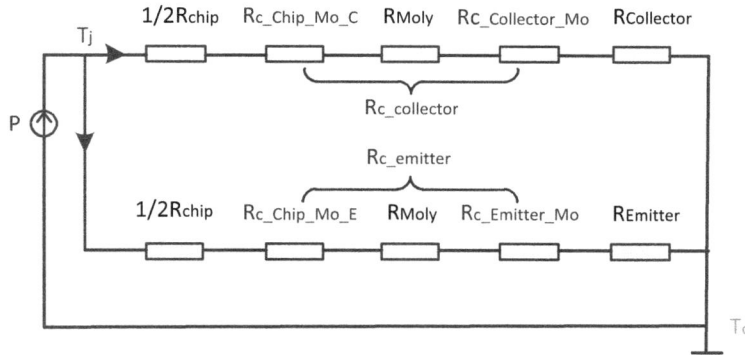

Fig. 13.11. Equivalent thermal network of a single FRD chip submodule.

Both sides of the FRD chip are covered with several μm of aluminium to realize the connection (aluminum metallization). The thermal contact resistance of the collector side and emitter side should be calculated separately because the surface roughness of the aluminium layer is quite different. According to the description above, a single FRD chip submodule contains four contact interfaces, which is shown in Fig. 13.10 and Fig. 13.11. The material properties of the single FRD submodule are listed in Table 13.2.

Table 13.2. Material parameters.

	Copper	**Moly**	**Alum**
Density [kg/m^3]	8930	10200	2700
Thermal conductivity [W/(m×K)]	394	145	210
Specific heat [J/(kg×K)]	385	217	900
Elasticity modulus [GPa]	110	330	68
Surface roughness [μm]	※	※	※
Surface slope	※	※	※

where ※ denotes that the material property should be measured via experiment.

The basic thermal differential equation in press pack IGBTs is shown in (13.9), where ρ is the material density, C_p is the constant pressure specific heat, t is the time, k is the thermal conductivity and Q is the volumetric heat source.

$$\rho \cdot C_p \cdot \frac{\partial T}{\partial t} - \nabla \cdot (k \cdot \nabla T) = Q. \tag{13.9}$$

The boundary conditions for this simulation are set as follows. The temperatures of the collector and emitter pole outer faces (case temperature) have been set to 60 °C, with the assumption that the external water-cooling system can sustain a constant case temperature. The heat generation in the active area of the FRD chip is set as 100 W. The equivalent clamping force of the single FRD chip submodule is assumed to be 1 kN according to press pack IGBTs' work condition (1.2 kN/cm^2) and the contact area.

13.2.3. Simulation Results and Analysis

As shown in the thermal contact resistance calculation formulas above, thermal contact resistance is mainly affected by surface roughness, surface slope, microhardness and contact pressure. Thermal contact resistance within all the interfaces of the single FRD submodule are calculated through different interface parameters based on the formulas aforementioned. In this Section, the interface between the collector side of the FRD chip and molybdenum is used as an example to show the influence of the previously mentioned parameters. Surface roughness ranges from 0.1 µm to 1 µm according to the manufacture craft for press pack IGBTs. Surface slope ranges from 0.02 to 0.2 according to the surface roughness. Microhardness ranges from 1 GPa to 10 GPa according to material property, and the equivalent clamping force ranges from 200 N to 2000 N based on the clamping force of a single FRD chip within presspack IGBTs. The calculation results are shown in Fig. 13.12 and Fig. 13.13. The unit of thermal contact resistance is K/W.

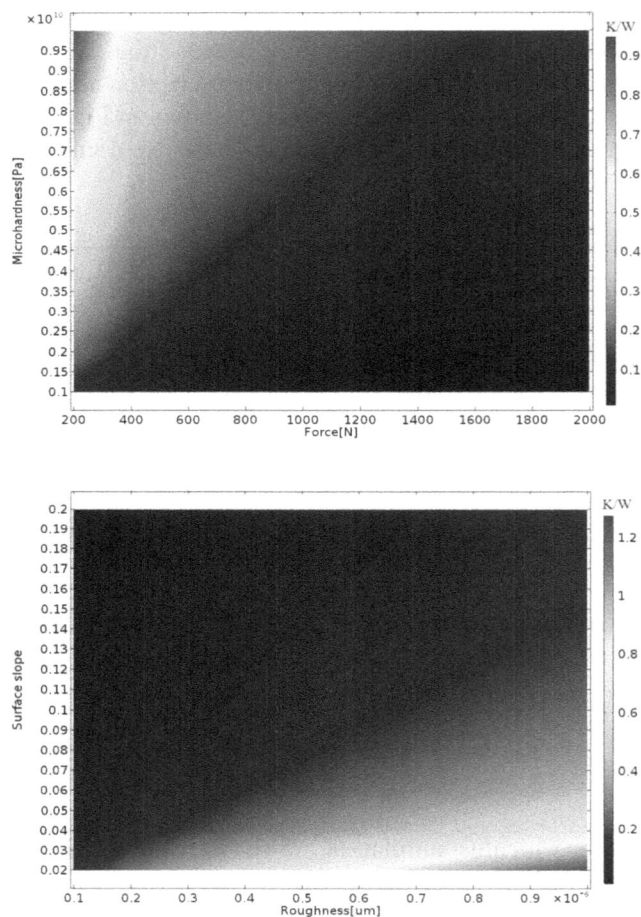

Fig. 13.12. Thermal contact resistance of the interface micro-contact: contact force and microhardness (*top*), surface roughness and slope (*bottom*).

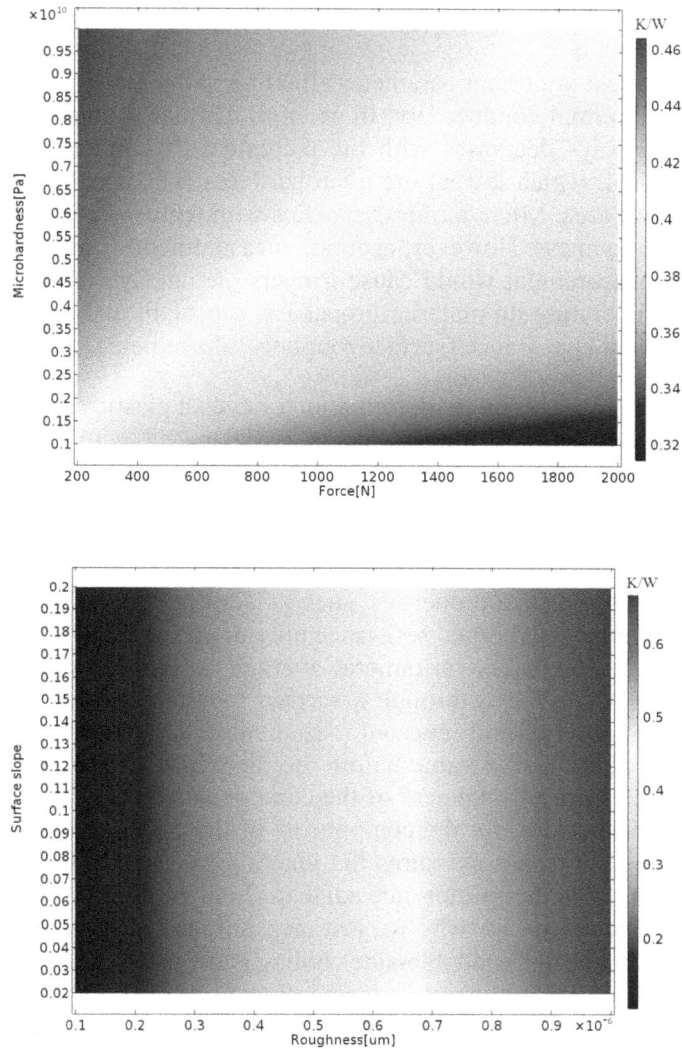

Fig. 13.13. Thermal contact resistance of the interface gap: contact force and microhardness (*top*), surface roughness and slope (*bottom*).

It is observed that as expected, a higher contact pressure (equivalent clamping force) and a lower microhardness H_{mic}, which accelerate the plastic deformation of the asperities and the formation of micro-contacts with a larger contact surface area, give arise to a lower value of the thermal micro-contact resistance. Likewise, a decrease in the surface roughness and an increase in the roughness slope, which result in higher values of the micro-contact surface area, also give rise to a lower thermal micro-contact resistance. The micro-gap thermal resistance is mainly affected by contact pressure, surface roughness and microhardness, with the exception of surface slope.

13.2.3.1. Influence of the Temperature

Temperature is the most important parameter affecting many material properties, such as microhardness and thermal conductivity. In general, thermal contact resistance between contact interfaces always decreases with the increment of temperature because that a temperature increment, which lowers the microhardness (i.e., it softens), gives rise to a larger contact surface area. Microhardness changes with temperature eventually lead to a change of surface roughness. However, accurate measurement of microhardness is quite difficult, and the measurement would cause irreversible damage to the chip. Therefore, the influence of temperature on material properties, especially the surface properties, is represented by its final appearance (surface roughness decrement).

Micro-contacts undergo an elasticity deformation or even a plastic deformation under the external clamping force as the microhardness is decreased with the augment of the temperature of the chip and other components during the thermal resistance test. Although the temperature of the chip and other components is reduced after the measurement is completed, the deformation is still unable to be fully restored to the state before measurement under the external clamping force and eventually leads to a decrease of surface roughness. The surface properties, such as surface roughness, would no longer change after several repeated thermal resistance measurements (equivalent to temperature cycling). The surface roughness (arithmetic average of absolute values R_a) of every contact surface can be measured through a surface roughness tester (such as Mitutoyo SURFTEST SJ-310) after several repeated measurements, as shown in Table 13.3, and compared with the experimental value before the thermal resistance test. It is observed that, as expected, the surface roughness of the components after testing is much smaller than before testing, especially for the components of the collector side. This is because the thermal resistance test only measures the junction-to-case thermal resistance of the collector-side cooling with the emitter-side adiabatic, and the heat flow only goes through the collector side. The reason for why we just measure the collector-side cooling is that thermal contact resistance under double-side cooling is the parallel relationship of the one under collector-side cooling and emitter-side cooling. So it is very difficult to distinguish the relationship of each interface and hard to verify the theoretical values. The calculated thermal contact resistance before and after the thermal resistance test based on the theoretical model mentioned before is shown in Table 13.4 with the clamping force of 1 kN.

Table 13.3. Comparison of the surface roughness.

Surface roughness σ [μm]	Before	After	Deviation
Collector	0.226	0.151	33.18 %
Collector-side molybdenum	0.344	0.268	22.09 %
Chip collector side	0.495	0.147	70.30 %
Chip emitter side	0.147	0.057	61.22 %
Emitter-side molybdenum	0.334	0.307	8.08 %
Emitter	0.638	0.617	3.29 %

Table 13.4. Thermal contact resistance comparison (@ 1 kN).

	Collector_Mo	Chip_Mo_C	Chip_Mo_E	Emitter_Mo
Before [K/W]	0.1024	0.1192	0.1211	0.1584
After [K/W]	0.086	0.09	0.117	0.156
Deviation [%]	16.02	24.50	3.39	1.52

13.2.3.2. Influence of the Clamping Force

Though thermal contact resistance can be reduced to some extent through the refinement of machining craft to decrease surface roughness, the effect is limited. The reason is that the surface slope, which is a function of surface roughness as shown in Table 13.1, reduces with the surface roughness decrement and leads to an increment of thermal contact resistance, therefore finally suppresses the decrement of thermal contact resistance. Most importantly, the refinement of machining craft will inevitably result in a sharp increase in cost. Microhardness is influenced by many external factors, such as temperature, but mainly determined by internal properties. Thus the effect of microhardness to reduce thermal contact resistance seems very small due to the material of press pack IGBTs that is assigned. As discussed above, the only parameter that can greatly change the thermal contact resistance in press pack IGBTs is the clamping force. The thermal contact resistance of each interface under different clamping forces is calculated based on the formulas aforementioned and is plotted in Fig. 13.14.

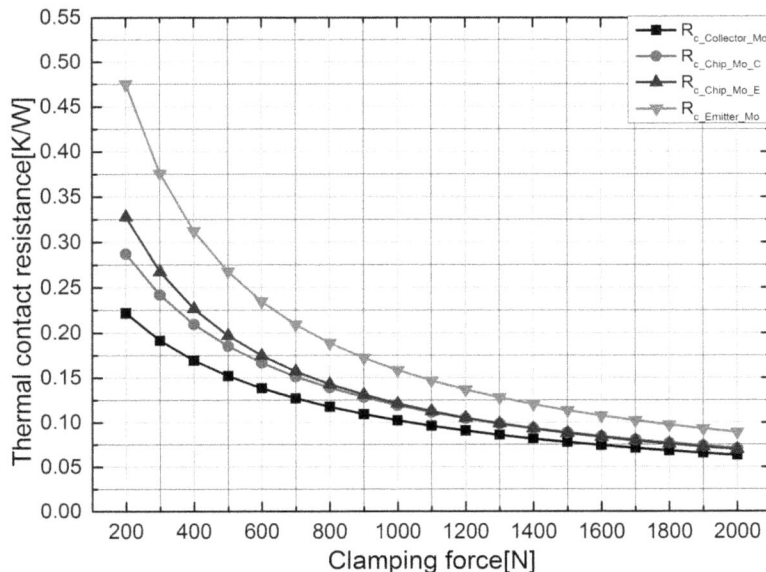

Fig. 13.14. The relationship between the clamping force and thermal contact resistance between four interfaces of the FRD chip submodule.

As shown in Fig. 13.14, the thermal contact resistance decreases drastically with the increase of the clamping force at an early stage; the rate then slows down in the middle stage and finally towards saturation. Due to the double-side cooling, the heat flow can go through both the collector and emitter sides, and the thermal contact resistance of each side is connected in series, as shown in Fig. 13.10 and Fig. 13.11. Fig. 13.15 shows the difference of thermal contact resistance of both the collector and emitter sides under different clamping forces with the rated clamping force of 1 kN. It can be seen that thermal contact resistance sharply increases when the clamping force is lower. And the increment of clamping force can reduce it to some extent, but with the clamping force continuing to increase the rate of thermal contact resistance reduction slows down. Meanwhile, the failure rate of the chip sharply increases. The thermal contact resistance increment caused by lower clamping force gives rise to a high junction temperature and eventually leads to low reliability. Thus, the strictly controlled distribution of clamping force within press pack IGBTs has a significant effect on the electrical and thermal characteristics and reliability of press pack IGBTs.

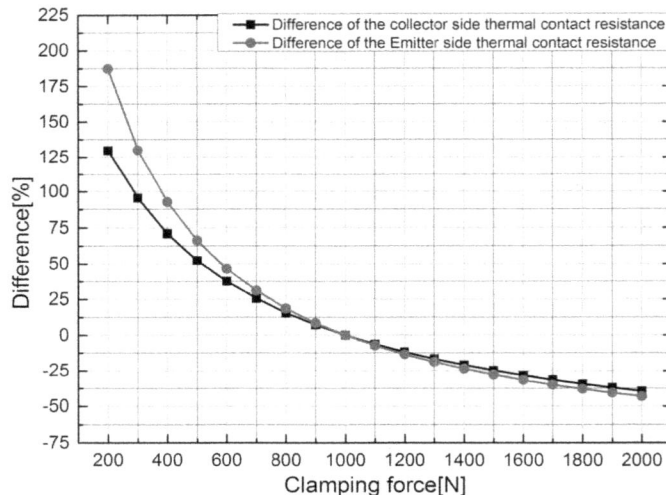

Fig. 13.15. The difference of the thermal contact resistance versus clamping force, with respect of the value at the nominal clamping force of 1 kN.

13.3. Experimental Measurement

Thermal contact resistance is a complex parameter affected by the material properties, surface morphology, contact pressure, temperature and so on [34, 35], and thus the precision of the test bench will be a significant challenge for measurement. At present, steady-state and transient methods are used to predict the thermal contact resistance, and the steady-state method is most commonly used.

In the steady-state method, the temperature difference between the contact interfaces can be obtained by a linear fitting after the temperature has been measured between two samples [36]. The schematic diagram is shown in Fig. 13.16 and usually the temperature is measured through thermocouples. The thermal flux Q can be measured by the flowmeter or calculated through thermal conductivity and temperature gradient. Then the thermal contact resistance can be obtained by formula (13.10).

$$R_c = \frac{T_1 - T_2}{Q}. \qquad (13.10)$$

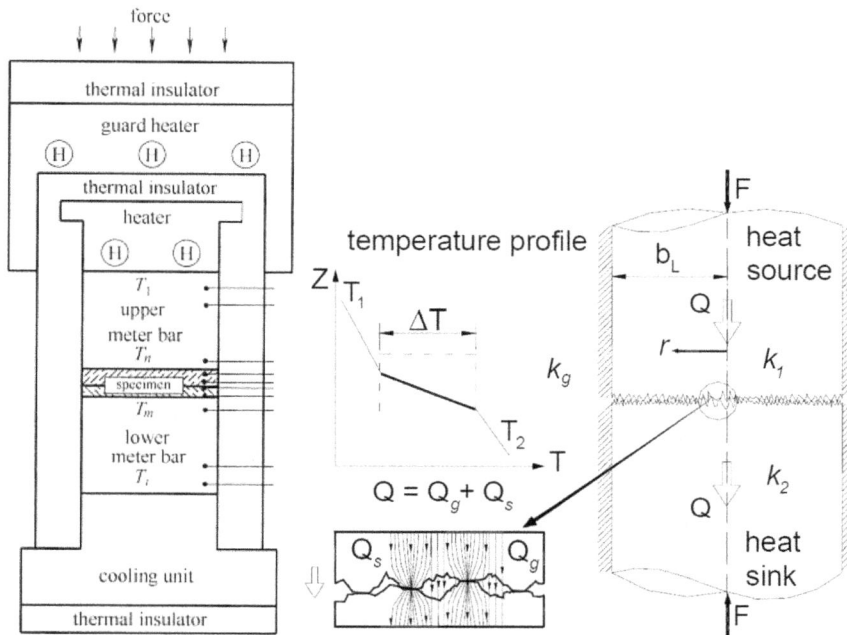

Fig. 13.16. Schematic diagram for the steady-state method.

This method has several disadvantages and the accuracy is limited. Not only will the temperature distribution of the specimen be disturbed by the embedded thermocouple, but the accuracy of the adjacent measured temperature will also be influenced since this steady method requires a thermocouple to measure the temperature. Most importantly, many thermocouples should be located in the samples, and thus the steady-state method is not suitable for a specimen whose geometry is small (in the millimeter ranger). To cover some disadvantages of the thermocouple method, an infrared imaging system with an accuracy of 0.1°C, instead of a thermocouple, is put forward by various scholars to record the temperature of the two-dimensional interface. Although the accuracy is greatly improved, an error of approximately 23 % is still presented [37].

The optical-thermal method is the widely used method among transient thermal contact resistance measurement methods [38, 39] and the schematic diagram is shown in Fig. 13.17. Thermal contact resistance is obtained by the phase difference of the heat wave and modulation wave after encountering the interface. However, the accuracy of the optical-thermal method is affected by the interface characteristic, as the heat wave is diffused at the contact interface, and it destroys their phase relationship [40].

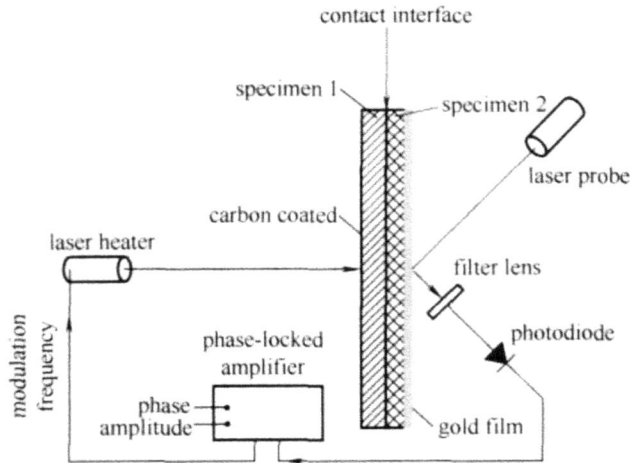

Fig. 13.17. Optical-thermal method for thermal contact resistance.

All the methods, including the steady-state and transient methods, so far are for the measurement of thermal contact resistance between power semiconductor devices and heatsinks or among copper bus bar junctions. Namely, all the methods are for the thermal contact resistance outside power semiconductors rather than within them. Furthermore, those methods are not suitable for the direct measurement of the thermal contact resistance within press pack IGBTs because that the components are too small. Until now there are two indirect methods for the measurement: the combination of the experimental results with a Finite Element Method (FEM) model called as *combination method* and the measurement of the junction-to-case thermal resistance called as *indirect experiment method* in this Chapter.

13.3.1. Combination Method

T. Poller et al. [41] measured the thermal contact resistance within press pack IGBTs through a combination of experiments and a FEM model. The measurement setup was developed to determine the contact resistances on a single chip assembly as shown in Fig. 13.18 and it is clamped with metallic stamps provided with drilled holes for installing thermal probes. The system also contains four thermal contact resistances:

- Interface between the emitter side (copper) and the Molybdenum plate;
- Interface between the Molybdenum plate and the emitter side of the chip (aluminum metallization);
- Interface between the collector side of the chip (aluminum metallization) and the Molybdenum plate;
- Interface between the Molybdenum plate and the collector pole (copper).

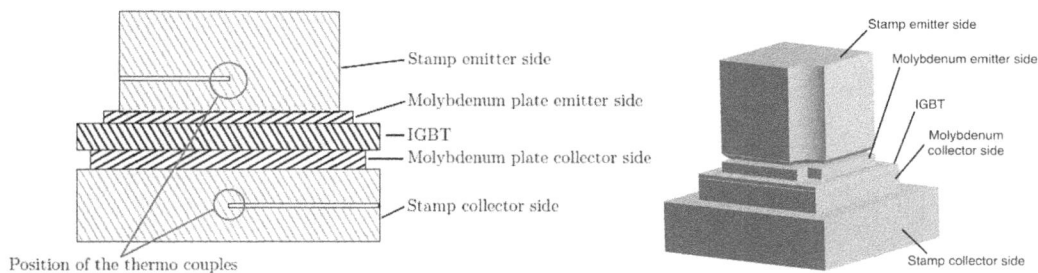

Fig. 13.18. A single chip assembly: measurement setup (*left*),
geometry of the FEM model (*right*).

The identification of the unknown parameters was performed with an iterative process based on MOGA (multi-objective genetic algorithm). This algorithm iteratively applies a variation on the unknown parameters followed by a solution of the FEM model. At each iteration, the match between the FEM solution and the experimental setup is verified by comparing some predefined target parameters calculated by the FEM with the actual measurement. And the flowchart of the iterative process is shown in Fig. 13.19. The obtained thermal contact conductance h_c is shown in Fig. 13.20. The TCC in the figure is the thermal contact conductance h_c.

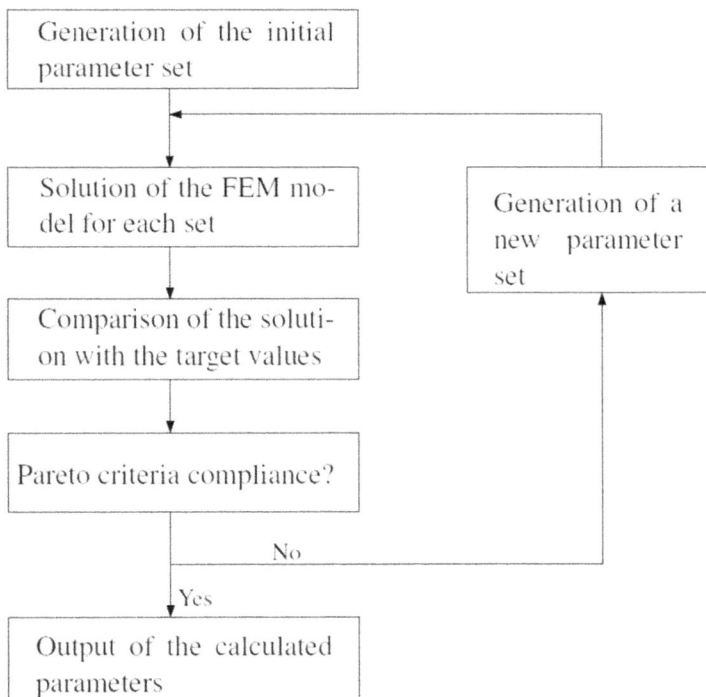

Fig. 13.19. Flowchart of the iterative process for parameter calculation.

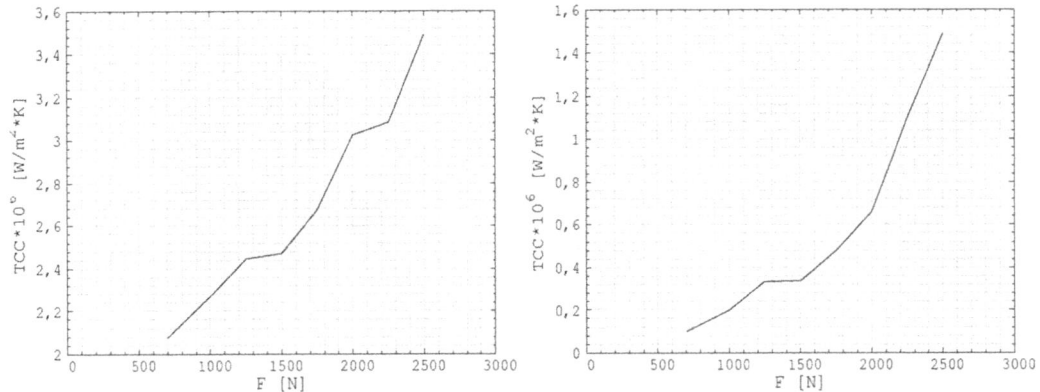

Fig. 13.20. Thermal contact conductance: the interconnection between aluminium and molybdenum (*left*), the interconnection between copper and molybdenum (*right*).

13.3.2. Indirect Experiment Method

E. Deng [15] also proposed an indirect method to measure the junction-to-case thermal resistance to reflect the thermal contact resistance change. And furthermore the thermal contact resistance can be accurate obtained by comparing the experimental results with a sintered value which will present in the optimization section. As shown in Fig. 13.11, the junction-to-case thermal resistance of the single FRD chip submodule with the collector/emitter side cooling consists of the bulk thermal resistance and the thermal contact resistance of specific contact interfaces. The variation of the thermal contact resistance under different conditions will lead to a variation of junction-to-case thermal resistance with the assumption that the bulk thermal resistance is not sensitive to the clamping force or temperature (the temperature is not very high). Generally, bulk thermal resistance is considered in constant and the change in thermal contact resistance will induce the change of the junction-to-case thermal resistance. It is therefore measured to indirect predict the thermal contact resistance. The junction-to-case thermal resistance can be obtained by formula (13.11) as the junction temperature T_j, case temperature T_c and the power dissipation P is given.

$$R_{jc} = \frac{T_j - T_c}{P}.$$

(13.11)

The power dissipation P can be easily obtained by the heating current and the voltage drop induced by the heating current. And the case temperature T_c can be obtained by the thermocouple located in the case surface. However, the junction temperature T_j cannot be accurately measured in a directly method and usually measured indirectly through an electrical method. The collector-emitter voltage drop caused by a notably small constant current is the most commonly used Temperature Sensitive Parameter (TSP) and its relationship with junction temperature is the most used method, which is defined as the K factor [42].

Thermal resistance tester (*phase11*) is used to measure the junction-to-case thermal resistance and special clamping force fixture is also needed for press pack IGBTs. Fig. 13.21 shows the clamping force fixtures for press pack IGBTs [15, 43].

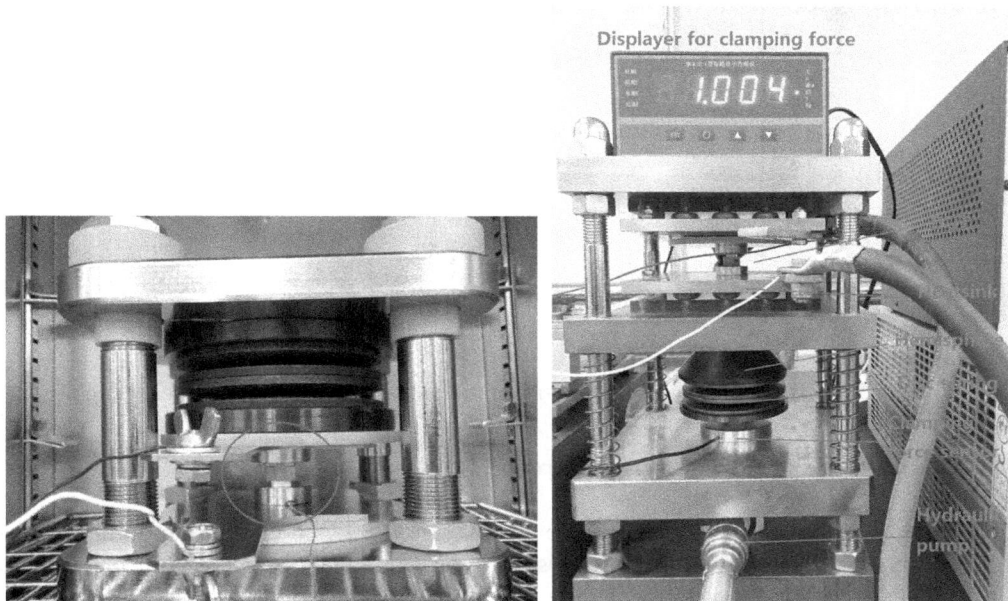

Fig. 13.21. Clamping force fixture for press pack IGBTs: K factor measurement (*left*), thermal resistance measurement (*right*).

The relationship between the external parameters, such as the clamping force and temperature, and the junction-to-case thermal resistance is considered to be equivalent to the relationship of the thermal contact resistance. Therefore the junction-to-case thermal resistance of the collector-side cooling with the emitter-side adiabatic is measured to deduce the behaviour of thermal contact resistance. The thermal contact resistance with the collector-side cooling $R_{c\_collector}$ is the sum of the $R_{c\_Collector\_Mo}$ and $R_{c\_Chip\_Mo\_C}$ as shown in Fig. 13.11.

13.3.2.1. Influence of the Temperature

The junction-to-case thermal resistance of the submodule under the rated clamping force of 1 kN is repeatedly measured by the thermal resistance tester. From the data shown in Fig. 13.22, we can see that the thermal resistance reduces quickly with the repeated measurements and tends to be approximately stable after the 11th test. The experimental results show that thermal contact resistance is changed with the temperature variation. The thermal contact resistance changes about 0.065 K/W and then trends to be stable.

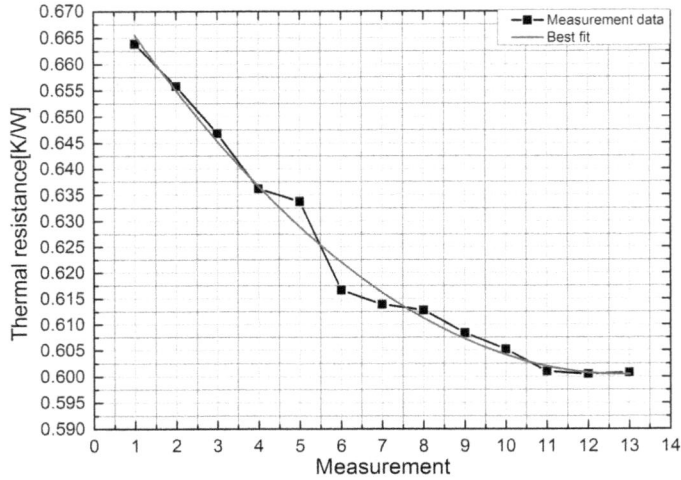

Fig. 13.22. The junction-to-case thermal resistance of the submodule (@ 1 kN).

13.3.2.2. Influence of the Clamping Force

The junction-to-case thermal resistance of the single FRD chip submodule is also measured under different clamping forces, ranging from 500 N to 1800 N. As shown in Fig. 13.23, the thermal resistance decreases along with the augment of clamping force. The thermal contact resistance changes about 0.185 K/W with the clamping force variation. The junction-to-case thermal resistance with the rated clamping force is about 0.6 K/W and the thermal contact resistance increases much with the clamping force reduction.

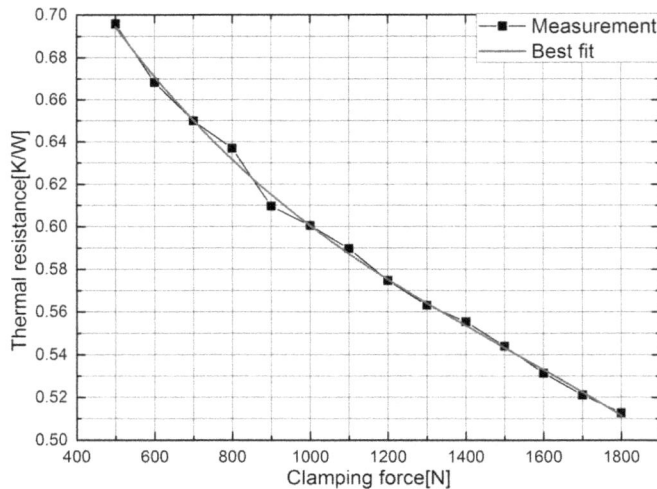

Fig. 13.23. The relationship between the clamping force and the junction-to-case thermal resistance.

13.4. Optimization

13.4.1. The Demand for Optimization

From the FEM simulation and experimental results above we can see that thermal contact resistance accounts for a considerable proportion within press pack IGBTs and it is very important to reduce it. The thermal contact resistance changes much with the clamping force variation and this is bad when multi-chips are connected in paralleled in press pack IGBTs. If the clamping force of one of the multi-chips is lower than other chips the thermal contact resistance increases a lot and then leads to a higher junction temperature. This phenomenon may cause the specific chip thermal damage as shown in Fig. 13.24 which is found in the press pack IGBTs.

Fig. 13.24. The thermal damaged silicon chip.

13.4.2. Nanosilver Sintering Technology

Although thermal contact resistance can be reduced via surface roughness refinement or increasing the clamping force as the simulation and experimental results present above, the effect of improvement is strictly limited by the machining craft and cost. As shown in Fig. 13.5, the thermal interface material is usually used to fill up micro-gaps to increase the contact surface area and reduce the thermal contact resistance because the heat flow mainly goes through micro-contacts rather than micro-gaps. Recently, the nanosilver sinetering technology has gradually been proposed to the die attach for its high electric conductivity, good thermal conductivity, high melting point [44].

In power electronics, the soft solders with melting points below 250 °C is the most commonly used die attach technology. An increase of the operation temperature above 150 °C leads to a significant decrease of the solder's reliability as failures due to solder fatigue or brittle fracture are accelerated. Two main technology trends became visible to overcome these problems: *sintering of nanosilver particles* and *diffusion soldering*. As the

351

melting point of Ag is 960 °C the Ag joint can suffer higher operation temperature. A compact Ag joint can be formed by sintering nanosilver particles under high pressure (5-30 MPa) and moderate temperatures (200-270 °C). Sintered silver layers show high temperature stability as well as excellent thermal and electrical conductivity and an improve stability during thermal and power cycling. Fig. 13.25 shows the power cycling results of samples with new die attach technologies, includes the *sintering of nanosilver particles* and *diffusion soldering* [45]. The LTJT in the picture is the abbreviation of the *Low Temperature Joint Technology* which is also the other name for the *sintering of nanosilver particles*. From the experimental results we can see that the samples with *sintering of nanosilver particles* show a higher reliability.

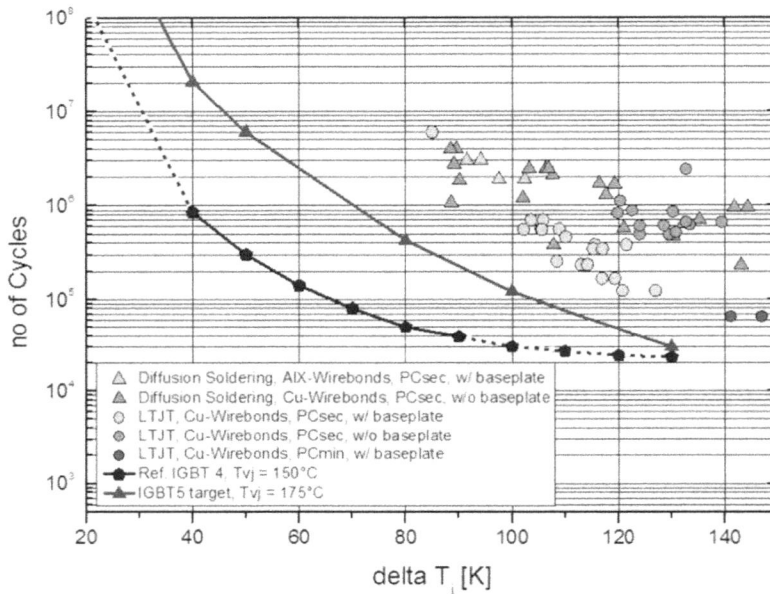

Fig. 13.25. Power cycling results of samples with new die attach technologies.

The nanosilver particles are suit for thermal interface materials to fill up the micro-gaps to reduce thermal contact resistance for its high electrical and thermal conductivity. The nanosilver sintering technology is therefore applied to the single FRD chip submodule and Fig. 13.26 shows the direct contact and nanosilver sintered submodule [15].

13.4.3. Experimental Results

The junction-to-case thermal resistance of the sintered submodule is measured through thermal tester (*phase 11*) and the clamping force fixture. All external conditions, such as the power dissipation, cooling conditions and clamping force, of the sintered submodule thermal resistance test remain the same with the direct contact submodule, with the consideration of just one experimental variable during the measurement. The bulk thermal resistance of sintered nanosilver layer can be neglected because the thickness is only 50 μm. The experimental result of the sintered submodule is shown in Fig. 13.27.

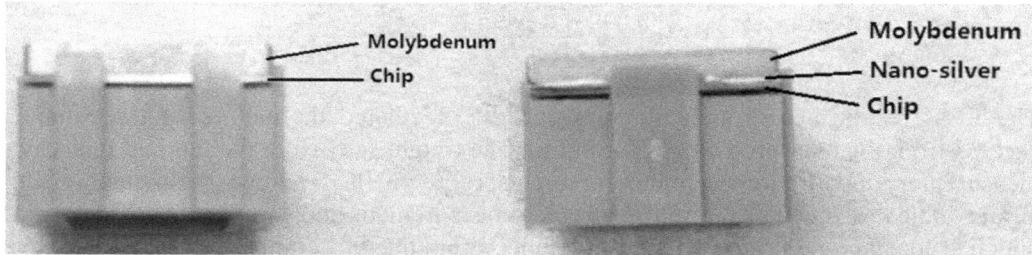

Fig. 13.26. Submodules for the thermal resistance measurement: direct contact submodule (*left*), sintered submodule (*right*).

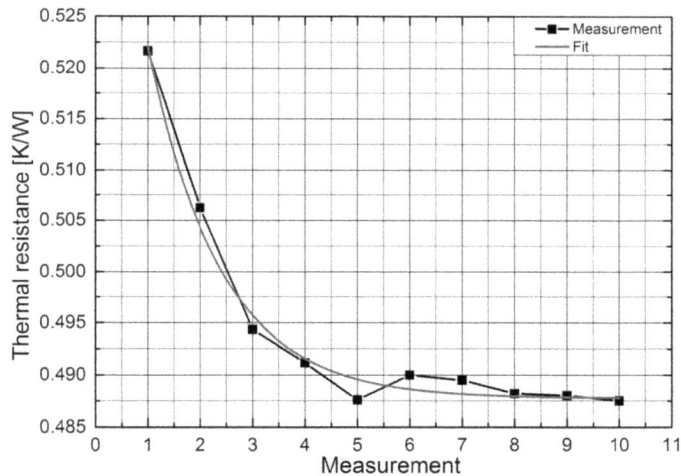

Fig. 13.27. The junction-to-case thermal resistance of the sintered submodule (@ 1 kN).

Similar to the direct contact submodule, the sintered submodule junction-to-case thermal resistance reduces with the repeated measurements and then tends toward a stable value, but it reaches stability more quickly than the direct contact one. The reason is that there only exists one contact layer with the collector side cooling. The final stable thermal resistance is much lower than the direct contact one, with the decrement of about 18.8 %. The experimental results given above shows that nanosilver sintering technology can greatly reduce the thermal contact resistance or even eliminate it. The thermal contact resistance $R_{c\_Chip\_Mo\_C}$ between the collector side of the chip and molybdenum can be obtained through the experimental results of the direct contact and sintered submodule. Meanwhile the value is also compared with the FEM simulation results in Section 1.2 as shown in Table 13.5.

Table 13.5. Thermal contact resistance comparison.

	FEM	Measurement	Deviation [%]
Before [K/W]	0.1192	0.142	16.06
After [K/W]	0.093	0.112	16.96

The FEM results before and after indicate the calculated thermal contact resistances derived from the measured values for the surface roughness before the thermal resistance measurement and after repeated measurements, respectively. And in experimental results, before indicates the first thermal resistance measurement and after indicates the value which is constant with repeated measurements. From the data comparison we can see that the theoretical model of thermal contact resistance is suitable for press pack IGBTs as the theoretical results agree with the experimental data well. The reason the experimental data are slightly larger is that the nanosilver layer cannot totally eliminate the thermal contact resistance, and the nanosilver layer will yield a bulk thermal resistance.

13.5. Conclusion and Outlook

Press pack IGBTs is a promising IGBT device to apply in the high voltage and high power density applications, such as flexible HVDC transmission system. Thermal contact resistance is a very important issue because that it connects the clamping force with the junction temperature among multi-chips within press pack IGBTs and influences the reliability. Various preliminary conclusions can be drawn via the calculation and measurement of the thermal contact resistance of the single FRD chip submodule.

1) The surface topography, such as the surface roughness and surface slope, has a great impact on the thermal contact resistance. Though reducing the surface roughness can greatly improve the thermal characteristics, the surface slope also decreases with the roughness which, in turn, increases the thermal contact resistance. At the same time, the cost will sharply increase. Thus, in actual applications, a comprehensive consideration should be taken regarding the reduction of thermal contact resistance.

2) The clamping force is the most important parameter that influences the thermal contact resistance. Thermal contact resistance sharply increases when the clamping force is reduced which, in turn, leads to a high junction temperature and eventually leads to the chip thermal damage and low reliability. Most importantly, sequential increment of the clamping force will increase the mechanical failure rate of the chip but with a slight thermal contact resistance reduction. Thus, the strictly controlled distribution of the clamping force among multi-chips within press pack IGBTs has a significant effect on the electrical and thermal characteristics and reliability of press pack IGBTs.

3) Thermal contact resistance within press pack IGBTs can be significantly reduced as the thermal interface material can effectively fill the gaps of the interface and increase the contact area. Nanosilver sintering technology is a promising connection technology in

power electronics and it is also good solution to reduce the thermal contact resistance within press pack IGBTs and improve its reliability.

Furthermore experiment works, especially the power cycling tests, should be done with the reliability of the sintered single chip submodule and multi-chips in press pack IGBTs. Meanwhile, the double-side nanosilver sintering of the silicon chip is also deserved to do.

References

[1]. H. Huang, Multilevel Voltage-Sourced Converters for HVDC and FACTS Applications, *Jabi*, 2007.

[2]. Q. Gao, Y. Lin, L. Huang, An overview of Zhoushan VSC-MTDC transmission project, *Power System & Clean Energy*, Vol. 31, Issue 2, 2015, pp. 33-38.

[3]. F. Wakeman, D. Hemmings, W. Findlay, G. Lockwood, Pressure contact IGBT, testing for reliability, *Westcode Semiconductors Ltd.*, March 2012.

[4]. H. Zeller, High Power Components from the State-of-the-Art to Future Trends, in *Proceedings of the Conference on Power Conversion (PCIM' 98)*, Nurmberg, Germany, May 1998, pp. 1-10.

[5]. C. Busca, R. Teodorescu, F. Blaabjerg, An overview of the reliability prediction related aspects of high power IGBTs in wind power applications, *Microelectronics Reliability*, Vol. 51, Issue 9-11, 2011, pp. 1903-1907.

[6]. A. Müsing, G. Ortiz, J. W. Kolar, Optimization of the current distribution in press-pack high power IGBT modules, in *Proceedings of the IEEE Power Electronics Conference (IPEC'10)*, June 2010, pp. 1139-1146.

[7]. A. Hasmasan, C. Busca, R. Teodorescu, Modelling the clamping force distribution among chips in press-pack IGBTs using the finite element method, in *Proceedings of the IEEE International Symposium on Power Electronics for Distributed Generation Systems (PEDG'12)*, June 2012, pp. 788-793.

[8]. A. Hasmasan, C. Busca, R. Teodorescu, Electro-thermo-mechanical analysis of high-power press-pack insulated gate bipolar transistors under various mechanical clamping conditions, *IEEE Journal of Industry Applications*, Vol. 3, Issue 3, 2014, pp. 192-197.

[9]. E. P. Deng, Z. B. Zhao, P. Zhang, Clamping force distribution within press pack IGBTs, *Transactions of China Electrotechnical Society*, Vol. 32, Issue 6, 2017, pp. 201-208.

[10]. C. Busca, R. Teodorescu, F. Blaabjerg, Dynamic thermal modelling and analysis of press-pack IGBTs both at component-level and chip-level, in *Proceedings of the Conference of IEEE Industrial Electronics Society (IECON'13)*, November 2013, pp. 677-682.

[11]. M. Sweet, E. M. Sankara Narayanan, S. Steinhoff, Influence of cassete design upon breakdown performance of a 4.5kV press-pack IGBT module, in *Proceedings of the International Conference on Power Electronics, Machines and Drives (PEMD'16)*, Glasgow, UK, April 2016, pp. 1-6.

[12]. L. Tinschert, A. R. Årdal, T. Poller, Possible failure modes in Press-Pack IGBTs, *Microelectronics Reliability*, Vol. 55, Issue 6, 2015, pp. 903-911.

[13]. T. Poller, S. D. Arco, M. Hernes, Influence of the clamping pressure on the electrical, thermal and mechanical behaviour of press-pack IGBTs, *Microelectronics Reliability*, Vol. 53, Issue 9-11, 2013, pp. 1755-1759.

[14]. H. Magnar, Determination of the thermal and electrical contact resistance in press-pack IGBTs, *Prenatal Diagnosis*, Vol. 32, Issue 13, 2013, pp. 1233-1241.

[15]. E. P. Deng, Z. B. Zhao, P. Zhang, Optimization of the thermal contact resistance within press pack IGBTs, *Microelectronics Reliability*, Vol. 69, 2017, pp. 17-28.

[16]. Z. Ping, Y. Xuan, L. I. Qiang, Development on thermal contact resistance, *Ciesc Journal*, Vol. 4, 2012, pp. 11-20.

[17]. M. Yovanovich, E. Marotta, Thermal Spreading and Contact Resistances, New Jersey, *Wiley*, 2003, pp. 261-395.

[18]. M. A. Lambert, L. S. Fletcher, Thermal contact conductance of spherical, rough metals, *Transactions of the Asme Serie C Journal of Heat Transfer*, Vol. 119, Issue 4, 1997, pp. 684-690.

[19]. M. Bahrami, Thermal resistances of gaseous gap for conforming rough contacts, in *Proceedings of the 42nd AIAA Aerospace Sciences Meeting and Exhibit*, 2004, pp. 1-11.

[20]. W. H. McAdams, Heat Transmission, *McGraw-Hill*, New York, 1954.

[21]. A. Bejan, D. Kraus, Heat Transfer Handbook, *John Willey*, New York, 2003.

[22]. J. B. P. Williamson, J. Pullen, R. T. Hunt, The shape of solid surfaces, *Surface Mechanics, ASME*, New York, 1969, pp. 24-35.

[23]. J. A. Greenwood, The area of contact between rough surfaces and flats, *Journal of Lubrication Tech*, Vol. 89, Issue 1, January 1967, pp. 81-91.

[24]. M. Bahrami, J. R. Culham, M. M. Yovanovich, Review of thermal joint resistance models for non-conforming rough surfaces in a vacuum, in *Proceedings of the ASME 2003 Heat Transfer Summer Conference*, 2003, pp. 411-431.

[25]. V. M. Antonetti, T. D Whittle, R. E. Simons, An approximate thermal contact conductance correlation, in *Proceedings of the 28th National Heat Transfer Conference*, Minneapolis, Minnesota, 1991, pp. 35-42.

[26]. L. H. Tanner, M. Fahoum, A study of the surface parameters of ground and lapped metal surfaces, using specular and diffuse reflection of laser light, *Wear*, Vol. 36, Issue 3, 1976, pp. 299-316.

[27]. M. A. Lambert, Thermal contact conductance of spherical, rough metals, PhD Thesis, Dept. of Mech. Eng., *Texas A&M University*, USA, 1995.

[28]. M. Bahrami, J. R. Culham, M. M. Yovanovich, G. E. Schneider, Thermal contact resistance of non-conforming rough surfaces. Part 1. Mechanical model, in *Proceedings of the 36th AIAA Thermophysics Conference*, Orlando, FL, June 2003, pp. 2003-4197.

[29]. M. Bahrami, J. R. Culham, M. M. Yovanovich, G. E. Schneider, Thermal contact resistance of non-conforming rough surfaces. Part 2. Thermal model, in *Proceedings of the 36th AIAA Thermophysics Conference*, Orlando, FL, June 2003, pp. 2003-4198.

[30]. S. Song, M. M. Yovanovich, Relative contact pressure: dependence upon surface roughness and vickers microhardness, *Journal of Thermophysics and Heat Transfer*, Vol. 2, Issue 1, 1987, pp. 43-47.

[31]. M. M. Yovanovich, New contact and gap correlations for conforming rough surfaces, in *Proceedings of the AIAA 16th Thermophysics Conference*, Palo Alto, CA., June 1981.

[32]. V. W. Antonetti, M. M. Yovanovich, Thermal Contact Resistance in Microelectronic Equipment, Thermal Management Concepts in Microelectronic Packaging From Component to System, *ISHM Technical Monograph Series 6984-003*, 1984, pp. 135-151.

[33]. E. P. Deng, Z. B. Zhao, P. Zhang, Influence of the temperature on the thermal contact resistance within press pack IGBTs, *Semiconductor Technology*, Vol. 41, Issue 12, 2016, pp. 906-912.

[34]. Y. Z. Li, C. V. Madhusudana, E. Leonardi, On the enhancement of the thermal contact conductance: effect of loading history, *Journal of Heat Transfer*, Vol. 122, Issue 1, 2000, pp. 46-49.

[35]. T. Mcwaid, E. Marschall, Thermal contact resistance across pressed metal contacts in a vacuum environment, *International Journal of Heat & Mass Transfer*, Vol. 35, Issue 11, 1992, pp. 2911-2920.

[36]. M. Bahrami, Thermal resistances of gaseous gap for conforming rough contacts, in *Proceedings of the 42nd AIAA Aerospace Sciences Meeting and Exhibit*, 2004, Paper 2002-0317.

[37]. A. Astm, D5470-06: standard test method for thermal transmission properties of thermally conductive electrical insulation materials, *America: Astm International*, 2006.

[38]. Y. Osone, T. Kubo, N. Nakazato, Optical measurement of thermal contact conductance between wafer-like thin solid samples, *Transactions of the ASME Serie C Journal of Heat Transfer*, Vol. 121, Issue 4, 1999, pp. 954-963.

[39]. L. Shi, Investigation of heat transport on the solid-solid contact interface at low temperature, PhD Thesis, *Huazhong University of Science & Technology*, Wuhan, 2006.

[40]. O. Kwon, L. Shi, A, Majumdar, Scanning thermal wave microscopy, *Journal of Heat Transfer*, Vol. 125, 2003, pp. 156-163.

[41]. T. Poller, J. Lutz, S. D'Arco, Determination of the thermal and electrical contact resistance in press-pack IGBTs, *Prenatal Diagnosis*, Vol. 32, Issue 13, 2013, pp. 1233-1241.

[42]. U. Scheuermann, R. Schmidt, Investigations on the $V_{ce(t)}$ - method to determine the junction temperature by using the chip itself as sensor, in *Proceedings of the International Exhibition and Conference for Power Electronics Intelligent Motion and Power Quality (PCIM'09)*, Europe, 2009, pp. 802-807.

[43]. E. P. Deng, Study on the methods to measure the junction-to-case thermal resistance of IGBT modules and press pack IGBTs, *Microelectronics Reliability*, 2017

[44]. G. Q. Lu, J. N. Calata, G. Lei, Low-temperature and pressureless sintering technology for high-performance and high-temperature interconnection of semiconductor devices, in *Proceedings of the International Conference on Thermal, Mechanical and Multi-Physics Simulation and Experiments in Microelectronics and Micro-Systems (EuroSimE'17)*, April 2007, pp. 1-5.

[45]. N. Heuck, K. Guth, M. Thoben, Aging of new interconnect-technologies of power-modules during power-cycling, in *Proceedings of the International Conference on Integrated Power Systems (CIPS'14)*, February 2014, pp. 1-6.

Chapter 14

Reliability Study on Nanoscale CMOS Devices

Jin He, Xingye Zhou, Mansun Chan, Xiaomeng He Haijun Lou

14.1. Introduction

Over the past decades, research on the reliability of devices and circuits has been one of the most important issues for the IC industry and IC study with the development of CMOS technology, such as the study on Fowler-Nordheim (FN) stress effect, hot-carrier injection (HCI) effect, negative bias temperature instability (NBTI) effect, current leakage, and variation in nanoscale devices. We would like to review the above aspects of traditional and non-traditional CMOS devices based on the research work of our group, focusing on the extraction of interface and oxide traps induced by high electrical field or high temperature, degradation modeling and simulation of devices and circuits, and the investigation of statistical variation in nanoscale devices.

Measurements of the interface traps in MOSFETs created by high electrical field or high temperature are of practical and scientific importance. These traps seriously affect the reliability of the devices and the aging of integrated circuits. For SOI devices, this phenomenon becomes more evident and complicated due to the multiple Si/SiO_2 interfaces and the front-back interface coupling effect. The interface traps are traditionally studied by the MOS capacitance-voltage (C-V) measurement, conductance technique, sub-threshold current, deep level transient spectroscopy (DLTS) and charge-pumping (CP) method [1-5]. A simple subthreshold analysis technique is presented for extraction of the interface traps in MOSFETs, which does not require the knowledge of subthreshold slope and device geometry parameters [6]. Following a forward gated-diode architecture frequently used to characterize the interface traps in small size devices [7-8], a refined forward gated-diode method is obtained, which enables the extraction of interface and oxide traps at both the front and back gate interfaces in SOI MOSFETs [9]. To determine the lateral distribution of the hot-carrier induced interface traps in the MOSFETs, a novel combined gated-diode technique is provided, making use of the modulation effect of an

Jin He
Peking University Shenzhen SoC Key Laboratory, PKU-HKUST Shenzhen —Hong Kong
Institution and Peking University Shenzhen Institution, Shenzhen, 518055 P. R. China

additional drain terminal on the R-G current peak of the common forward gated-diode method [10-11]. Compared with other techniques such as the CP and DLTS, this method has the advantages of simplicity, high sensitivity and wide application range to the different device structures. In addition, when p-type MOSFETs are stressed under a negative bias or high temperature, interface states are generated owing to the breakdown of Si-H bonds at the Si/SiO$_2$ interface. This so called NBTI effect induces a degradation of threshold voltage, subthreshold slope and circuit switching speed [12-18]. To bridge the gap between simulation and fabrication, NBTI prediction modeling has been widely studied [12]. By solving the reaction equation combined with the diffusion theory, a physics-based unified R-D model is demonstrated and an accurate description of NBTI degradation over a wide range of stress times is obtained [19]. With the CMOS technology scaling down to the 32 nm technology node and beyond, non-traditional MOSFET structures are proposed due to their superior gate controllability, improved short channel effect (SCE) immunity and enhanced current drive, such as FinFETs [20], nanowire transistors (NWTs) [21], and junctionless multi-gate transistors (JMTs) [22]. The reliability issues, such as HCI, NBTI and tunneling leakage current, directly affect the performance of these novel devices and circuits [23-27], which needs to be investigated carefully. Moreover, in nano-scale transistors, the statistical variation has been becoming a serious reliability issue. In nano-scale novel structure devices such as FinFET and nanowire transistors, Random Dopant Fluctuation (RDF) effect, lien edge roughness (LER) effect or fin sidewall angle effect could result in significant variation [22, 26, 28-31]. Therefore, the investigation of statistical variation in these novel structure devices is necessary.

Interface trap extraction, modeling and simulation for predicting the degradation of devices and circuits, investigation of statistical variation in nano-scale devices are reviewed for traditional and non-traditional CMOS devices in this Chapter.

14.2. Reliability Study on Nanoscale Traditional CMOS Devices

The interface traps in MOSFETs seriously affect the device reliability [1-2]. In this Section, different extraction methods of interface and oxide traps in traditional CMOS devices are described. Additionally, a physics-based unified Reaction-Diffusion (R-D) model for predicting NBTI degradation is presented.

14.2.1. Extraction of Interface Traps

14.2.1.1. Extraction of the Average Density of Interface Traps

It is reported that a mean density of interface traps (D_{it}) can be estimated from the subthreshold characteristics [32-33]. In this Subsection, a new subthreshold analysis technique, namely, the linear cofactor difference method, is presented for the extraction of interface traps induced by electrical stress in MOSFETs.

The theory of this method is described as follows. If a function $f(x)$ is strictly monotonic, nonlinear and continuous over (x_0, x_1) and differentiable on $[x_0, x_1]$, then there definitely exists point x_p, $x_0 < x_p < x_1$, such that

$$G(x_p) = \frac{\partial G}{\partial x}\bigg|_{x=x_p} = 0 \quad , \tag{14.1}$$

where

$$G(x) = \Delta f(x) \equiv b + Kx - f(x) \tag{14.2}$$

is the linear cofactor difference of $f(x)$, $\Delta f(x)$ is the linear cofactor difference operator, and b and K are the linear cofactor operator intersection and factor, respectively.

The constants b and K can be determined by the choice of the linear cofactor difference operation region (x_0, x_1) via

$$G(x_1) = b + Kx_1 - f(x_1) \quad , \tag{14.3}$$

$$G(x_0) = b + Kx_0 - f(x_0) \quad . \tag{14.4}$$

For the monotonic increased function $f(x)$, constant $K > 0$. Otherwise, $K < 0$. Here, the linear cofactor difference operation region (x_0, x_1) is chosen in the subthreshold region of MOSFETs for the experimental data, at which the gate voltage V_{gs} changes from 0 to 0.6 V.

According to the device physics of MOSFETs, the drain current in the subthreshold region can be shown to have a form [34]

$$I_{ds} = \beta(\eta - 1)V_t^2 \exp\left[\frac{(V_{gs} - V_{th})}{\eta V_t}\right] \cdot \left[1 - \exp\left(-\frac{V_{ds}}{V_t}\right)\right] \quad , \tag{14.5}$$

where $\beta = \mu_s W C_{ox}/L$, $V_t = kT/q$, and $\eta = 1 + (qD_{it}/C_{ox}) + (C_d/C_{ox})$ are the transistor gain, thermal voltage, and ideal factor of the subthreshold slope, respectively. All other symbols have common meanings. For example, D_{it} is the interface trap density and C_d is the depletion layer capacitance.

The subthreshold transfer characteristics of MOSFETs can be rewritten

$$V_{gs} = V_{th} + \eta V_t \cdot \left\{ \ln I_{ds} - \ln\left[\beta(\eta - 1)V_t^2\right] - \ln\left[1 - \exp\left(-\frac{V_{ds}}{V_t}\right)\right] \right\} \quad . \tag{14.6}$$

Therefore, according to Eq. (14.2), the linear cofactor difference gate voltage can be expressed as

$$\Delta V_{gs} = V_{th} + \eta V_t \cdot \left\{ \ln I_{ds} - \ln\left[\beta(\eta-1)V_t^2 \right] - \ln\left[1 - \exp\left(-\frac{V_{ds}}{V_t} \right) \right] \right\} - KI_{ds} \quad , (14.7)$$

where the linear cofactor difference intersection b is set to zero because of the small drain current at zero gate voltage, and K is so chosen such as a fixed 2×10^5 V/A as to keep the linear cofactor difference maximum of the gate voltage always in the subthreshold region.

The linear cofactor difference operator condition has been fully satisfied for Eq. (14.6), and then ΔV_{gs} has an extreme maximum as the drain current I_{ds} is varied to I_{dP}, where

$$\frac{d\Delta V_{gs}}{dI_{ds}}\bigg|_{I_{ds}=I_{dP}} = 0 \qquad (14.8)$$

It is evident that I_{dP} is the linear cofactor difference gate voltage peak position. From Eq. (14.7), we have

$$\eta = \frac{I_{dP} \cdot K}{V_t} = 1 + \frac{qD_{it}}{C_{ox}} + \frac{C_d}{C_{ox}} \qquad (14.9)$$

We know the interface traps increase with a rise of the gate oxide stress time. According to Eq. (14.9), the increase of the interface traps is demonstrated by the shift of the peak position I_{dP} of the linear cofactor difference gate voltage. As a result, for the same linear cofactor difference factor K_P, the relationship between the increased interface trap and the linear cofactor difference gate voltage peak position can be obtained as

$$q\Delta D_{it} = \frac{KC_{ox}}{V_t} \Delta I_{dP} \qquad (14.10)$$

The above formulation constitutes a new technique to extract the increased D_{it} without the knowledge of the subthreshold slope, the threshold voltage, and device geometry parameters.

During the stress measurements, the nMOSFETs are stressed by F-N tunneling current in the conditions of constant $V_g = 5$ V with different stress time. Fig. 14.1 shows the variation of subthreshold characteristics of a MOSFET with different gate oxide stress time. As seen in this figure, the subthreshold current shows slight reduction with the increase in the stress time. The current variation is too small to make the traditional slope technique fit the extraction of interface traps induced by the F-N stress.

Fig. 14.2 shows the typical $\Delta V_{gs}-I_{ds}$ characteristics of the NMOSFET with increasing accumulated gate oxide stress time for the same linear cofactor difference factor K. Not only the peak position I_{dP} of the linear cofactor difference gate voltage but also its peak height shift significantly. The variation of the subthreshold linear cofactor difference characteristics is due to the stress-generated SiO_2/Si interface traps.

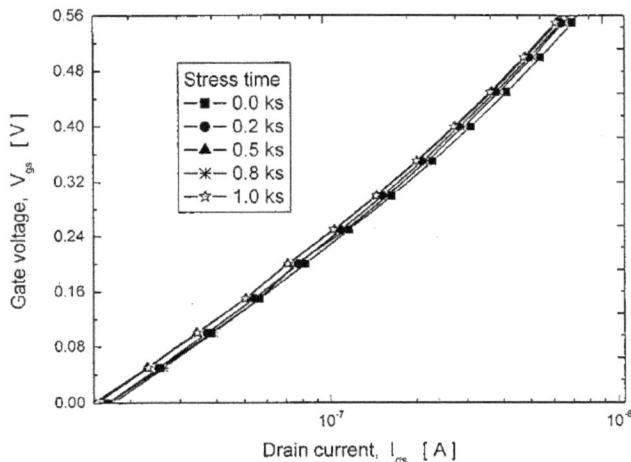

Fig. 14.1. The variation of subthreshold characteristics of MOSFET with different gate oxide stress time [6].

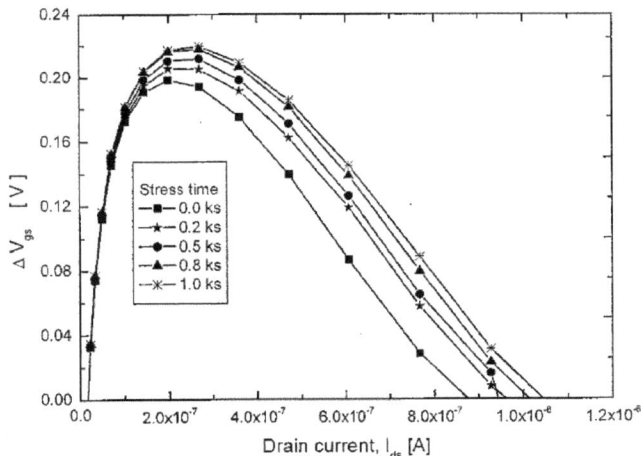

Fig. 14.2. Typical $\Delta V_{gs} - I_{ds}$ characteristics of nMOSFET with increasing stress time [6].

The relationship between the gate oxide stress-induced interface trap density and the accumulated stress time is obtained with Eq. (14.10), as shown in Fig. 14.3. As seen here, the gate oxide stress-induced interface trap density increases in a power law with the relationship of $\Delta D_{it} \sim t^n$, here the factor n is fitted to 0.414. This result is consistent with the earlier observation [35-36].

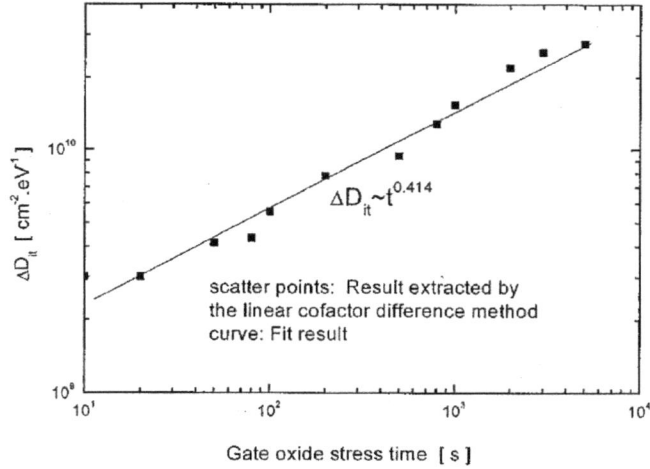

Fig. 14.3. Dependence of the increased interface traps and the gate oxide stress test time obtained by the linear cofactor difference method [6].

14.2.1.2. Front and Back Gate Interface and Oxide Traps in SOI MOSFETs

It is very important to extract or measure the interface and oxide traps accurately for modeling the performance and designing the device structure. However, the analysis of hot-carrier induced degradation of SOI MOSFETs is complicated due to the dual channels and the front-back interface coupling effect. Here, a refined forward gated-diode method which enables the extraction of interface and oxide traps at both the front and back gate interfaces is proposed [9].

The hot-carrier-stress induced interface traps will contribute to an increase in the gated-diode R-G current peak, which can be used to determine the hot-carrier induced interface trap density.

With the hot-carrier stress, the increased R-G current component is simply modeled by

$$\Delta I_{pure} = I_{peak\text{-}stress,j} - I_{peak\text{-}origin},$$

(14.11)

where j represents the jth hot-carrier-stress experiment, $I_{peak\text{-}origin}$ and $I_{peak\text{-}stress,j}$ denote the gated-diode R-G current peak before and after the hot-carrier-stress, respectively.

The increased R-G current peak values thus can be extracted from Eq. (14.11). Based on the SRH statistics theory, the relationship between the increased R-G current peak values and the interface trap density is described by

$$\Delta I_{pure} = \frac{1}{2} q n_i (c_n c_p)^{1/2} \Delta D_{it} A_G e^{qV_b/2kT},$$

(14.12)

where n_i is the intrinsic carrier density, ΔD_{it} is the density of the induced interface traps, A_G is the effective gate area, V_b is the diode forward voltage and $c_n = c_p = 10^{-8} \text{cm}^{-3} \text{s}^{-1}$.

The shift of the critical gate voltage associated with R-G current peak also directly indicates the amount of generated interface and oxide traps, which can be expressed as

$$\Delta V_{\text{g-peak}} = V_{\text{g-peak-stress},j} - V_{\text{g-peak-origin}} = \Delta V_{it} + \Delta V_{ot},$$ (14.13)

where

$$\Delta V_{it} = \frac{q\Delta D_{it} \cdot e\phi_F}{C_{ox}},$$

$$\Delta V_{ot} = \frac{qN_{ot}}{C_{ox}}.$$

$V_{\text{g-peak-origin}}$ and $V_{\text{g-peak-stress},j}$ represent the gated-diode R–G current peak position before and after the hot-carrier stress. ΔV_{it} and ΔV_{ot} are the R–G current peak voltage shifts due to the interface and oxide traps induced by the hot-carrier stress, respectively.

Since the generated interface trap density D_{it} is calculated from Eq. (14.12), the oxide trap density N_{ot} can be determined from Eq. (14.13). Thus, via the R–G current peak shift of the forward gated-diode, the induced oxide traps are readily measured either at the front or back gate interface. To avoid the influence of the back interface, the strong accumulation case of the back interface must be used in measuring the front gated-diode current. Similarly, once the front gate interface is forced into the strong accumulation via a negative front gate voltage, the measured back gated-diode current only indicates the damage of the back interface due to the front channel hot-carrier-stress effect.

Fig. 14.4 plots the measured front gated-diode current against the front gate voltage, which exhibits the evolution of R-G current with different hot-carrier-stress time. As shown, the maximum front gated-diode R-G current peak occurs at the critical front gate of $V_{g1} = -0.22$ V before the hot-carrier stress. It is observed that the sharp R-G current peak grows in magnitude and the hot-carrier-stress causes a positive shift of the critical voltage corresponding to the front gate R-G current peak position, which can be explained by the electrons trapped at the front gate oxide layer. This result is used for the extraction of the front gate interface and oxide traps.

Fig. 14.5 shows the induced front gate interface and oxide traps for different hot-carrier-stress time. As shown in Fig. 14.5(a), the front gate interface trap density significantly grows with the accumulated stress time, exhibiting logarithmic time dependence of $\Delta D_{it} \sim t^n$, where n is fitted to 0.7. The induced oxide traps show a trend similar to that of the interface traps [see Fig. 14.5 (b)]. The fitted relationship $\Delta N_{ot} \sim t^n$ has a factor n of 0.85, slightly larger than that of interface traps.

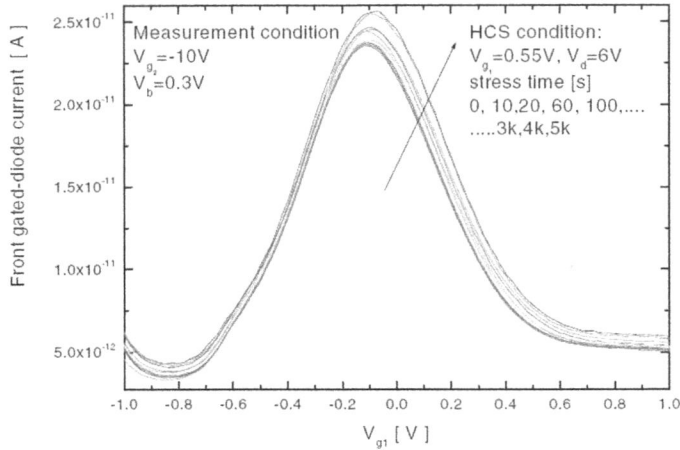

Fig. 14.4. Measured front gated-diode current versus front gate voltage of an SOI MOSFET [9].

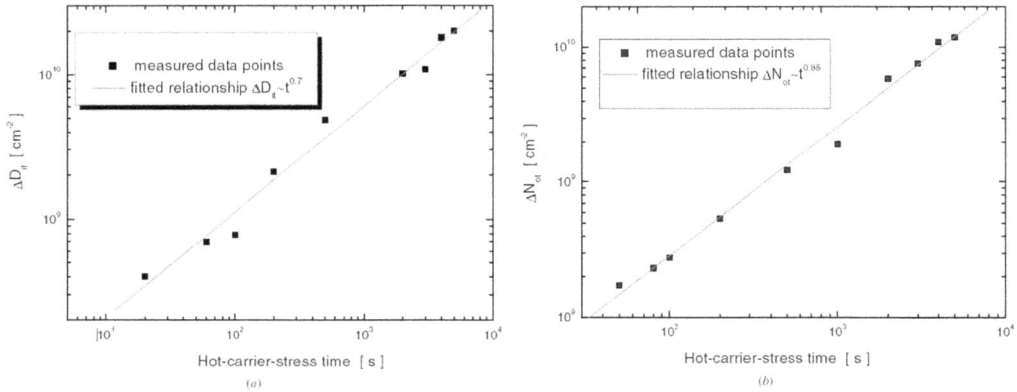

Fig. 14.5. Extracted front gate interface trap density (a), and oxide trap density (b) in an SOI MOSFET [9].

Similarly, by setting the front gate voltage of $V_{g1} = -1.5$ V to make the front interface into strong accumulation, the back gated-diode current is measured and the results are illustrated in Fig. 14.6. The result verifies the previous report on the charge-pumping method [37] that one-channel stress will lead to great damage in the opposite channel in SOI devices, in terms of the interface traps and oxide trap charges.

The induced back gate interface and oxide traps are extracted and shown in Fig. 14.7. The back gate interface trap density also has a logarithmic dependence on the stress time and increases as a power law of $\Delta D_{it} \sim t^n$ with $n = 0.4$. This shows a lower rate of increase in the back gate interface traps compared with that in the front interface. This result is consistent with [35]. The induced back gate oxide traps increase exponentially with stress time and the fitted relationship $\Delta D_{it} \sim t^n$ has a factor n of 0.39, indicating a close connection between the induced back gate interface and oxide traps.

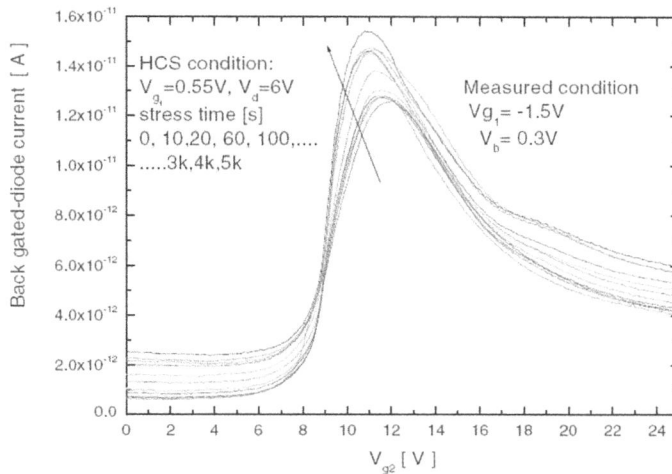

Fig. 14.6. Measured back gated-diode current versus back gate voltage of an SOI MOSFET [9].

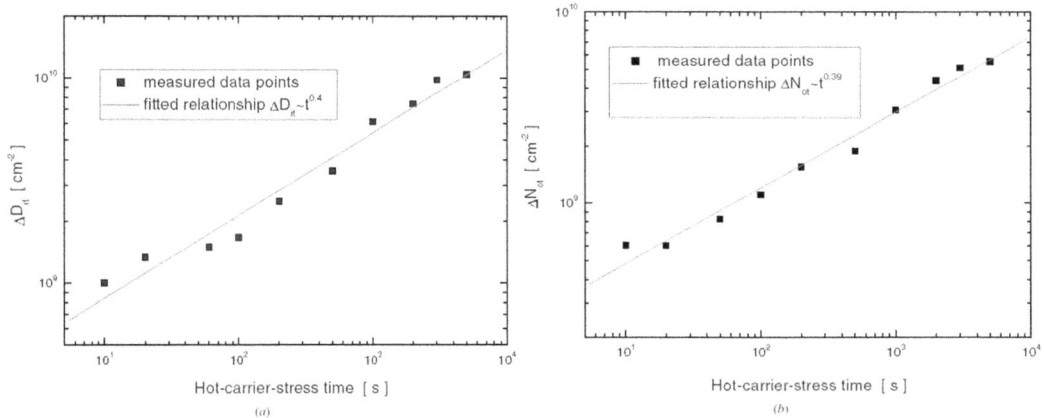

Fig. 14.7. Extracted back gate interface trap density (a), and oxide trap density (b) in an SOI MOSFET [9].

14.2.1.3. Extraction of the Lateral Spatial Distribution of Interface Traps

A novel combined gated-diode technique is provided to determine the lateral distribution of the hot-carrier induced interface traps in the MOSFETs [10]. The measurement structure used in this method is shown in Fig. 14.8. As seen in this figure, this novel technique makes use of four terminals of NMOSFETs (i. e. source, drain, gate and substrate). The source/bulk pn junction is forward biased while the drain/bulk pn junction is with a reverse bias for extracting the interface trap distribution from the drain edge to the channel region. Similarly, interface traps at the source side can be determined by interchanging the source and drain terminals with the drain/bulk pn junction forward-biased while the source/bulk pn junction reverse-biased.

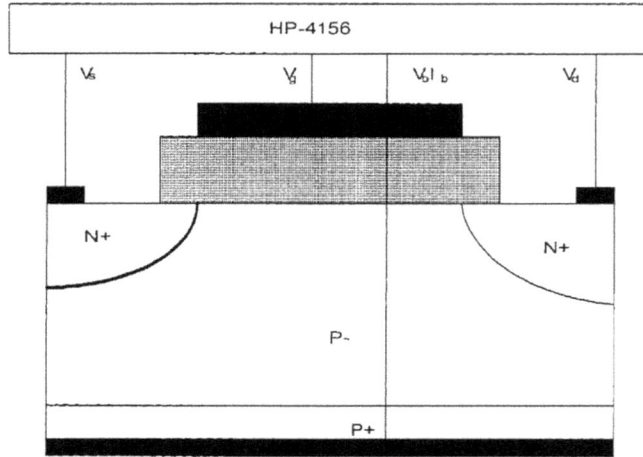

Fig. 14.8. Cross-section view of the MOSFET and the combined gated-diode experimental set-up using HP 4156 parameter analyzer [10].

In the extraction of interface traps at the drain side, the source/bulk junction of the MOSFET is an effective forward gated diode which operates in the threshold region ($V_{b, (s)} \leq 0.6$ V) and its maximum R-G rate is reached by getting the gate voltage $V_{g\text{-peak}}$, so that the R-G current predominates in the measured diode current. Since the reverse gated-diode R-G current is always far less than that of the forward one, the contribution of the drain region to the R-G current peak could be approximated as zero at the first-order approximation. Because the amplitude of the reverse drain voltage modulates the effective channel length with the maximum recombination rate F_{max}, the measured R-G current peak will decrease with the increase of drain voltage. Therefore, differentiating the measured R-G current by the drain voltage or the effective channel length can represent the lateral distribution of the interface traps.

In the common forward gated-diode case, the source and drain terminals are grounded while the bulk terminal is fixed at a constant voltage less than 0.6 V, not only the excess carrier concentration but also the contributing R-G current's depletion layer boundary are the function of the gate voltage V_g. From the SRH theory, the R-G originating from the interface trap recombination can be expressed by

$$I_{R\text{-}G}(V_i, x) = qW \int N_{it}(x) F(V_i, x) dx \,,$$

(14.14)

with

$$F(V_i, x) = \upsilon_{th}\sigma\frac{n_1(x)p_1(x) - n_i^2}{n_1(x) + p_1(x) + 2n_i},$$

(14.15)

$$n_1(x)p_1(x) \equiv n_i^2 \exp\left[\frac{qV_{b,s(d)}}{kT}\right].$$

(14.16)

Here, υ_{th} and σ are the electron thermal velocity and capture cross-section, respectively. All other symbols have the common physical meanings. The F is the R-G current factor modulated by the gate and bulk terminal voltages.

With a constant bulk terminal voltage and the source/drain grounded, as the gate voltage is swept from the inversion threshold to the strong accumulation, the surface potential and thus the excess carrier concentration will be strongly changed. As a result, the diode diffusion current will keep constant, but the R-G current will slowly increase to a peak value at the condition of $V_{g\text{-peak}}$ and then decrease to a constant. As shown in Eq. (14.15), the maximum R-G recombination rate, calculated by $\partial F(V_i, x)/\partial n = 0$, occurs when $n_1(x) = p_1(x)$ or when the surface potential coincides with the intrinsic Fermi level. From Eq. (14.16), we have

$$n_1(x) = p_1(x) = n_i \exp\left[\frac{qV_{b,s(d)}}{2kT}\right],$$

(14.17)

and then, the maximum factor F is expressed as

$$F_{max} = \frac{1}{2}\upsilon_{th}\sigma \exp\left[\frac{qV_{b,s(d)}}{2kT}\right].$$

(14.18)

At the first-order approximation, F_{max} can be assumed the same in the contributed R-G current channel region. Thus, the whole channel always satisfies Eq. (14.18) and then the R-G current peak is rewritten

$$I_{\text{R-G(peak)}} = qWF_{max}\int_{x_1}^{x_2} N_{it}(x)dx.$$

(14.19)

From the above discussion, Eq. (14.19) is differentiated along the channel direction to obtain the local interface trap density $N_{it}(x)$ as

$$N_{it}(x) = (qWF_{max})^{-1}\frac{dI_{\text{R-G(peak)}}}{dL_i}.$$

(14.20)

For the determination of interface traps at the drain side, $V_s = 0$, $V_d > 0$ and $V_{b,s} = V_b \leq$ 0.6 V, the reverse V_d will deplete the injected excess carrier from the drain edge into the channel region and then the maximum F factor will exponentially decrease in this region according to Eq. (14.18). Therefore, the R-G current contributed by this depleted region can be neglected and its F approximates zero. As a result, the effective channel length that contributes the R-G current will decrease with the increase of V_d, causing the R-G current peak decreased.

The effective channel length that contributes the R-G current at V_{peak} or $n_1(x)=p_1(x)$ can be written as

$$L_i = L - \sqrt{2\varepsilon V_d / q n_1(x)}.$$ (14.21)

Here, L is the original channel length defined by the gate mask.

As a result, the lateral distribution of the interface traps expressed by Eq. (14.20) is re written as

$$N_{it}(x) = (qWF_{max})^{-1} \frac{dI_{R\text{-}G(peak)}}{dV_d} \left(\frac{dL_i}{dV_d} \right)^{-1}.$$ (14.22)

Similarly, the interface trap distribution from the source edge into the channel region can be obtained based on Eqs. (14.23) and (14.24).

$$L_i = L - \sqrt{2\varepsilon V_s / q n_1(x)},$$ (14.23)

$$N_{it}(x) = (qWF_{max})^{-1} \frac{dI_{R\text{-}G(peak)}}{dV_s} \left(\frac{dL_i}{dV_s} \right)^{-1}.$$ (14.24)

The experimental verification of this interface extraction method is performed in both surface-channel NMOSFETs and partially SOI NMOSFETs. Fig. 14.9 shows the bulk diode current versus gate voltage of NMOSFET with different drain voltages and sharp current peaks with different amplitudes are observed. The sharp current peak $I_{b\text{-}peak}$ is due to the contribution of R-G process via interface trap recombination, which is just the R-G current $I_{R\text{-}G(peak)}$. Thus, the R-G peak current ($I_{b\text{-}peak}$) directly reflects the magnitude of the interface trap density.

Fig. 14.10 plots the raw R-G current peak $I_{b\text{-}peak}$ versus the drain voltage V_d of a NMOSFET measured by the combined gated-diode method. By fixing the bulk voltage V_b and the gate voltage $V_{g\text{-}peak}$, only the drain voltage V_d gradually increases from zero to a positive value. In this case, the $I_{b\text{-}peak}$-V_d curve describes the variation of the forward R-G current peak with the reverse drain voltage. As seen in the figure, the R-G current peak $I_{b\text{-}peak}$ decreases with an increase of the drain voltage, which confirms the theoretical analysis presented here.

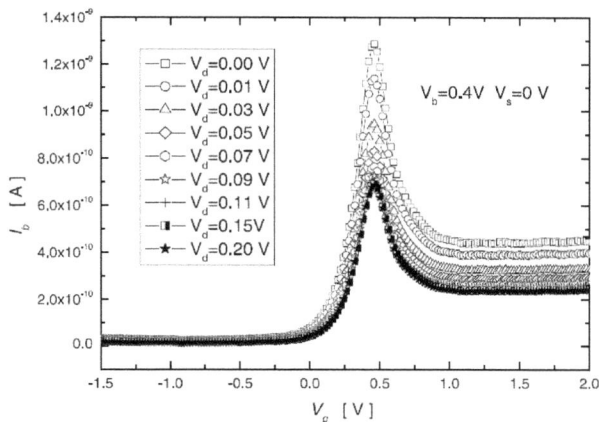

Fig. 14.9. The bulk terminal current of a NMOSFET obtained by the combined gated-diode method in the determination of the interface traps at the drain side [10].

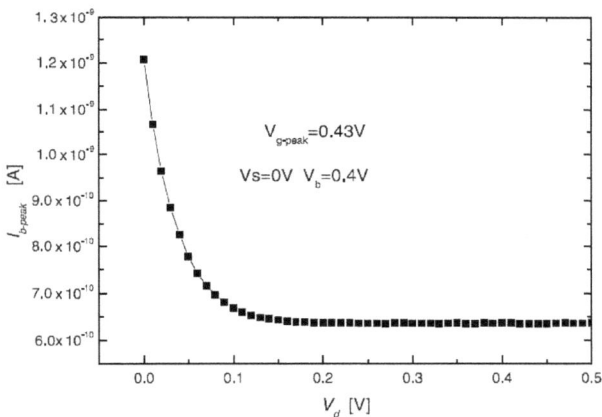

Fig. 14.10. The raw R-G current peak versus the drain voltage of a NMOSFET for the determination of the interface traps at the drain side [10].

According to the presented principle, the $I_{b\text{-peak}}$-V_d curve can be transferred into the $I_{b\text{-peak}}$- L_i curve, and then the lateral distribution of interface traps near the drain edge can be determined via Eq. (14.22), which is shown in Fig. 14.11. It is observed that the high-density interface traps indeed localize in the drain region edge. The deeper the lateral distance extends into the channel region, the lower the interface trap density. This obtained result qualitatively coincides with the previous reports on MOSFETs [3].

Fig. 14.12 gives the lateral distribution of the hot-carriers induced interface traps in an SOI NMOSFET with different stress time. The lateral distribution of the interface traps shows a sharp peak near the drain region and then gradually decreases toward the channel region, which is similar to that of other methods such as the CP technique. Additionally, with an increase of the accumulated stress time, the lateral distribution of the interface

371

traps gradually extends into the channel region and the peak of the interface trap profile increases significantly. This picture demonstrates the dynamic of the interface trap generation caused by the hot-carriers, which can be used in study of the degradation mechanism of the MOSFET devices and circuits.

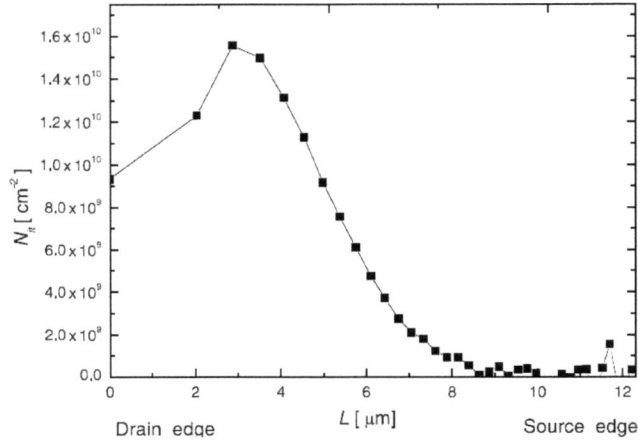

Fig. 14.11. Interface traps in a NMOSFET at the drain side [10].

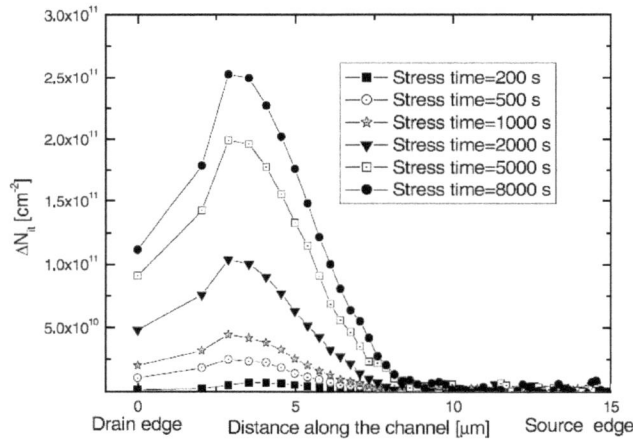

Fig. 14.12. Extracted interface trap distribution in an SOI NMOSFET with different accumulated stress time [11].

14.2.2. Negative Bias Temperature Instability Effect

The reaction-diffusion (R-D) of hydrogen species and hole-trapping are two of the basic possible explanations for the NBTI effect. As shown in Fig. 14.13, the classical R-D model ($\Delta V_{th} \sim t^n$) shows discontinuous time exponents between short- and long-term stress regions for the H/ H_2 induced threshold voltage shift (ΔV_{th}) in p-type MOSFETs [38]. On the other hand, an NBTI model based on the hole-trapping theory ($\Delta V_{th} \sim \log t$) shows good agreement with measured data in the short-term stress region but is unable to describe the V_{th} degradation during long-term stress processes [39]. To obtain a better overall fitting capability, another empirical model was developed by simply combining different degradation models ($\Delta V_{th} \sim \log t$ and $\Delta V_{th} \sim t^n$) [40].

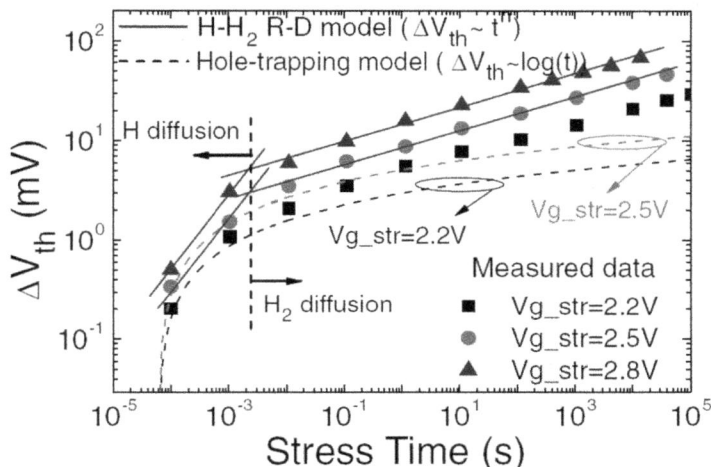

Fig. 14.13. ΔV_{th} measured using ultrafast technique compared with different NBTI models [19].

In this Subsection, we demonstrate a physics-based unified R-D model by solving the reaction equation combined with the diffusion theory, and an accurate description of NBTI degradation over a wide range of stress times is obtained [19]. The principle of the developed R-D model is illustrated in Fig. 14.14. By applying a negative gate stress bias to the p-MOSFET, a chemical reaction occurs between Si-H bonds and injected holes in the short initial period of stress process [Fig. 14.14 (a)]. H atoms are released from the broken Si-H bonds and diffuse into the poly-Si gate where they form molecular H_2. Since the diffusion constant in SiO_2 is much larger than that in poly- Si, most of the H atoms remain in SiO_2 and only some of them diffuse into poly-Si during long-term stress [Fig. 14.14(a)]. Therefore, the most significant ΔV_{th} shift occurs at the start of the short-term stress region owing to the reaction process, and ΔV_{th} increases slowly during the long term H_2 diffusion. After the stress is removed, the H atoms remaining in the SiO_2 can repair Si dangling bonds within a very short time [Fig. 14.14 (b)] and cause a fast recovery of ΔV_{th}. However, H_2 molecules located in the poly-Si take a long time to diffuse back into SiO_2 or even flow out into the electrode, which results in a slow diffusion-like recovery [Fig. 14.14(b)]. H_2 molecules that escape via the electrode cause the non-recoverable degradation observed experimentally.

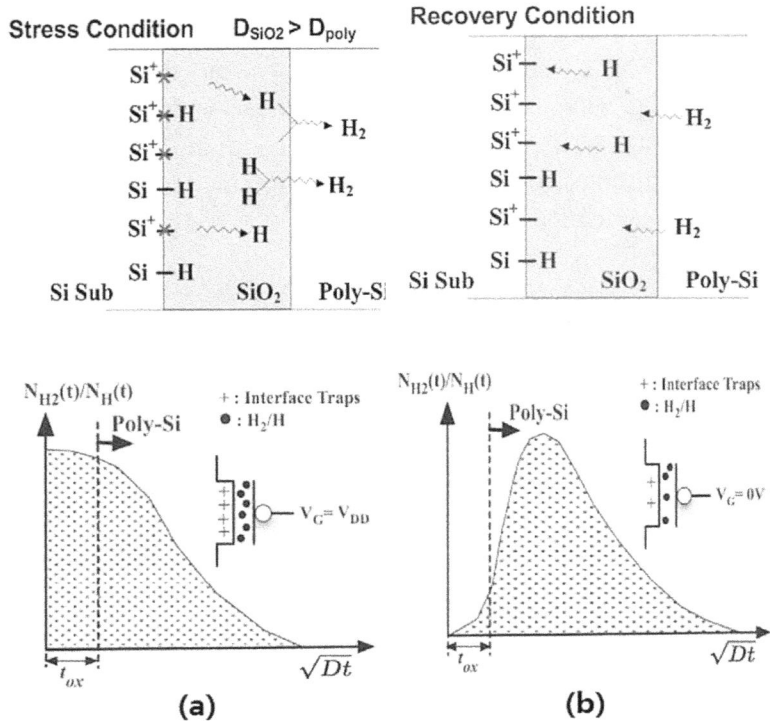

Fig. 14.14. Illustration of (a) stress process and (b) recovery process [19].

The classical R-D model is usually described by the reaction function Eq. (14.25) and the diffusion function Eq. (14.26) [12-13]:

$$\frac{dn_T}{dt} = k_f(N_T - n_T) - k_r N_{H0} n_T,$$
(14.25)

$$\frac{dN_H}{dt} = D_H \frac{d^2 N_H}{dx^2},$$
(14.26)

where n_T is the interface state density generated by the NBTI effect. k_f and k_r are the forward and reverse reaction rates, respectively. The total number of Si-H bonds is denoted by N_T and by H-atom diffusion constant D_H. N_{H0} and N_H are the diffusing hydrogen concentration at the Si/SiO$_2$ interface and in the gate oxide, respectively.

With the reverse reaction (Si dangling bonds reparation during the stress process) neglected and $N_T \gg n_T$ assumed, Eq. (14.25) is simplified to

$$\frac{dn_T}{dt} = k_f N_T.$$
(14.27)

Since the net trap generation rate (dn_T / dt) as well as the reaction rate (k_f) is suppressed with increasing stress time, k_f is assumed to be an inverse function of the stress time t ($k_f \sim 1/t$). k_f is exponentially related to the electric field in the gate oxide [$k_f \sim \exp(E_{ox})$] [12], so it is possible to express k_f by gate current density (j_g) [41]. Finally, the reaction rate k_f can be written as a function of stress time (t) and gate current density (j_g) as

$$k_f = \frac{j_g k}{t} ,$$

(14.28)

where a reaction proportion coefficient k is introduced considering that not all the injected holes have a chance to react with Si-H bonds.

The implementation of Eq. (14.28) into Eq. (14.27) leads to rewriting dn_T / dt as

$$\frac{dn_T}{dt} = \frac{j_g k N_T}{t} .$$

(14.29)

The interface state density generated during the reaction process can be derived from Eq. (14.29) as

$$n_T(t) = j_g k N_T \log t ,$$

(14.30)

and the stress-induced V_{th} degradation (ΔV_{ths_rea}) dominated by the reaction can therefore be written as

$$\Delta V_{ths\_rea}(t) = \frac{q}{C_{ox}} \cdot n_T(t) = \frac{q}{C_{ox}} \cdot j_g k N_T \log t .$$

(14.31)

With increasing stress time, diffusion begins to dominate the V_{th} degradation. Following the classical H$_2$ R-D model with the time exponent $n = 1/6$ [42], the diffusion-induced ΔV_{th} (ΔV_{ths_diff}) is written analytically as

$$\Delta V_{ths\_diff}(t) = \frac{q}{C_{ox}} n_T(t) \cdot (D_{f\_poly} t)^{1/6} ,$$

(14.32)

where D_{f_poly} is the forward H$_2$ diffusion constant in poly-Si. The final stress-induced ΔV_{th} is written as the sum of both the reaction part and the reaction-limited diffusion part:

$$\Delta V_{th}^{S} = \Delta V_{th\_max} - \Delta V_{th0}^{R} =$$

$$= \frac{q}{C_{ox}} \cdot j_g k N_T \log t \cdot [1 + (D_{f\_poly} t)^{1/6}] \tag{14.33}$$

After stress is removed, the H atoms existing in the SiO_2 diffuse back to repair Si dangling bonds within a very short time and induce a fast recovery (ΔV_{thr_rea}). On the other hand, H_2 molecules located in poly-Si need a much longer time to diffuse back into the oxide, which induces a slow recovery (ΔV_{thr_diff}). On the basis of the stress model shown in Eq. (14.33), the corresponding recovery model is developed as

$$\Delta V_{th}(t) = \Delta V_{ths\_max} - \Delta V_{thr\_rea} - \Delta V_{thr\_diff}$$

$$= (\Delta V_{th\_max} - R \log t) \cdot [1 - (D_{r\_poly} t)^{1/6}], \tag{14.34}$$

where ΔV_{ths_max} is the maximum ΔV_{th} at the end of the stress process, R is the recovery rate, and D_{r_poly} is the reverse H_2 diffusion constant in the poly-Si gate.

The improved R-D model is implanted into the advanced surface potential MOSFET model HiSIM for circuit simulation. In the developed R-D model, the gate current density (j_g) is the main concern for ΔV_{th} calculation and is directly obtained using the advanced surface potential MOSFET model HiSIM.

Fig. 14.15 shows the measured data obtained by the ultrafast technique [43] on a log–log scale. Different from the previous models (classical R-D model and hole-trapping model), the developed model realizes a smooth transition from the short term region to the long-term region under different stress conditions.

The developed R-D model is verified using dynamic NBTI data with 1000 s stress and 1000 s recovery durations [44], as shown in Fig. 14.16. Good agreement between the simulation and measurement results is obtained during several stress and recovery cycles. Based on the experimental result shown in Fig. 14.16(a), the shifts in gate current density (I_g) and drain current (I_{ds}) under dynamic NBTI stress are simulated in Figs. 14.16(b) and 14.16(c) by implementing the developed R-D model into the compact MOSFET model HiSIM. I_g and I_{ds} are solved consistently, and their time-dependent degradation is due to the V_{th} degradation during each stress period.

Table 14.1 shows a list of the parameter values of the developed R-D model [Eqs. (14.33) and (14.34)], as extracted from the experimental data [Figs. 14.15 and 14.16(a)].

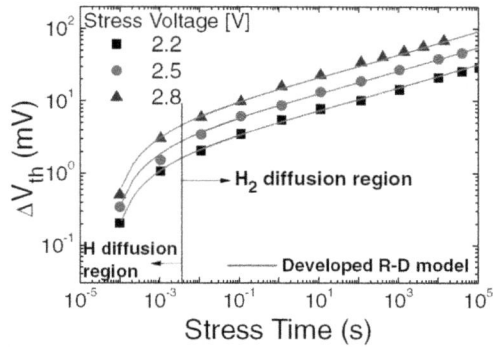

Fig. 14.15. Comparison between the developed model and measured data [19].

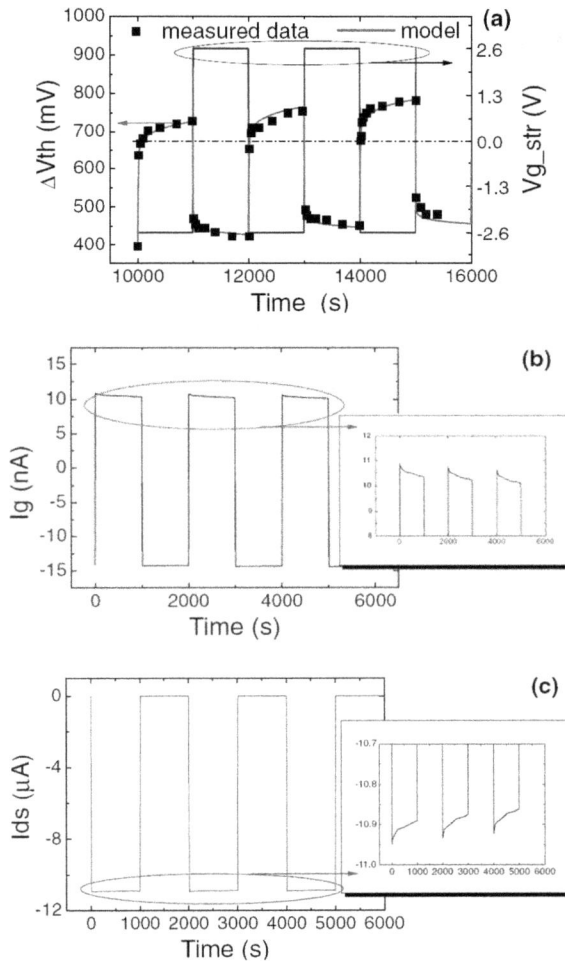

Fig. 14.16. (a) Shift in the threshold voltage under dynamic NBT stress with 1000 s stress and 1000 s recovery durations; (b) NBTIinduced gate current (I_g) degradation; (c) NBTI-induced drain current (I_{ds}) degradation [19].

Table 1. Parameters of the developed R-D model in Eqs. (14.33) and (14.34) as extracted from the experimental data [Figs. 14.15 and 14.16 (a)].

Parameters	Values
H_2 forward diffusion constant in poly-Si (D_{f_poly})	1×10^{-1} nm/s
H_2 reverse diffusion constant in poly-Si (D_{r_poly})	1×10^{-5} nm/s
Reaction rate (k)	2×10^{-10} cm^2/s
Maximum Si–H bond density (N_T)	1×10^{11} cm^{-2}
Gate oxide thickness (T_{ox})	2.2 nm
Recovery rate (R)	0.015

14.3. Reliability Study on Nanoscale Non-Traditional CMOS Devices

14.3.1. Reliability of FinFETs

14.3.1.1. Generation of Interface State Traps in FinFETs

In this Subsection, the degradation of a FinFET is studied by using the gated-diode G–R method [45]. It is identified that the degradation is due to two different mechanisms: oxide interface states generation and oxide traps formation. The impacts of both the mechanisms are investigated in this work to identify the dominant mechanism that causes the degradation of the FinFET characteristics.

The schematic structure of a FinFET that can be operated as a gated diode is shown in Fig. 14.17.

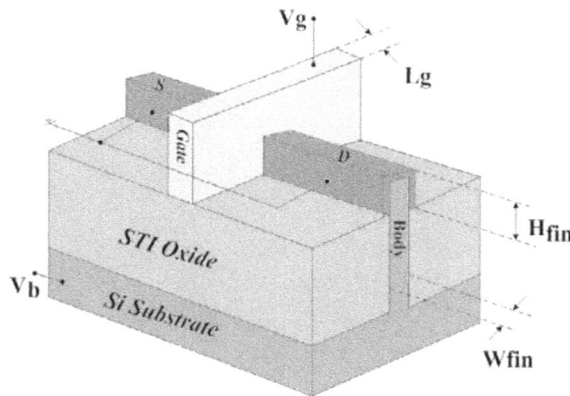

Fig. 14.17. Schematic structure of a FinFET [45].

To characterize the interface states distribution, the forward gated-diode current method is used to measure the generation–recombination (G–R) current as described in [10-11]. By keeping a constant bias for the gate diode, the G–R current changes with the gate voltage. It is observed that the G–R current rises to a peak with increasing gate voltage, and then decreases to a constant level, as potted in Fig. 14.18.

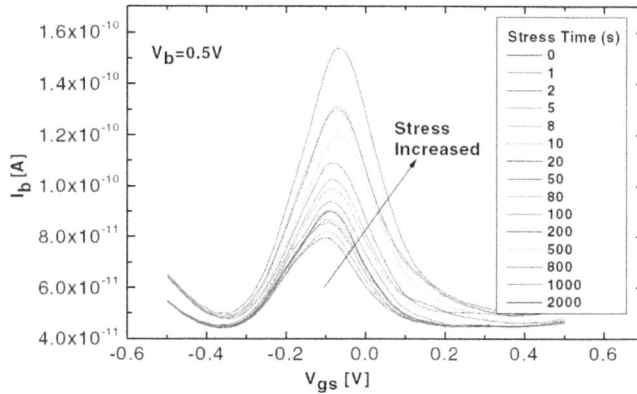

Fig. 14.18. The substrate current I_b versus V_{gs} characteristics of a FinFET under different F-N stress time [45].

The maximum G–R current (I_{peak}) could be expressed as [45]

$$I_{peak} = \frac{1}{2} q n_i (c_n c_p)^{1/2} N_{it} WL \exp(\frac{qV_b}{2kT})$$
$$c_n = c_p = 10^{-8} \text{ cm}^{-3}\text{s}^{-1}$$

(14.34)

Here, all symbols have their common physical meanings.

Fig. 14.19 shows the extracted interface state density in FinFETs as a function of the F-N stress time t. Assuming an exponential relationship, the increase in ΔN_{it} ($\Delta N_{it} = N_{it} - N_{it0}$) can be expressed in terms of stress time as $\Delta N_{it} \propto t^{0.39}$, which shows an evident superior of the FinFET reliability than SOI MOSFET ($\Delta N_{it} \propto t^{0.7}$) [9].

14.3.1.2. Temperature Dependence of Interface States in FinFETs

Due to the ultra-short channel and thin Fin body, FinFET suffers from a large electrical field, leading to inevitable HCI effect [23]. The high density of interface states induced by HCI seriously affected the device performance. Here, the distribution of interface states along the FinFET channel at various temperatures is extracted and investigated by using the forward gated-diode method, and an empirical model is proposed for predicting the distribution of interface states in FinFETs [46].

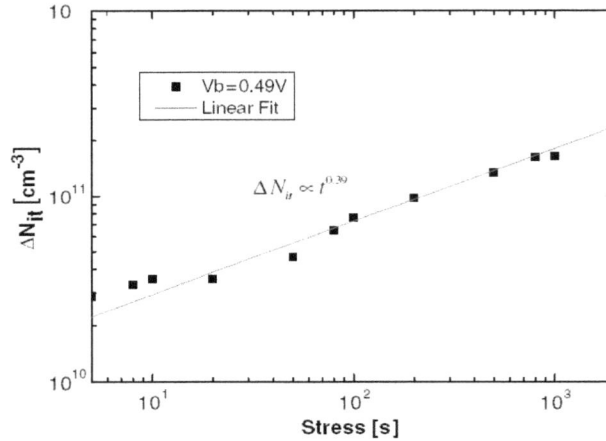

Fig. 14.19. The extracted interface state density in a FinFET as a function of the F-N stress time [45].

The interface state density can be derived as a function of position y along the channel as follows

$$N_{it}(y) = \frac{\dfrac{dI_{\text{G-R,peak}}}{dV_{ds}} \left(\dfrac{dy}{dV_{ds}}\right)^{-1}}{\dfrac{1}{2} qn_i (c_n c_p)^{1/2} W \exp\left(\dfrac{qV_b}{2kT}\right)}$$

$$\frac{dy}{dV_{ds}} = \sqrt{\frac{\varepsilon_0 \varepsilon_{si}}{2qn_i \exp\left(\dfrac{V_b}{2kT}\right)}} \sqrt{\frac{1}{V_{ds}}}, \tag{14.35}$$

where $dI_{\text{G-R,peak}} / dV_{ds}$ is the differential coefficient of the maximum G-R current which ban be experimentally measured with the forward gated-diode [10].

Fig. 14.20 shows the extracted HCI-induced interface states in FinFETs. As shown in Fig. 14.20 (a), the distribution of interface states along the channel is obtained with the temperature ranging from 28 °C, 78 °C to 128 °C. The maximum interface state density increases not only with longer HCI stress time, but also with elevated temperature. The maximum interface state density $N_{\text{it\_peak}}$ at different stress time and temperatures is plotted in Fig. 14.20 (b). $N_{\text{it\_peak}}$ increases linearly with the stress time in Log-Log scale, but the change of generation rate slows down with higher temperatures.

Additionally, a Gaussian-like empirical model for interface state distribution under different stress time is proposed as Eq. (14.36) based on the measured data, which can be implemented into the HSPICE simulator to predict the FinFET-based circuit degradation.

As shown in Fig. 14.21, the result shows a good agreement between the empirical model and the experimental measurement data at 28 °C.

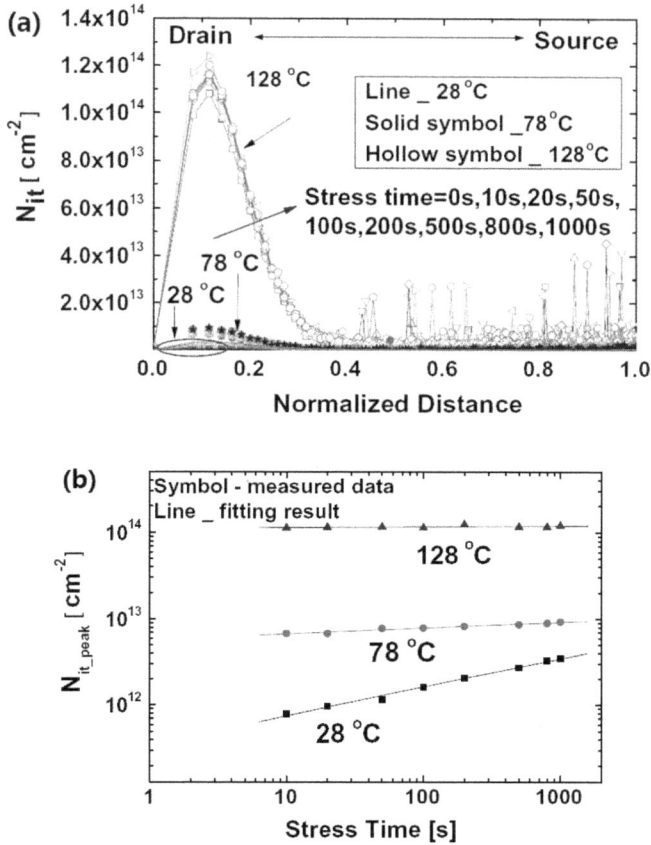

Fig. 14.20. The extracted interface state density induced by HCI effect in a FinFET [46].

$$N_{it}(y) = N_{it\_ch} + \frac{A}{\delta\sqrt{2\pi}}\exp\left(-\frac{(y-\xi)^2}{2\delta^2}\right).$$

(14.36)

Here N_{it_ch} is the interface state density in the channel, and parameters of δ, ξ and A are fitting parameters determined by measured data.

14.3.1.3. Asymmetric Issues of FinFETs

In this Subsection, with the above obtained empirical interface state distribution model [Eq. (14.36)], a physical-based FinFET device degradation model is developed considering the effect of non-uniform distribution of interface states [47], which is verified by the experimental data.

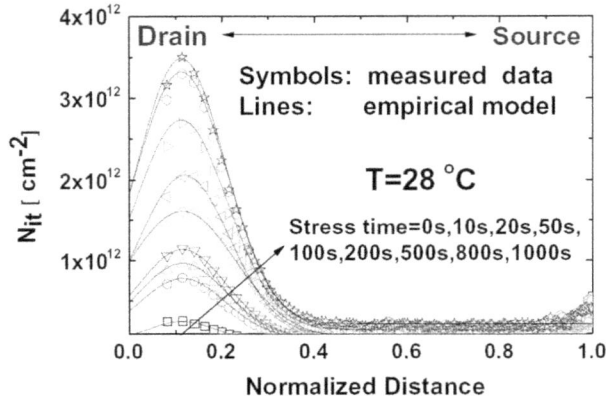

Fig. 14.21. Comparison of the empirical model and measured data for the interface states in a FinFET [46].

For simplicity, FinFET is considered to be a double-gate MOSFET. The interface states are supposed to be acceptors located along the channel with the density of $N_{it}(y)$. The inversion charge concentration $Q_n(y)$ is expressed as [48]

$$Q_n(y) = -C_{ox}[V_{gs} - V_{th0} - V(y)] + qN_{it}(y) ,$$
(14.37)

where C_{ox} is the gate oxide capacitance, V_{th0} and $V(y)$ are threshold voltage and quasi Fermi level, respectively. The drain current is obtained from the inversion charge as

$$I_{ds} = -W\mu Q_n(y)\frac{dV}{dy} = W\mu\left\{C_{ox}[V_{gs} - V_{th0} - V(y)] - qN_{it}(y)\right\}\frac{dV}{dy} ,$$
(14.38)

where $\mu = \mu_0/(1 + \kappa N_{it})$ is the carrier mobility following the work in [48], μ_0 is the electron mobility of fresh device and κ is the mobility degradation factor.

Based on the compact model for double-gate devices [49], the drain current considering the impact of interface states is finally obtained by combining Eqs. (14.36) and (14.38) as

$$I_{ds} = \frac{I_{ds0} - \dfrac{qW\mu_0}{L_g}\displaystyle\int_0^{V_{ds}} N_{it}(y)\frac{dV}{dy}dy}{1 + \dfrac{\kappa}{L_g}\displaystyle\int_0^{L_g} N_{it}(y)dy}$$

$$= \frac{L_g}{L_g + \kappa\displaystyle\int_0^{L_g} N_{it}(y)dy}I_{ds0} - \frac{qW\mu_0 V_{ds}\displaystyle\int_0^{L_g} N_{it}(y)dy}{L_g\left[L_g + \kappa\displaystyle\int_0^{L_g} N_{it}(y)dy\right]} ,$$
(14.39)

where I_{ds} and I_{ds0} are the drain current of damaged and fresh device, respectively. The model obtains a good agreement with the experimental results, as shown in Fig. 14.22.

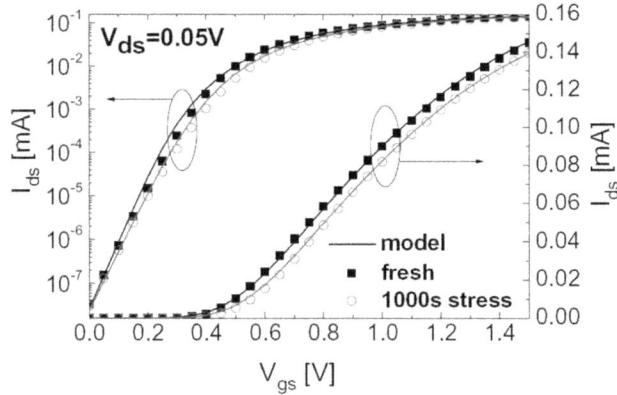

Fig. 14.22. The transfer characteristics of a FinFET before and after 1000 s HCI stress [47].

Fig. 14.23 exhibits an evident asymmetric HCI-induced degradation for the saturation drain current of FinFETs in forward operation and reverse operation mode ($\Delta I_{ds} = I_{ds} - I_{ds0}$), and Fig. 14.24 gives the diagram for the corresponding degradation mechanism. Eq. (14.39) is still available by replacing L_g with effect channel length $L_{eff} = L_g - l$. Here l is the pinch-off length calculated in [50]. In the forward operation mode, large amount of interface states are generated in the pinch-off region and influence the pinch-off length l [see Fig. 14.24 (a)]. As seen from Fig. 14.23 (a), the saturation current not only shows a larger decline but also a more evident channel length modulation effect. Since the pinch-off region is not influenced much by the low density of interface states in the reverse mode [see Fig. 14.24(b)], the saturation current shows a less degradation [see Fig. 14.23 (b)].

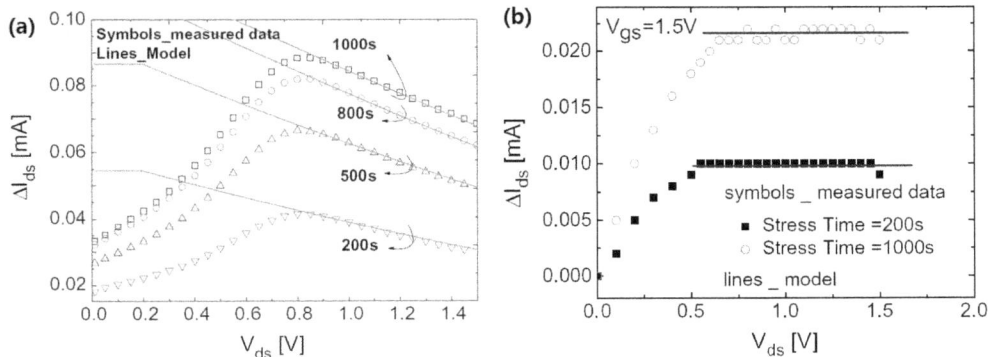

Fig. 14.23. $I_{ds} - V_{ds}$ degradation due to HCI stress in (a) forward operation mode, and (b) reverse operation mode [47].

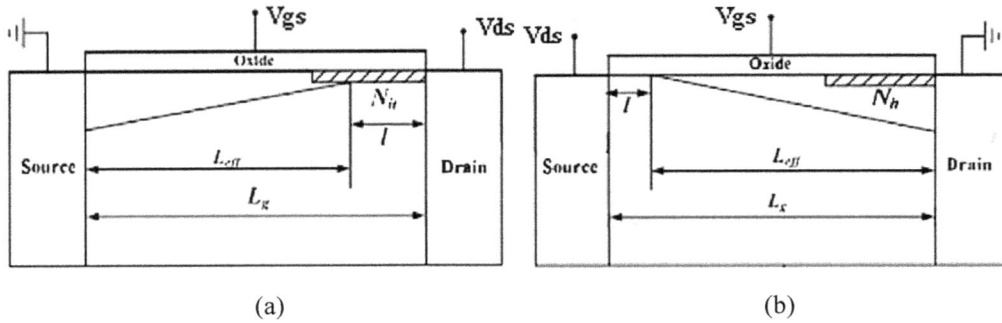

(a)   (b)

Fig. 14.24. Saturation region in (a) forward operation mode and (b) reverse operation mode [47].

As shown in Fig. 14.25, 6-T SRAM is simulated to demonstrate the impact of HCI on the circuit performance. The Static Noise Margin (SNM) and the Write Margin (WM) are used to mentor the degradation. The access devices may suffer from the asymmetric degradation with source and drain reversal during operation.

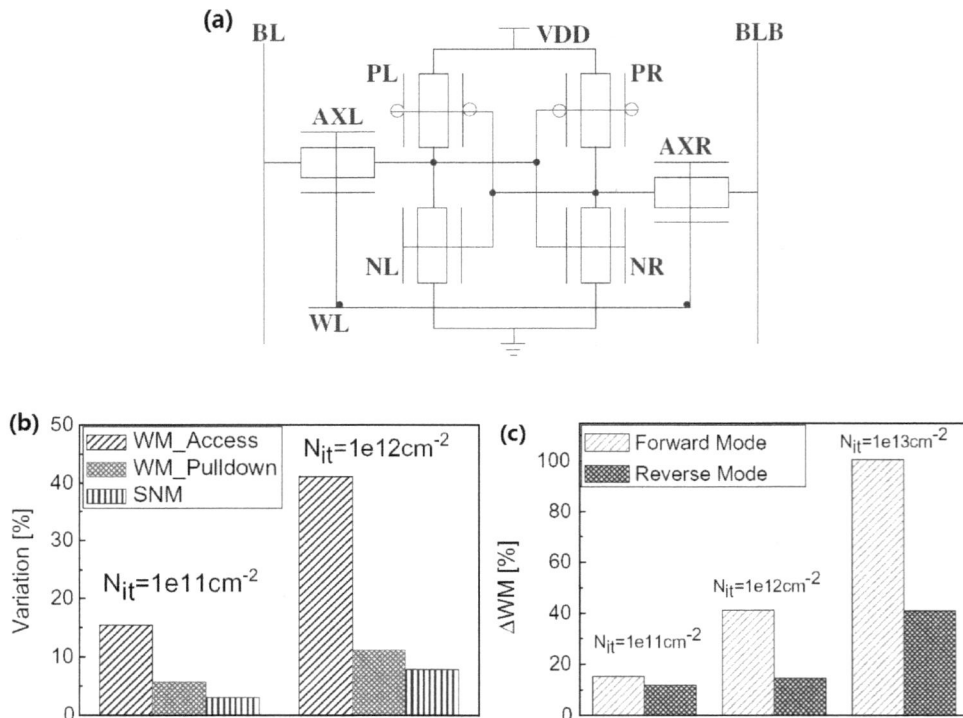

Fig. 14.25. Impact of HCI stress on the 6-T FinFET SRAM circuit performance: (a) 6-T FinFET SRAM cell; (b) comparison of WM and SNM variation due to HCI in different devices; (c) variation of WM with Access transistors operated on the forward and reverse mode [47].

Fig. 14.25 (b) compares the variation of SNM and WM with HCI stress applied to the Access and Pull-down transistors. The maximum degradation appears on the Access transistor while the SNM remains relatively stable. When Access transistors operated on the forward and reverse mode, the WM exhibits an obvious asymmetric degradation as shown in Fig. 14.25 (c). The comparison implies that HCI causes more serious WM degradation in the forward mode than that in the reverse mode.

14.3.2. Reliability of Nanowire MOSFETs

14.3.2.1. NBTI Model for Nanowire MOSFETs

As discussed in Section 14.2.2, NBTI remains as one of the most important reliability issue. Here, the common reaction-diffusion (R-D) theory is used to develop a first-order compact NBTI model for SNWTs [51]. Moreover, the reaction probability effect is taken into account for the random nature of Si-H bond breaking.

The device structure is based on the Twin Silicon Nanowire MOSFET, as shown in Fig. 14.26 (a). To simplify the model derivation, the SNWT is assumed to be a gate-all-around MOSFET with a circular channel [see Fig. 14.26 (b)]. The channel radius is R and gate oxide thickness is t_{ox}. Fig. 14.26 (c) illustrates the NBTI induced hydrogen diffusion along the radius of the nanowire, and λ is the diffusion constant.

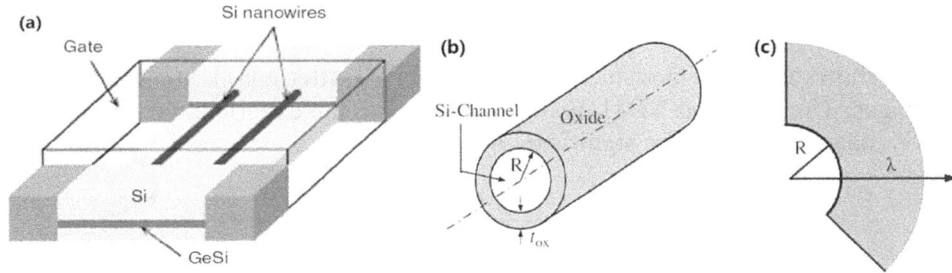

Fig. 14.26. Schematic view of the silicon nanowire MOSFET [51].

Based on the conventional R-D theory, the degradation of threshold voltage ΔV_{th}^0 is derived as [51]

$$\Delta V_{th}^0(t) = \frac{qt_{ox}}{\varepsilon_{ox}\varepsilon_0\sqrt{R}}\left[\sqrt[2n]{K^2\left(\exp\frac{E_{ox}}{E_0}\right)^2 C_{ox}(V_{gs}-V_{th})}\right]$$
$$\times \left(\frac{R\sqrt{Ct}}{2}+\frac{Ct}{6}\right)^{2n} , \qquad (14.40)$$

where $C = (1/T_0)\exp(-E_a/(kT))$ is the diffusion constant, $K = 5.2 \times 10^4$ is the field acceleration prefactor, E_{ox} is the oxide electrical field, $T_0 = 10^{-8}$ s/nm². With the assumption of diffusing specie H, we have $n = 1/4$, $E_a = 0.12$ eV, and $E_0 = 2$ V/nm.

Since not all the H atoms diffusing into gate oxide will induce Si-H bond breaking, considering the Si-H bond breaking probability $R_b = 1 - \sigma e^{-t/\tau}$, the modified degradation model of threshold voltage can be written as

$$
\begin{aligned}
\Delta V_{th}^0(t) &= \Delta V_{th\_max} \cdot R_b \\
&= \frac{qt_{ox}}{\varepsilon_{ox}\varepsilon_0\sqrt{R}}\left[\sqrt[2n]{K^2\left(\exp\frac{E_{ox}}{E_0}\right)^2 C_{ox}(V_{gs} - V_{th})}\right. \\
&\quad \times \left.\left(\frac{R\sqrt{Ct_1}}{2} + \frac{Ct_1}{6}\right)^{2n} \cdot (1 - \sigma e^{-t/\tau})\right]
\end{aligned}
\tag{14.41}
$$

where $\tau = \tau_0 \exp\left(\dfrac{1.5 - 0.056(V_{gs} - V_{th})/t_{ox}}{kT}\right)$, $\sigma \approx 1$, $T_0 = 3 \times 10^3$ s/nm², $K = 5.2 \times 10^4$,

$t_1 = 1000$ s and $\tau_0 = 1.26e^{-7}$ s.

Fig. 14.27 shows the characteristics of a p-type SNWT after NBTI stress with different gate voltages. The modified NBTI model matches with the experimental data very well, and the V_{th} degradation is more significant at higher stress voltage.

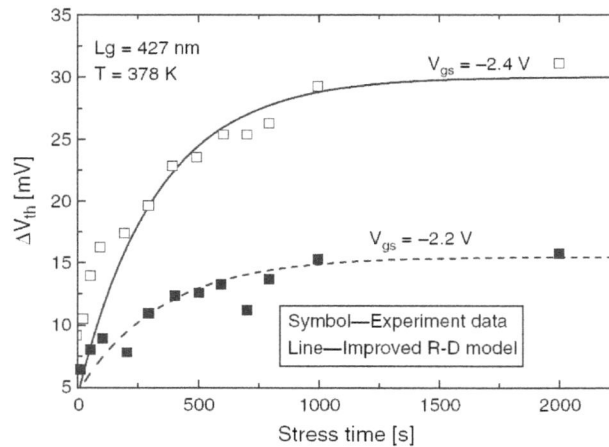

Fig. 14.27. Threshold voltage degradation due to NBTI in p-type SNWT [51].

In the recovery process, the change in interface trap density is supposed to be the same as the stress process, the degradation of threshold voltage during the recovery process is given by [51]

$$\Delta V_{\mathrm{th}}^{0}(t) = \Delta V_{\mathrm{th\_max}} \left[1 - \frac{(R\sqrt{\zeta D(t-t_1)}/2) + (\zeta D(t-t_1)/6)}{(R\sqrt{Dt}/2) + (Dt/6)} \right],$$

(14.42)

where t_1 is beginning of the recovery process, D is the hydrogen diffusion constant in SiO_2, and ζ is the ratio of diffusion constants in the stress and recovery process.

Most experimental measurements for NBTI stress indicate that the threshold voltage cannot recover completely but saturates at a certain value due to the partial back diffusion of H atoms. The maximum of threshold voltage recovery has the form as [51]

$$\Delta V_{\mathrm{th0}}^{R} = \frac{q t_{\mathrm{ox}}}{\varepsilon_{\mathrm{ox}} \varepsilon_0} \sqrt{\frac{k_{\mathrm{f}} N_0}{R k_{\mathrm{r}}}} C_{\mathrm{ox}} (V_{\mathrm{gs}} - V_{\mathrm{th}}) \left(\frac{R t_{\mathrm{ox}}}{2} + \frac{t_{\mathrm{ox}}^2}{6} \right)^{1/2},$$

(14.43)

where N_0 is the initial concentration of Si-H bonds for a fresh device, k_{f} and k_{r} are the reaction rates of the forward and reverse reactions, respectively.

Similarly, the bond reformation probability can be expressed as

$$R_{\mathrm{r}} = 1 - R_{\mathrm{b}} = \sigma e^{-(t-t_1/\tau)}.$$

(14.44)

The saturation threshold voltage degradation $\Delta V_{\mathrm{th}}^{S}$ is the difference between the maximum threshold voltage degradation $\Delta V_{\mathrm{th\_max}}$ and the maximum threshold voltage recovery $\Delta V_{\mathrm{th0}}^{R}$

$$\Delta V_{\mathrm{th}}^{S} = \Delta V_{\mathrm{th\_max}} - \Delta V_{\mathrm{th0}}^{R}.$$

(14.45)

Combining Eqs. (14.43)-(14.45), the threshold degradation $\Delta V_{\mathrm{th}}^{R}(t)$ in recovery process is

$$
\begin{aligned}
\Delta V_{\mathrm{th}}^{R}(t) &= \Delta V_{\mathrm{th0}}^{S} + \Delta V_{\mathrm{th0}}^{R} \cdot R_{\mathrm{r}} \\
&= (\Delta V_{\mathrm{th\_max}} - \Delta V_{\mathrm{th0}}^{R}) + \frac{q t_{\mathrm{ox}}}{\varepsilon_{\mathrm{ox}} \varepsilon_0} \sqrt{\frac{k_{\mathrm{f}} N_0}{R k_{\mathrm{r}}}} C_{\mathrm{ox}} (V_{\mathrm{gs}} - V_{\mathrm{th}}) \\
&\quad \times \left(\frac{R t_{\mathrm{ox}}}{2} + \frac{t_{\mathrm{ox}}^2}{6} \right)^{1/2} \cdot \sigma e^{-t(t-t_1/\tau)}
\end{aligned}
$$

(14.46)

Fig. 14.28 shows the comparison between experimental data and the R-D model. In the stress process, both the conventional and improved R-D models agree well with the experimental data in the beginning, but the conventional R-D model produces poor saturation results. When $\zeta = 0.53$, the recovery process fails to match the experimental data until $t = 1500$ s. With the increase in ζ, such as $\zeta = 0.8$, the prediction with the R-D model produces a sharp drop in the beginning of the recovery process but never saturates. By taking into account of the reaction probability, the improved R-D model has improved accuracy in matching the experimental data, especially in the recovery process.

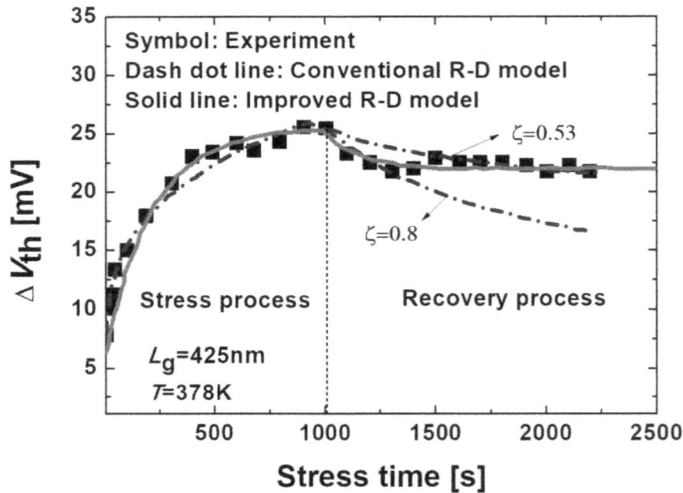

Fig.14.28. Threshold voltage degradation due to NBTI in p-type SNWT.

14.3.2.2. Suppression of Tunneling Leakage Current in Nanowire MOSFETs

Junctionless nanowire transistor (JNT) has been considered as one of the candidates to overcome severe challenges on doping techniques [21]. However, the band-to-band tunneling (BTBT) current between the channel and drain region will significantly raise the OFF-state leakage current [26].

In this Subsection, the dual-material gate (DMG) structure is proposed and incorporated into JNTs [52], and the consequent suppression of the BTBT effect is verified by using numerical simulation tools [53]. Fig. 14.29 (*a*) shows the architecture of the proposed DMG-JNT including two metal gates with different work functions. The gate on the source and drain side are denoted as 'control gate' and 'screen gate', respectively. The conventional JNT with one single gate material is also provided for comparison. Fig. 14.29 (*b*) gives the energy band diagrams of the DMGJNT and conventional JNT at different positions. For the OFF-state and negative gate bias ($V_{gs} \leq 0$ V) in JNTs, no matter whether at the channel center or channel surface, the valence band in the channel is higher than the conduction band near the drain, resulting in the valence band electrons tunneling from the channel into the drain. Hence, the BTBT is triggered and degrades the OFF-state characteristics of JNT. As for the DMG-JNT, the unique energy band step greatly reduces the overlap of valance and conduction band and increases the potential tunneling path length on the drain side. Therefore, the tunneling effect will be effectively suppressed by DMG structure.

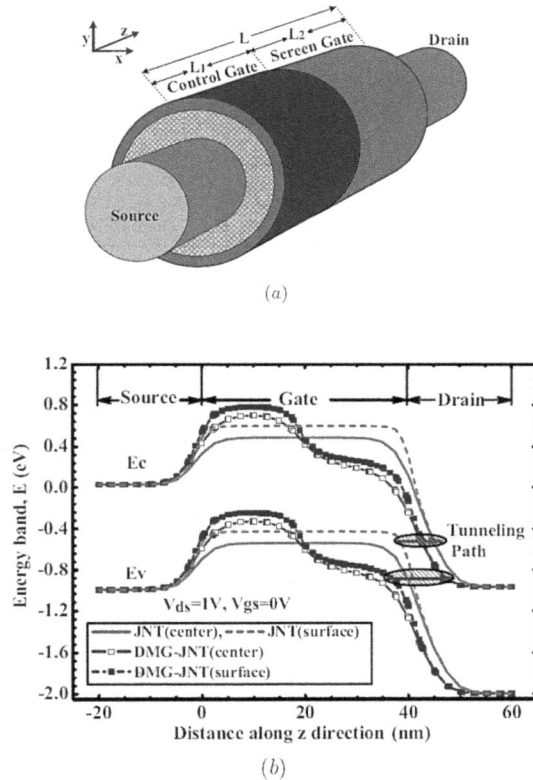

(a)

(b)

Fig. 14.29. (a) Schematic view of the proposed DMG-JNT; (b) energy band diagrams of the DMG-JNT and conventional JNT, where gate voltage $V_{gs} = 0$ V and drain voltage $V_{ds} = 1$ V [52].

With the simulation model calibrated with the experimental results [26], the suppression of tunneling leakage current is investigated for DMG-JNTs. Fig. 14.30 shows the transfer characteristics of DMG-JNT and conventional JNT with different channel lengths. It is observed that the OFF-state current of the DMG-JNT is decreased by almost two orders of magnitude as compared with the conventional JNT.

In Fig. 14.31, the tunneling current I_T characteristics with scaling channel length for the DMG-JNT are shown. Simultaneously, the impacts of radius R and doping N_d are included. As shown in Fig. 14.31 (a), the I_T increases with N_d for both DMG-JNT and conventional JNT. Larger R induces raised I_T due to the lower tunneling barrier height and width [see Fig. 14.31(b)]. As the channel length is larger than 15 nm, the I_T of DMG-JNT is much smaller than that of conventional JNT. As channel length scales down to 15 nm and below, the I_T of both JNT and DMG-JNT increases drastically.

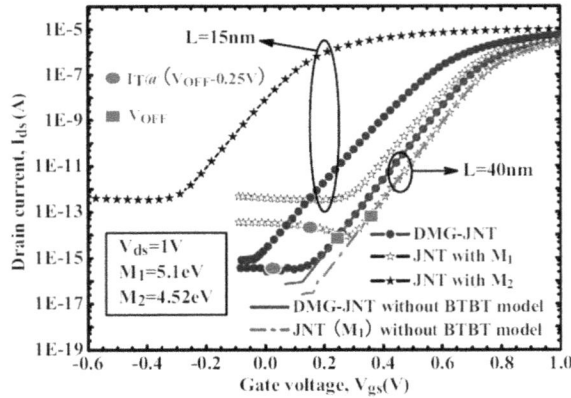

Fig. 14.30. (*a*) Schematic view of the proposed DMG-JNT; (*b*) energy band diagrams of the DMG-JNT and conventional JNT, where gate voltage V_{gs} = 0 V and drain voltage V_{ds} = 1 V [52].

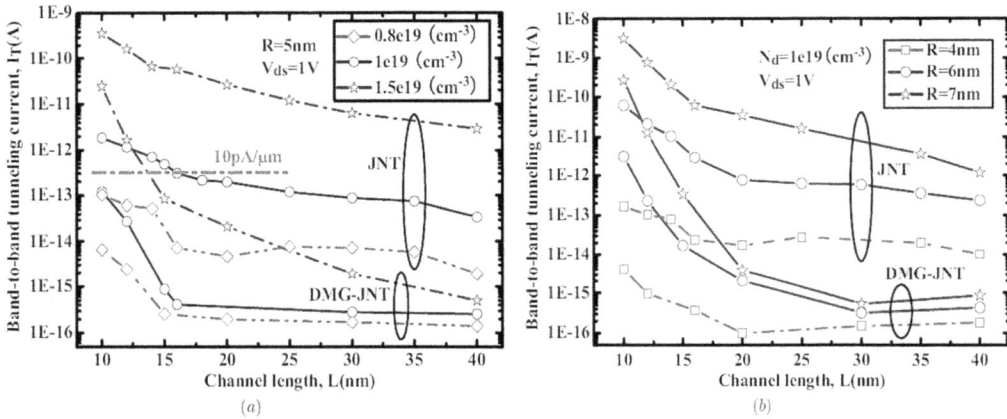

Fig. 14.31. The BTBT current characteristics for the scaling DMG-JNT and conventional JNT [52].

Fig. 14.32 shows the dependence of I_T characteristics on R_a (L_1/L) for fixed workfunction M_1 and M_2. It is evident that as R_a approaches 0 or 1, which means, the control gate length in DMG-JNT is 0 nm or L, the I_T increases drastically up to the tunneling current level of the JNT. However, when Ra is between 0.2 and 0.825, the I_T of DMG-JNT remains at a lower level than the corresponding JNT even as the channel length is scaled down to 15 nm, which implies the energy band step in DMG-JNT has non-negligible influences on BTBT effect.

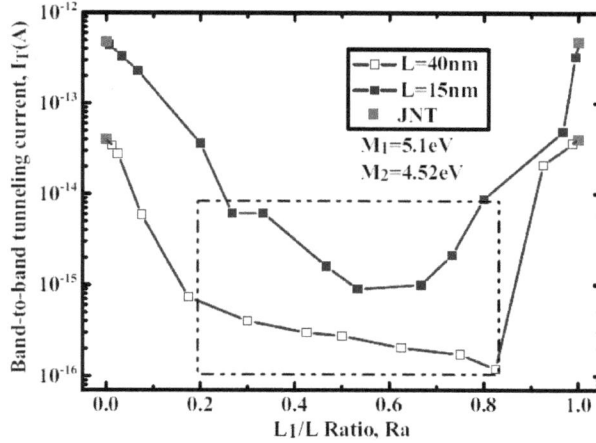

Fig. 14.32. The BTBT current characteristics versus various L_1/L ratio [52].

14.4. Variation Study on Nanoscale Novel MOSFET Devices

14.4.1. Random Dopant Fluctuation Effect

Based on the atomic level simulation, a general method is provided to study the impact of RDF effect on the surrounding gate MOSFET device and circuit performances [54].

Threshold voltage is taken as the representation to reveal the impact of RDF on the device performance. Monte Carlo method [55] is used to generate the dopant atom distribution in the volume of interest by following the random dopant placement algorithm [56], as shown in Fig. 14.33. In the expression $P\_abs = N*v$, N is the expected doping concentration with any doping profile, and v is the volume owned by each atom in the pseudo lattice. The equivalent charge density corresponding to a single dopant is given as [57]

$$\rho(r) = \frac{ek_c^3}{2\pi^2} \frac{\sin(k_c r) - (k_c r)\cos(k_c r)}{(k_c r)^3}, \tag{14.47}$$

where k_c and r are the inverse of screening length and the distance from the single dopant atom for the grid point, respectively.

Fig. 14.34 (a) shows an example of random dopant atom placement. The low density of dopant atoms in nano-scale devices lead to significantly discretized doping profile. Dopant atoms located in different positions have distinct influences on the device performance [see Fig. 14.34 (b)]. Fig. 14.35 plots a statistical simulation result of threshold voltage variation in surrounding gate MOSFET, which is fitted into Gaussian distribution with a standard deviation of 53.1 mV.

391

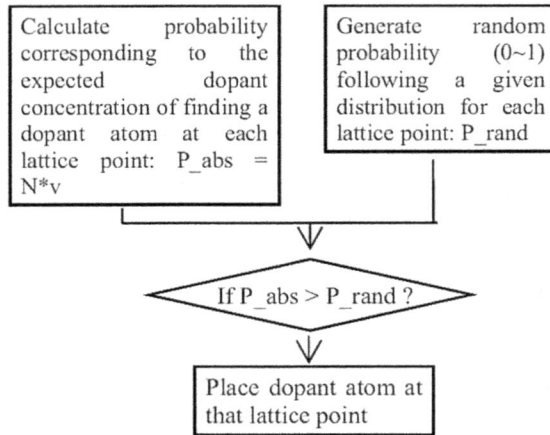

| Calculate probability corresponding to the expected dopant concentration of finding a dopant atom at each lattice point: P_abs = N*v | Generate random probability (0~1) following a given distribution for each lattice point: P_rand |

If P_abs > P_rand ?

Place dopant atom at that lattice point

Fig. 14.33. Random dopant placement algorithm [54, 56].

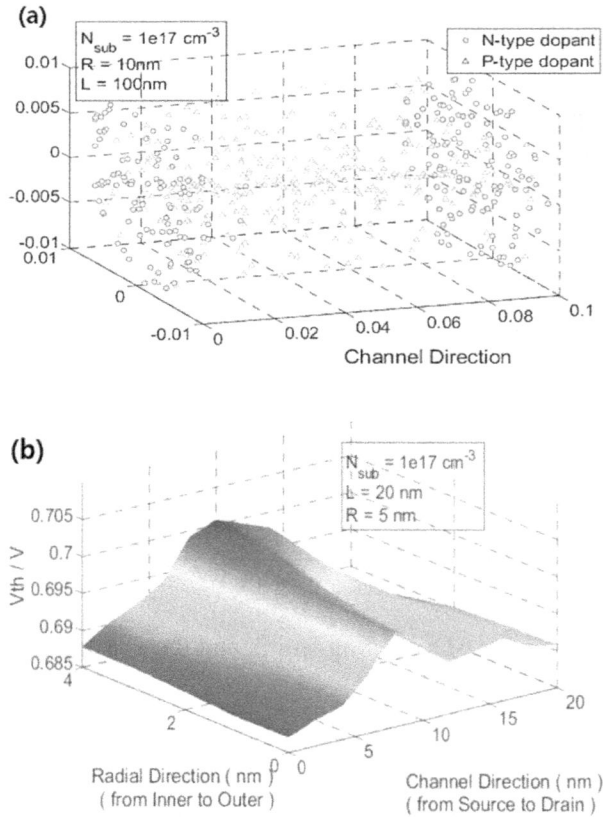

Fig. 14.34. (a) An example of random dopant placement in a surrounding gate MOSFET; (b) Threshold voltage variation due to single acceptor dopant at different locations [54].

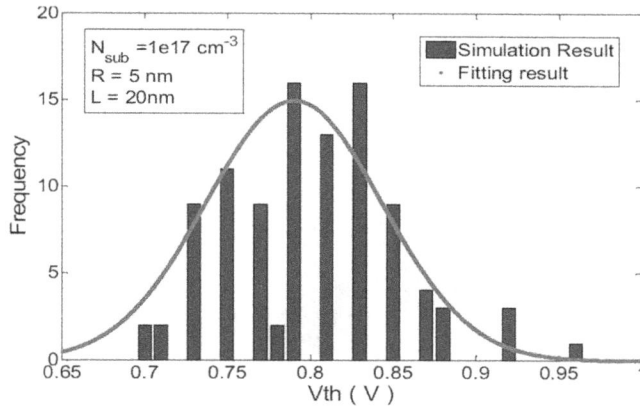

Fig. 14.35. Statistical threshold voltage variation of surrounding gate MOSFET due to RDF [54].

In order to investigate the influence of RDF effect on the circuit performance, the RDF effect is implanted into the compact model of surrounding gate MOSFET [58]. Three thousand devices with RDF effect variations are simulated in SRAM cell circuit. Fig. 14.36 shows the statistical distribution of static noise margin (SNM) of SRAM cell during hold time. In spite of the aggressive fluctuation of threshold voltage, the SRAM cells do not amplify the variation, which indicates that surrounding gate MOSFET can be a promising device due to its resistibility to RDF effect.

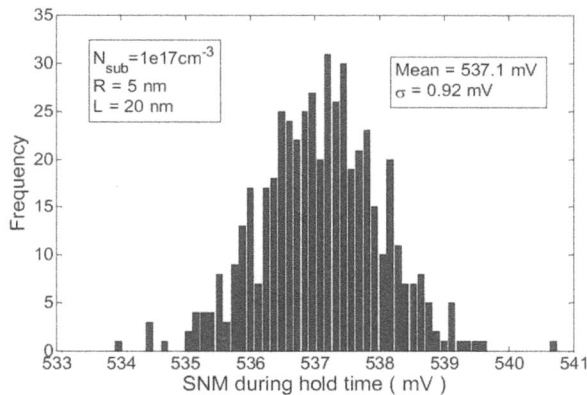

Fig. 14.36. Statistical variation in SNM of SRAM cell during hold time [54].

14.4.2. Fin-Width Line Edge Roughness Effect

Line edge roughness (LER), which means the random deviation of line edge from its ideal pattern, plays a significant role on the nanoscale transistor performance [59]. In this Subsection, the LER issue in FinFETs is investigated through a full 3-D statistical simulation approach [60]. The fin-width roughness modeling approach is based on a sequence of fixed root-mean-square amplitude by Matlab. The parameters used to

393

generate the edges are root-mean-square amplitude and a cut-off frequency, which reflects the nature of LER in the simulation. The generated fin channel with fin-width LER for double-gate FinFETs studied here is shown in Fig. 14.37.

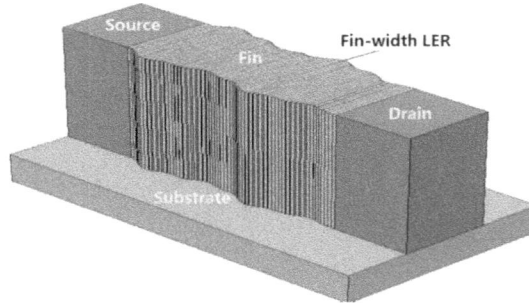

Fig. 14.37. The generated fin channel with fin-width LER for double-gate FinFETs [60].

The root-mean-square amplitude of fin-width LER is assumed 0.71 nm, which means the variation fluctuations to be 10% of its critical dimension (fin width W_{fin} = 12 nm). Fig. 14.38 shows the typical distributions of transconductance, subthreshold slope and threshold voltage in 30 nm gate length FinFETs. The variation of all the three parameters is almost close to a Gaussian distribution, and it infers that the fin-width LER is a critical source of intrinsic parameter fluctuations in nanometer FinFETs, especially for the threshold voltage.

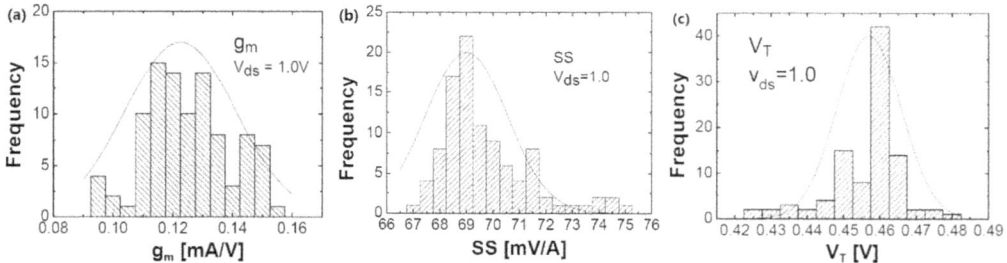

Fig. 14.38. The fin-width LER induced variation in transconductance, subthreshold slope and threshold voltage of double-gate FinFETs [60].

14.4.3. Fin Sidewall Angle Fluctuation Effect

Non-ideal cross section such as a trapezoidal shape is often obtained for the fin channel in the fabrication of multi-gate MOSFETs [26], and the vertical non-uniformity significantly affects the device characteristics. Here, the effect of trapezoidal shape on the subthreshold characteristics of junctionless multi-gate transistors (JMTs) is analyzed based on the 3-dimensinal device simulation and compared with those of the inversion mode multi-gate MOSFETs (IM-MuGFETs) [61]. A three-dimensional schematic view of a JMT with trapezoidal fin body is shown in Fig. 14.39, and the key parameter, fin sidewall angle (Θ), is normally kept between 65° and 90° according to the fabrication process.

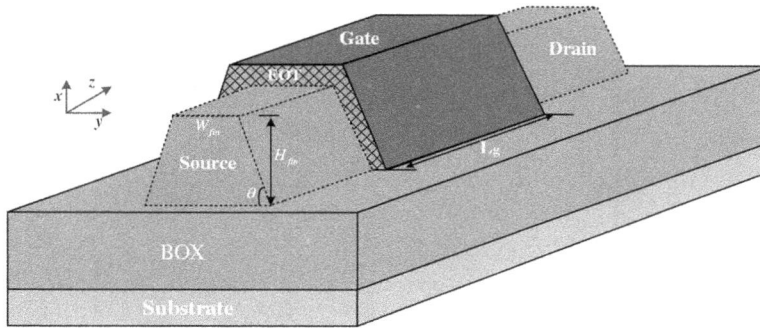

Fig. 14.39. Schematic view of JMT with trapezoidal fin body [61].

Fig. 14.40 (a) shows the SS difference between the case with a trapezoidal cross section and the ideal case (Θ = 90°) for both the JMT and IM-MuGFET. It is shown that the SS difference increases almost linearly with the reduction in Θ . The slopes of SS difference for the JMT and IM-MuGFET are 0.43 and 0.93, respectively. For the JMT, SS increases with increasing fin height H_{fin} and decreasing Θ [see Fig. 14.40 (b)].

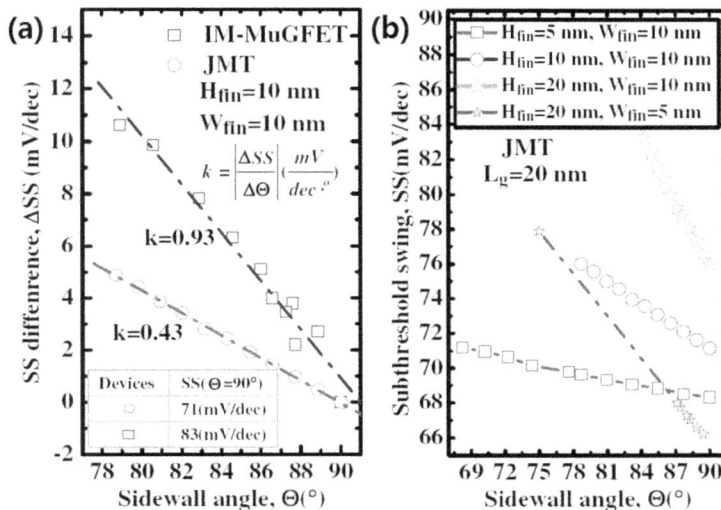

Fig. 14.40. Variation in SS difference and SS for both JMT and IM-MuGFET [61].

Fig. 14.41 plots the characteristics of DIBL for both the JMT and IM-MOSFET. The DIBL voltage increases almost linearly with the reduction in Θ for both the devices. As

H_{fin} of the JMT increases from 5 to 20 nm, DIBL is enhanced and the slope of the DIBL vs Θ increases from 0.5 to 6.83.

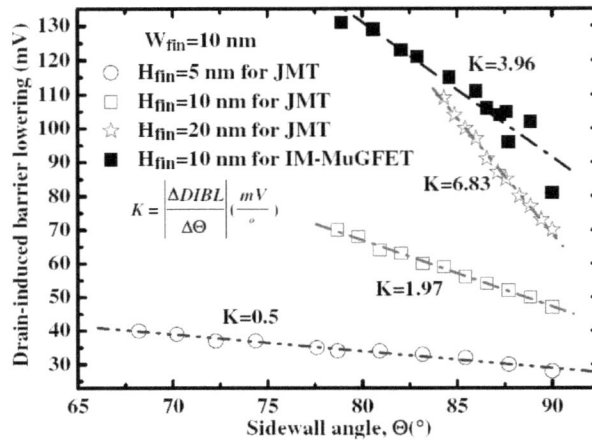

Fig. 14.41. DIBL variation vs. sidewall angle for both JMT and IM-MuGFET [61].

The doping concentration change will result in important variation for JMTs. We evaluate the influence of sidewall angle on SS and DIBL as compared with that of doping variation (see Fig. 14.42). The values of SS and DIBL for different doping concentrations and sidewall angles are normalized to the ideal JMT with doping of 1×10^{19} cm^{-3} and $W_{fin} = H_{fin} = 10$ nm. As seen from the figure, DIBL is more sensitive to the variations in both doping concentration and sidewall angle than SS. In addition, fin sidewall angle impact the SS and DIBL variation more severely, indicating that Θ should be carefully designed and limited.

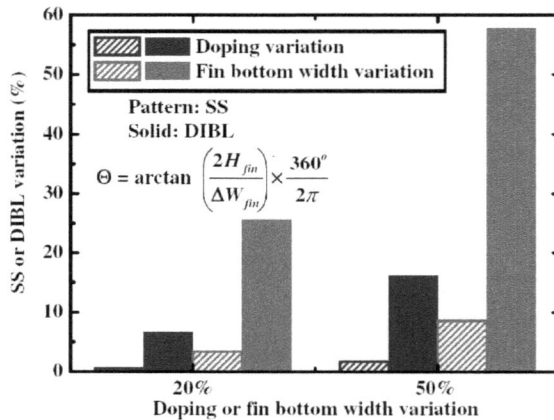

Fig. 14.42. DIBL variation vs. sidewall angle for both JMT and IM-MuGFET [61].

To suppress the subthreshold characteristics variation due to the non-ideal sidewall angle, the high-k spacer is introduced for JMTs [62]. Fig. 14.43 gives the proposed device architecture of the JMT with high-k spacer.

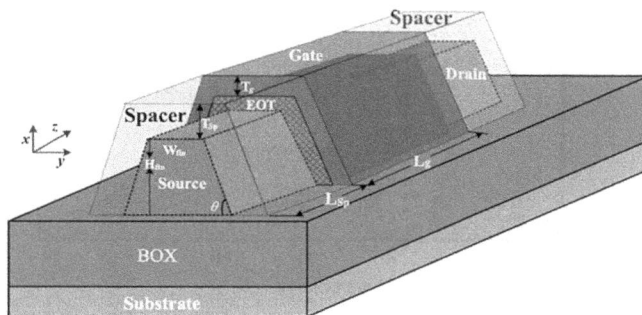

Fig. 14.43. Schematic view of JMT with the trapezoidal cross section and high-k spacer [62].

The fluctuation of SS difference and DIBL difference induced by the sidewall angle Θ is obtained, as shown in Fig. 14.44. As for $\Theta = 90°$, the SS value in JMTs degrades from 91.72 mV/dec to 81.73 mV/dec when the dielectric constant ξ_{sp} of spacer increases from 1 to 21 [see Fig. 14.44(a)]. The DIBL values for the case of $\Theta = 90°$ are 145 mV, 115 mV and 83 mV, respectively [see Fig. 14.44 (b)]. Both the SS and the DIBL value in JMTs increases almost linearly with the decrease of the sidewall angle. However, the short channel immunity of JMTs can be enhanced by the stronger fringe capacitance applied on the source/drain region through the spacers

Fig. 14.45 shows the electrical field distribution along z direction for explaining the variation suppression mechanism by high-k spacers, and the value is extracted under the same electron density with the minimum in the x-y cross section. Due to the volume depletion in JMTs, the lower electrical field represents the stronger depletion and the better gate control of the channel under the same electron density. As the sidewall angle Θ decreases from 90° to 78.7°, the electrical field increases, indicating the reduced gate control. As the high-k spacer is introduced, the electrical field is effectively reduced with the broader depletion region, which explores more superior subthreshold characteristics.

14.5. Conclusions

In this Chapter, the reliability study on the nanoscale traditional and non-traditional CMOS devices is discussed. With the presented extraction methods of interface traps at the SiO_2/Si interface, the reliability issues about FN effect, HCI effect and NBTI effect are studied and the corresponding models for predicting the degradation of devices and circuits are provided. Finally, the statistical variation in the nano-scale novel devices such as FinFET or surrounding gate FET is investigated, including the influences of random dopant fluctuation effect, line edge roughness effect and sidewall angle variation effect.

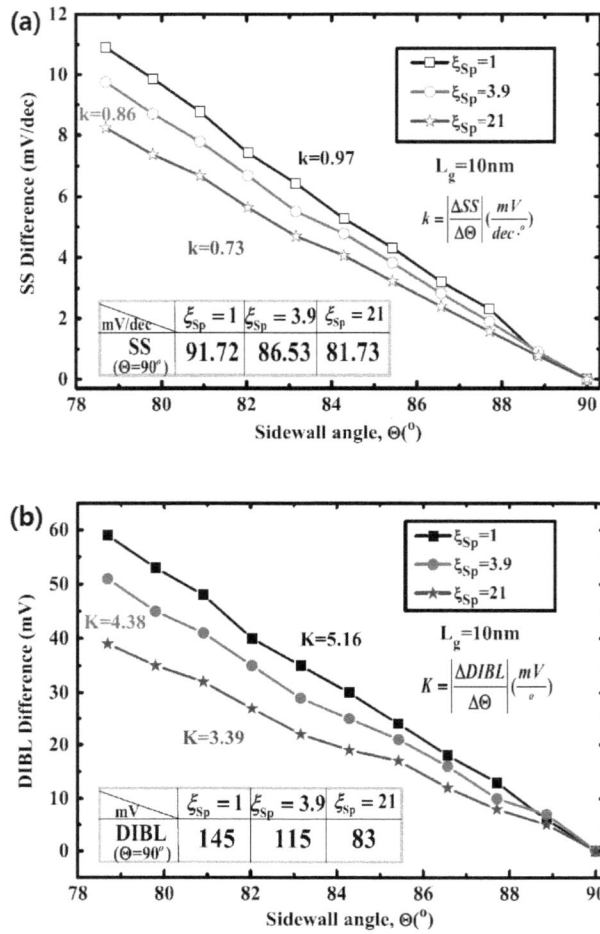

Fig. 14.44. The dependence of SS difference and DIBL difference on the sidewall angle [62].

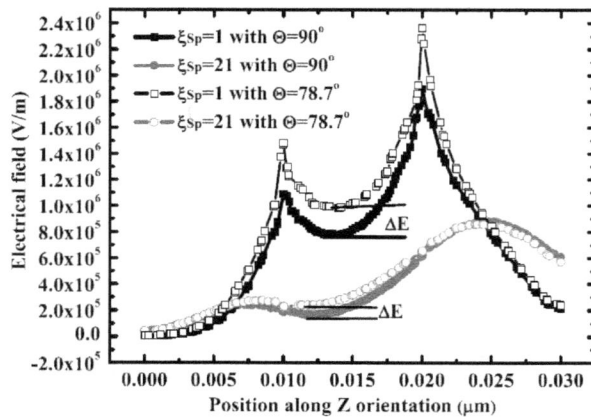

Fig. 14.45. Electrical field distribution along z direction [62].

Acknowledgements

This work is funded by National Natural Science Foundation of China under Grant (61574005), by Fundamental Research Project of Shenzhen Sci. & Tech. Fund (JCYJ20160329161334453, JCYJ20170412153812353, JSGG20170414140411874, Graphene based device and circuit SPICE model for chip applications, Research on key issues of millimeter-wave 5G wireless transmission, Fundamental research on Space Division Multiplexing (SDM) optical switching network for data center optical interconnection). It is also supported by IER Funding of PKU-HKUST Shenzhen-Hong Kong Institution.

References

[1]. E. Nicollian, G. Goetzberger, The Si-SiO$_2$ interface-electrical properties as determined by the metal-insulator-silicon conductance technique, *Bell Syst. Tech. J.*, Vol. 46, Issue 6, 1967, pp. 1055-1066.

[2]. A. Pacelli, A. L. Lacita, S. Villa, Reliable extraction of MOS interface traps from low frequency CV measurements, *IEEE Electron. Dev. Lett.*, Vol. 19, Issue 5, 1998, pp. 148-150.

[3]. C. H. Ling, S. E. Tan, D. S. Ang, A study of hot carrier degradation in NMOSFET's by gate capacitance and charge pumping current, *IEEE Trans. Electron Devices*, Vol. 42, Issue 7, 1995, pp. 1321-1328.

[4]. T. Wang, L.-P. Chiang, N.-K. Zous, T.-E. Chang, C. Huang, Characterization of various stress-induced oxide traps in MOSFET's by using a sub-threshold transient current technique, *IEEE Trans. Electron Devices*, Vol. 45, Issue 8, 1998, pp. 1791-1796.

[5]. C.-H. Wang, A. Neugroschel, New technique for life and surface or interface recombination velocity measurement in thin semiconductor layer, *IEEE Trans. Electron Devices*, Vol. 39, Issue 3, 1992, pp. 662-670.

[6]. J. He, X. Zhang, R. Huang, Y. Wang, Linear cofactor difference method of MOSFET subthreshold characteristics for extracting interface traps induced by gate oxide stress test, *IEEE Trans. Electron Devices*, Vol. 49, Issue 2, 2002, pp. 331-334.

[7]. T. Ernst, S. Cristoloveanu, A. Vandooren, T. Rudenko, J.-P. Colinge, Recombination current modeling and carrier lifetime determination in dual-gated fully-depleted SOI devices, *IEEE Trans. Electron Devices*, Vol. 46, Issue 7, 1999, pp. 1503-1509.

[8]. J. He, X. Zhang, R. Huang, Y. Wang, Application of forward gated-diode R–G current method in extracting F–N stress-induced interface traps in SOI NMOSFETs, *Microelectron. Reliab.*, Vol. 42, 2002, pp. 145-148.

[9]. J. He, X. Zhang, R. Huang, and Y. Wang, A refined forward gated-diode method for separating front channel hot-carrier-stress induced front and back gate interface and oxide traps in SOI NMOSFETs, *Semicond. Sci. Technol.*, Vol. 17, 2002, pp. 487-492.

[10]. J. He, X. Zhang, R. Huang, Y. Wang, Extraction of the lateral distribution of interface traps in MOSFETs by a novel combined gated-diode technique, *Microelectron. Reliab.*, Vol. 41, 2001, pp. 11953-11957.

[11]. J. He, X. Zhang, A.-H. Huang, R. Huang, A novel experimental technique: combined gated-diode method for extracting lateral distribution of interface traps in SOI NMOSFETs, *Solid-State Electron.*, Vol. 45, 2001, pp. 1107-1113.

[12]. A. E. Islam, H. Kufluoglu, D. Varghese, S. Mahapatra, M. A. Alam, Recent issues in negative-bias temperature instability: initial degredation, field dependence of interface trap generation, hole trapping effects, and relaxation, *IEEE Trans. Electron Devices*, Vol. 54, Issue 9, 2007, pp. 2143-2154.

[13]. J. B. Yang, T. P. Chen, S. S. Tan, L. Chan, Analytical reaction-diffusion model and the modeling of nitrogen-enhanced negative bias temperature instability, *Appl. Phys. Lett.*, Vol. 88, 2006, pp. 172109-1 – 172109-3.

[14]. S. Mahapatra, M. A. Alam, Defect generation in p-MOSFETs under negative-bias stress: an experimental perspective, *IEEE Trans. Device Mater. Reliab.*, Vol. 8, Issue 1, 2008, pp. 35-46.

[15]. V. Huard, M. Denais, C. Parthasarath, NBTI degradation: from physical mechanisms to modeling, *Microelectron. Reliab.*, Vol. 46, Issue 1, 2006, pp. 1-23.

[16]. M. A. Alam, S. Mahapatra, A comprehensive model of PMOS NBTI degradation, *Microelectron. Reliab.*, Vol. 45, Issue 1, 2005, pp. 71-81.

[17]. S. V. Kumar, C. H. Kim, S. S. Sapatnekar, A finite-oxide thickness-based analytical model for negative bias temperature instability, *IEEE Trans. Devices Mater. Reliab.*, Vol. 9, 2009, pp. 537-556.

[18]. R. Fernandez, B. Kaczer, A. Nackaerts, S. Demuynck, R. Rodrıguez, M. Nafria, G. Groeseneken, AC NBTI studied in the 1 Hz - 2 GHz range on dedicated on-chip CMOS circuits, in *Proceedings of the International Electron Devices Meeting (IEDM'06)*, San Francisco, USA, 11-13 December 2006, pp. s12p4.

[19]. C. Ma, H. J. Mattausch, M. Miyake, K. Matsuzawa, T. Lizuka, S. Yamaguchi, T. Hoshida, A. Kinoshita, T. Arakawa, J. He, M. M.-Mattausch, Unified reaction-diffusion model for accurate prediction of negative bias temperature instability effect, *Jpn. J. Appl. Phys.*, Vol. 51, 2012, pp. 02BC07-1 - 02BC07-5.

[20]. D. Hisamto, W.-C. Lee, J. Kedzierski, H. Takeuchi, K. Asano, C. Kuo, E. Andeson, T.-J. King, J. Bokor, C. Hu, FinFET-a self-aligned double-gate MOSFET scalable to 20 nm, *IEEE Trans. Electron Devices*, Vol. 47, Issue 12, 2000, pp. 2320-2325.

[21]. J.-P. Colinge, C.-W. Lee, A. Afzalian, N. D. Akhavan, R. Yan, I. Ferain, P. Razavi, B. O'Neill, A. Blake, M. White, A.-M. Kellehe, B. McCathy, R. Murphy, Nanowie transistors without junctions, *Nature Nanotechnol.*, Vol. 5, 2010, pp. 225-229.

[22]. C. W. Lee, I. Ferain, A. Afzalian, R. Yan, N. D. Akhavan, P. Razavi, J. P. Colinge, Performance estimation of junctionless multigate transistors, *Solid-State Electron.*, Vol. 54, 2010, pp. 97-103.

[23]. S.-Y. Kim, J. H. Lee, Hot carrier-induced degradation in bulk FinFETs, *IEEE Electron. Dev. Lett.*, Vol. 26, Issue 8, 2005, pp. 566-568.

[24]. W.-S. Liao, S.-S. Chen, S. Chiang, W.-T. Shiau, Reliability investigation upon 30 nm gate length ultra-high aspect ratio FinFETs, in *Proceedings of the IEEE 43rd Annual Int. Reliability Physics Symposium (IRPS'05)*, San Jose, USA, 17-21 April 2005, pp. 541-544.

[25]. G. Kapila, B. Kaczer, A. Nackaets, N. Collaert, G. V. Groeseneken, Direct measurement of top and sidewall interface trap density in SOI FinFETs, *IEEE Electron. Dev. Lett.*, Vol. 28, Issue 3, 2007, pp. 232-234.

[26]. S. J. Choi, D. I. Moon, S. Kim, J. P. Duarte, Y. K. Choi, Sensitivity of threshold voltage to nanowire width variation in junctionless transistors, *IEEE Electron. Dev. Lett.*, Vol. 32, Issue 2, 2011, pp. 125-127.

[27]. S. Gundapaneni, M. Bajaj, R. K. Pandey, K. V. R. M. Murali, S. Ganguly, A. Kottantharayil, Effect of band-to-band tunneling on junctionless transistors, *IEEE Trans. Electron Devices*, Vol. 59, Issue 4, 2012, pp. 1023-1029.

[28]. M.-H. Chiang, J.-N. Lin, K. Kim, C.-T. Chuang, Random dopant fluctuation in limited-width FinFET technologies, *IEEE Trans. Electron Devices*, Vol. 54, Issue 8, 2007, pp. 2055-2060.

[29]. E. Baravelli, M. Jurczak, N. Speciale, K. De Meyer, A. Dixit, Impact of LER and random dopant fluctuations on FinFET matching performance, *IEEE Trans. Nanotechnol.*, Vol. 7, Issue 3, 2008, pp. 291-298.

[30]. C.-H. Park, M.-D. Ko, K.-H. Kim, S.-H. Lee, J.-S. Yoon, J.-S. Lee, Y.-H. Jeong, Investigation of low-frequency noise behavior after hot-carrier stress in an n-channel junctionless nanowire MOSFET, *IEEE Electron Device Lett.*, Vol. 33, Issue 11, 2012, pp. 1538-1540.

[31]. C.-H. Park, M.-D. Ko, K.-H. Kim, R.-H. Baek, C.-W. Sohn, C. K. Baek, S. Park, M. J. Deen, Y.-H. Jeong, J.-S. Lee, Electrical characteristics of 20-nm junctionless Si nanowire transistors, *Solid-State Electron.*, Vol. 73, 2012, pp. 7-10.

[32]. E. Kameda, T. Matsuda, M. Yasuda, and T. Ohzone, Interface state density in n-MOSFETs with Si-implanted gate oxide measured by subthreshold slope analysis, *Solid-State Electron.*, Vol. 43, 1999, pp. 565-573.

[33]. Z. Lun, D. S. Ang, C. H. Ling, A novel subthreshold slope technique for the extraction of the buried-oxide interface trap density in fully depleted SOI MOSFET, *IEEE Electron Device Lett.*, Vol. 21, Issue 8, 2000, pp. 411-413.

[34]. N. Arora, MOSFET Models for VLSI Circuit Simulation-Theory and Practice, *Springer-Verlag*, 1993.

[35]. C. Hu, S. C. Tam, F.-C. Hsu, P.-K. Ko, T.-Y. Chan, K. W. Terrill, Hot-electron-induced MOSFET degradation-model, monitor, and improvement, *IEEE J. Solid-State Circuit.*, Vol. 20, Issue 1, 1985, pp. 295-305.

[36]. B. S. Doyle, K. R. Mistry, J. Faricelli, Examination of the time power law dependencies in hot carrier stressing of n-MOS transistors, *IEEE Electron Device Lett.*, Vol. 18, Issue 2, 1997, pp. 51-53.

[37]. Y. Li, T.-P. Ma, A front-gate charge-pumping method for probing both interfaces in SOI devices, *IEEE Trans. Electron Devices*, Vol. 45, Issue 6, 1998, pp. 1329-1335.

[38]. A. E. Islam, H. Kufluoglu, D. Varghese, M. A. Alam, Critical analysis of short-term negative bias temperature instability measurements: explaining the effect of time-zero delay for on-the-fly measurements, *Appl. Phys. Lett.*, Vol. 90, 2007, pp. 083505-1 – 083505-3.

[39]. V. Huard, C. R. Parthasarathy, C. Guerin, M. Denais, Physical modeling of negative bias temperature instabilities for predictive extrapolation, in *Proceedings of the IEEE Int. Reliability Physics Symposium (IRPS'06)*, San Jose, USA, 26-30 March 2006, pp. 733-734.

[40]. H. Reisinger, O. Blank, W. Heinrigs, A. Muhlhoff, W. Gustin, C. Schlunder, Analysis of NBTI degradation- and recovery- behavior based on ultra fast VT-measurements, in *Proceedings of the IEEE Int. Reliability Physics Symposium (IRPS'06)*, San Jose, USA, 26-30 March 2006, pp. 448-453.

[41]. N. Sadachika, D. Kitamaru, Y. Uetsuji, D. Navarro, M. M. Yusoff, T. Ezaki, H. J. Mattausch, M. Miura-Mattausch, Completely surface-potential-based compact model of the fully depleted SOI-MOSFET including short-channel effects, *IEEE Trans. Electron Devices*, Vol. 53, Issue 9, 2006, pp. 2017-2024.

[42]. H. Kufluoglu, M. A. Alam, A generalized reaction-diffusion model with explicit H-H_2 dynamics for negative-bias temperature-instability (NBTI) degradation, *IEEE Trans. Electron Devices*, Vol. 54, Issue 5, 2007, pp. 1101-1107.

[43]. C. Shen, M.-F. Li, C. E. Foo, T. Yang, D. M. Huang, A. Yap, G. S. Samudra, Y.-C. Yeo, Characterization and physical origin of fast V_{th} transient in NBTI of pMOSFETs with SiON dielectric, in *Proceedings of the International Electron Devices Meeting (IEDM'06)*, San Francisco, USA, 11-13 December 2006, pp. s12p6.

[44]. Z. Q. Teo, D. S. Ang, K. S. See, Can the reaction-diffusion model explain generation and recovery of interface states contributing to NBTI? In *Proceedings of the International Electron Devices Meeting (IEDM'09)*, Baltimore, USA, 7-9 December 2009, pp.737-740.

[45]. C. Ma, B. Li, Y. Wei, L. Zhang, J. He, X. Zhang, X. Lin, M. Chan, FinFET reliability study by forward gated-diode generation-recombination current, *Semicond. Sci. Technol.*, Vol. 23, 2008, pp. 075008-1 - 075008-8.

[46]. C. Ma, H. Wang, C. Zhang, X. Zhang, J. He, X. Zhang, Temperature dependence of the interface state distribution due to hot carrier effect in FinFET device, *Microelectron. Reliab.*, Vol. 50, 2010, pp. 1077-1080.

[47]. C. Ma, L. Zhang, C. Zhang, X. Zhang, J. He, X. Zhang, A physical based model to predict performance degradation of FinFET accounting for interface state distribution effect due to hot carrier injection, *Microelectron. Reliab.*, Vol. 51, 2011, pp. 337-341.

[48]. H. Kang, J.-W. Han, Y.-K. Choi, Analytical threshold voltage model for double-gate MOSFETs with localized charges, *IEEE Electron Device Lett.*, Vol. 29, Issue 8, 2005, pp. 927-930.

[49]. L. Zhang, J. He, F. Liu, J. Zhang, Y. Song, A unified charge-based model for symmetric DG MOSFETs valid for both heavily doped body and undoped channel, in *Proceedings of the International Conference Mixed Design of Integrated Circuits & Systems (MIXDES'08)*, Poznan, Poland, 19-21 June 2008, pp. 367-372

[50]. P. Ratnam, C. A. T. Salama, A new approach to the modeling of nonuniformly doped shot-channel MOSFETs, *IEEE Trans. Electron Devices*, Vol. 31, Issue 9, 1984, pp. 1289-1298.

[51]. C. Ma, B. Li, H. Wang, X. Zhang, J. He, Compact negative bias temperature instability model for silicon nanowire MOSFET (SNWT) and application in circuit performance simulation, *J. Comput. Theor. Nanosci.*, Vol. 7, 2010, pp. 107-114.

[52]. H. Lou, D, Li, Y. Dong, X. Lin, J. He, S. Yang, M. Chan, Suppression of tunneling leakage current in junctionless nanowire transistors, *Semicond. Sci. Technol.*, Vol. 28, Issue 12, 2013, pp. 5016-5021.

[53]. TCAD Sentaurus Device Users Manual, *Synopsys*, Mountain View, CA, 2010.

[54]. H. Wang, C. Ma, C. Zhang, F. He, X. Zhang, X. Lin, Impact of random dopant fluctuation effect on surrounding gate MOSFETs: from atomic level simulation to circuit performance evaluation, in *Proceedings of the Int. Nanoelectron. Conference (INEC'10)*, Hong Kong, China, 3-8 January 2010, pp. 1136-1137.

[55]. D. J. Frank, Y. Taur, M. Ieong, H.-S. P. Wong, Monte Carlo modeling of threshold variation due to dopant fluctuations, in *Proceedings of the Symposium on VLSI Technology*, Kyoto, Japan, 14-16 June 1999, pp. 169-170.

[56]. V. Vidya, Thin-body silicon FET devices and technology, PhD Thesis, *University of California*, Berkeley, 2007.

[57]. N. Sano, K. Matsuzawa, M. Mukai, N. Nakayama, Role of long-range and short-range Coulomb potentials in threshold characteristics under discrete dopants in sub-0.1 um Si-MOSFETs, in *Proceedings of the International Electron Devices Meeting (IEDM'2000)*, San Fransico, USA, 10-13 December 2000, pp.275-278.

[58]. J. Yang, J. He, F. Liu, L. Zhang, X. Zhang, M. Chan, A compact model of silicon-based nanowire MOSFETs for circuit simulation and design, *IEEE Trans. Electron Devices*, Vol. 55, Issue 11, 2008, pp. 2898-2906.

[59]. T. Linton, M. Chandhok, B. J. Rice, G. Schrom, Determination of the line edge roughness specification for 34 nm devices, in *Proceedings of the International Electron Devices Meeting (IEDM'02)*, San Francisco, USA, 8-11 December 2002, pp. 303-306.

[60]. X. Guo, S. Wang, C. Ma, C. Zhang, X. Lin, W. Wu, F. He, W. Wang, Z. Liu, W. Zhao, S. Yang, A novel approach to simulate fin-width line edge roughnesss effect of FinFET performance, in *Proceedings of the International Conference Electron Dev. Solid-State Circ. (EDSSC'10)*, Hong Kong, China, 15-17 December 2010, pp. 1-4.

[61]. H. Lou, D. Li, Y. Dong, X. Lin, S. Yang, J. He, M. Chan, Effect of fin sidewall angle on subthreshold characteristics of junctionless multigate transistors, *Jpn. J. Appl. Phys.*, Vol. 52, 2013, pp. 104302 .

[62]. H. Lou, B. Zhang, D. Li, X. Lin, J. He, M. Chan, Suppression of subthreshold characteristics variation for junctionless multigate transistors using high-k spaces, *Semicond. Sci. Technol.*, Vol. 30, 2015, pp. 015008.

Chapter 15

Adaptive Routing for Fault Tolerance and Congestion Avoidance for 2D Mesh and Torus NoCs in Many-Core Systems-on-Chip

Mounir Benabdenbi, Lorena Anghel, Michael Dimopoulos and Yi Gang

15.1. Introduction

The quest for higher-performance and low-power consumption has driven the microelectronics' industry race towards aggressive technology scaling and multicore chip designs. We are in the many-core era and chips like TILE-Gx72 [1] by Tilera, with 72 cores or the Intel Xeon Phi™ coprocessor 5110P [2] with 60 cores are just some examples that show the technology trend. All these chips are highly complex homogeneous and heterogeneous systems on chip and need a fast and scalable infrastructure for supporting their communication demands. The Networks-on-chip (NoCs) offer a promising scalable and modular solution to the increasing demands for higher speed and lower power.

According to the International Technology Roadmap for Semiconductors (ITRS) [3], as we move towards the post-CMOS era with the continuous downscaling of the feature sizes, the lowering of power supply voltages and the increasing of operating frequencies, the reliability of circuits is threatened [4-6] by the increased process-voltage and frequency variations. In these environments, the designs should be able to provide either full functionality (e.g. full critical systems), or degraded one in the case of consumer applications even in the presence of high failure rates. To accomplish this, the systems should be able to adapt to manufacturing and runtime failures and continue their functioning. A holistic solution would be the most efficient way to cope with reliability

Mounir Benabdenbi
University Grenoble Alpes, TIMA, F-38031
Grenoble, France

issues mentioned before [4] which will be encountered in post-CMOS era. However, such an effective solution has yet to come.

Error resilience in on-chip networks is addressed in this work. Although, the presented solution tackles the problem at the network level, it is orthogonal to other developed solutions at other levels (e.g. link-level, end to end recovery) [7, 8] and can work complementarily to them.

The proposed solution works by exploiting local information about the state of links and routers. It consists of an adaptive fault tolerant routing algorithm named CAFTA (Congestion Aware Fault Tolerant Algorithm) enhanced with neighbor fault-aware knowledge. In order to improve transmission latencies and to effectively guide the routing decisions towards load-balanced network traffic, a new metric is introduced. In the presence of runtime errors, packet retransmission combined with novel message recovery mechanisms are utilized in order to provide fault tolerance under high failure rates. Extensive simulation experiments under various fault types (permanent, transient, intermittent) commonly encountered in advanced technology nodes demonstrate the effectiveness of the proposed fault tolerant scheme.

This Chapter is organized as follows. Section 15.2 describes the related work. The proposed fault tolerant routing is detailed in Section 15.3. The experimental setup and various results for the performance of the proposed method are presented in Section 15.4. Finally, Section 15.5 concludes the Chapter.

15.2. Related Work

In the literature, there is a wealth of work regarding fault tolerance on Networks on Chip. The focus here is on methods that deal with routing in the presence of link/router failures. So, the algorithms in literature can be classified according to the targeted type of fault: permanent [8, 10, 12, 15-17, 25, 27], transient [8, 13, 17, 27], intermittent [10] and according to the number of faults they can sustain [11, 14, 18]. Some algorithms use virtual channels to avoid deadlock conditions [10, 12, 15, 16, 20], others avoid to use virtual channels and use turn restrictions [11, 14, 19, 24]. Another classification is based on the used routing function [9]: there are deterministic algorithms that always make their routing decisions without considering the state of the network, and adaptive algorithms that use information about network traffic and/or channel status to avoid congested or faulty regions of the network.

There are also some recent proposals [11, 14, 18] that work on-chip and tackle the problem of unconstrained faults by reconfiguring the network in the presence of faults. However, in [11] fault tolerant and deadlock-free routing is not guaranteed.

The authors in [13] present a fault tolerant mechanism for soft error handling. Their routing algorithm is named DOR XY [9] and it cannot handle cases where propagation path is blocked due to a fault affecting the link of the output port. They propose the usage of packet fragmentation in cases where a virtual channel is faulty as a mean to release occupancy of network resources.

The authors in [12] propose an adaptive fault tolerant algorithm that does not use routing tables. In [15] a fault tolerant algorithm FTCAR is presented with congestion awareness that is able to tolerate partial permanent faulty link routers.

In [25] the authors present an adaptive algorithm called MiCoF that works on shortest paths. In order to achieve this, the architecture of the routers is modified so as to include extra wires that connect each router with its neighbor routers. These wires are used as by-pass links when the router fails and therefore the connectivity of the network remains intact. BFT-NoC [27] provides dynamically reconfigured bidirectional links with the purpose to retain the connectivity between the routers in the case of link failures. For this purpose, the remaining healthy unidirectional links between the routers are configured so as to support bidirectional communication (transmission and reception).

In the presence of faults, the communication fabric changes from regular to irregular. So, in order to cope with dynamic (runtime) failures, an adaptive routing algorithm has been selected here. It is deadlock-free, livelock-free, and does not use routing tables. In comparison to the works of [12, 15, 16] the proposed algorithm uses a new congestion-avoidance metric and, unlike the others, it also provides a novel error-recovery mechanism to cope with runtime failures.

It should be noted that a preliminary version of CAFTA was briefly presented in [18]. However, the computation of the congestion metric (*FR* metric) was not included and no comparative results were provided to assess its effectiveness. In addition, the error recovery mechanism presented in [18] is generalized here to support two different recovery schemes that will be described later on (Section 15.3.5) namely *SMR (Split Message Recovery)* and *BMR (Bypass Message Recovery)*. Results for the SMR scheme will be presented in the experiments (Section 15.4) whereas in [18] the *BMR* scheme has been used.

Most of the relative works presented in literature focuses mainly on permanent faults (some consider only router faults others only link faults) and only a few of them study the influence of permanent and transient [8] or permanent and intermittent [10]. In advanced technology nodes where it is expected to encounter an unconstrained number of various fault types (permanent, transient, intermittent), methods and schemes are needed to cope efficiently with this situation. In this respect, this work and the work presented in [18] represent, to the best of the authors' knowledge, the first attempts to provide some insight about the efficiency of a fault tolerant routing algorithm under the simultaneous occurrence of an unconstrained number of different types of faults: permanent, transient and intermittent.

15.3. Fault Tolerant Routing with CAFTA

In the following subsections the proposed fault tolerant routing algorithm CAFTA will be presented. It uses one-hop fault look ahead to find valid paths through faulty areas (see Section 15.3.2). It also exploits congestion-awareness (Section 15.3.4) to efficiently route packets. In order to provide fault-tolerance under runtime failures (temporary and permanent) it includes packet-retransmission and a new recovery scheme (Section 15.3.5). As CAFTA is firstly dedicated to NoC with 2D Mesh topology, an extension is proposed for 2D Torus topology in Section 15.3.6.

15.3.1. Router Architecture

The basis for our router architecture, depicted in block form in Fig. 15.1a, is an input-buffered router with credit-based wormhole flow control. A typical 5-stage pipeline router architecture is used. The five stages are: Buffer Write (BW), Route Computation (RC), Virtual Channel Allocation (VA), Switch allocation (SA) and finally the link traversal (LT). These functions are combined into 4 stages: (BW/RC), (VA/SA), (ST), (LT). The router contains also:

- The fault management unit which is responsible for the propagation (reception) of fault status signals to (from) the neighbor routers (fault awareness information). This information is used by the RC logic to dismiss possible output port candidates after being invalidated due to the presence of faults.

- The congestion management unit that performs the computation of a metric that grades the network state with respect to the communicated traffic. Again, this information is employed as a secondary criterion to rank the possible output port candidates which have been selected by the RC logic.

- The recovery unit consisting of the message retransmission logic and the recovery logic employed in each virtual channel.

15.3.2. Fault Model and Assumptions

The routers in a NoC are connected through pairs of unidirectional links (bi-directional connection) and each unidirectional link is composed of a group of parallel wires whose number is equal to the size of the phit (the smallest unit of information that travels on the NOC links). The used fault model is a coarse-grain fault model under which a whole unidirectional link could be faulty. Partially faulty links where only a partial set of link wires could be faulty will not be considered in this study. For a whole link to fail, faults on both of its constituent's unidirectional links should be present. Under this fault model, a faulty node is considered as a node where all of its links are faulty. Also, a fault in a unidirectional link can be seen as if the output port of the upstream router and the input port of the downstream router, which are connected through this link, are faulty. So with this model we may also model faults in router's ports. Faults can be permanent (either pre-existing due to manufacturing/design errors or runtime due to wear-out, aging effects, etc.), transient (due to soft errors, crosstalk, etc.), and intermittent [6].

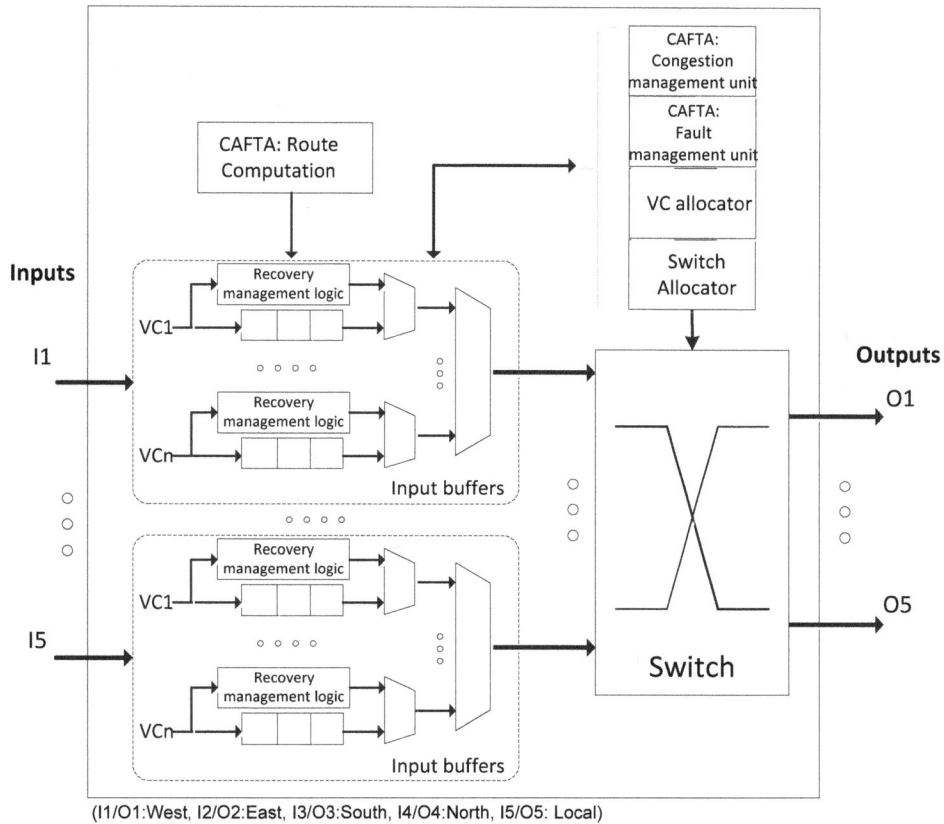

(I1/O1:West, I2/O2:East, I3/O3:South, I4/O4:North, I5/O5: Local)

Fig. 15.1 (a). The proposed router architecture.

Although our focus is not on the fault detection mechanisms, in our fault tolerant scheme the following assumptions are made for fault tolerant purposes:

- In the presence of permanent static faults, each source-destination pair of healthy nodes is reachable; i.e. there exists at least one path that connects the given source with the given destination. This assumption is necessary in order to be able to evaluate a routing algorithm. However, after a certain number of faults, there is high probability that the network becomes partitioned (disjoint clusters of healthy nodes) and there can be no packet transfer between different partitions. In such cases, the operating system selects a partition of healthy nodes to assign the application execution, performs the thread migration and marks the other partitions as unreachable. Solutions for recovering from such cases with dedicated emergency links have been proposed in the literature [26-28].

- The messages are protected by ECC [30]. In case of error, flit retransmission occurs from the upstream router.

- Software or hardware self-checking can be used to detect faults at processing elements (PEs), routers, and links [4]. For example, Software-Based Self-Test (SBST) methods is one of the option to be used for detecting errors at PEs. Permanent faults may be detected by self-tests, whereas for temporary faults on links, routers, and PEs a low-cost double sampling technique like GRAAL [4] may be employed. For routers, a failure detection mechanism like heart beats [31] may be utilised. According to this, each router exchanges heart beat messages with its neighbors. If a particular router fails to send these heart beats, then it is considered by its neighbors as being faulty.

- Once a router becomes faulty it stops providing services to the network (is considered as a fail-silent node [32]).

- All the healthy neighbors of a defective router are informed of its status and are reconfigured so that they inhibit the traffic towards this router (e.g. they deactivate the output port leading to this router). In the case where the defective router is the destination for some message, the transmission will fail. If the defective router is an intermediate node in the path set-up by the header flit then the flits that are already in the buffers of the faulty router are lost and the source will need to re-send the message. Generally, if the defective router is partially functioning one might use design techniques such as [26-28] to recover the data from this router otherwise one might use higher level techniques such as checkpointing and rollback recovery [4] to increase the fault tolerance of the system.

- The end routers of a defective link are informed of its status and they are reconfigured in the case of a permanent failure so that they prohibit the traffic through this link (e.g. the upstream router disables the output port connected to this link whereas the downstream router disables its respective input port).

It should be noted that, during the time period between the detection of a fault, the fault-status propagation and (eventually) the network reconfiguration, there are packets in transit and the routing paths selected by the routing computation logic may be no longer valid. This situation might create message dependencies between old packets (before the fault) and new packets (after the fault) which, generally, might lead to deadlock. The techniques proposed in Section 15.3.5 help to resolve from such situations.

Each router maintains the health status of its links and of the links of its neighbor routers (routers at a distance one hop away) in a fault status register similar to the work presented in [17], and contrary to [17] where the link failures are assumed to be bidirectional, the link failures here may be unidirectional. In addition, two bits are used for the status of a link {00 (no fault), 11 (permanent fault), 01 (transient fault)}. Dedicated wires are utilised (Fig. 15.1 b) to support the mechanism of propagating the fault status information between the routers of a NoC. To facilitate the understanding of this mechanism, let us consider the router $T_{i,j}$ in Fig. 15.1 b. This router keeps information about the health status of all

410

the 16 numbered links. Therefore, the size of the fault status register is 64 bits (= 32 unidirectional links x 2-bits for link status). Router $T_{i,j}$ transmits (receives) the fault status information to (from) all its neighbor routers. For example, let us consider fault status propagation between the routers $T_{i,j}$ and $T_{i,j+1}$ in Fig. 15.1 b: Router $T_{i,j}$ sends to router $T_{i,j+1}$ seven bits composed by: one bit (1 for faulty, 0 for fault free) for the status of its outgoing link (link 4 in Fig. 15.1 b) whereas the remaining six bits (three 2-bit pairs) correspond to the fault status information regarding the pairs of links 1, 2, 3. Quite analogously, router $T_{i,j}$ receives seven bits of information from router $T_{i,j+1}$ which amount to: one bit for the status of its incoming link 4 (= outgoing link for router $T_{i,j+1}$) and the 6 bits cover the fault status of the links 14, 15, 8. To discriminate between permanent and temporary failures and to avoid misdiagnosing a temporary failure as permanent, threshold values [17, 29] are kept for the status of the links. For example, a threshold value of k means that a link will be declared as being permanent faulty after having been found faulty for $k+1$ times. For the permanent faulty links, the network is reconfigured as it was described earlier in order to isolate the defective link.

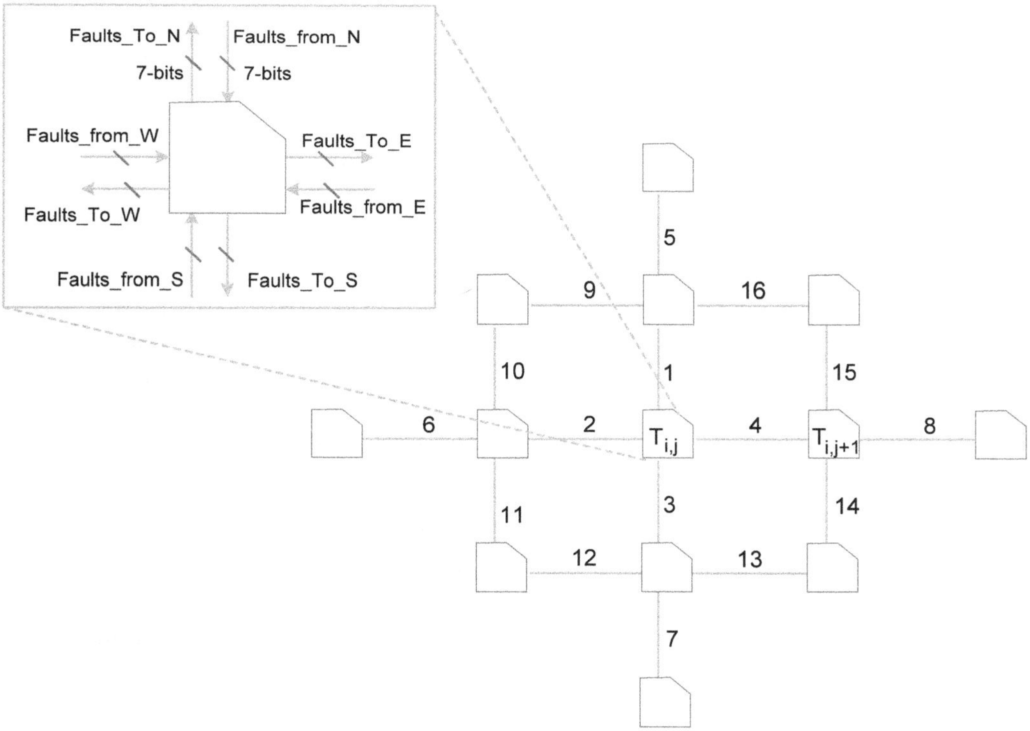

Fig. 15.1(b). The region of fault-awareness of a router $T_{i,j}$ and the fault propagation mechanism.

15.3.3. Routing Algorithm

Generally, any routing algorithm may benefit from the proposed improvements presented later on in Sections 15.4 and 15.5. The routing algorithm employed here as a case study, is a version of the algorithm presented in [12], named Variant B. It is an adaptive fault-tolerant and deadlock-free routing algorithm developed for 2D mesh NoCs that does not use routing tables. It combines North-Last and South-Last adaptive routing algorithms which use turn restrictions to avoid deadlocks [19]. Two separate virtual networks (VN) are utilized, one for North-Last and another for South-Last routing. Four virtual channels (VCs) are used per port and two of them are assigned for each VN. The selection of which VN to use, depends on the relative position of the destination with respect to the source. If the destination lies in the north (south), the South-Last (North-Last) VN is used.

Let us suppose that a message is sent from a source node S to a destination D using only routing algorithm Variant B and a runtime fault corrupts the message. We discriminate between the following cases:

- The header flit had reached the destination before the runtime fault, therefore D, after a timeout is asserted, will signal the message corruption to S by the return of negative acknowledgment (NACK).

- The header flit has not yet reached the destination. In this situation the NACK message may be delayed further depending also on the network traffic status (no congestion metric is used resulting in increased latencies).

- The header flit becomes corrupted or the message path is interrupted due to an intermediate link/node runtime failure. This case will lead to message loss and transmission failure and neither the source nor the destination are aware of this situation.

From the above discussion it becomes obvious that the original Variant B algorithm as well as its improved version Variant C with echo mode [12], are not able to handle efficiently or even they may fail in the presence of runtime (dynamic) faults.

To cope with more ambitious fault tolerant goals and overcome Variant B limitations, the following additions/extensions have been introduced to the basic Variant B:

- Neighbor fault awareness: Each router knows the health status of its own links and of the links of its neighboring routers [17] (routers at a distance of one hop away).

- Network traffic regulation by the use of a new congestion-avoidance metric (Section 15.3.4).

- Packet-retransmission combined with a new message recovery scheme (Section 15.3.5).

The proposed algorithm CAFTA is presented in pseudo code in Fig. 15.2. Being based on Variant B, CAFTA is also proven being deadlock-free and always terminates.

Routing Algorithm(x,D,M):

For a given source x and destination D
VN_i : the is the virtual network currently in use by the message M
x : the current node, *w* is the next node
Initial:
w= $\varnothing$
Q : the list of available output nodes // with Neighbor fault awareness

If x=D then eject to local and return *success* //destination is reached!
else begin
If $(Q \neq \varnothing)$ *then begin*

  If (sizeof(Q) = 1) // only one available choice
then set w= $Q_i \in Q$.

   else begin // more than one available choices
        set w= $Q_i' \in Q$, where Q_i' has minimal distance. To break possible ties for the
        cases where all $Q_i' \in Q$, have either minimal or non-minimal distances, use as
        a secondary ranking the *FR value* (preference to smaller values) // section 3.4.
   end
end
  If w ≠ $\varnothing$ then begin // Route message to node *w*
   If w is not in VN_i *then begin*
        eject M to *Network Interface Controller* (NIC) of local router
        re-inject M to network and try to route to *w*.
   end
   If w is blocked by a fault *then begin//runtime* faults
   try *retransmission* and *if it fails try either SMR or BMR* // section 3.5
   If w is still blocked then return *failure* // destination *unreachable*!
   end
  end
  else return *failure* // destination *unreachable*!.
end

Fig. 15.2. The CAFTA algorithm in pseudo-code form.

15.3.4. Network Traffic Regulation with the FR Metric

Various metrics have been presented in the literature on how to dynamically regulate the network traffic flow in order to have improved performance (low latency). Thus, there are metrics which make use of local information, like the number of free VCs [20], the buffer availability of adjacent nodes [21], and metrics like RCA [22], DBAR [23] which utilize regional information. These metrics have been proposed and used mainly in a fault-free environment. However, their usefulness can be exploited to guide the selection of propagation paths also in faulty environments. In a faulty environment, the primary goal for a fault-tolerant routing algorithm is the careful selection of fault-free paths towards the destination. The congestion metric may be only used as a secondary criterion for path selection in case of several available outputs available.

For an online fault tolerant algorithm like CAFTA, which relies on local (limited region) information, the safe region of applicability of a congestion metric is the region of fault awareness. It should be noted that any of the proposed congestion metrics in the literature may be used to guide a fault tolerant routing algorithm provided that the region of fault awareness is taken into consideration.

Here, a simple congestion metric is introduced, namely Flits-remain (FR). In fact, in a wormhole-based flow control, the header flit of a packet defines the path while it traverses the network from the source towards the destination, followed by the packet data (payload). This path remains active until the tail flit of the packet passes and releases the resources. A measure of occupancy of the resources (buffers, VCs) in a router is the number of flits that remain to pass through (*flits-remain* or FR). Usually, the header flit carries the size of the packet which is attached to. This size represents the number of flits that will follow the header flit. In our router architecture, a counter is added at each port for keeping track of the *FR* value. When the header flit of a packet leaves a router from one of its output ports, it increases the FR value of that port by an amount equal to the value of its packet-size (in flits). After that, each time a data flit of the packet exits the router through one of its ports, the FR value in the respective exit port is decreased by one. The computation of the FR metric will be better understood with the following example. Let us consider the situation appearing in Fig. 15.3a where we have three input buffered routers I, J, and K with I being the current router. For illustration purposes we consider two virtual channels per port of two flits depth each. The header flit from a message with a size of 3 flits has entered router I through port W (Fig. 15.3 (a)). Also, we have a message entering from north port of router I propagating towards the east through port W of router J and a message entering from west port of router I propagating towards the south through port N of router K. During routing computation stage (RC-stage) for the header flit, the possible output port candidates for header flit are router J on the east and router K on the south. By considering the *FR* metrics of these two output port candidates we have: $FR(N)_K = 2 < 5 = FR(W)_J$; the *FR* value of the north port of router K is smaller than the *FR* value of the west port of router J. So, the router port with the smaller *FR* value will be chosen for message propagation. In Fig. 15.3 (b) we have the network status after the header flit has propagated to router K and the new (updated) value of $FR(N)_K = 5$. In the case of ties, where all port candidates have the same *FR* value, an output port is randomly selected for the next propagation.

Let us suppose we have the situation appearing in Fig. 15.4 where we want to route a message from S (0, 1) to D (3, 3). At the same time, there is some network traffic, which is denoted by the size and color of the links; the greater the link size, the higher the traffic that passes through that link. Generally, we have an increased traffic in the central area of the network. Application of the original Variant B algorithm gives the following routing path: {(0, 1), (0, 2), (0, 3), (1, 3), (1, 2), (2, 2), (3, 2), (3, 3)}; the message is routed through the congested area. The proposed algorithm of Section 15.3.3 by using the FR metric selects the route: {(0, 1), (1, 1), (2, 1), (3, 1), (3, 2), (3, 3)} thus it avoids the congested area. It should be noted that, in the routing algorithm of Section 15.3.3, minimal routes are favored; packets proceed to progressive routes. To break possible ties, the FR metric is used as a secondary criterion and the choice with the smaller FR value is given a higher preference. Thus, the proposed metric provides opportunities for the algorithm to travel through less congested areas as it will be presented in the experimental results section.

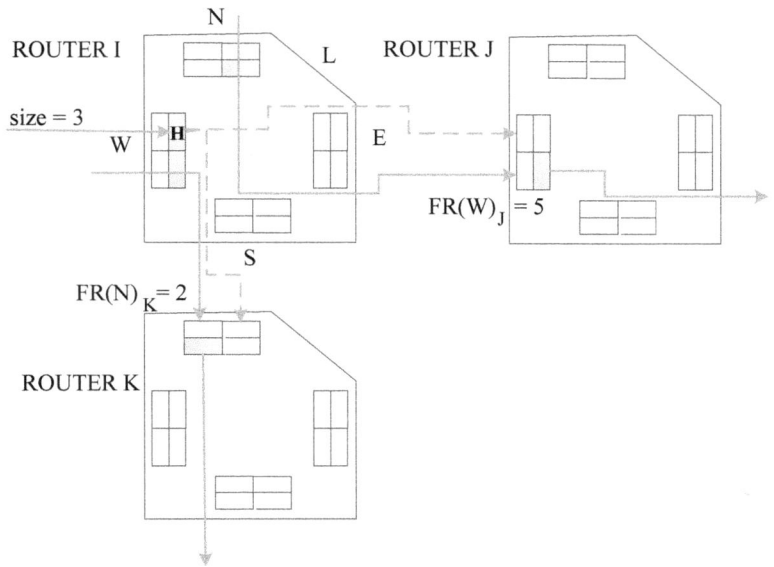

(a)

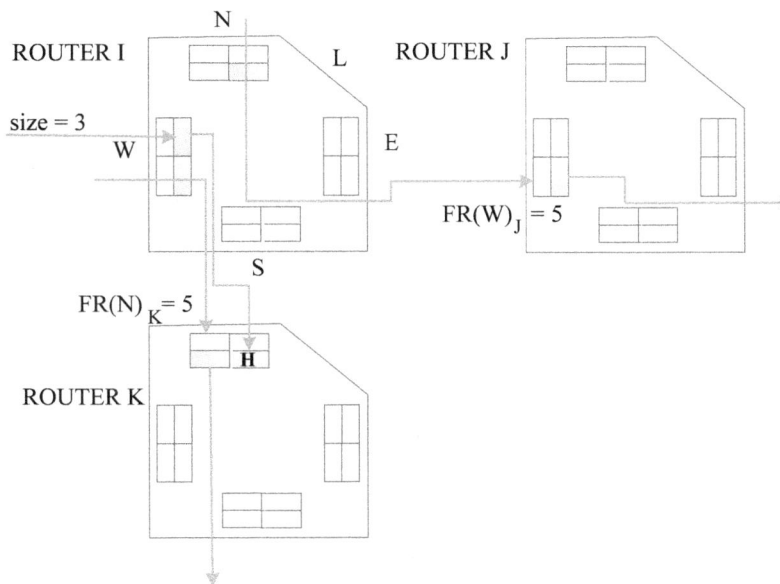

(b)

Fig. 15.3. Example of FR metric computation and updating: (a) Selection for the next propagation based on the FR metric; (b) Updating of FR value after header flit propagation.

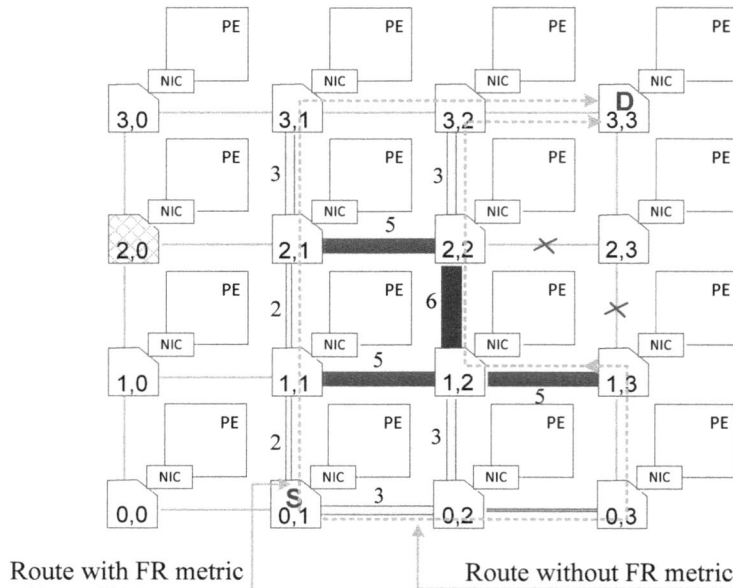

Fig. 15.4. Example of routing a packet around a congested area using FR metric. Node (2, 0) is faulty, faulty links: $\{(2, 2)\leftrightarrow(2, 3)\}$, $\{(1, 3)\leftrightarrow(2, 3)\}$. The numbers on the links denote FR values.

15.3.5. Message Recovery

During circuit operation, a situation may arise where a runtime fault (permanent, transient, intermittent) may cause the corruption of the message, or lead to message deadlocks and eventually message loss. A k-retry-retransmission mechanism is employed at the router and for the output port as selected by the route computation stage (RC). This remedies cases where runtime faults, temporary in nature, lead to errors which are uncorrectable by the ECC mechanism (e.g. multi-bit error). For situations where the message propagation path towards the destination is blocked due to a permanent link or node failure, and after k unsuccessful retries (for our studies we have set k=2) the proposed adaptive fault-tolerant routing algorithm provides a solution by selecting another output port to route the message. Last, for the case of worm split, where an intermediate link/node in the path set by the header flit fails, the message (payload) cannot progress towards the destination. For such cases, a special mechanism is adopted here for message recovery. In order to facilitate the understanding of the proposed recovery schemes an example will be presented next.

Let us consider the situation appearing in Fig. 15.5 (a) where a message is routed from S (0, 1) to D (3, 3). The header flit has reached the destination D and the payload is guided to D via the path $\{(0, 1), (0, 2), (1, 2), (2, 2), (2, 3), (3, 3)\}$. A run-time failure at link $(0, 2) \leftrightarrow (1, 2)$ splits the worm. As a result, the rest of the message cannot reach the destination D and eventually the transmission will fail (for a permanent failure). To resolve from such situation, a pseudo tail flit is created in router (1, 2) and a timer is set to *k1* cycles (*k1* a user defined number). At the same time, in router (0, 2) a second retransmission is attempted. If the output link is still unusable, a pseudo header flit is created in router (0, 2) and send through another path {(0, 2), (0, 3), (1, 3) (1, 2)} with the purpose to by-pass the faulty link and re-establish the path-connection. If the pseudo header flit fails to re-connect the broken path within *k1* cycles, the pseudo tail flit which was created in router (1, 2) is send immediately to the destination (3, 3) thus releasing the occupied resources.

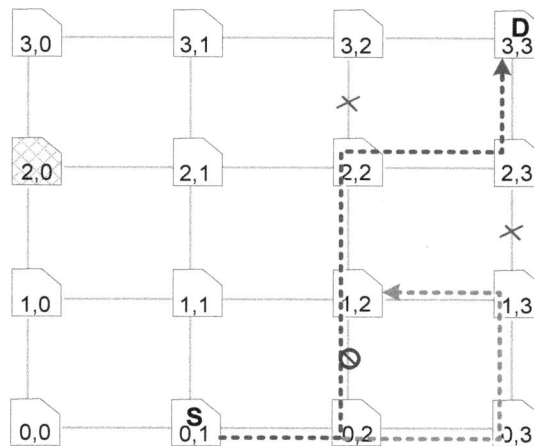

(a)

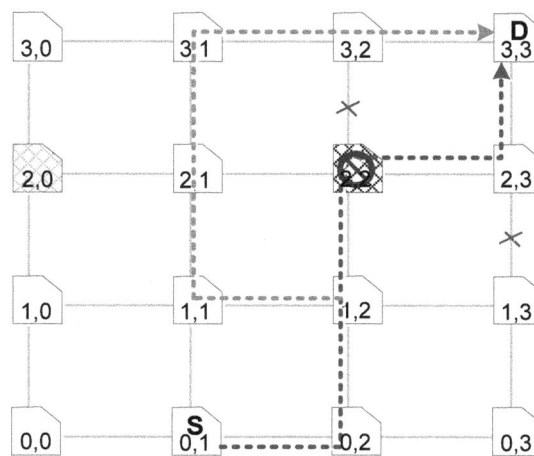

(b)

Fig. 15.5. Examples of message recovery in the case of a worm split.
Example of message recovery: (a) by BMR mechanism, (b) by SMR mechanism.
Node (2, 0) is faulty, faulty links: {(1, 3) ↔ (2, 3)}, {(2, 2) ↔ (3, 2)}.
A runtime failure appears in: (a) link {(0, 2) ↔ (1, 2)}, (b) node (2, 2).

In Fig. 15.5 (b) an alternative recovery scheme is presented for the message routed from S (0, 1) to D (3, 3) via the path {(0, 1), (0, 2), (1, 2), (2, 2), (2, 3), (3, 3)}. A run-time failure at node (2, 2) splits the worm and blocks the rest of the message from reaching the destination. In the case of a permanent run-time failure the transmission will eventually fail. To resolve from such situation, a pseudo tail flit is created in router (2, 3) and a timer is set to *k1* cycles (*k1* a user defined number). At the same time, in router (1, 2) a second retransmission is attempted. If the output link is still unusable, a pseudo header flit is created in router (1, 2) and is send towards the destination through an alternative path selected by the adaptive routing algorithm: {(0, 1), (0, 2), (1, 2), (1, 1), (2, 1), (3, 1), (3, 2), (3, 3)}. Furthermore, after *k1* cycles, the pseudo tail flit from router (2, 3) is send immediately to the destination (3, 3) thus releasing the occupied resources.

For the management of the pseudo flits (header, tail) which are used in the recovery schemes, the mechanism depicted in Fig. 15.6 is adopted which is an extension of [13]. According to this mechanism, when a new packet arrives at a router port, a copy of the header flit is kept in a 1-flit buffer (Fig. 15.6) until the tail flit releases the occupied resources. In case of a runtime fault, where we have a worm split, the occupied resources cannot be freed because the data and tail flits are forced to follow the path set by the header flit which is now blocked, therefore leading into a deadlock. To recover from this deadlock, a new pseudo header flit is created from the copy of the original header flit and requests a new output port (RC/VC/SW allocation) for propagation. After the request is granted, the rest of the flits follow the new path set up by the pseudo header flit. The architecture of Fig. 15.6 with the help of the pseudo-flit modification logic: a) allows the creation of pseudo-tail flits by changing the flit tag from "header" to "tail", b) supports the modification of the destination node ID of the pseudo-flit. By setting as destination node ID the ID of the downstream router (router (1, 2) in Fig. 15.5 (a) of the failed link/node, we have the situation of Fig. 15.5 (a) termed as *Bypass Message Recovery* or *BMR*, where an attempt is made to bypass the blocked path and re-join the message. If the bypass is successful, the pseudo-head flit is consumed at the downstream router of the failed link/node. Thus, the propagation of the rest of the message continues towards the destination node which is actually unaware of the instantaneous message rupture (no pseudo flit has reached the destination).

When the downstream router of the failed link/node is unreachable by the pseudo header flit, either due to multilink faults or due to the whole router being faulty, *BMR* might fail. An example where the downstream router (2, 3) is unreachable by the pseudo header flit is given in Fig. 15.5 (b). Alternatively, or in cases such as the one described previously, the *SMR* mechanism, described next, may be utilized.

If the destination node ID of the pseudo-flit remains unaltered, then the situation of Fig. 15.5 (b) arises which is termed as *Split Message Recovery* or *SMR*. In such case, which is similar to the concept of packet fragmentation [13], the remaining part of the message may continue the propagation towards the destination through another new path. The release of the currently occupied resources is ensured by the creation and propagation of a pseudo tail flit from the downstream router of the failed router link/node. Now, the destination node becomes more involved in the management of the pseudo flits by considering the pseudo tail flit as the end of current packet and the pseudo head flit as the beginning of a new packet. Thus, the complexity of the packet reassembly at the end node is increased.

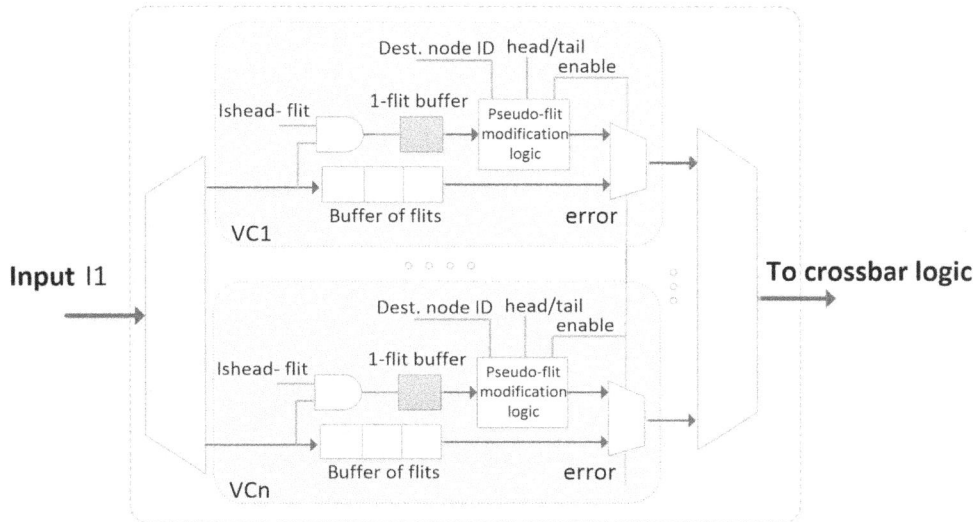

Fig. 15.6. Detailed diagram of an input port with the modifications of router logic to accommodate for message recovery.

Depending on the status of the network, either one of the *BMR* and *SMR* mechanisms may be used. For example, under situations where the neighborhood of the runtime fault does not impede the reachability of healthy resources, *BMR* may be the preferred mechanism, since it avoids the extra management during the packet reassembly at the destination node, whereas when the neighborhood of the failed link/node becomes problematic to reach or totally unreachable (Fig. 15.5 (b)) then the *SMR* may be selected. In the experimental results (Section 15.4) the *SMR* scheme has been used, whereas in [18] are presented results by CAFTA+*BMR*.

In the situation where a runtime node is faulty the above-described recovery mechanism may be complemented by a task migration mechanism to allocate the tasks of the failed node to another healthy node. The above error recovery mechanism may be proved useful in the cases of high volume data transfer between selected nodes e.g. in multimedia applications.

15.3.6. Extension of CAFTA for Torus Topology

To evaluate the impact of the network topology on CAFTA efficiency, we decided to extend the proposed algorithms to a 2D torus Network-on-Chip. Due to its higher connectivity, torus topology offers more connectivity in terms of potential paths between a source and destination pair. The main difference between Mesh and Torus topology is that for torus vertical and horizontal wrapping links are added.

The Torus network can be partitioned in two virtual networks (VN). As for CAFTA in 2D Mesh topology, the routing algorithm SouthLast (SL) and NorthLast (NL) are applied to each VN. In Fig. 15.7 the algorithm leading to the Virtual Network selection is given. However, although the routing algorithm is the same as for 2D Mesh, having additional links may provoke deadlocks.

Given a pair of nodes S : source (xs,ys), D : destination(xd, yd), in a network of radix k

VN: virtual network to be used to reach the destination node

dx=xd-xs and dy=yd-ys;

(1) $0 < dy < k/2$ or $dy < -(k/2)$ => select SouthLast VN

(2) $-(k/2) < dy < 0$ or $dy > k/2$ => select NorthLast VN

(3) $dy = 0$ => select South-Last VN or North-Last VN, depending on the available channels

Fig. 15.7. Path selection in a 2D torus network.

15.3.6.1. Routing and Deadlocks in 2D torus NoC

Before extending CAFTA to this topology, we must guarantee that the routing algorithm is deadlock free. Based on Dally theorem [20], a routing algorithm is deadlock free if and only if its channel dependency graph is acyclic. Variant B algorithm for 2D Mesh is proved deadlock free but is it still the case when introducing the additional Torus wrapping links?

To answer this question, we will consider the impact of adding vertical and horizontal wrapping links. However, we assume in the following that the NorthLast and SouthLast routing algorithms are deadlock free for the mesh topology. **Thus we will only consider the potential cycles added by the wrapping links.**

15.3.6.2. Vertical Wrapping Links

Let us consider the vertical links of a 3 x 3 Torus network (see Fig. 15.8). In Variant B, two distincts virtual networks coexist with two different routing schemes: NorthLast and SouthLast. Fig. 15.8 describes channels that are used within the NorthLast VN. When adding two unidirectional vertical wrapping links, a packet may go for example through the channels C06 -> C63 -> C30 -> C06. The corresponding dependency graph introduces cycles, thus the routing is not deadlock free. The same reasoning can be applied to the SouthLast VN.

To break the cycles we introduce some restrictions: in Fig 15.9 (a) we remove the dependency between C30 and C06 and in Fig. 15.9 (b) between C36 and C60. Thus, the channel dependency graph becomes acyclic and the routing is guaranteed as deadlock free.

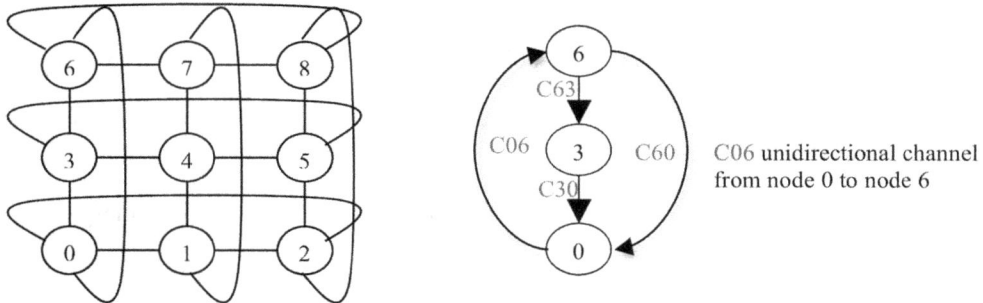

Fig. 15.8. A 3x3 torus network and the connexion graph for a column.

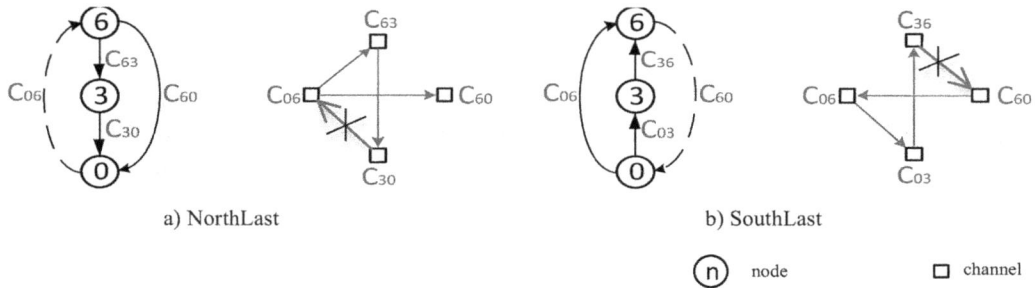

a) NorthLast   b) SouthLast

Fig. 15.9. Restrictions applied to vertical wrapping links in the Virtual Networks (a) NorthLast, and (b) Southlast.

The example showed in Fig 15.9 (a), removes dependency between node 3 and 6, by preventing node 3 to send a paquet to node 6 with the same VN. The only way to allow this transfer consists in making use of the virtual source feature proposed in Variant B [12]. A virtual source consists in consuming locally a packet in a node and creating a new packet with the same destination but injected in the dual VN. This technique is used when according to a turn restriction, faulty links prevent the propagation of a packet in the NoC, while preserving the deadlock freeness of the algorithm. Here, to allow the transfer between node 3 and node 6, a virtual souce (VS) is created in node 0 and new packets are routed to node 6 using the dual VN, in this case the SouthLast VN. However, a paquet can be sent directly from node 0 to node 6 without creating a VS.

Symetrically, the same reasoning applies for packets in the SouthLast VN (see Fig. 15.9 (b)), with similar corresponding restriction.

We can extrapolate this restriction to the vertical wrapping links of a 2D Torus NxN Network on Chip. This restriction can be defined as follows:

Restriction 1:

    o   For the Virtual Network NorthLast, for each node at the south boundary:

        ■ If the node is a source node, then the packet can be sent through the south port in the vertical wrapping link, using the NorthLast virtual network;

        ■ If the node is a transit node, a virtual source must be created and the new packet is sent through the south port by using SouthLast virtual network.

    o   For the Virtual Network SouthLast, for each node at the north boundary:

        ■ If the node is a source node, then the packet can be sent through the north port in the vertical wrapping link, using the SouthLast virtual network;

        ■ If the node is a transit node, a virtual source must be created and the new packet is sent through the north port by using NorthLast virtual network.

15.3.6.3. Horizontal Wrapping Links

If we consider now the horizontal wrapping links, similar issue arise. We apply the same deadlockness verification and the corresponding restriction. The NorthLast and SouthLast routing algorithms having the same restrictions for horizontal links, we will not differentiate the two cases.

When adding without restrictions horizontal wrapping links to the mesh network, two cycles are introduced in the channel dependency graphs (see Fig. 15.10): e.g. $C20 \rightarrow C01 \rightarrow C12 \rightarrow C20$ and $C02 \rightarrow C21 \rightarrow C10 \rightarrow C02$. To break these cycles, we have to remove at least two dependencies introduced by the wrapping links, one for each VN. We can chose to remove C12-C20 or C20-C01 and C10-C02 or C02-C21. Here C12-C20 and C10-C02 are removed.

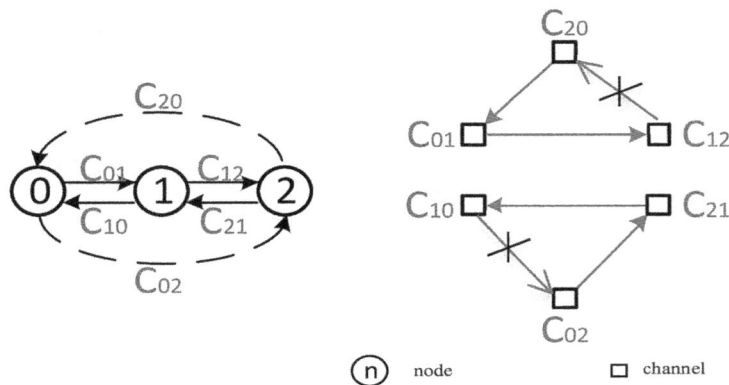

Fig. 15.10. Restrictions applied to horizontal wrapping links for NorthLast and SouthLast VNs.

As for the previous Subsection for a NxN Torus NoC we can extrapolate and formalize this new restriction as follows:

Restriction 2:

For each node located at east or west boundary:

- If the node is a source node, then the packet can be sent using any of the virtual networks, through the horizontal wrapping link;

- If the node is a transit node, a virtual source must be created in this node. The new packet is sent through the corresponding port (east or west) using the horizontal wrapping link but switching to the other virtual network.

15.3.6.4. Vertical and Horizontal Wrapping Links

The last case to check relates to the wrapping links connected to the four routers at the NoC boundaries.

In Fig. 15.11, one can see that the channel dependency graph exhibit cycles. However applying the two restrictions previously described implies that dependencies C06-C68 and C82-C20 are also removed. The graph becomes acyclic, thus no more restrictions are needed.

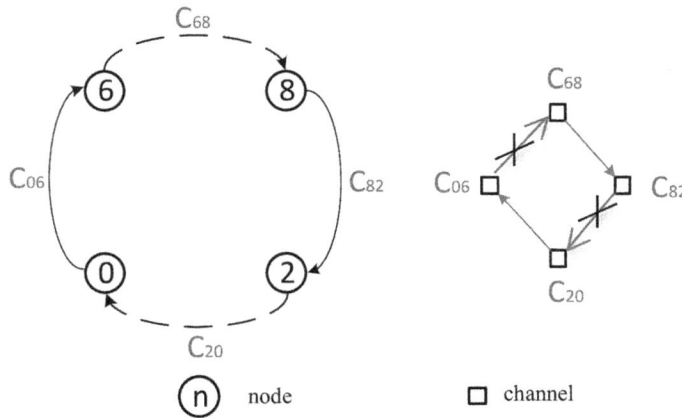

Fig. 15.11. Restrictions applied to surrounding wrapping links for NorthLast and SouthLast VNs.

The CAFTA algorithm presented in Fig. 15.2 dedicated to Mesh topology is slightly modified to take into account these two restrictions. This new CAFTA algorithm applies then to torus NoC providing adaptive deadlock free routing, neighbor fault awareness, network traffic regulation and packet-retransmission combined with a new message recovery scheme.

15.4. Experimental Setup and Results

15.4.1. Experimental Setup

CAFTA and the microarchitecture described in Section 15.3, the necessary framework to perform the fault injection campaigns, have been implemented in C++ on top of IRIS [33] which is a cycle-accurate network simulator. For power estimations the ORION [34] power model has been incorporated into IRIS. Three different 2D-mesh and 2D-torus networks sizes (10×10, 12×12, 16×16) are studied. Table 15.1 presents the simulated configurations. CAFTA's performance is analyzed under three synthetic traffic patterns namely uniform random, transpose and bit complement.

Two different fault types are examined, namely unidirectional link faults and node faults whereas the fault patterns being considered are: permanent faults, transient faults and intermittent faults. More specifically, we have mixed type of fault injections and for a specific number n_f of injected faults the ratios of permanent, transient and intermittent faults are: 15 % permanent faults and 5 % permanent runtime faults to account for failures due to wear-out, electromigration, aging effects etc., 40 % transient faults (due to single event effects, cross-talk, etc.) with duration of one cycle and 40 % intermittent faults with random duration from 200 cycles up to 800 cycles. Since intermittent faults manifest in bursts, we set for each intermittent fault the burst length [6] to a random value between 1 and 10.

Seven different fault injection ratios (fir) are investigated (see Table 15.1) and for each fir the chosen network size was simulated until 5000 injected packets either reach their destination or they are dropped. Every fault injection experiment was repeated 1000 times (Table 15.1). In the simulations, a warm-up period of 2000 cycles has been utilized to preload the network with sufficient traffic. This is necessary to evaluate the impact of the fir on the CAFTA expected benefits.

Table 15.1. Simulated network configurations for Mesh topology.

Topologies	10×10 Mesh (100 nodes), 12×12 Mesh (144 nodes), 16×16 Mesh (256 nodes)
Traffic pattern (Synthetic)	Uniform random, Transpose, Bit complement
Flit size (in bits)	128
Arbitration	Round-robin
Router latency (cycles)	4
VNs	2
VCs/port	4 / 2 per VN
Buffer size (in flits)	8
Packet Size (in flits)	10
Fault types:	Faulty unidirectional links, faulty routers
Fault injection ratios (%)	0, 2, 5, 10, 20, 30, 40
Total simulation Time	until 5000 injected packets are received or dropped
Warm-up time (cycles)	2000
Iterations/fault injection pattern	1000

It should be noted that, the total number of possible link faults for a *nxn* Mesh network using bi-directional links, is given by the relation: $Total\_faults = 4n(n-1)$. Therefore, for the networks under study we will have: 360 possible link faults for a network of 10×10, 528 possible link faults for 12×12 network and 960 possible link faults for 16×16 network.

15.4.2. Performance of FR Metric Under Synthetic Traffic

In the first experiments, the impact of using the FR metric (Section 15.3.4) will be evaluated. Figs 15.12 (a), 15.12 (b) and 15.12 (c) present results for a fault-free 16×16 NoC regarding the average latency variation under three different traffic patterns: Uniform Random, Transpose, Bit complement. The routing algorithm CAFTA employs three different metrics for congestion management. The case "free VC" corresponds to a metric that takes into account the availability of free VCs in the downstream routers [20], and "FR" is the proposed metric from Section 15.3.4. The case "no metric" refers to CAFTA without the use of a congestion metric (baseline configuration). In this situation, the actual routing algorithm is Variant B so a direct comparison can be made with Variant B with respect to the average latency. For each congestion metric we developed a corresponding version of the CAFTA algorithm. In all these figures (Figs. 15.12 (a), 15.12 (b), and 15.12 (c)) we see that by not taking into account the traffic status of the network ("no metric" case) the average latency of the routing algorithm increases very much and under heavy network traffic tends to saturate at lower values of packet injection ratio: 0.046 (flits/node/cycle) for Uniform Random traffic (Fig. 15.12 (a)), 0.041 (flits/node/cycle) for Transpose traffic (Fig. 15.12 (b)) and 0.036 (flits/node/cycle) for Bit complement traffic (Fig. 15.12 (c)).

We should note that the latency measured when using the FR metric is always smaller when compared to the other two cases. More specifically we have the following cases: For the Uniform Random traffic (Fig. 15.12 (a)) the use of FR metric results in a latency improvement (at the saturation point) of 10.8 % with respect to the free-VC metric, and of 65.2 % with respect to the baseline. For the Transpose traffic (Fig. 15.12 (b)) the latency improvement when using FR metric is 20.16 % compared to free-VC, and 30 % compared to the baseline. Finally, for the Bit Complement traffic (Fig. 15.12 (c)) the latency improvement when using FR metric is 28.75 % with respect to the free-VC metric, and 66.67 % with respect to the baseline.

Generally, in all studied cases, the CAFTA version implementing the use of FR metric exhibits the best behavior (smaller average latency, saturates at higher injection rates). Thus, it manages to guide the routing algorithm through paths with lower average latencies. For the remaining experiments we will consider CAFTA with the FR metric.

a) Uniform random trafic

b) Transpose trafic

c) Bit complement trafic

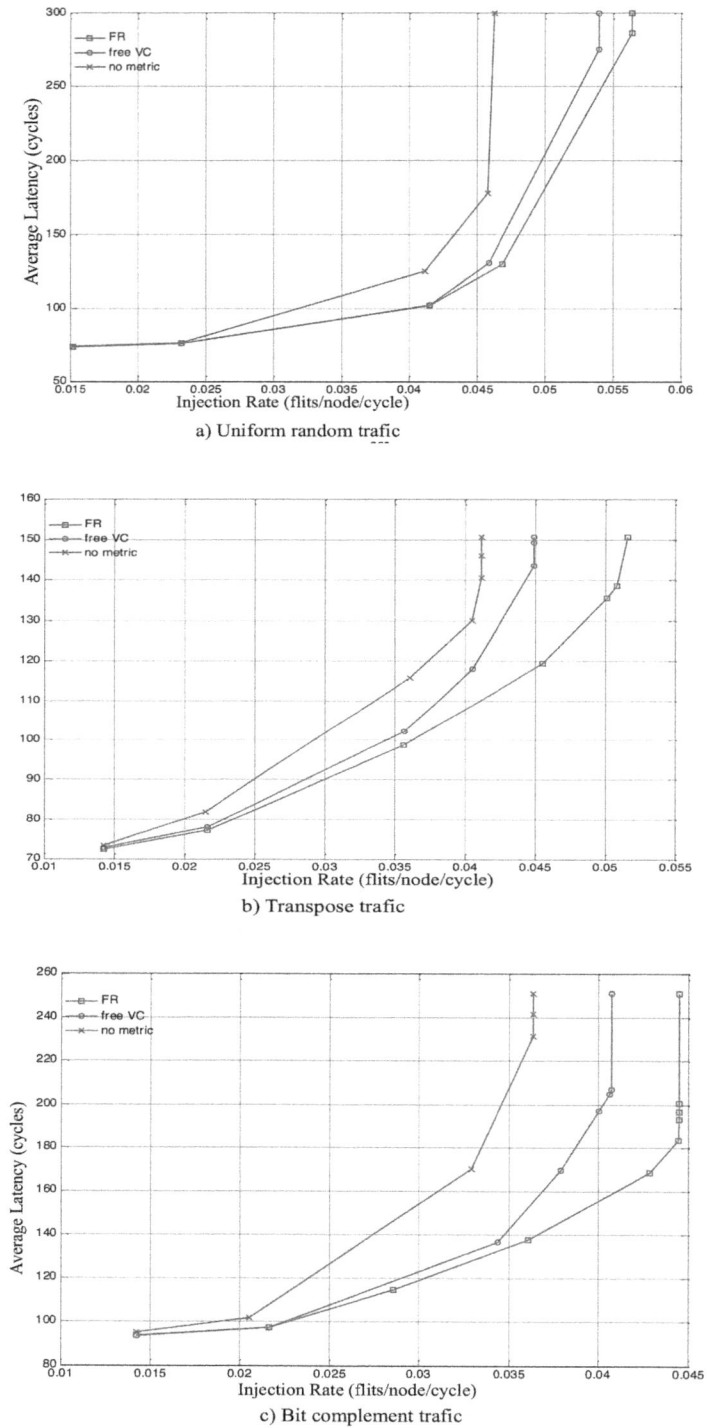

Fig. 15.12. Average latency variation for different traffic metrics and for different traffic types, for a fault-free 16×16 NoC.

15.4.3. Variation of Average Latency Due to the Presence of Faults Under Synthetic Traffic

In the following, we are assessing variation of the average latency due to the presence of faults. Two different sets of experiments are considered. The first one evaluates the effect of faulty links (Figs 15.13 (a), 15.13 (b) and 15.13 (c)) under three different synthetic traffic patterns in a 16×16 NoC. More specifically, three different fault injection ratios are considered: a) 0 % of links are faulty which corresponds to the fault-free case (Figs. 15.13 (a)), b) 10 % of links are faulty (Fig. 15.13 (b)) and c) 40 % of links are faulty (Fig. 15.13 (c)).

As expected, the injection of faults affects the network's average latency which tends to increase with the increase in the percentage of faulty links as the network traffic increases (injection rate). The second set of experiments evaluates the effect of faulty routers (Figs 15.14 (a), (b) and (c)). Again, we see an increase of the average latency as the percentage of faulty routers increases. It should be noted that in the case of faulty routers, as it is expected, the average latency increase is higher than the corresponding case of faulty links. This is attributed to the model used for the faulty routers: a faulty router is a router where all its links are faulty. Therefore, a specific percentage of faulty routers actually amounts to a bigger percentage of link faults.

15.4.4. Reliability Estimation

The percentage of packets successfully delivered is related to the reliability of a fault-tolerant routing algorithm. From Figs. 15.15 ((a), (b) and (c)) and Figs. 15.16 ((a), (b), and (c)) we see that the percentage of packets delivered decreases as the failure rate increases.

Overall, it is seen that CAFTA maintains high levels of reliability. Let us consider for example the case of a 16x16 mesh NoC and the faulty link model (see Figs. 15.15 (a), (b), and (c)). For this network, CAFTA succeeds in delivering more than 99.9 % of the transmitted messages when the percentage of faulty links is up to 10 %; i.e. 96 simultaneous link faults and under all the examined traffic patterns. The success delivery rate reduces to 99.38 % for Uniform Random traffic (Fig. 15.15 (a)), to 99.03 % for Transpose traffic (Fig. 15.15 (b)) and to 97.68 % for Bit complement traffic (Fig. 15.15 (c)) when the percentage of faulty links increases to 40 %; i.e. 384 simultaneous link faults. Similar observations can be made if we consider faulty routers instead of faulty links (see Figs. 15.16 (a), (b), and (c). In this case, as it was previously explained, the faulty routers are expected to have a stronger effect on the reliability measures of the NoC. However, even in the scenario where 40 % of routers in a 16x16 NoC are faulty, the success delivery rate is still high, close to 93.40 % for Uniform Random traffic (Fig. 15.16 (a)), to 95.98 % for a Transpose traffic (Fig. 15.16 (b)) and 93.53 % for Bit complement traffic (Fig. 15.16 (c)).

a) Uniform random trafic

b) Transpose trafic

c) Bit complement trafic

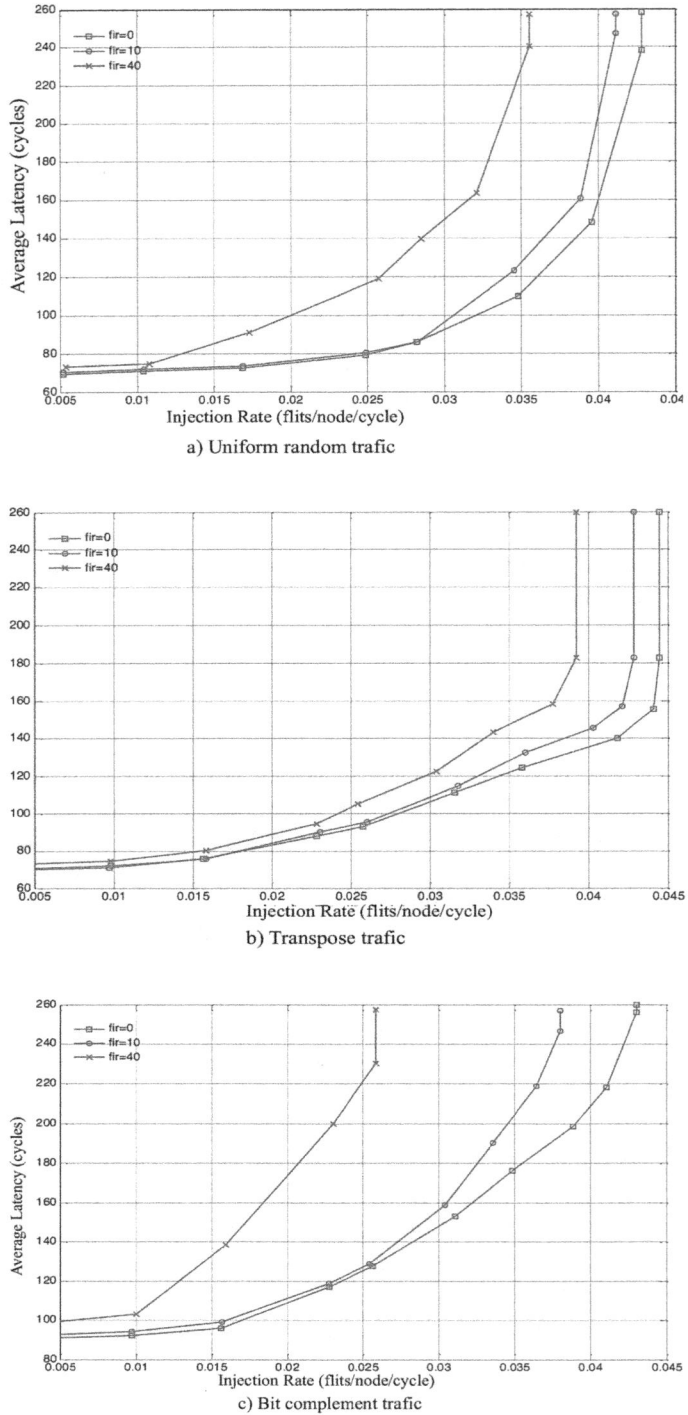

Fig. 15.13. Average latency variation of CAFTA for various fault injection ratios (faulty links) for different traffic types for a 16×16 NoC.

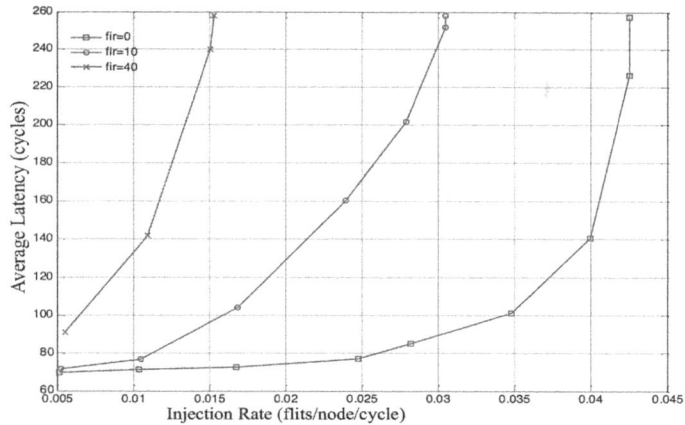

a) Uniform random trafic

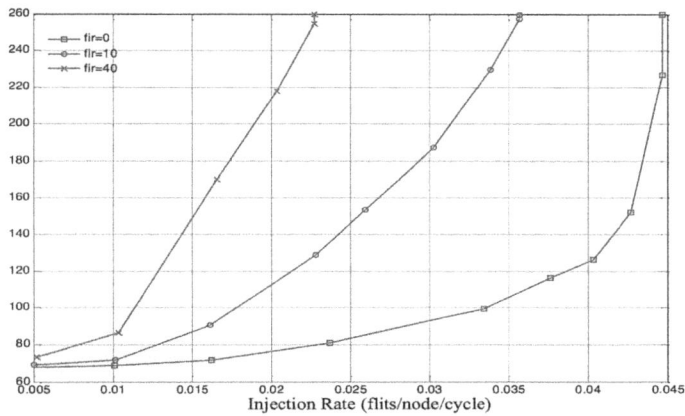

b) Transpose trafic

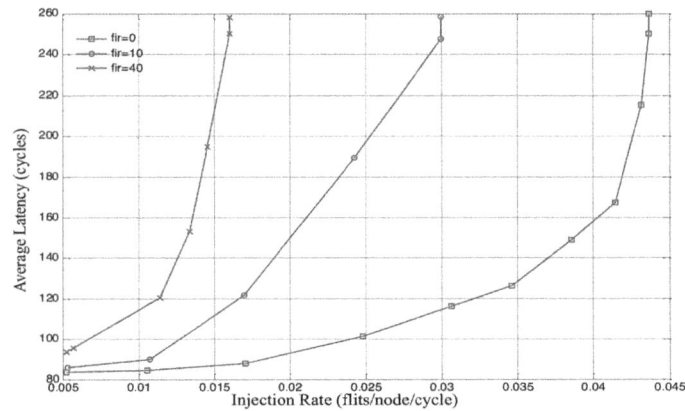

c) Bit complement trafic

Fig. 15.14. Average latency variation of CAFTA for various fault injection ratios (faulty routers) for different traffic types, for a 16x16 NoC.

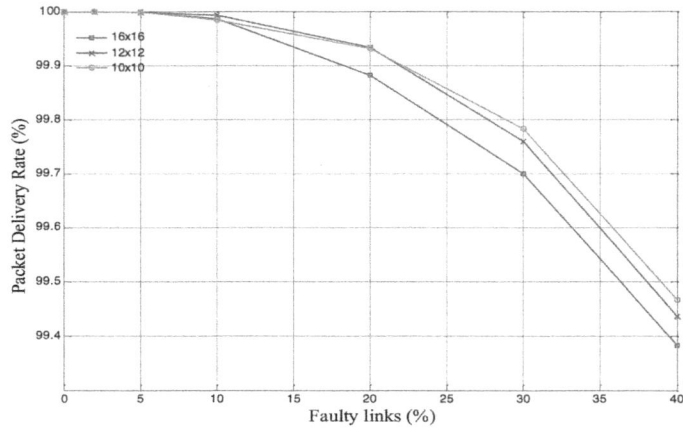

a) Uniform random trafic

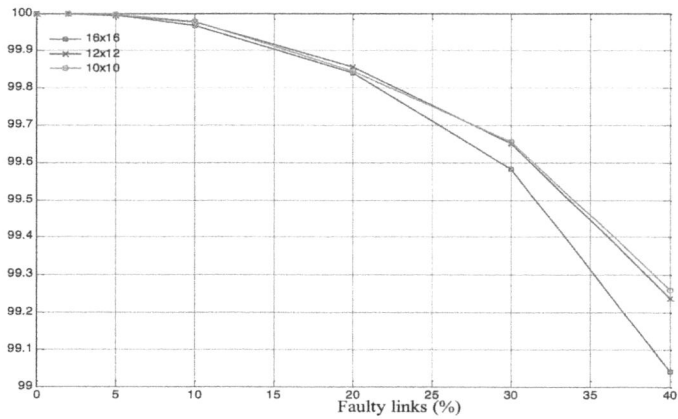

b) Transpose trafic

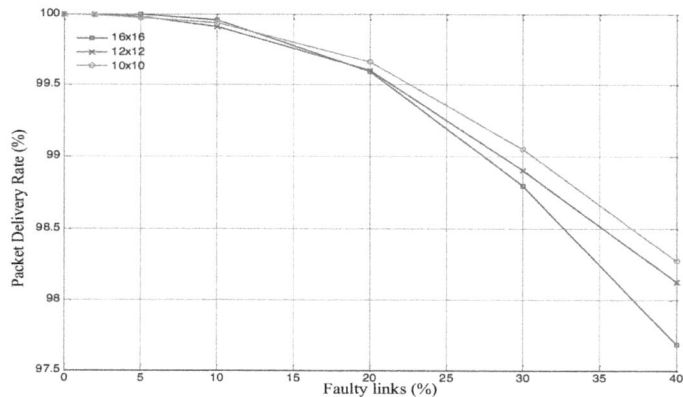

c) Bit complement trafic

Fig. 15.15. Packet delivery rate for mesh topologies with different sizes for various fault injection ratios (faulty links) and for different traffic types.

a) Uniform random trafic

b) Transpose trafic

c) Bit complement trafic

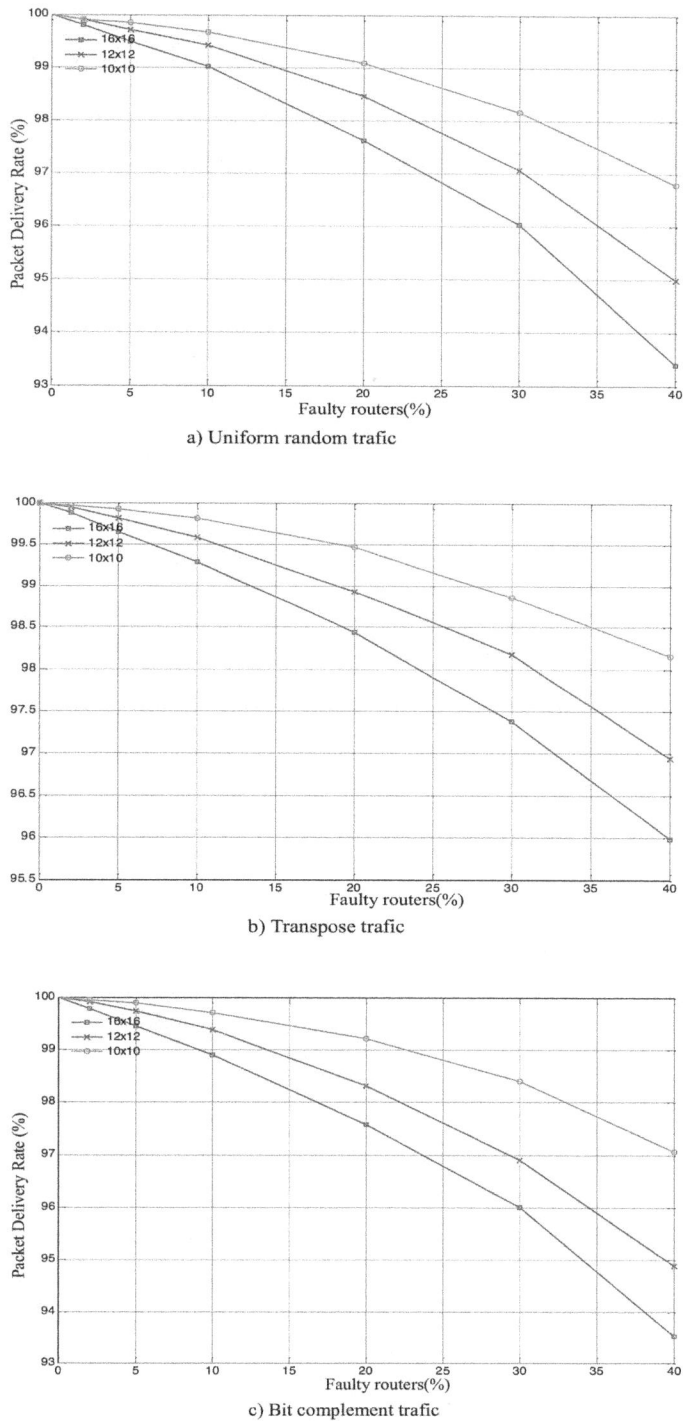

Fig. 15.16. Packet delivery rate for mesh topologies with different sizes for various fault injection ratios (faulty routers) and for different traffic types.

It should be noted that the reliability decrease, is attributed mainly to the employed routing algorithm (Variant B) which sometimes fails to reach the destination even though the destination is reachable. In such cases, if it is required, the reliability may be further improved either by substituting the current routing algorithm by a more powerful one or by considering additional techniques like those presented in [27, 28].

15.4.5. CAFTA for Torus NOC Topology

To evaluate the impact of the topology, we applied the CAFTA algorithm to 2D-torus networks, using the same simulation platform as described in Section 15.4.1. We conducted the same experiments as in Section 15.4.3. We measured the average latency for different faulty link injection rate and for different traffic types (as in Fig. 15.12). Compared to the mesh topology, the experimental results shows, that the saturation threshold is improved on average of 20 % and up to 25 % for all traffic types and for all fir.

If we compare the packet delivery rate for torus and mesh topologies (Fig. 15.17) and for a random uniform traffic, we can see that more than 99, 6 % of the packets reach their destination even in the presence of 40 % of faulty links. This ratio is higher than 98 % for the mesh topology. As also expected, the size of the network is less impacting in torus topology than in mesh topology.

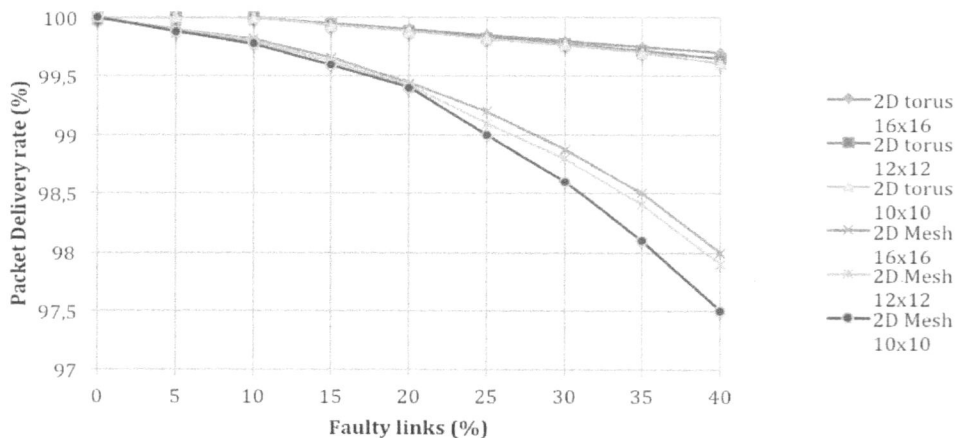

Fig. 15.17. Packet delivery rate with faulty links: torus vs. mesh.

Due to its higher connectivity, torus topology is much more efficient than the mesh topology in terms of latency, fault tolerance capability and traffic congestion management.

Thus if the target application is critical and the reliability is the higher priority, torus topology implementing the CAFTA algorithm can be used. However more power consumption and area is expected for this topology, due to the wrapping links and the additional cost of the routing, and above defined constraints implementation.

15.5. Conclusions

An adaptive fault tolerant routing algorithm, CAFTA, has been presented. It utilizes a new load balancing metric (FR metric) which aids in routing the messages through less congested areas. Under different traffic patterns, it was observed that the FR metric results in a latency improvement (at the saturation point) of 53.95 % on the average with respect to the baseline case where a congestion metric is not utilised and 19.9 % with respect to the free-VC metric. Furthermore, FR saturates at higher injection rates.

An enhanced fault recovery mechanism has been introduced to CAFTA to cope with run-time failures. Extensive simulations have been carried out under different failure rates, traffic patterns, fault models (faulty links, faulty routers), network sizes and under different types of faults (permanent, transient, intermittent). As the experiments have shown, CAFTA maintains high reliability of more than 97.68 % for a 2D mesh network of 16x16 and in the presence of 384 simultaneous link faults (out of 960 total links). For the extreme scenario of 103 out of 256 routers being simultaneously faulty, the reliability attained by CAFTA is more than 93.40 %. An extension of CAFTA to the torus topology is also proposed. We demonstrate the deadlock freeness of the extension and showed that the topology improves the latency and the reliability when compared to the mesh network. For this topology, the network saturation threshold is improved by an extra 20 % on average. For the extreme scenario where 40 % of the links are faulty, the packet delivery rate is more than 99.6 %.

References

[1]. http://www.mellanox.com/page/press_release_item?id=1681

[2]. http://www.intel.com/content/www/us/en/processors/xeon/xeon-phi-detail.html

[3]. International technology roadmap for semiconductors, Semiconductor industry association 2012, http://public.itrs.net/Links/2012ITRS/Home2012.htm

[4]. M. Nicolaidis, L. Anghel, N.-E. Zergainoh, Y. Zorian, T. Karnik, K. Bowman, J. Tschanz, Lu Shih-Lien, C. Tokunaga, A. Raychowdhury, M. Khellah, J. Kulkarni, V. De, D. Avresky, Design for test and reliability in ultimate CMOS, in *Proceedings of the Design, Automation & Test in Europe Conference & Exhibition (DATE'12)*, 2012, pp. 677-682.

[5]. D. Park, C. Nicopoulos, J. Kim, N. Vijaykrishnan, C. R. Das, Exploring fault-tolerant Network-on-Chip architectures, in *Proceedings of the International Conference on Dependable Systems and Networks (DSN'06)*, Philadelphia, PA, USA, 2006, pp. 93-104.

[6]. D. Gil, J. Gracia, J. Baraza, L. Saiz-Adalid, P. Gil Vicente, Analyzing the impact of intermittent faults on microprocessors applying fault injection, *IEEE Design & Test of Computers*, Vol. PP, No. 99, pp. 66-73.

[7]. S. Shamshiri, A. Ghofrani, K.-T. Tim Cheng, End-to-end error correction and online diagnosis for on-chip networks, in *Proceedings of the IEEE International Test Conference (ITC'11)*, Sept. 2011, Paper 10.3.

[8]. Q. Yu, J. Cano, J. Flich, P. Ampadu, Transient and permanent error control for high-end multiprocessor systems-on-chip in *Proceedings of the 6th ACM/IEEE Intl. Symposium on Networks-on-Chip (NOCS'12)*, May 2012, pp. 169-171.

[9]. J. Duato, S. Yalamanchili, L. Ni, Interconnection Networks: an Engineering Approach, *Morgan Kaufmann Publishers*, 2003.

[10]. S. Pasricha, Y. Zou, NS-FTR: a fault tolerant routing scheme for networks on chip with permanent and runtime intermittent faults, in *Proceedings of the 16th Asia and South Pacific Design Aut. Conference (ASPDAC'11)*, Colorado State Univ., Fort Collins, CO, USA, Jan. 2011, pp. 443-448.

[11]. A. DeOrio, D. Fick, V. Bertacco, D. Sylvester, D. Blaauw, J. Hu, G. Chen, A Reliable Routing Architecture and Algorithm for NoCs, *IEEE Trans. on Computer-Aided Design*, Vol. 31, Issue 5, May 2012, pp. 726-739.

[12]. F. Chaix, D. Avresky, N. Zergainoh, M. Nicolaidis, Fault-tolerant deadlock-free adaptive routing for any set of link and node failures in multi-cores systems, in *Proceedings of the 9th IEEE International Symposium on Network Computing and Applications (NCA'10)*, Cambridge, Massachusetts, USA, July 2010, pp. 52-59.

[13]. Y. Kang, T. Kwon, J. Draper, Fault-tolerant flow control in On-chip networks, in *Proceedings of the Fourth ACM/IEEE International Symposium on Networks-on-Chip (NOCS'10)*, Washington, DC, USA, 2010, pp. 79-86.

[14]. K. Aisopos, A. DeOrio, L. Peh, V. Bertacco, ARIADNE: Agnostic reconfiguration in a disconnected network environment, in *Proceedings of the International Conference on Parallel Architectures and Compilation Techniques (PACT'11)*, Galveston Island, TX, October 2011, pp. 298-309.

[15]. M. Valinataj, S. Mohammadi, J. Plosila, P. Liljeberg, A fault-tolerant and congestion-aware routing algorithm for Networks-on-Chip, in *Proceedings of the 13th IEEE Symposium on Design and Diagnostics of Electronic Circuits and Systems (DDECS'10)*, 2010, pp. 139-144.

[16]. M. Ebrahimi, M. Daneshtalab, J. Plosila, H. Tenhunen, MAFA: Adaptive fault-tolerant routing algorithm for Networks-on-Chip, in *Proceedings of the 15th Euromicro Conference on Digital System Design (DSD'12)*, 2012, pp. 201-207.

[17]. C. Feng, Z. Lu, A. Jantsch, M. Zhang, Z. Xing, addressing transient and permanent faults in NoC with efficient fault-tolerant deflection router, *IEEE Transactions on Very Large Scale Integration (VLSI) Systems*, Vol. 21, No. 6, June 2013, pp. 1053-1066.

[18]. M. Dimopoulos, Y. Gang, M. Benabdenbi, L. Anghel, N. Zergainoh, M. Nicolaidis, Fault-tolerant adaptive routing under permanent and temporary failures for many-core systems-on-chip, in *Proceedings of the 19th International On-Line Testing Symposium (IOLTS'13)*, Chania, Crete, Greece, 2013, pp. 7-12.

[19]. C. Glass, L. Ni, The turn model for adaptive routing in *Proceedings of the 19th Annual International Symposium on Computer Architecture (ISCA'92)*, New York, NY, USA, 1992, pp. 278-287.

[20]. W. Dally, H. Aoki, Deadlock-free adaptive routing in multicomputer networks using virtual channels, *IEEE Trans. Parallel Distrib. Syst.,* Vol. 4, No. 4, April 1993, pp. 466-475.

[21]. J. Kim, D. Park, T. Theocharides, N. Vijaykrishnan, C. Das, A low latency router supporting adaptivity for On-Chip interconnects, in *Proceedings of the Design Automation Conference (DAC'05)*, June 2005, pp. 559-564.

[22]. P. Gratz, B. Grot, S. W. Keckler, Regional congestion awareness for load balance in networks on chip, in *Proceedings of the International Symposium on High Performance Computer Architectures (HPCA'08)*, 2008, pp. 203-214.

[23]. S. Ma, N. Jerger, Z. Wang, DBAR: an efficient routing algorithm to support multiple concurrent applications in networks-on-chip, in *Proceedings of the 38th Annual International Symposium on Computer Architecture (ISCA '11)*, 2011, pp. 413-424.

[24]. C. Glass and L. Ni, Fault-tolerant wormhole routing in meshes without virtual channels, *IEEE Trans. Parallel Distrib. Syst.*, Vol. 7, No. 6, June 1996, pp. 620-636.

[25]. M. Ebrahimi, M. Daneshtalab, J. Plosila, H. Tenhunen, Minimal-path fault-tolerant approach using connection-retaining structure in Networks-on-Chip, in *Proceedings of the 7th IEEE/ACM International Symposium on Networks on Chip (NoCS'13)*, April 2013, pp. 1-8.

[26]. A. DeOrio, K. Aisopos, V. Bertacco, L.-S. Peh, DRAIN: Distributed recovery architecture for inaccessible nodes in multi-core chips, in *Proceedings of the 48th ACM/EDAC/IEEE Design Automation Conference (DAC'11)*, June 2011, pp. 912-917.

[27]. W. Tsai, D. Zheng, S. Chen, Y. Hu, A fault-tolerant NoC scheme using bidirectional channel, in *Proceedings of the 48th ACM/EDAC/IEEE Design Automation Conference (DAC'11)*, June 2011, pp. 918-923.

[28]. K. Latif, A.-M. Rahmani, E. Nigussie, T. Seceleanu, M. Radetzki, H. Tenhunen, Partial virtual channel sharing: a generic methodology to enhance resource management and fault tolerance in Networks-on-Chip, *Journal of Electronic Testing Theory and Applications*, Vol. 29, Issue 3, June 2013, pp 431-452.

[29]. A. Ghofrani, R. Parikh, S. Shamshiri, A. DeOrio, K. Cheng, V. Bertacco, Comprehensive online defect diagnosis in on-chip networks, in *Proceedings of the IEEE 30th VLSI Test Symposium (VTS'12)*, April 2012, pp. 44-49.

[30]. Kohler, G. Schley, M. Radetzki. Fault tolerant network on chip switching with graceful performance degradation, *Trans. Comp.-Aided Des. Integ. Cir. Sys.*, Vol. 29, No. 6, June 2010, pp. 883-896.

[31]. M. Aguilera, W. Chen, S. Toueg, Heartbeat: a timeout-free failure detector for quiescent reliable communication, Technical Report 97-1631, Department of Computer Science, *Cornell University*, May 1997.

[32]. A. Avizienis, J. Laprie, B. Randell, C. Landwehr, Basic concepts and taxonomy of dependable and secure computing, *IEEE Trans. Dependable Secur. Comput.*, Vol. 1, No. 1, January 2004, pp. 11-33.

[33]. Manifold, http://manifold.gatech.edu

[34]. A. Kahng, B. Li, Li-Shiuan Peh, K. Samadi, ORION 2.0: A power-area simulator for interconnection networks, *IEEE Transactions on Very Large Scale Integration (VLSI) Systems*, Vol. 20, No. 1, 2012, pp. 191-196.

Chapter 16

Wearable Data Collecting and Processing Hub for Personal Vital Signs Measurement

Eliasz Kańtoch and Piotr Augustyniak

16.1. Introduction

A considerable effort in recent development of electronics has been made towards intelligent environments for aging well. Particular needs first come from cardiology with advent of Holter systems, providing circadian records for caregivers and a feeling of safety for patients. Next step was made with digital communication, its ubiquitous access allows for a pervasive telemedicine instead of limiting the range of medical services to in-hospital patients.

Although millions of people may be satisfied with watching or hearing a digital content with a standardized software, making calls or taking pictures in a similar way, a medical or scientific use of the wearable computers requires their adaptability to interpersonal differences and progress of disease. Such adaptation may easily be done with the software and technically is not much different than downloading and installing of an app on smartphone. However, from the functional point of view there is one important difference: the patient (or the healthy monitored person) should not be involved in the adaptation.

For this reason we designed a wearable data collecting and processing hub being a multi-purpose recording, stimulating, networking and processing device of flexible architecture. The idea is similar to a smartphone, however instead of voice or picture messaging this small, lightweight and power efficient computer is a remotely programmable device for biological research and health-related measurements. It is friendly for patients for its human-like adaptation, favoring personal needs and easy operation. It is a complicated, yet powerful distributed measurement computer with easy programming for medics allowing them for wide-range personalization of acquisition, interpretation and reporting. Finally it is a complex measurement, processing and communication device for

Eliasz Kańtoch
AGH-University of Science and Technology, 30, Mickiewicz Ave.
30-059 Kraków, Poland

manufacturers, suitable for mass-production due to a common hardware platform, easy software adaptation and compliance with various needs from the health care.

16.2. Related Work

Collecting data from living subjects (humans or animals) was recognized since decades as the most versatile tool for studying their physiology and behavior. Recent approaches required the presence of specimen in measurement laboratory, what influenced the data acquisition process and contradicted the fundamental assumption of measurement not influencing the observed processes. Recent advances of electronic and communication technologies allowed for designing measurement systems as networks of intelligent sensors applied and programmed according to variable measurement conditions. Most relevant paradigms of such measurement systems are presented in following sections.

16.2.1. Intelligent Sensors for Physiological Signals

Sensors usually produce a huge amount of raw data which are vulnerable to distortion and unauthorized access. The idea of 'intelligent sensor' addresses all the issues: to be called 'intelligent', the sensor should process and protect the data consequently it is a kind of digital microprocessor-based measurement system running a dedicated software and optimized for communication and autonomy (i.e. power saving). All designs presented here assume the cooperation within a measurement system supervised by a management computer system responsible for signal processing and storage.

Tseng et al. [1] proposed a biosignal-dedicated ultra-low power front-end amplifier built in 0.18 μm CMOS technology and featuring the SNR value of over 54 dB at a supply voltage of 0.4 V (0.09 μW). Its architecture included a chopper-stabilized instrumental amplifier with cancellation of common mode interference and an amplifier with programmable gain (0–30 dB in frequency bands 0.5–100 or 10–400 Hz). Another monitoring system with wireless programmability and telemetry interface suitable for various electrophysiological recordings like ECG, BCI, EMG etc. was presented by Morrison [2]. The gain can be selected on the run between 43 and 80 dB and the bandwidth between 230 and 2000 Hz. It consists of a 4-channel low-noise front-end, 8-bit ADC, MICS/ISM transmitter and infrared (935 nm) programming input (gain/bandwidth and transmission channel selection). The prototype was built in 0.13 μm CMOS technology, weights 0.6 g and consumes 1.07 mW from a 1.2 V supply. Recording of a wide range of biopotentials was also possible with a configurable 0.35 μm CMOS integrated circuit front-end proposed by Teng [3]. Its topology consists of a single-differential or double-differential recording channels with adjustable gain in a range of 37 to 66 dB with a bandwidth of 2.8 kHz. Besides the transmission and storage the system needs between 110 and 324 μW for operation. A highly-flexible miniature multi-purpose biosignal data sensor and recorder (weighting ca. 2.3 g) was also proposed by Bailey [4]. It has programmable parameters sampling frequency and sensitivity to comply with wide range of biosignal types (ECG, EEG, EMG, EOG) and a huge (8 GB) flash memory for storage of weeks of raw signals.

A comparative analysis suitable as a departure point for studying and designing of intelligent biosensors and sensing front-ends may be found in the paper by Wang et al. [5], who also proposed their own general-purpose low noise, low current biopotential amplifier with configurable bandwidth and gain.

16.2.2. Computer Systems for Biosignals Measurements

Several computer systems were conceived for laboratory setups for biosignal measurements in human and animals. These solutions usually include a multi-purpose analog front-end and a collection of signal-oriented interpretation procedures. In some products the procedures are implemented in a high level programming language for easy customization, in some others they are embedded in a locked accompanying software. The first category offer high flexibility and therefore is worth mentioning here. Peterek [6] proposed a setup composed of a general-purpose biosignal front-end g.BSamp from Gugger and acquisition mode-dependent analysis software in Matlab. A typical processing chain with configurable parameters for ECG, EEG and PPG signals can be modified or completed accordingly to the experiment needs. An alternative laboratory system for acquisition, processing and evaluation of ECG signal, blood and perfusion pressure dedicated to recording from animals was presented in [7]. Similarly in this case an analog front-end consisting of specialized instrumental amplifiers with adjustable gain and 12-bit ADC sampling at a rate up to 1 kHz per channel is accompanied by LabView- or Matlab-based custom-developed software. Third example from a paper by Chmelar et al. [8] also consists of specialized signal-related inputs, ADC with processing unit and a storage (SD card) and transmission (USB) module. The system is conceived for recording of cardiovascular signals, such as electrocardiogram (ECG) and photoplethysmogram (PPG), in this case, however, a new non-invasive method is used for blood pressure measurement without a cuff.

16.2.3. Dedicated Portable Systems for Biosignal Measurements

A common drawback of the computer-based measurements is a limited portability of the setup, being a price to pay for the flexibility of data processing. Consequently some experimental scenarios are hardly feasible or require specialized (i.e. less flexible) instrumentation. An example of such system was proposed by Liu [9] as a wireless data logger with EMG, ECG, and accelerometer transducers for field measurements. It consists of a 4-channel amplifier, a 16-bit ADC with sampling frequency up to 1 kHz and a microSD card-based storage, or ZigBee wireless interface of a range up to 100 m with 11 dBi/9 dBi directional antennas. The device weights 102 g, and requires 430 mW at full power (6 h with two batteries of 860 mAh). These parameters are clearly different to their counterparts from intelligent sensors (e.g. [3]), but this time a complete measurement system is considered. Another system for ECG, EMG, EOG, and EEG measurement [10] consists of a custom-designed digital controller to support alternately the network management and biosignal measurements. A mixed-signal system-on-chip analog front-end has programmable gain and bandwidth (45–63 dB, 0.5–1000 Hz) and complies with Intelligent Electrode and Active Cable concept [11]. The 8-bit ADC and digital core was made in 0.18 μm CMOS technology and consumes 20 μW from a 1.2 V supply. A wireless digital communication with signal storage equipment is performed by a separate low-energy Bluetooth chip. A typical configuration of an analog front-end, ADC and digital signal processing unit and a wireless transmitter shows unprecedented configurability for matrix recording. A system designed to high-density recordings of brain signals (up to 400 channels) was proposed in [12]. It was fabricated in 0.65 μm CMOS technology and a single amplifier of 2.5 μV noise and 10 kHz bandwidth consumes 17.2 μW from a 1 V supply.

In some portable systems the digital core is used for supervising of data acquisition, transmission and storage. More sophisticated designs also allow for spike detection, feature extraction and data clustering, however in most cases the procedures applied are related to usage with particular biosignals or to work with a particular scenario (e.g. fall detection). The embedded algorithms are hardly customizable and face several resources limitations that compromise the quality and reliability of results. Moreover, usage of more sophisticated algorithms requires high computational power and limits the autonomy of the acquisition system. The group of dedicated portable recording systems fills the gap besides all-purpose computer-based systems and has opposite functional characteristics.

16.2.4. Wearable Sensor Networks

The advancement in wearable sensor technology enables to build wearable, low-power acquisition systems for biosignal measurements. The most commonly used architecture of such systems consists of the Wearable Sensor Hub (WSH) and a set of wearable sensors which are placed on the body. Data from wearable sensors are acquired, processed and analyzed with dedicated algorithms by the Wearable Sensor Hub (WSH). System elements should be small enough to have high wearability and high biocompatibility. Sensor Hub generally contains a microcontroller, mass storage, a wireless communication module and a battery module. Microcontroller controls the data acquisition from sensors. Wireless communication module is responsible for communication between Wearable Sensor Hub (WSH) and computer while the battery module is responsible for providing energy for entire system. The microcontrollers used for wearable sensor networks operate at frequencies ranging from several MHz to several GHz. They are equipped with peripherals like analog-to-digital converters (ADC) and include a number of interfaces including I²C and SPI. The most common wireless communication standards for exchanging data over short distance for application in wearable sensor networks are IEEE 802.15.1 (Bluetooth and Bluetooth Low Energy), IEEE 802.15.4 (ZigBee) and IEEE 802.11 (Wi-Fi).

The subject of the application wearable sensors for vital signs measurement, was raised by many researchers. T. Klingeberg et al. developed mobile wearable device for monitoring of ECG, blood pressure and skin temperature [13]. S. Coyle discussed the signals that can be measured using smart textiles. He indicated the following: breathing, heart activity, muscle activity, blood oxygen saturation and body movement [14]. M. J. Abreu investigated the integration of off-the-shelf electronic components into textile products [15]. The number of previous studies have focused on human monitoring using various wearable sensors including heart rate monitors, body temperature sensors or accelerometers [16]. It was also demonstrated that house embedded sensors can be successfully used to monitor human behavior [17]. Qi Wang et al. investigated different wearable systems that are used mostly for the monitoring and provision of feedback on posture and upper extremity movements in stroke rehabilitation. The results indicated that accelerometers and IMUs are the most frequently used sensors, in most cases attached to the body through ad hoc contraptions for the purpose of improving range of motion and movement performance during upper body rehabilitation [18].

16.2.5. Programmable Mobile Platforms for Biosignal Measurements

The survey made in previous sections suggests that a wearable data collecting hub with flexible programming of data interpretation and transmission is highly welcome by natural physiology and behavior experimentators.

A mixed signal design proposed by Kim et al. [19] focuses on low power consumption, but it also provides high degree of configurability thanks to a custom digital signal processor. The prototype was built in a wrist-mounted housing and with power consumption of 32 μW (at 1.2 V) it allows for long time continuous ECG monitoring in various scenarios including life record, emotion monitoring and fitness. The embedded software performs ECG-specific operations like motion artifacts removal, heart beats detection, classification and arrhythmia detection based on general signal processing procedures (including Least Mean Square, Principal Component Analysis and Wavelet Transform). The innovative character of this design is also expressed by monitoring of electrode-tissue impedance, adaptive sampling, program memory loop buffer and low energy Bluetooth transmission with AES encoding. Despite the advantageous approach, implementation of advanced ECG interpretation procedures was not possible due to limited computational power.

Another proposal by Augustyniak [20] is a prototype of wearable computer with measurement-oriented operating system, biomedical signal-oriented hardware, three-ways programmable data transmission system and BSN control ability. The prototype and its programming rules will be presented in details in Chapter 16.3.

16.3. Central Hub Hardware and Management Rules

In this section we take a closer look into an example programmable data acquisition, processing and communication platform for mobile experiments with human biosignals. Such platform was prototyped in our laboratory [20] and is currently under industrial investigation as possible general-purpose wearable assisted living hub. The device has four functional blocks accordingly to generally defined purposes of application:

- Biosignal acquisition and electrostimulation module, capable to gather direct signals from the body (ECG, EMG etc.) and from transducers (light, pressure, acceleration etc.) and to produce bias current for measurements (galvanic skin response or impedance rheography) and stimuli for various therapy.

- Local wireless transceiver for communication with wearable sensors and management of body sensor network (initializing, programming, status and data querying) and connecting to other intelligent devices in active environment.

- Data processing platform with continuous resources management and reporting, dynamic distant update/modification of executable code, hardware and communication configurator and simplified user interface.

- Mass storage and communication module (wired and wireless) with optimized reporting timing (synchronous, asynchronous or event-related) and content, configurable data encryption and removable storage media.

16.3.1. Biosignal Interface

The direct measurement interface uses analog front-end ADAS1000 (Analog Devices) with five configurable gain single-ended channels, and a 24-bit resolution analog-to-digital converter with programmable data rate up to 128 kHz [21]. The chip was designed for ECG, but works equally well with other high-amplitude biomedical signals like electromyogram or electrohysterogram (signal from uterine contraction activity). Acquisition of weaker signals (as electroencephalogram or electrooculogram) is problematic due to bit resolution of only 3.27 μV/LSB, but such measurements are avoided in mobile health care due to high interferences and noise level. Differential measurements (e.g. surface electromyogram) are performed with a pair of single ended leads and mathematically transformed to a sole differential data stream. The ADAS1000 was selected for the biosignal interface for its compliance with safety requirements of Association for the Advancement of Medical Instrumentation (AAMI EC11: 1991/2001/2007, AAMI EC38 R2007, EC13:2002/2007) and International Electrotechnical Commission (IEC60601-1 ed. 3.0 b:2005 and others related) for medical electronic equipment.

Analog inputs of the ADAS1000 can also be configured for capturing electrical signals from wired body sensors of pressure, acceleration or position (e.g. magnetometers) located on the body. The connected sensors have to deliver an analog voltage signal in the range of 0 – 1.3 V with the bandwidth of 0 – 2 kHz. This is a typical specification for wearable accelerometers, other sensors usually require adaptation of interface (Fig. 16.1).

Several modern MEMS sensors, suitable for physiology-related measurements embed a digital converter and provide a direct serial data stream. For communication with wired body sensors with such output, the device provide digital serial interfaces in I2C or SPI standards. An intrinsic advantage here is a robust data transmission in presence of electrical fields from other sources.

The recorder provides a single output line of programmable voltage to supply power necessary for operation of the external wired sensors.

The ADAS1000 chip also provides four general-purpose digital I/O lines which are used as inputs from buttons of the user interface.

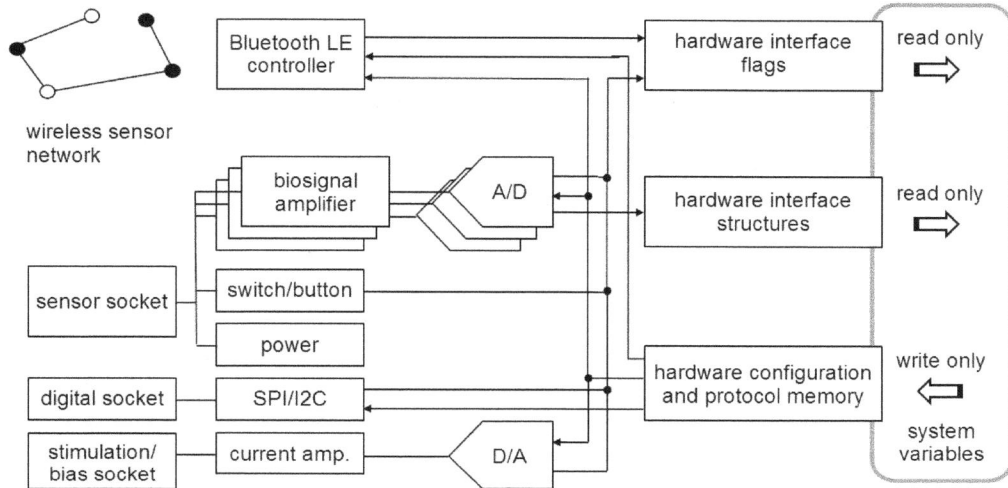

Fig. 16.1. The flow diagram of biosignal and sensor network interface.

16.3.2. Sensor Network Management Module

A wireless body sensor network (BSN) handled by the recorder, is another source of physiological data, but also improves the extensibility of the recorder. The measurement capability can include a broad range of data from the human and his or her environment including processed images or sounds and records of interaction with the surrounding equipment. Moreover, the BSN can be built adaptively of new sensors, not known at the recorder's design stage. These sensors have to be engineered accordingly to common communication rules, but including new measurement nodes to the BSN requires only updating of data exchange protocols in recorder's memory.

The address and respective communication protocol are to be configured in the recorder for correct identification, initialization, synchronization and interrogation of all devices participating in data acquisition process. Each remote sensor has a unique address identifying it in the sensor network, but operates independently and is powered with an individual battery. Sensors deployed on the body are strictly personal, but environmental sensors are not restricted to work within a unique network. The sensors have their own data processors dynamically configured by the hub in course of the acquisition.

The example BSN is organized in a star topology and uses the Bluetooth Low Energy (4.0) technology. The biosignal recording device is the central point (hub) of the BSN. Its supervisory tasks include: identification, initialization, control and reading of all subordinated sensor nodes accordingly to the programmed schedule. Three data transfer modes are provided accordingly to the function and properties of the sensors:

- Short data packets are used for identification and initialization of sensors, occasional verification of their status (e.g. battery level) or time synchronization. Due to fast connection setup the recorder synchronizes and checks subordinated sensors in short time and with very low energy expense.

- Asynchronous signal packets are employed for transmission of results locally collected and processed by the sensor. Unless a real-time interpretation is required, the difference between data collection and transfer speeds shortens connection duty cycle and thus improves the data security and reduces the power dissipation.

- Continuous signal transmission is used rarely only for seamless or delay-sensitive data.

Combining modern communication technology with power saving-oriented transmission protocols allows to design sensors operating on a single coin-cell battery for several months (e.g. [2, 3]). However, even in low-energy designs the security of wireless data transmission within the BSN is crucial. We applied an AES-CCM encryption algorithm which first establishes short-term individual channels to each remote sensor with a unique key of a length of 6 digits, and then uses this channel for broadcasting long time keys to each node of the network [22]. The configuration of the body sensor network lasts for few seconds and occurs randomly, therefore attacks like man-in-the-middle have little chance to success.

16.3.3. Device Management Kernel

The device management kernel consists of bootable, management, dataflow and security control procedures programmed at the manufacturing stage to the separate permanent storage memory area. This area is accessible uniquely for a cable-connected computer with dedicated software. This area is not accessible for data or signal processing procedures and cannot be updated remotely via wireless transmission channel. The management kernel plays an essential roles in coordinating of principal device's functions:

- Initialization and monitoring of the hardware, interpretation of configuration codes and definitions of initialization and operation status codes of the recorder and cooperating sensors.

- Handling of data transmission between successive procedures of data interpretation chain.

- Supervision of external data connectivity and file allocation on local storage media.

- Cryptographic procedures and device network identification.

- Backup procedures in compliance with Active Fault Tolerance paradigm [23].

In the recording device with unprecedented remote programmability and dynamically linked software, the issues of fault tolerance were particularly addressed in engineering of management kernel. Several factors may have impact to the system causing either malfunctioning either suspension of operating impossible to recover remotely. Among others, the management kernel has to be robust to errors in hardware configuration, including the BSN nodes, power sources and connectivity, faulty execution of interpretation software or dynamic linking, over usage of resources, erroneous interpretation result or limited availability of data recipient.

Based on initial values of hardware configuration vector the device is set up at the beginning of the monitoring session. Reconfiguration is possible as many times as needed in the course of measurement by simply uploading of new values. The present configuration is reported in a similar vector form suitable for a direct comparison. Each configuration parameter is accompanied by a decision switch allowing for a priori selection of system behavior in case of unsuccessful action (retrying, omitting, breaking). Similar mechanism was applied for supervising of execution of interpretation procedures chain. Each procedure is followed by a vector of limit values (e.g. execution time, number of communication attempts, memory and processor usage etc.) accompanied by fault decision switches. The present values of these parameters are gathered in the respective format allowing for comparison and reporting. Regular automatic inspections of the hardware and software are driven by an interrupt procedure of the kernel. This procedure updates a set of status words every 10 ms and software selected values are included to a general status report.

Another security mechanism buffers the raw data during the dynamic linkage of new interpretation procedure. All signals are first interpreted by previously used procedure set, next the interpretation is repeated with new procedures and both results with appropriate time stamps are transmitted to the server for acceptance. If the acknowledgement is received, further processing is performed with updated procedures, otherwise the obsolete code is reloaded to the RAM and executed. The memory map of the wearable data collecting and processing hub is presented in Fig. 16.2.

For the evaluation of the quality of interpretation locally performed by the recorder, selected strips of raw signals are appended to the report for a redundant interpretation in almost unconstrained computational environment. To this point the configuration attributes include parameters of time interval and length of signal excerpts transmitted directly to the server for the ground truth interpretation. Accordingly to the configuration, the evaluation of recorder-side interpretation is triggered as time-dependent, event-dependent or on the demand from the server.

Two alternative data recipients: the Secure Digital card and the wireless transmission (WiFi) make the operation cheap and reliable or immediate and interactive. In case of inefficient wireless transmission the kernel employs a procedure of local backup and prioritized asynchronous transmission of all outgoing messages. Interpretation results, raw data strips and status reports are automatically redirected to the storage card and queued for immediate transmission when the connection is restored. During the backup storage, the recorder is not controllable by the server. The absence of outgoing acknowledgment informs the server that reprogramming eventually initialized in this time was ineffective.

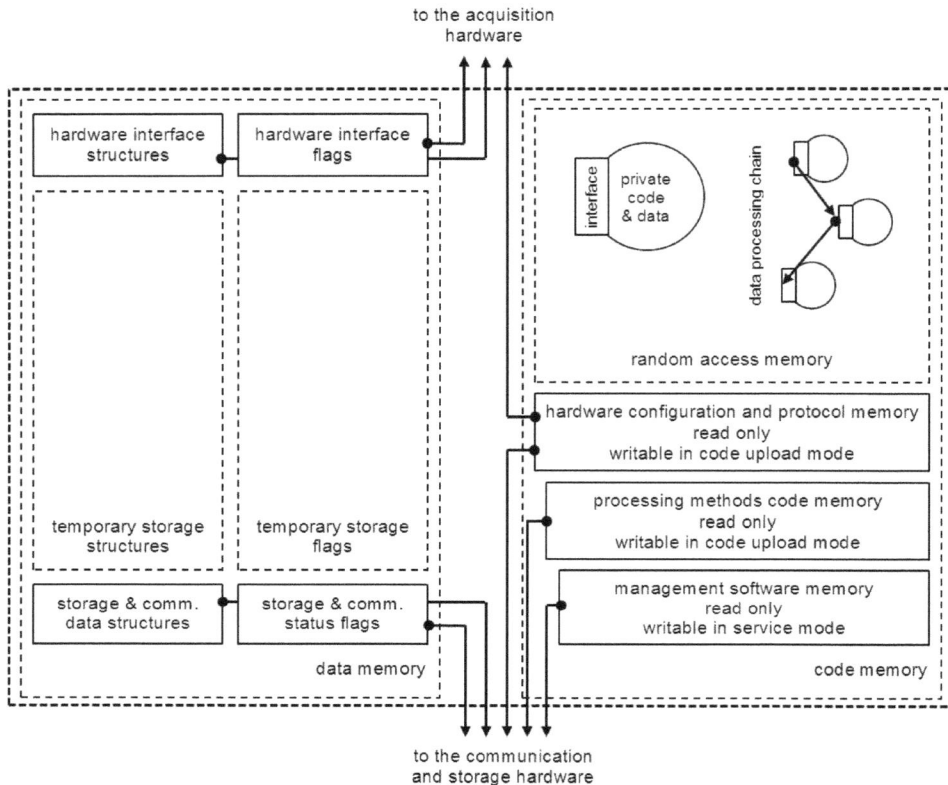

Fig. 16.2. Memory map of the wearable data collecting and processing hub.

16.3.4. Storage and Communication Module

The storage and communication module consists of specialized circuitry and cooperating embedded software for external data communication. Three independent channels are provided for immediate communication and storage of results of physiological measurement in software-selected form (raw or processed).

Long-distance wireless communication is main communication channel for immediate reporting of the wearer health status, environment and device status as well as for remote programming and adjustment of recorders function (e.g. procedures of interpretation chain). This channel uses the Wi-Fi protocol and has software-controlled limited access to the program memory and read-only access to the content of data containers defined as output from data processing procedures (Fig. 16.3).

Removable data storage card can be selected as a data recipient for long-term recordings or as a data buffer for time-delayed reporting in case when a wireless connection is inefficient. The storage uses a micro Secure Digital (SDHC) standard flash memory card of maximum capacity of 32 GB. The content of the card is stored in a standard FAT32 file format and can easily be accessed by an external card reader or via wired connection.

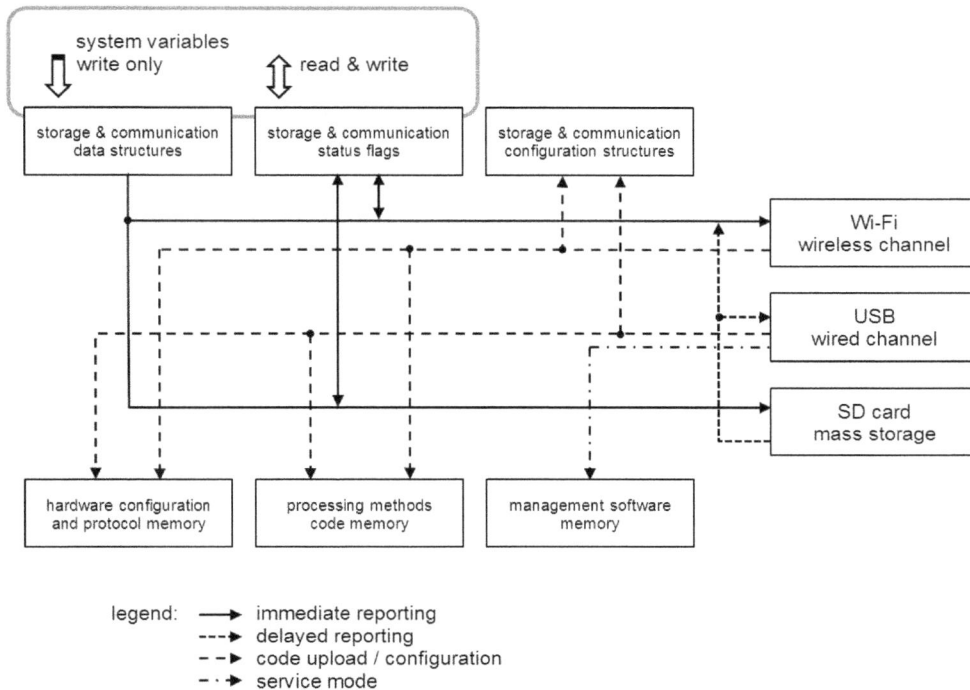

Fig. 16.3. Diagram flow of the storage and communication module.

Wired digital communication is used for read out of time-delayed results of the measurement, but also to remote programming and adjustment of recorder's interpretation procedures. This is the only way for updating the procedure of recorder's management kernel. The wired connection uses a USB 2.0 standard (in MTP mode) and offers a direct read-write access to the program memory and removable data storage. These areas are mountable as independent hard disks in a host operating system. Simultaneous operation of the analog front-end and the wired connection is not possible due to safety regulations on interconnecting medical devices.

Although all three data recipients may be programmed for independent usage, a scenario of simultaneous usage of the wireless reporting and local storage of raw signals is particularly advantageous. The diagnostic parameters calculated by on-board procedure chain and immediately available to the caregiver may be verified by later off-line interpretation based on original raw signal stored in high capacity flash card.

16.4. Building a Network of Cooperating Sensors

We designed multi-sensor system architecture for vital sign measurement. The system consists of: wearable sensor hub (WSH) connected to wearable sensors and infrastructure sensors, which are located in different everyday objects in human environment i.e. TV etc.

The system was designed to simultaneously acquire signals from multiple sensors. If a user starts to interact with the object (infrastructure) data are exchanged between the object sensor and the WSH. We investigated the following sensors for integration with the system: accelerometers, body and environment temperature meters, pressure and light intensity sensors. The system architecture was shown in Fig. 16.4.

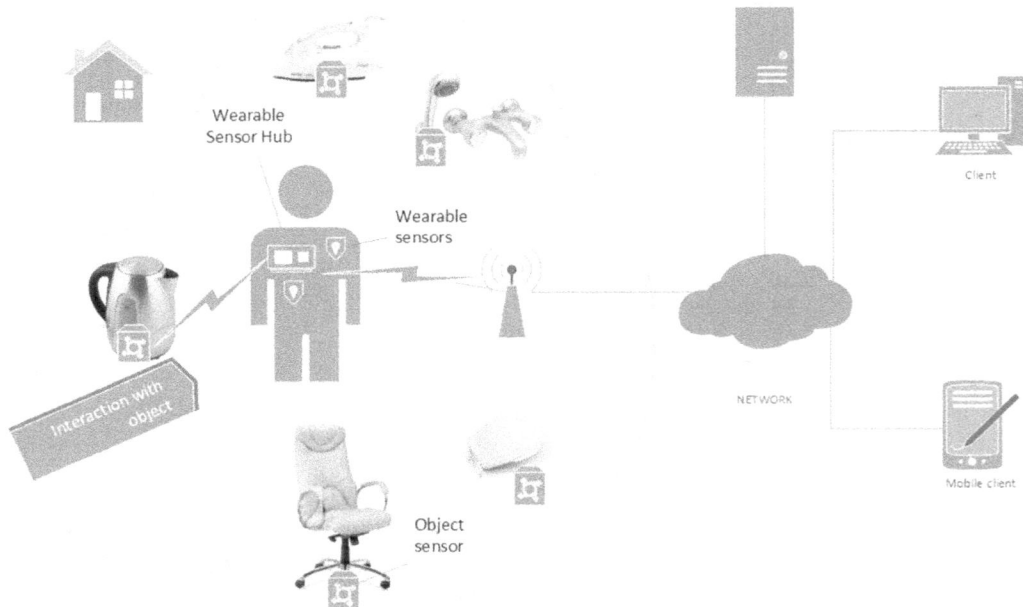

Fig. 16.4. System architecture.

One of prototype implementation of wearable sensor hub was implemented in our laboratory. The device consists of MEMS and analog sensors connected to the microcontroller and the circuits supporting the power supply from Li-Poly battery which can be easily charged with USB port. The prototype is embedded inside the plastic casing. The prototype is equipped with the universal IDC connector supporting the ISP programming and connection of external wearable sensors. The prototype takes advantage of the advanced MEMS sensors dedicated to the specific measurement of environmental parameters. The MPU-6050, used for motion detection, is the integrated 6-axis

MotionTracking device combining a 3-axis gyroscope, 3-axis accelerometer and Digital Motion Processor. It features high resolution and can be used to perform advanced motion analysis with the use of motion processor. The BMP-085 sensor performs the barometric pressure measurement and in conjunction with altitude calculations may be used in the systems monitoring the height above sea level. The DS18b20 digital thermometer is utilized for ambient temperature measurements. A volunteer during measurement process was shown in Fig. 16.5.

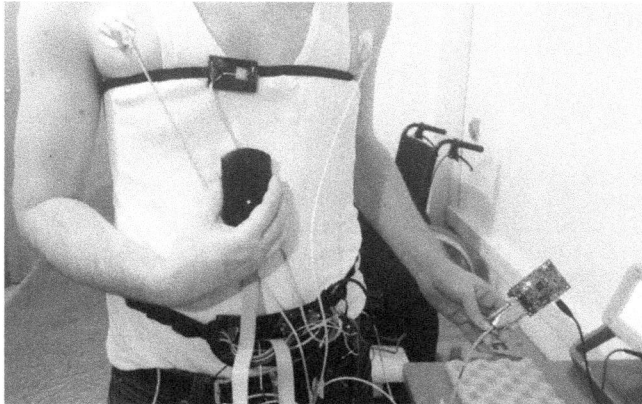

Fig. 16.5. The volunteer during measurement process.

16.5. Sensor Management Protocols

The Wearable Sensor Hub (WSH) uses the sensor pooling method to control data acquisition from the sensor network and available object sensors [24]. The WSH requests data from both wearable and identifies nearby object sensors. The WSH performs data acquisition, processing and analyzing tasks. The presence of object sensor triggers interaction between the WSH and object sensor. The WSH collects also information about the following states: sensor state, link state. Sensor management system performs a variety of management control tasks based on collected network state such as changing the sampling frequency or switching the sensor on/off. The WSH, wearable sensors and object sensors work under resource constrains such as limited battery power, memory and computational power. The communication protocol operates in the Application Layer of the Internet Protocol Suite (Layer 7 of the OSI model). The sensors interaction diagram was shown in Fig. 16.6.

16.6. Data Processing Platforms

The software of wearable data collecting and processing hub was designed accordingly to a functional layer paradigm [20] and consists of: data acquisition, data processing and reporting layers. Separate layers are independently modified by the configuration commands accordingly to their specificity, and subject to further individual development.

Operation of the layers is supervised by the management kernel being in charge of resources allocation, data flow and security and active fault tolerance.

16.6.1. Wearable Data Processing Platform

Wide range programmability of patient-side interpretation of acquired biosignals and accompanying data is main novelty but also challenge of the wearable hub. This functionality was carefully designed with respect to unconstrained programming and processing efficiency.

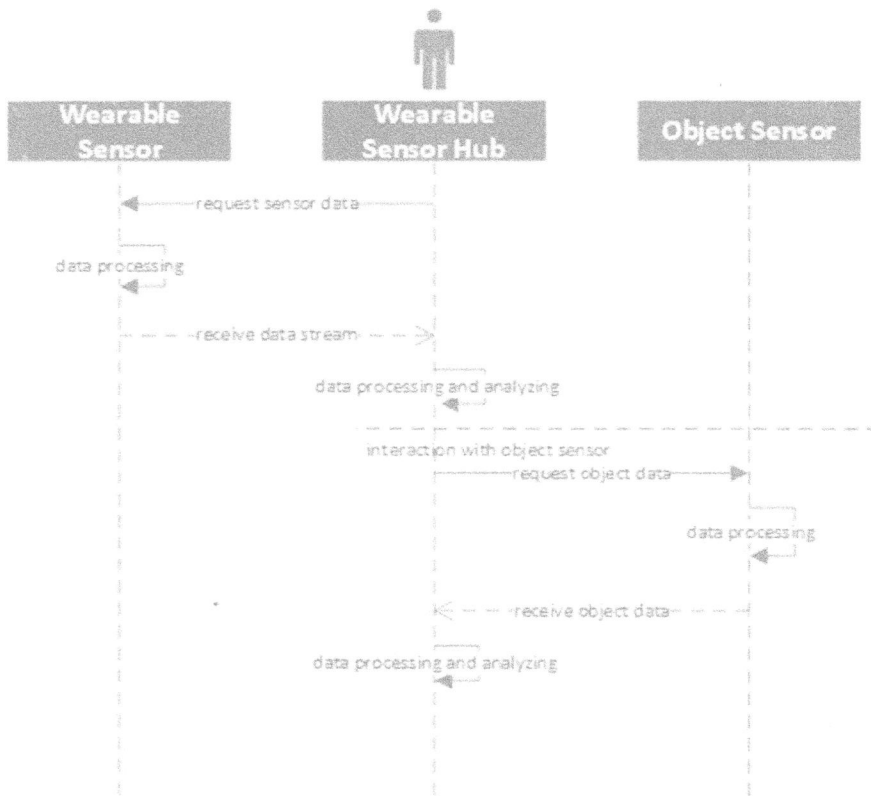

Fig. 16.6. Sensors interaction diagram (sequence diagram).

The data representation uses three kind of variables: numbers, vectors and frames. Predefined variables are used for interfacing between the layers and for communicating of the acquisition layer and reporting layer with corresponding hardware. Examples of such default structures bridging between the acquisitions and processing layers are: acquisition configuration and acquisition status. Additional variables may be randomly defined for temporary storage of the data in all layers. The data memory can be dynamically allocated when new variables are declared or the existing ones are deleted.

Data structures accessed by active procedures are locked until respective procedure is deactivated.

Data processing procedures are programmed accordingly to the object programming paradigm as methods with public-mode access to its interface. The variables used as elements of interfaces are accessible in public memory area and thus are useful for building of procedure chains or splitting of the data flow. Besides the interface, each procedure is attributed with the public vector for configuration and status report. The configuration vector is modified by the reporting layer (working in code configuration mode) but is protected from being overwritten by the procedure code. The status vector is written by the procedure code and read by the error tracking routine being part of management kernel. The executable code and internal storage space of each method are private and hidden for external calls. The prototype executable code of each method is stored in a separate program memory and copied to the random access memory during the software linking. This allows for activating several independent instances of each method within a single interpretation chain. The content of program memory can be modified by the reporting layer (working in code upload mode), but it is protected from being overwritten during the execution of processing code. In result of the code upload, a new method may be dynamically linked to the processing chain. The obsolete method is unlinked once the program counter no longer points to its code.

Analysis of the usage of certain data processing procedures most commonly used with biomedical signal led us to distinguishing three categories of methods:

- Simple and versatile procedures (such as moving window average, digital filtering or heart beat detector) loaded by default to the recorder's program memory.

- Compound procedures of moderate complexity and resources requirement (such as ST-segment investigation or time-domain HRV assessment) used less often and loaded from a remote repository to the recorder's program memory on a specific demand.

- Compound procedures of high complexity or high resources requirement (such as respiration wave reconstruction or frequency-domain HRV assessment) not suitable for run on the wearable data processing platform, and thus performed by a central server software thread (see 16.6.2).

Execution of user's software is supervised by a specialized routine of the management kernel. It is launched by default at the beginning of the recorder's operation, and performs tasks related to resources allocation and monitoring, dynamic code linking, data consistency and security and active fault tolerance.

16.6.2. Central Data Processing Platform

Although the wearable hub can acquire and process biosignals and collect data into the internal storage, revealing all advantages of the reconfigurable health monitoring system requires the appropriate server software. We assume that every wearable hub is subordinated to a caregiver's server, which is a multi-user machine of high computational power to guarantee the scalability of the monitoring service for several thousands of individuals. Since our design assumes the independence of each client with separate memory areas and processes, adequate description of distant cooperation between the server and the recorder is based on a one-to-one model. The role of the server is twofold: firstly it collects all health-related data from the monitored individual and participates in the interpretation process and secondly it maintains the control over the recorder functions and configuration (Fig. 16.7).

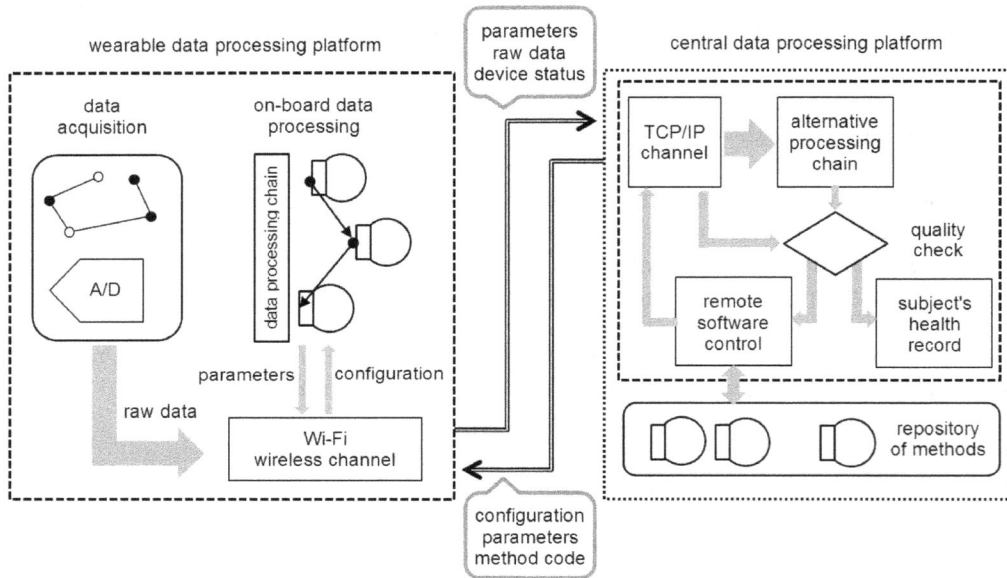

Fig. 16.7. Cooperation of wearable and central processing platforms.

Collecting of the data is based on the setup of data acquisition and processing layer and uses respective data structures declared as components of individual's longitudinal record. These structures are defined as continuous (e.g. for seamless signal storage) or occasional (for storage of description of irregular events). Since the wearable hub and the server are both capable of biosignals interpretation, this process can be distributed between the two accordingly to requirements and external conditions:

1. The recorder captures and collects raw signals, as a typical long-term (e.g. Holter) recorder. The acquired raw signal is available for the interpretation software once the data acquisition and transfer are completed.

2. The recorder captures and transmits raw signals, as a typical interactive distant recorder. The acquired raw signal is immediately available for the interpretation software provided there is a continuous access to the wireless network.

3. The recorder processes the acquired signals and collects the diagnostic results, as a typical independent remote recorder. Thanks to immediate diagnosis and recorder-side storage of diagnostic data, continuous availability of data carrier is not required. However, the quality of diagnostic results may be compromised by limitations of interpretative software running on a wearable platform.

4. The recorder processes the acquired signals and transmits the diagnostic results along with the representative strips of source raw signals. The outcome of the recorder-side processing may be verified or corrected by complementary server-side software. This mode combines the advantages of immediate results availability and unrestricted processing quality.

Supervision of the wearable recorder is maintained continuously by a client-related instance of server-side software, whenever the monitoring is activated. The server is reading the recorder's status even if health-related reports are suspended by configuration and builds a functional mirror of recorder's signal processing software. The human operator can thus design his or her own data processing chain, compile and send it for a remote execution and then verify the data flow and processing results with the help of incoming status reports. Similar mechanism was also used in the research on automatic optimization of remote recorder software with use of libraries of precompiled methods uploaded upon necessity to the recorder [25].

16.7. Discussion

The idea of the wearable data collecting and processing hub was implemented in an experimental platform based on a development kit of the PXA-270 CPU [26] and custom-built analog circuitry (ADAS1000 [21]) and storage/wireless communication unit (CC3100MOD [27]). This platform provides all the functional features except for the size of $100 \times 104 \times 12$ mm, which is about 5 times larger than the intended target product (Fig. 16.8).

Technical details along with conditions and results of various tests are presented in [20]. Actual evaluation with volunteers playing out realistic behavioral scripts was based on cardiac and motion signals. Besides the correct functionality of the prototype, it confirmed several unprecedented features as:

- Hardware reconfiguration, connectivity of wired and wireless sensors on the run;
- Recognition and data interchange with beacons mounted in active environment;
- Programmability, uploading and linking of data interpretation procedures on the run;
- Adaptive, conditional and prioritized communication and report format;
- Multicriterion power saving options.

The prototype was also tested in an auto-adaptive scenario. The programmed measurement protocol used one ECG and three accelerometer channels to monitor the heart rate at rest, but in case a certain subject's mobility was detected, two additional ECG channels were enabled to monitor exercise-related changes of ST-segment and avoid motion artefacts. This scenario was found particularly interesting by medical scientists, who would use the recorder in its target small form factor in several *in-vivo* research programs for human behavior and functional health.

The prototype was built with an aim to bridge a gap between intelligent sensors for biosignals and computer-based biosignal recording systems. Without loss of generality it was tested mainly with ECG signals which is a handy example for general investigations. Among other biosignals, for the ECG there exist best clinically-evaluated methods for automatic detection of components, best knowledge on physiological backgrounds and best documented annotated databases of reference records.

Fig. 16.8. An example prototype of wearable data collecting and processing hub.

For the lack of medical procedures of remotely controlled transcutaneous electrical nerve stimulation (TENS) we could not practically evaluate this feature. Nevertheless, with offering a technical background to fully control the stimulation, we hope that therapeutics from several domains will soon show their interest for using this effective form of electroanalgesia in patients' homes.

Several platforms are developed each year for distant acquisition of biosignals for both: telemedical and assisted living applications. Most of them are combined of data acquisition and transmission modules with only occasional on-board data processing. The processing routines are always embedded in the read only memory with only few adjustable parameters (e.g. age, weight). Main novelty of our approach consists in making the processing remotely programmable. In result the wearable data collecting and processing hub can be dynamically adapted to the needs of each particular wearer. This makes the telemedicine a pervasive tool for supervising several health aspects and ameliorate the quality of life of most citizens.

Apart from popular smartphone operating systems like Google Android or Apple iOS there are operating systems dedicated for smartwatches like Android Wear (released March 2014) and WatchOS (released April 2015). Future works will include integrating our system with the Android Wear. The Android Wear is an operating system based on modified Linux kernel. It supports 32-bit ARM, MIPS and x86 platforms as well as Bluetooth and Wi-Fi connectivity. In context of monitoring daily activities, we focused on the following advantages of using Android Wear OS: acceleration of software development using Android Wear API, acceleration of computation using multi-core processors, custom user interfaces (UI) for wearable devices, a large number of available devices from different hardware vendors and a support for Bluetooth and Wi-Fi connectivity.

Acknowledgment

The scientific work supported by the AGH University of Science and Technology in Krakow, under the grant No. 11.11.120.612.

References

[1]. Y. Tseng, Y. Ho, S. Kao, C. Su, A 0.09 µW low power front-end biopotential amplifier for biosignal recording, *IEEE Trans. Biomed. Circ. and Syst.*, Vol. 6, Issue 5, 2012, pp. 508-516.

[2]. T. Morrison, M. Nagaraju, B. Winslow, A. Bernard, B. P. Otis, A 0.5 cm² four-channel 1.1 mW wireless biosignal interface with 20 m range, *IEEE Trans. Biomed. Circ and Syst.*, Vol. 8, Issue 1, 2014, pp. 138-147.

[3]. S.-L. Teng, R. Rieger, Y.-B. Lin, Programmable ExG biopotential front-end IC for wearable applications, *IEEE Trans. Biomed. Circ and Syst.*, Vol. 8, Issue 4, 2014, pp. 543-551.

[4]. C. Bailey, G. Hollier, A. Moulds, M. Freeman, J. Austin, A. Fargus, T. Lampert, Miniature multisensor biosignal data recorder and its evaluation for unsupervised Parkinson's disease data collection, in *Proceedings of the 5th International Conference on Sensor Device Technologies and Applications (SENSORDEVICES'14)*, 2014, pp. 84-92.

[5]. T.-Y. Wang, M.-R. Lai, C. M. Twigg, S.-Y. Peng, A fully reconfigurable low-noise biopotential sensing amplifier with 1.96 noise efficiency factor, *IEEE Trans. Biomed. Circ and Syst.*, Vol. 8, Issue 3, 2014, pp. 411-422.

[6]. T. Peterek, M. Augustynek, P. Žůrek, M. Penhaker, Global courseware for visualization and processing biosignals, in *Proceedings of the World Congress on Medical Physics and Biomedical Engineering*, 2009, pp. 404-407.

[7]. P. Kalavský, V. Rosík, S. Karas, M. Tyšler, Measuring system with compound software architecture for measurement and evaluation of biosignals from isolated animal hearts, in *Proceedings of the 8th International Conference MEASUREMENT*, Smolenice, Slovakia, 2011, pp. 379-382.

[8]. M. Chmelar, R. Ciz, O. Krajsa, J. Kouril, Biosignal data acquisition and its post-processing, advances in biology, bioengineering and environment: http://www.wseas.us/e-library/conferences/2010/Vouliagmeni/BIOLED/BIOLED-14.pdf

[9]. Y-P. Liu, H-C. Chen, P-C. Sung, Wireless Logger for Biosignals, *International Journal of Applied Science and Engineering*, Vol. 8, Issue 1, 2010, pp. 27-37.

[10]. G. Yang, Hybrid integration of active bio-signal cable with intelligent electrode. Steps toward wearable pervasive-healthcare applications, PhD Thesis, *KTH Information and Communication Technology*, Stockholm 2012, http://kth.diva-portal.org/smash/get/diva2:610512/ FULLTEXT01.pdf

[11]. G. Yang, J. Chen, Y. Cao, H. Tenhunen, L.-R. Zheng, A novel wearable ECG monitoring system based on active-cable and intelligent electrodes, in *Proceedings of the 10th*

International Conference on e-health Networking, Applications and Services (Healthcom'08), 2008, pp. 156-159.

[12]. R. J. Chandler, A system-level analysis of a wireless low-power biosignal recording device, UCLA Electronic Theses and Dissertations 2012, http://escholarship.org/uc/item/1836k3z4

[13]. T. Klingeberg, Mobile wearable device for long term monitoring of vital signs, *Computer Methods and Programs in Biomedicine*, Vol. 106, Issue 2, 2012, pp. 89-96.

[14]. S. Coyle, Medical Applications of Smart Textiles. Multidisciplinary Know-How for Smart-Textiles Developers, *Woodhead Publishing Ltd*, Cambridge, UK, 2013, pp. 420-443.

[15]. M. J. Abreu, H. Carvalho, A. Catarino, A. Rocha, Integration and embedding of vital signs sensors and other devices into textiles, in *Proceedings of the International Conference and Exhibition on Healthcare & Medical Textiles (THE MEDTEX'07)*, 2007.

[16]. E. Kantoch, BAN-based health telemonitoring system for in-home care, in *Proceedings of the Computing in Cardiology Conference (CinC'15)*, Nice, 2015, pp. 113-116.

[17]. P. Augustyniak, E. Kantoch, Turning domestic appliances into a sensor network for monitoring of activities of daily living, *Journal of Medical Imaging and Health Informatics*, Vol. 5, Issue 8, 2015, pp. 1662-1667.

[18]. Qi Wang, Interactive wearable systems for upper body rehabilitation: a systematic review, *Journal of Neuro Engineering and Rehabilitation*, Vol. 14, 2017, p. 20.

[19]. H. Kim, S. Kim, N. Van Helleputte, A. Artes, M. Konijnenburg,, J. Huisken, C. Van Hoof, R. F. Yazicioglu, A configurable, low-power mixed signal SoC for portable ECG monitoring applications, *IEEE Trans. Biomed. Circ and Syst.*, Vol. 8, Issue 2, 2014, pp, 257-267.

[20]. P. Augustyniak, Remotely programmable architecture of a multi-purpose physiological recorder, *Microprocessors and Microsystems*, Vol. 46, 2016, pp. 55-66.

[21]. Analog Devices Web Portal, http://www.analog.com/

[22]. J. Padgette, K. Scarfone, L. Chen, Security Guide to Bluetooth - Recommendations of the National Institute of Standards and Technology, Special Publication 800-121, Revision 2, http://nvlpubs.nist.gov/nistpubs/SpecialPublications/NIST.SP.800-121r2.pdf

[23]. Y. Zhang, J. Jiang, Bibliographical review on reconfigurable fault-tolerant control systems, *Annual Reviews in Control*, Vol. 32, 2008, pp. 229-252.

[24]. F. Sanfilippo, K. Y. Pettersen, A sensor fusion wearable health-monitoring system with haptic feedback, in *Proceedings of the 11th IEEE International Conference on Innovations in Information Technology (IIT'15)*, 2015, pp. 262-266.

[25]. P. Augustyniak, Autoadaptivity and optimization in distributed ECG interpretation, *IEEE Transactions on Information Technology in Biomedicine*, Vol. 14, Issue 2, 2010, pp. 394-400.

[26]. Toradex Web Portal, https://www.toradex.com/computer-on-modules/colibri-arm-family

[27]. TI Web Portal, http://www.ti.com/

Chapter 17

Challenges and Approaches for Assurance of Safety-Critical Programmable Hardware

Jaspal S. Sagoo

17.1. Introduction

Rapid technological advances within the semiconductor industry have meant that complex logical functionality can be implemented on programmable electronic hardware devices (such as CPLDs, FPGAs and ASICs). These devices can not only implement complex combinatorial logic functions but also complete systems using single or many core processors [1, 2].The main benefit of increased programmable capacity is the reduction in the overall component count, which has had an impact on reducing the power consumption and weight of a system whilst increasing its reliability. Also the availability of many design and simulation tools [3] coupled with well-defined design methodologies [4] has meant that the process of programmable hardware design has become automated and easier to apply. This is in contrast to the previous era when relatively simple logic functionality was implemented on discrete logic devices but which required considerable effort to design and test. Thus the versatility and flexibility of programmable electronic hardware has meant that these devices are ubiquitous as they are used in everyday computer based applications as well as specialised systems such as within military and avionics applications [5].

Within the avionics and military domains, where electronic systems tend to be safety critical, a greater emphasis is placed on providing assurances that these systems function as intended and in a safe manner [6]. In producing these systems, a developer needs to provide several levels of assurances in their development. Firstly, the developer needs to ensure that the system meets all the customer requirements. The developer would also need to ensure that a safety process has been applied throughout the system development which has identified and mitigated (or removed) all known hazards. Using the developmental and safety assessment evidence, generated during the systems development, the developer would need to formulate a reasoned argument to show these systems are safe. Typically, this argument can be encapsulated within a structured

Jaspal S. Sagoo
QinetiQ, St Andrews Road, Malvern, Worcestershire WR14 3PS, United Kingdom.

argument framework known as a safety case [7].The developer would need to provide assurances to the customer and a certification, or regulatory, authority, that the system is 'fit for purpose' and is safe [8]. In particular the customer (or certification authority) may conduct an independent technical review (such as an audit) of the developmental and safety evidence in order to gain sufficient assurances in the system. Due to the various levels of assurance required in a system development, assurance tends to be a rigorous and costly, but necessary, activity.

This chapter examines the challenges that are faced in the assurance of programmable electronic hardware for safety applications. Specifically, several approaches are presented to illustrate how assurance is provided for systems which contain programmable electronic hardware that have been developed using different approaches. The type of systems considered are those that are developed 'in-house', where a developer can provide a greater level of assurance, and systems that are constructed from 3rd party Commercial-Off-The-Shelf (COTS) or Intellectual Property (IP) items, were limited or no developmental evidence exists to provide assurance. This chapter presents a case study which shows how assurance may be provided for the latter category. The case study considers a fuel pump control system that uses programmable electronic hardware to perform the control function.

17.2. Challenges in Assurance

The process of providing assurance involves using a set of planned and systematic actions to gain confidence that a product or process satisfies its requirements [9]. From a system developer's perspective, this involves ensuring that the requirements have been satisfied within the various phases of the system development and that evidence (in the form of documentation) exists to show the satisfaction of these requirements. From a customer's perspective, assurance involves gaining enough confidence from the developer, often by reviewing developmental evidence, that the system requirements have been satisfied.

The assurance of modern day programmable electronic hardware is a challenging activity [10] and this is due to factors such as:

1. The designer no longer uses discrete logic (such as AND or gates) to create a logic function; instead programming constructs from hardware description languages (such as VHDL) are used. These languages create a design abstraction (which is known as the Register Transfer Logic (RTL)) that models a digital synchronous circuit in terms of the flow of digital signals within the hardware registers and the logical operations that are performed on these signals. Hence, the process of logic design has become very high level (i.e. abstracted from the electronic components) and is more akin to software programming.

2. The actual 'gate level' description of the logic design is obtained from the RTL description using automated tools via a process called synthesis. Although these tools are well-established and tend to have a successful track record of use, the designer has limited or no control of how the RTL is translated into the gate level

description. Hence, the designer is reliant on activities such as simulation and testing to verify the logic design.

3. The logic designs may not be fully testable because it may be impossible to set test points in all parts of the design. Thus, the logic design verification may need to be achieved by a combination of testing, simulation and design reviews.

4. The abstract nature of logic design means that a greater emphasis is placed on ensuring that the logic functionality is developed in accordance to a well-defined methodology that is compliant to accepted national or international standards. Typically, a well-defined methodology should contain:

   - Distinct development stages such as requirements, design, detailed design, implementation and testing.

   - Processes that control the hardware development such as configuration management and quality assurance.

   - Verification and validation processes that are applied throughout the development stages.

   - Activities that require the generation of documentation which show that the development methodology has been defined and the system requirements have been satisfied by the hardware development.

5. The logic development needs to ensure that a safety process has been applied throughout the development that allows:

   - Identification of the hazards associated with the logic functionality. Based on the criticality of these hazards the safety level of the hardware needs to be defined.

   - Mitigation (or removal) of hazards throughout the various stages of the development.

The above factors show that detailed information about a system's development processes and documentation is needed to provide assurance. Although this information may be readily available for systems that a developer has constructed, it presents challenges for systems built using COTS devices and IP. Hardware COTS are those items that are purchased as pre-programmed devices and IP are purchased as pre-developed hardware designs that could be inserted into the hardware device. Whilst these items are desirable because they reduce the time-to-market and developmental costs, no evidence may be available to show what methodology had been used in their development. When these items are used in a system, a developer would need to satisfy themselves (as well as the customer and certification authority) that sufficient evidence is available to show that the system is safe to use and satisfies its requirements, without any undesirable emergent behaviours.

17.3. Assurance Standards

In the avionics domain, guidance for showing assurance in programmable electronic hardware devices is detailed in the standard DO 254 [11]. This guidance was developed not only due to the increased programmable capacity of these devices but also because developers were opting to implement complex functionality on these devices (for which there was no guidance and standards) rather than in software (the development for which was regulated by standards such as DO 178 since the 1980s; the latest version being DO 178C [12]).

Although DO 254 was developed in 2000, it was considered as means of compliance by the Federal Aviation Administration (FAA) in 2005 and subsequently it has gained widespread use. Within the UK military avionics domain, DO 254 is considered by standards such as DEF STAN 00-970 Part 13 [13] to be an acceptable means of compliance for providing design assurance in airborne hardware when supported by a robust, documented and auditable safety assessment (as described in DEF STAN 00-056 Issue 7 [14]).

An examination of the guidance detailed in DO 254 shows that it was essentially the culmination of the best practice available, at the time of publication, in the development of airborne electronic hardware. The standard is based on the premise that a system level hazard analysis has been performed from which the hazards for the hardware functionality can be identified. These hazards are used to allocate a hardware design assurance level (DAL) and define the hardware safety requirements. Within DO 254, the DALs are defined as A, B, C, D and E which correspond to failure conditions of catastrophic, hazardous/severe major, major, minor and no effect, respectively.

DO 254 does not stipulate that a specific methodology should be followed by a programmable electronic hardware development, but defines a generic hardware lifecycle process (see Fig. 17.1) that should be followed and a set of certification objectives that should be satisfied. The generic hardware lifecycle defines:

1. Hardware Design Processes (requirements, conceptual design, detailed design, implementation and production transition).

2. Hardware Planning Process.

3. Validation and Verification Process.

4. Configuration Management Process.

5. Process Assurance Process.

6. A documentation set that needs to be generated for each of the DALs.

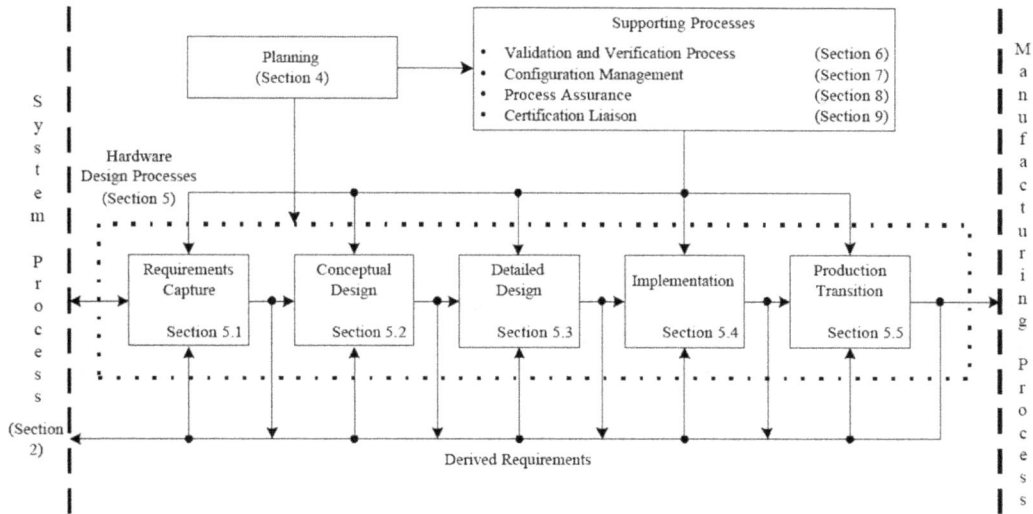

Fig. 17.1. DO 254 Hardware Lifecycle Processes (diagram reproduced from DO 254 [11] - the section numbers shown in this diagram are references to the sections with DO 254).

Additional guidance on the interpretation and application of DO 254 is provided by documents such as Certification Authorities Software Team (CAST) 27 [15], FAA Complex Electronic Hardware (CEH) Job Aid [16] and EASA SW CEH 001 document [17].

When following the DO 254 guidance, a developer would need to form a hardware design assurance strategy to ensure that the device will be developed to the assigned DAL. This strategy would firstly require the DAL for the device to be defined and a classification of whether the hardware device is simple or complex. Within DO 254, a device is classified as simple if a comprehensive combination of deterministic tests and analysis can ensure that correct functional behaviour is performed under all foreseeable operating conditions with no anomalous behaviour. A complex device is defined as a one that is not simple. Although DO 254 can be applied to the development of simple devices, it is essentially applicable to complex devices (which are referred to as complex electronic hardware). For devices that are classified as DAL A or B (i.e. higher levels of criticality) DO 254 requires additional design assurance strategies to be applied. These are listed in DO 254 and can include the use of:

1. Architectural Mitigation.

2. Product Service Experience.

3. Elemental Analysis (such as code coverage).

4. Safety Specific Analysis.

5. Formal Methods.

461

The guidance of DO 254 requires that certain developmental activities are performed with independence from the design process and these include process assurance, configuration management and verification of DAL A and DAL B hardware. Hence, the developer would need to ensure that this independence is achieved within the project's organisational structure and that documentary evidence exists to show this independence.

17.4. Assurance of Developer's Hardware Development

When implementing DO 254 [18, 19] a developer commences with the formation of a plan that shows how the certification objectives of this standard will be satisfied. This plan is referred to as the Plan for Hardware Aspects of Certification (PHAC) and provides:

- Overview descriptions of the system and hardware.

- A declaration of the hardware DAL with justifications based on safety assessments.

- Definition of the hardware lifecycle processes that will be used to develop the programmable hardware.

- Definition of the hardware lifecycle data (such as hardware plans, hardware designs and test data) that will be generated as evidence to meet the DAL allocated to the hardware.

- Definition of any additional considerations that will be taken such as the use of COTS or legacy hardware.

- A schedule for achieving certification.

Once the PHAC has been finalised within the developer's organisation it would need to be sent to the certification authority or the customer for approval.

On approval of the PHAC, the developer would commence the hardware development by executing the phases of specification, design, implementation and testing, whilst ensuring that certification objectives of DO 254 are satisfied. However, during this hardware development, the developer would need to involve the customer (or the certification authority) in order to provide assurances that it is progressing as specified in the PHAC. In this assurance task, the certification authority or customer may decide to conduct their own independent reviews of the hardware development. Typically, these reviews are based on the guidance provided by the FAA CEH Job Aid [16], which presents a series of reviews known as Stage of involvement (SOI) reviews. These reviews are summarised as follows:

1. SOI#1 Planning Review – this review is performed after the completion of the hardware planning stage, when the hardware planning documentation has been generated and placed under configuration control.

2. SOI#2 Design Review – this review is performed when the hardware requirements have been documented and reviewed, and shown to be traceable to the system requirements. Additionally, the detailed design and implementation stage must have been completed and traceability to the hardware requirements must be shown.

3. SOI#3 Verification and Validation Review – this review is performed when test procedures and test results have been documented and reviewed. Also the detailed design must be shown to satisfy the hardware requirements and the test coverage analysis should have been performed.

4. SOI#4 Final Review – this review occurs when the conformity of the hardware development has been shown to all hardware life cycle processes.

In conducting these reviews an independent assessor (such as the customer or certification authority):

- Has visibility of the hardware development;

- Can review the application of the hardware lifecycle processes and data;

- Can provide feedback to the developer on the CEH development.

For each SOI review, the CEH Job Aid [16] provides a set of questions, which are meant for guidance only, that the independent assessor could consider when performing the review. The most useful aspect of these reviews is that the assessor is closely involved with the hardware development and the developer can receive early indications of when the development may not be in compliance to DO 254, so that remedial action can be taken.

17.5. Assurance of Hardware COTS and IP Items

COTS and IP items are typically supplied by the manufacturer with only a description on their functionality, so these items can have an unknown pedigree. This is because it may not be clear how these items were developed and the item's manufacturer may not provide development or lifecycle data. Hence, unless the COTS or IP manufacturer provides evidence to show compliance of these items to the development standard that has been used, the developer (using these items) cannot provide the same level of assurance, as would be provided if the hardware was developed in-house.

Since assurance needs to be provided for COTS or IP items, for use in safety related applications, the developer needs to find an approach for showing this assurance. The standards of DO 254 [11] and EASA Certification Memo [17] provide some guidance on the type of assurance evidence needed for COTS and IP items and this includes:

- Ensuring the verification of the item through the overall design process.

- Determining the technical suitability of the item.

- Determining the track record of the items usage.

- Determining the track record of the item's manufacturer.

- Use of in-service evidence.

This guidance can be used to form a systematic approach which aims to determine what evidence can be obtained from the COTS and IP manufacturer and what evidence can be generated by the developer of the COTS and IP item. The form of a plausible systematic approach is shown as follows:

1. COTS and IP manufacturer - the developer would need to establish:

    (a) How has the manufacturer of the COTS and IP items demonstrated a track record for the production of high quality components?

    (b) What quality control procedures have been used in the manufacture of the COTS or IP items?

    (c) What evidence has been provided by the manufacturers of the COTS or IP items to show that these items have been qualified and that component quality has been controlled?

The above questions may be addressed by a combination of examining the manufacturer's datasheets and liaising with the manufacturer. Moreover, if the items are produced and used in large volumes, particularly in the application area that is relevant to the developer, then a level of hardware pedigree can be established for these items. However, if questions (such as the above) cannot be addressed then the developer would need to determine what additional measures (such as additional testing) or mitigations need to be adopted for assurance.

2. COTS and IP developer – the developer would need to:

    (a) Document the rationale for the technical suitability of the item. For this task the developer needs to have sufficient understanding of the COTS or IP item's functionality.

    (b) Document the track record of usage, if the developer has previously used the item.

    (c) Understand the amount of developmental evidence that is missing for showing compliance to DO 254. A useful approach is to conduct a gap analysis between the evidence that is available on the COTS or IP item against that required to meet the certification objectives of DO 254. Also the gaps in evidence will inform the developer on what additional approaches (such as testing or providing additional mitigations) are required to provide assurance.

(d) Consider whether reverse engineering could be used to generate the developmental evidence (based on the gap analysis). However, this is mostly likely to be a long and costly activity, which may not provide all the compliance evidence that is required.

(e) Perform a safety assessment on the overall system in which the COTS and IP item is used. This would be used to: determine the safety integrity or DAL for the system and show the hazards that the hardware functionality may contribute to.

(f) Use verification activities to show that the COTS or IP items satisfy their system requirements. This evidence could include test plans, test procedures and test reports.

(g) Use verification to show that the system (which integrates COTS or IP items) satisfies its system requirements.

The above approach is not meant to be definitive or comprehensive; the variety of COTS and IP items and their uses would require an approach to be tailored for the specific application. However, these types of approaches have been applied (from an independent assessor's perspective) to determine what assurance evidence is available and required. Thus if large parts of the evidence is not available and cannot be generated then the COTS or IP item may be deemed to carry a high risk. In this case other mitigations (such as additional testing or architectural mitigations) may need to be considered.

17.6. Formal Approaches to Hardware Assurance

A system can be constructed from COTS items (such as the MIL-STD 1553 interface cards and single processor boards [20, 21]) which themselves contain software and programmable electronic hardware components. Typically, for these COTS items, no information is available from the manufacturer on how the constituent components were developed. However, since there is an increasing trend to use these types of system in safety related applications, then assurances in their functionality and safety behaviour must be provided. Although an assurance approach may be developed based on the guidance of DO 254, approaches based on formal modelling can be used for providing systems assurance [18].

Due to the discrete event nature of software and programmable electronic hardware based systems, formal techniques such as CSP [22], Z [23] and Petri nets [24] have often been used for modelling these systems. Since formal notations can be used to create a model at different levels of abstraction, formal techniques can be used to model systems for which components are partially or fully described. This situation is particularly suited to modelling a system that is constructed from COTS components, where varying degrees of information is available from manufacturers on the COTS components. Hence, these descriptions can be used to create a top-level model that shows the interactions between the various components of the system. The formal model can then be executed and the resulting state-space and state sequences examined to identify the undesirable behaviours.

An illustration of the above formal approach is shown by considering a case study in which assurance is provided on the functionality of a fuel pump control system, which could be used in avionics or automotive applications.

17.6.1. Case Study

This case study considers part of a pump and valve control system that is used to transfer fuel within a system. An overview of the pump and valve control system is provided in Fig. 17.2, which is represented in the terms of a plant and a controller. The plant contains the physical equipment (such the pumps and valves) and the controller is an embedded system that comprises an FPGA and associated control software. The control system shown in Fig. 17.2 is a simplified form of an industrial scale system which contains a range of pumps and valves that are interconnected by a large array of signals to the controller.

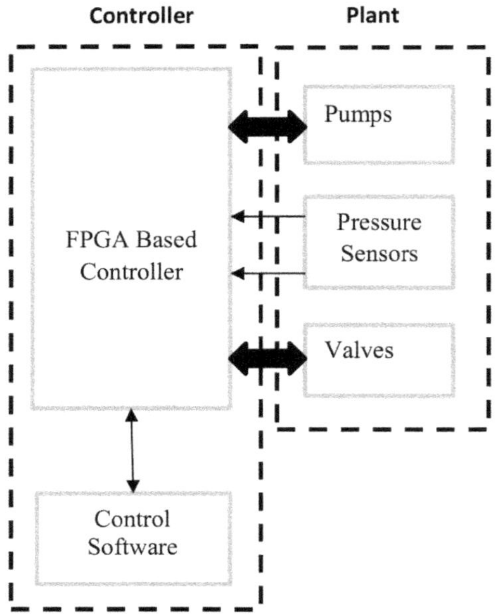

Fig. 17.2. Overview of the FPGA based Embedded Pump Control System.

17.6.1.1. Plant

Due to complexities in presentation, this case study does not consider the functionality of the valves. Also the plant contains two pumps per tank, which are located in different areas of a tank, to ensure that all the fuel can be removed when the tank is displaced at an angle. However, this case study only considers the operation of a single pump.

The operation of the pumps is monitored by pressure switches, which are positioned in close proximity to the pumps and indicate whether pressurized fuel is present at the pump outlet. The signal produced by the pressure switch (which is referred to as the Pressure Switch Signal (PSS)) is such that:

1. Low pressure signal indicates that either the pump is faulty or the tank has no fuel;

2. High pressure signal indicates that the tank has fuel and the pump can be activated. However, if the pump has been commanded off and PSS indicates a high pressure then the pressure switch is deemed to be faulty.

Pressure switches are used with the pumps to allow the controller to monitor the continuity of operation and determine the 'health' status of the pump. However, the failure of a pressure switch does not necessarily indicate that it's corresponding pump has failed because pump failure would be detected by the controller's monitoring of the fuel level.

17.6.1.2. Controller

The overall functionality of the FPGA based controller is known, however, no detailed information is available about the components that form this controller or the design details of the FPGA. Design documentation is available for the control software and this documentation can be used to determine the functionality of the FPGA based system. Hence, from an examination of the control software documentation and the plant's functionality it is understood that signals connected with the FPGA based system are:

1. PSS – this signal is received by the controller. This signal indicates whether the pressure switch is pressurised or unpressured and is assigned a validity flag;

2. Pump Command (PC) - this signal is received by the controller. PC is generated by control software and indicates the commanded state of the pump (i.e. whether the pump should be commanded on or off);

3. A signal that is sent by the controller to cause the pump to be activated on or off.

A further examination of the functionality of the control software, pumps and pressure switches shows that a state machine (as shown in Fig. 17.3) is implemented within the FPGA based system. The state machine of Fig. 17.3 shows that the 'pump on' functionality (represented by S1, S3, S4 and S5 in Fig. 17.3) has the same sequence of operations as the 'pump off' functionality (represented by S2, S6, S7 and S8 in Fig. 17.3). A detailed view of the controller's functionality can be deduced by understanding the behaviour arising from state S1 (which represents the state in which the controller is checking whether the pump is commanded on). This functionality is given by:

1. In state S1, the controller performs a check to determine whether:

(a) The control software requires the pump to be commanded off (S2),

(b) The PSS and control software requires the pump to be switched on (S3);

(c) The PSS and control software are set to default values requiring the pump to be switched off and set to failed (S4). Default values are set when the controller cannot establish with certainty the value of the measured variable;

2. In state S4, the pump could either:

(a) Remain in S4 or

(b) Transition to S5. This transition occurs if the pump has been commanded on and a valid PSS is received. State S5 is referred to as 'fail off timing' because it represents the delay that is required for the physical system to change state.

3. In state S5, the pump can either:

(a) Remain in S4, if the PSS and control software are set to default values or

(b) Transition to S3, if the pump is commanded on and a valid PSS is received.

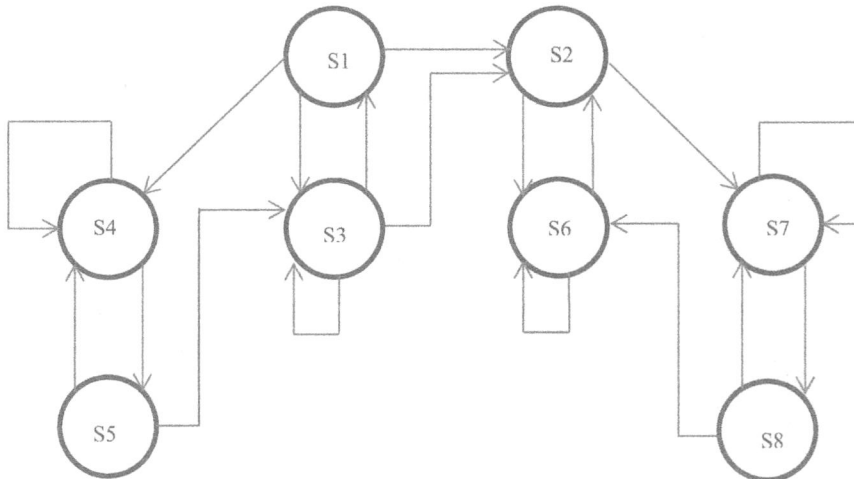

State	Semantics
S1	Check is Pump Commanded On
S3	Pump On
S4	Pump Fail Off
S5	Pump Fail Off Timing

State	Semantics
S2	Check is Pump Commanded Off
S3	Pump Off
S7	Pump Fail On
S8	Pump Fail On Timing

Fig. 17.3. State Machine Implemented within the FPGA based Controller for the Pump Control System.

The above description provides sufficient logical details for a Petri net model of the pump control system to be generated. Due to space limitations the full Petri net model of the pump control system cannot be shown, however, an extract of the model for a single pump is shown in Fig. 17.4.

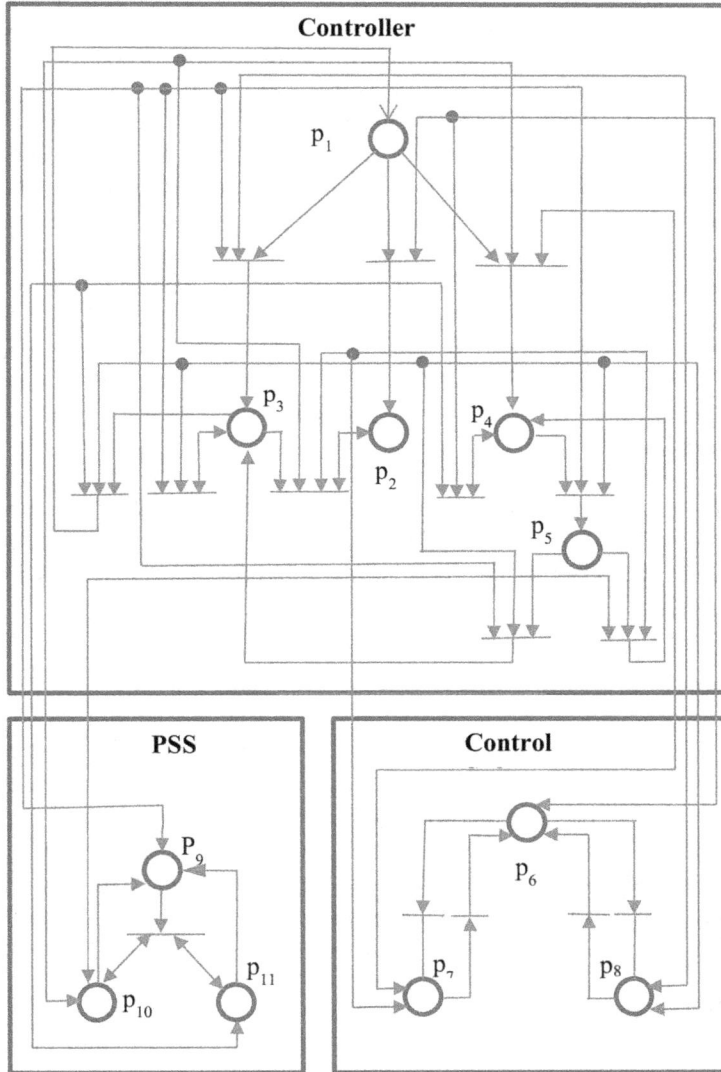

Place	Semantics	Place	Semantics	Place	Semantics
p_1	Check PC On	p_5	Pump Fail Off Timing	P_9	PSS is True
P_2	Check PC Off	p_6	PC is off	p_{10}	PSS is Default
P_3	PC On	P_7	PC is Default	p_{11}	PSS is False
p_4	Pump Fail Off	P_8	PC is On		

Fig. 17.4. Petri Net Model of the Pump Control System.

17.6.1.3. Analysis of the Controller Petri Net Model

Traditional methods of analysing a Petri net model are based on performing an exploration of the state reachability graph and identifying undesirable behaviours. State enumeration is easily performed using automated tools (such as CPN [25]) and undesirable behaviours that are typically detected include conditions such as deadlock (where processes reach an indefinite wait state and unable to reach another state) and livelock (where a process executes in an indefinite loop without interacting with other processes).

In the context of safety related embedded control systems, undesirable behaviours are concerned with those states and sequences that can lead to the occurrence of hazards. Since a global state of a system can be represented by the conjunction of the local states of the individual processes, which form the system, an examination of this state can show the combination of local states that are undesirable or can lead to a hazardous situation. Furthermore, if the set of local states of processes that can occur in parallel with a particular process state could be identified then an assessment can be performed to determine whether these states can cause a hazard. To this end, in this chapter, the set known as the Concurrent States Set (CSS) is defined, which for a particular state, identifies the states of the processes, within a system, that can exist concurrently with it. This set is defined as follows:

Concurrent States Set (CSS) - Let a system C be composed of a set of concurrent processes $(c_1, c_2, c_3, ..., c_n)$ and s_m be an arbitrary local state of process c_m. The CSS of s_m is the set of all local states of the other processes within C that can be concurrent with s_m. This set is denoted by $CSS(s_m)$.

17.6.1.4. CSS Analysis for the Pump Control System

Since an analysis needs to be performed to determine whether the controller does not create a hazardous situation, it is most useful to generate the CSS for the controller. This approach would allow the identification of the local states of the system that are potentially concurrent with a local state of the controller. Hence, an examination of these concurrent states could be performed to identify those states that represents a hazardous situation or states that can lead to a hazardous situation. The CSS for the controller and Petri net model of Fig. 17.4 is shown in Table 17.1.

For an example of reading Table 17.1 consider the controller state of P_1. When the controller is in state P_1 (which represents the state in which the controller is checking whether the pump command is on) the concurrent states shown in rows 1-9 of Table 17.1 are:

$CSS(P_1) = \{(P_1, P_9, P_6), (P_1, P_9, P_7), (P_1, P_9, P_8), (P_1, P_{10}, P_6), (P_1, P_{10}, P_7), (P_1, P_{10}, P_8), (P_1, P_{11}, P_6), (P_1, P_{11}, P_7), (P_1, P_{11}, P_8)\}$

Table 17.1. Concurrent States Set (CSS) for controller state P_1.

Controller State	PSS State	PC State	Overall State
P_1	P_9	P_6	Check Pump Command On; PSS is True; PC is off
		P_7	Check Pump Command On; PSS is True; PC is Default
		P_8	Check Pump Command On; PSS is True; PC is On
	P_{10}	P_6	Check Pump Command On; PSS is Default; PC is off
		P_7	Check Pump Command On; PSS is Default; PC is Default
		P_8	Check Pump Command On; PSS is Default; PC is On
	P_{11}	P_6	Check Pump Command On; PSS is False; PC is off
		P_7	Check Pump Command On; PSS is False; PC is Default
		P_8	Check Pump Command On; PSS is False; PC is On
P_2	P_9	P_6	Check Pump Command Off; PSS is True; PC is off
		P_7	Check Pump Command Off; PSS is True; PC is Default
		P_8	Check Pump Command Off; PSS is True; PC is On
	P_{10}	P_6	Check Pump Command Off; PSS is Default; PC is off
		P_7	Check Pump Command Off; PSS is Default; PC is Default
		P_8	Check Pump Command Off; PSS is Default; PC is On
	P_{11}	P_6	Check Pump Command Off; PSS is False; PC is off
		P_7	Check Pump Command Off; PSS is False; PC is Default
		P_8	Check Pump Command Off; PSS is False; PC is On
P_3	P_9	P_6	Pump Command On; PSS is True; PC is off
		P7	Pump Command On; PSS is True; PC is Default
		P_8	Pump Command On; PSS is True; PC is On
	P_{10}	P_6	Pump Command On; PSS is Default; PC is off
		P_7	Pump Command On; PSS is Default; PC is Default
		P_8	Pump Command On; PSS is Default; PC is On
	P_{11}	P_6	Pump Command On; PSS is False; PC is off
		P_7	Pump Command On; PSS is False; PC is Default
		P_8	Pump Command On; PSS is False; PC is On
P_4	P_9	P_6	Pump Fail Off; PSS is True; PC is off
		P_7	Pump Fail Off; PSS is True; PC is Default
		P_8	Pump Fail Off; PSS is True; PC is On
	P_{10}	P_6	Pump Fail Off; PSS is Default; PC is off
		P_7	Pump Fail Off; PSS is Default; PC is Default
		P_8	Pump Fail Off; PSS is Default; PC is On
	P_{11}	P_6	Pump Fail Off; PSS is False; PC is off
		P_7	Pump Fail Off; PSS is False; PC is Default
		P_8	Pump Fail Off; PSS is False; PC is On
P_5	P_9	P_6	Pump Fail Off Timing; PSS is True; PC is off
		P_7	Pump Fail Off Timing; PSS is True; PC is Default
		P_8	Pump Fail Off Timing; PSS is True; PC is On
	P_{10}	P_6	Pump Fail Off Timing; PSS is Default; PC is off
		P_7	Pump Fail Off Timing; PSS is Default; PC is Default
		P_8	Pump Fail Off Timing; PSS is Default; PC is On
	P_{11}	P_6	Pump Fail Off Timing; PSS is False; PC is off
		P_7	Pump Fail Off Timing; PSS is False; PC is Default
		P_8	Pump Fail Off Timing; PSS is False; PC is On

An interpretation of CSS(P_1) is that in state p1: the state of PSS can be either True (P_9), Default (P_{10}) or False (P_{11}) and the pump command can be either off (P_6), default (P_7) or on (P_8).

An examination of the states in Table 17.1 shows that some states represents behaviours that are desirable and thus do not present a hazardous situation whereas other states are undesirable and can lead to a hazard. Examples of these states are given in the following:

1. Desirable state - state is $\{(P_1, P_9, P_6)$ is desirable, where:

   - The controller is checking whether the pump should be commanded on (P_1),

   - The PSS measurement indicates that the fuel is pressurised (P_9),

   - The control software indicates that the pump should be commanded off (P_6).

From the state reachability graph of the Petri net model of Fig. 17.3 (which has been generated using CPN [15] but not shown due to space limitations), it is apparent that the above state leads to the pump being commanded off, and this is not a hazardous state.

2. Undesirable state – state (P_1, P_{11}, P_8) is undesirable, where:

   - The controller is checking whether the pump should be commanded on (P_1),

   - The PSS measurement indicates that the fuel is unpressurised (P_{11}),

   - The control software indicates that the pump should be commanded on (P_8).

In this state, the control software requires the pump to be on (P_8) but the PSS indicates that either there is no fuel in the tank or that the PSS is faulty. From the state reachability graph, it is apparent that this state can lead to a desirable state if either the PSS indicates False or the software commands the pump to be off. However, state (P_1, P_{11}, P_8) becomes hazardous if the system remains in this state, for a longer duration; in this case the pump can be running without fuel and at worst case become damaged.

In the above manner an inspection of all the states in Table 17.1 was performed and the states that have the potential to cause a hazard were identified (as shown in Table 17.2).

Although the states listed in Table 17.2 are considered as causing a potential deadlock, they need to be examined in the context of the overall fuel system to ascertain whether a deadlock or hazard can occur. This is because the overall system may have mechanisms (such as architectural redundancy or watchdog timers) that can mitigate against the hazards caused by these states. Hence, if appropriate mitigations are in place then greater assurance is provided in the system. On the other hand, if no mitigations are provided, then greater assurance can be provided by adding mechanisms that prevent the hazard from occurring.

Table 17.2. States from the CSS of Table 17.1 that can cause a hazardous situation.

Global State			Comments
P_1	P_{11}	P_8	Potential deadlock state that can be removed if PSS = True or PC = Off.
P_3	P_9	P_6	Potential deadlock state that can be removed if PC = on, PSS = False or PSS = default.
P_3	P_9	P7	Potential deadlock state that can be removed if PC = on or PSS = default.
P_3	P_{10}	P_6	Potential deadlock state that can be removed if PSS = True.
P_3	P_{10}	P_8	Potential deadlock state that can be removed if PSS = True or False.
P_3	P_{11}	P_6	Potential deadlock state that can be removed if PC = default or on.
P_4	P_9	P_6	Potential deadlock state that can be removed if PSS = False or PC = on.
P_4	P_9	P_7	Potential deadlock state that can be removed if PSS = False or PC = off.
P_4	P_{10}	P_6	Potential deadlock state that can be removed if PSS = False or PC = on.
P_4	P_{10}	P_7	Potential deadlock state that can be removed if PSS = False and PC = On, or PC = off and PSS = True.
P_4	P_{10}	P_8	Potential deadlock state that can be removed if PSS = True.
P_4	P_{11}	P_7	Potential deadlock state that can be removed if PC = off PC = On and PSS = True.
P_4	P_{11}	P_8	Potential deadlock state that can be removed if PC = Off or PSS = True
P_5	P_9	P_6	Potential deadlock state that can be removed if PC = On
P_5	P_9	P_7	Potential deadlock state that can be removed if PC = On or PSS = Default.
P_5	P_{10}	P_6	Potential deadlock state that can be removed if PC = Default or PSS = True and PC = On.
P_5	P_{10}	P_8	Potential deadlock state that can be removed if PSS = True.
P_5	P_{11}	P_6	Potential deadlock state that can be removed if PSS = True and PC = On.
P_5	P_{11}	P_7	Potential deadlock state that can be removed if PSS = True and PC = On.
P_5	P_{11}	P_8	Potential deadlock state that can be removed if PSS = True.

17.7. Conclusion

Due to their great flexibility, ease of programming and large programmable capacity, programmable hardware devices are used in a wide range of electronic systems. However, since complex logic functionality can be implemented in these devices, they may not be fully testable, and require well-defined development methodologies to ensure they are programmed correctly.

These issues become crucial when these devices are used in safety critical systems (such as in the avionics applications). Specifically, the developer must ensure that the device is 'fit for purpose', safe to use and satisfies regulatory standards. Moreover, the developer needs to provide assurances on the device development to the customer and a certification authority. Hence, providing assurances in programmable hardware is a rigorous and costly process.

In the avionics domain, DO 254 provides the main guidance for a developer to show assurance in the development of programmable hardware. For the assurance of hardware that is developed in-house by a developer, the guidance provided by DO 254 is well understood and may be considered to be standard practice.

For hardware developments that use COTS or IP items, the main challenge in assurance is that the developer may not know how these items were developed. However, DO 254 does provide guidance that may be tailored to produce an assurance approach.

Systems that are constructed from many COTS devices (such as interface cards and boards) provides the most challenges in assurance. Each COTS device may contain several programmable hardware items, IP and software items and no developmental evidence may be available on these items. As illustrated by the case study, formal modelling has provided a means of analysing these systems and provides the basis for forming an assurance argument. However, these approaches are not standard and require an initial study to determine their applicability.

Since the programmable capacity of hardware devices is ever increasing and there is great motivation in using these devices, the assurance of these devices within safety critical systems will remain a challenging area.

References

[1]. S. Z. Ahmed, G. Sassatelli, L. Torres, L. Rouge, Survey of new trends in Industry for programmable hardware: FPGAs, MPPs, MPSoCs, structured ASICs, eFPGAs and the new wave of innovation in FPGAs, in *Proceedings of the International Conference on Field Programmable Logic and Applications (FPL'10)*, 2010, pp 291-297.

[2]. K. Pocek, R. Tessier, A. DeHon, Birth and adolescence of reconfigurable computing: a survey of the first 20 years of field programmable custom computing machines, in *Proceedings of the IEEE International Symposium on Field Programmable Custom Computing machines Commemorative Book (FCCM'13)*, 2013, pp. 3-19.

[3]. R. Nane, V. Sima, C. Pilato, J. Choi, B. Fort, A. Canis, Y. Chen, H. Hsiao, S. Brown, F. Ferrandi, J. Anderson, K. Bertels, A survey and evaluation of FPGA high level synthesis tools, *IEEE Transactions on Computer-Aided Design pf Integrated Circuits and Systems*, Vol. 35, Issue 10, 2016, pp 1591-1604.

[4]. Standard Cell ASIC to FPGA Design and Methodology and Guidelines, Altera, April 2009, https://www.altera.com/content/dam/altera-www/global/en_US/pdfs/literature/an/an311.pdf

[5]. E. Monmasson, M. N. Cirstea, I. Bahri, A. Tisan, M. W. Naousar, FPGAs in industrial control applications, *IEEE Transaction on Industrial Informatics*, Vol. 7, No. 2, May 2011, pp. 224-243.

[6]. A. Bain, Safety certification in the defence sector, Assuring the safety of systems, in *Proceedings of the 21ˢᵗ Safety-Critical Systems Symposium (SSS'13)*, Bristol, UK, February 2013.

[7]. J. L. De La Vera, M. Borg, K. Wnuk, L. Moon, An industrial survey of safety evidence change impact analysis practice, *IEEE Transactions on Software Engineering*, Vol. 42, No. 12, December 2016, pp 1096-1117.

[8]. A. Wassyng, T. Maibaum, M. Lawford, On software certification: we need product-focused approaches, *Lecture Notes in Computer Science*, Vol. 6028, 2010, pp 250-274.

[9]. D. G. Raheja, M. Alloco, Assurance Technologies Principles and Practices: A Product, Process and System Safety Perspective, 2ⁿᵈ Edition, *Wiley*, 2006.

[10]. I. Kuon, R. Tessier, J. Rose, FPGA architecture: survey and challenges, *Foundations and Trends in Electronic Design Automation*, Vol. 2, No. 2, 2008, pp. 135-253.

[11]. Design Assurance Guidance for Airborne Electronic Hardware, *EUROCAE/RTCA, DO 254 Training*, 2000.

[12]. Software Considerations in Airborne Systems and Equipment Certification, *EUROCAE/RTCA, DO 178C*, 1992.

[13]. Design and Airworthiness Requirements for Service Aircraft part 13: Military Common Fit Equipment, *UK Ministry of Defence Standard 00-970*, Issue 12, September 2016.

[14]. Safety Management Requirements for Defence Systems. Part 1: Requirements and Guidance, *Defence Standard 00-056*, Issue 7, February 2017.

[15]. Clarifications on the Use of RTCA Document DO-254 and EUROCAE ED-80, Design Assurance Guidance for Airborne Electronic Hardware, *CAST 27*, 2006.

[16]. Conducting Airborne Electronic Hardware Reviews, *FAA Job Aid*, 28th February 2008.

[17]. Development Assurance of Airborne Electronic Hardware, *EASA Certification Memo CM-SWCEH-001*, 2011.

[18]. A. J. Kornecki, J. Zalewski, Hardware certification for real-time safety-critical systems: state of the art, *Annual Reviews in Control*, Vol. 24, 2010, pp 163-174.

[19]. V. Hilderman, T. Baghai, Avionics Certification: A Complete Guide to DO 178 (Software) DO 254 (Hardware), *Avionics Communications Inc.*, 2008.

[20]. MIL-STD-1553 Interface Datasheet, https://www.astronics.com/ballard-technology/test-simulation-interfaces/lx1553-5-pci-pcie-avionics-interface-cards-for-mil-std-1553

[21]. 2nd Generation Intel Core, http://www.adl-usa.com/wp-content/uploads/files/d2a3fe3c4b0dad29981fcb638346d179.pdf

[22]. A. W. Roscoe, The Theory and Practice of Concurrency, *Prentice-Hall*, 1998.

[23]. A. Diller, Z: An Introduction to Formal Methods, *Wiley*, 1994.

[24]. T. Murata, Petri nets: properties, basic concepts, analysis methods and practical use, *Proceedings of the IEEE*, Vol. 77, 1989, pp 541-577.

[25]. AIS Group, CPN Tools Version 4.0, http://cpntools.org/

Chapter 18

Reliability-Aware Energy Management for Weakly Hard Real-Time Systems

Linwei Niu and Wei Li

Abstract Aggressive scaling in technology size has dramatically increased the power density and degraded the reliability of real-time embedded systems. In this chapter, we study the problem of reliability-aware energy minimization for scheduling fixed-priority realtime embedded systems with weakly hard QoS-constraint. The weakly hard QoS-constraint is modeled with (m, k)-constraint, which requires that at least m out of any k consecutive jobs of a task meet their deadlines. We first propose a technique that can balance the static and dynamic energy consumption for real-time jobs with better speed determination than the classical strategies during their feasible intervals. Then based on it, we propose an adaptive fixed-priority scheduling scheme to reduce the energy consumption for the system while preserving its reliability. Through extensive simulations, our experiment results demonstrate that the proposed techniques can significantly outperform the previous research in energy performance while satisfying the weakly hard QoS-constraint under the reliability requirement.

18.1. Introduction

As transistor density continues to grow, the power/energy conservation problem becomes more and more critical in the design of pervasive real-time embedded systems. In the meantime, system reliability has become increasingly important in ubiquitous and pervasive computing and reliability-aware scheduling has attracted extensive attentions for the purpose of enhancing the energy efficiency while preserving the system reliability.

Many real-time scheduling based techniques, *e. g.* [21, 11, 36, 22, 17, 4, 2, 6], have been proposed to reduce energy dissipation. Among them dynamic voltage scaling (DVS) has been a widely adopted technology in the past two decades. Although DVS is very effective in reducing the energy consumption, it has been shown that voltage scaling has a direct and negative effect on system reliability as the probability of faults could be much higher

Linwei Niu
Department of Math and Computer Science, West Virginia State University, USA

at lower supply voltages [14, 39, 49]. For safety critical real-time systems such as avionics and industrial controls, catastrophical consequences may occur if system faults are not handled in a timely manner [18]. Generally, computing system faults can be classified into transient and permanent faults [20]. Transient faults are mainly caused by temporary factors such as electromagnetic interference and cosmic ray radiations while permanent faults could be caused by permanent damage in processing unit(s). It is shown in [9] that transient faults occur much more frequently than permanent faults. So in this chapter we focus on recovery from transient faults for reliability consideration.

With system reliability in mind, some previous works (e. g. [48, 50, 45]) have been proposed to reduce the energy consumption for real-time systems. Most of them are based on Earliest Deadline First (EDF) scheme. Although EDF schemes can support systems with higher utilization than fixed-priority (FP) schemes, they are difficult to implement in commercial kernels which do not provide explicit support for timing constraints, such as periods and deadlines [8]. Moreover, when the system is overloaded, EDF can produce unbounded and unpredictable deadline misses while FP has better stability in such cases because it is always known in advance that which task will miss its deadline. Due to its high predictability, low overhead, and ease of implementation, FP schemes are widely adopted in real-time embedded applications [25].

On the other hand, most of the works above are for *hard* real-time systems. However, few practical applications in real world are truly *hard* real-time. Many applications can allow some deadline misses provided that user's perceived quality of service (QoS) levels can be satisfied. The *weakly hard QoS-constrained model* is more appropriate to meet the requirement of such type of systems. In the weakly hard QoS-constrained model, tasks have both firm deadlines (*i. e.,* deadline missing is useless) and a throughput requirement (*i. e., sufficient* task instances must meet their deadlines to provide the required quality levels) [30]. Ramanathan *et al.* [37] proposed a deterministic weakly hard QoS-constraint model called the (m, k)-model. In the (m, k)-model, a periodic task is associated with a pair of integers, i. e., (m, k), such that among any k consecutive instances of the task, at least m of the instances must finish by their deadlines for the system behavior to be acceptable. A *dynamic failure* occurs, which implies that the weakly hard QoS-constrain is violated and the scheduler is thus considered failed, if within any k consecutive jobs more than $(k-m)$ job instances miss their deadlines. With the (m, k) requirement in mind, Ramanathan *et al.* [38] proposed to partition the jobs into *mandatory* and *optional* jobs. So long as all the mandatory jobs can meet their deadlines, the (m, k)-constraints can be satisfied.

In this chapter, we study the problem of reliability-aware energy minimization for scheduling fixed-priority real-time systems with (m, k)-constraint.

The rest of the chapter is organized as follows. Section 18.2 introduces the system models and some preliminaries. Section 18.3 presents our general algorithm of reliability-aware energy minimization for fixed-priority real-time systems. Section 18.4 further explores dynamic slack reclamation. The effectiveness of our approaches is demonstrated using experimental results in Section 18.5. Section 18.6 talks about the related work. In Section 18.7, we offer the conclusions.

18.2. Preliminary

In this section, we first introduce the system/fault-recovery models and power model. Then we provide some motivations.

18.2.1. System Model

The real-time system that we are interested in consists of N independent periodic tasks, $\mathcal{T} = \{\tau_1, \tau_2, \cdots, \tau_N\}$, scheduled according to the fixed-priority (FP) scheme. Each task τ_i is characterized using five parameters, *i. e.*, $(C_i, D_i, T_i, m_i, k_i)$. C_i, D_i, and T_i represent the worst case execution time (WCET), the deadline, and the period for τ_i, respectively. We assume $D_i \leq T_i$. The (m, k)-constraint for τ_i is represented by a pair of integers, *i. e.*, (m_i, k_i) $(0 < m_i \leq k_i)$, which require that, among any k_i consecutive jobs of τ_i, at least m_i jobs meet their deadlines. Without loss of generality, we assume that τ_i has a higher priority than τ_j if $i < j$ in the FP scheme. And according to [5], deadline-monotonic scheduling (DMS) scheme has been shown to be the optimal approach among the different FP schemes for periodic task sets.

Each task contains an infinite sequence of periodically arriving instances called jobs. The jth job of task τ_i is represented with J_{ij}. It is not hard to see that J_{ij}'s worst case execution time $w_{ij} = C_i$. In addition, we represent the arrival time, actual execution time under the highest processor speed, and absolute deadline of J_{ij} with r_{ij}, aw_{ij} and d_{ij}, respectively. And the recovery job for job J_{ij} is represented as R_{ij}.

18.2.2. Fault and Recovery Models

Similar to [31, 18], we focus on transient faults in this chapter since transient faults occur much more frequently than permanent faults in modern semi-conductor devices [15]. In this chapter we assume that faults can be detected using sanity (or consistency) checks [33] when a job finishes its execution. Meanwhile, the overhead for detection can be integrated into the job's worst case execution time. Moreover, when transient fault occurs and is detected at the end of a job's execution, the affected job can be addressed by re-executing a recovery job with the same worst-case execution time and deadline as the original one [33]. Once released, the recovery job will be executed like a normal job.

Following the fault model in [49], we assume that the transient faults will present Poisson distribution [43] and the average transient fault rate for systems running at speed s (and the corresponding supply voltage) is [49]:

$$\lambda(s) = \lambda_0 \cdot g(s), \tag{18.1}$$

where λ_0 is the average fault rate corresponding to the maximum speed s_{max}. That is, $g(s_{max}) = 1$. Considering the negative effect of DVS on the transient fault rate, in general, we have $g(s) > 1$ for $s < s_{max}$ [49].

18.2.3. Power Model

The power consumption on a DVS processor can be divided into two parts: the speed-dependent part $P_{dep}(s)$ and the speed-independent part P_{ind} [46]. The speed-dependent power $P_{dep}(s)$ mainly comes from the dynamic power consumption and short-circuit power consumption. And the speed-independent power P_{ind} mainly comes from the standby leakage power which is mainly due to the subthreshold leakage current and the reverse bias junction current in the CMOS circuit. The dynamic power mainly consists of the switching power for charging and discharging the load capacitance and can be represented [10] as $P_{dyn} = C_L V^2 f$, where C_L is the load capacitance, V is the supply voltage, and f is the system clock frequency. So the total power consumption when the processor is in its active status, *i. e*, $P_{act}(s)$, is thus $P_{act(s)} = P_{dep}(s) + P_{ind}$. The speed dependent power $P_{dep}(s)$ can be modeled as a strictly convex and increasing function of s. For example, usually the speed dependent power can be expressed as $P_{dep}(s) = \alpha s^3$, where α is a constant.

The processor can be in one of three states: *active, idle* and *sleeping* states. When the processor is idle, the major portion of power consumption comes from the static power P_{ind}. Shutting-down strategy, *i. e.,* putting the processor into its sleeping state, can greatly reduce the static energy. However, it has to pay extra energy and timing overhead to shut down and later wake up the processor. As shown in [51], the timing overhead is usually a very small value for contemporary microprocessors if properly designed for power awareness and thus can be safely ignored in simulation frameworks. However, the energy overhead is considerable [21, 12]. Assume that the power consumption of a processor in its idle state is P_{idle} and the energy overhead of shutdown/wakeup the processor is E_o. Then the processor can be shut down with positive energy gains when the length of the idle interval is larger than the *break even* time $t_{be} = \dfrac{E_0}{P_{idle}}$.

18.2.4. Practical Critical Speed

When the processor is active, without consideration of the static power P_{ind}, it would be most energy-efficient to run the tasks with voltages as low as possible. However, reducing the supply voltage will stretch the execution time of the task and increase the static energy from P_{ind}. Considering a job with workload w to be executed with total power $P_{act}(s)$, the total energy $(E_{act}(s))$ consumed to finish this job with speed s can be represented as $E_{act}(s) = P_{act}(s) \times \dfrac{w}{s}$. Hence, to minimize $E_{act}(s)$, let $\dfrac{dE_{act}(s)}{ds}$. When $P_{dep}(s) = \alpha s^3$, the derived speed is $s = \sqrt[3]{\dfrac{P_{ind}}{2\alpha}}$, which is the theoretical optimal speed to minimize the active energy for executing the job. We therefore call this speed the *critical speed*, and denote it as s_{crit}.

However, although seemingly reasonable, the theoretical critical speed s_{crit} calculated above might not always be achievable in practical processor models. With the continued scaling of CMOS technology, the leakage power, which is the major part of static power P_{ind}, has been increasing very fast and could be higher than the dynamic power in modern processors. For processor models in which leakage power is extremely high, the theoretical critical speed s_{crit} calculated above could be even greater than the maximal available speed s_{max} and thus might not be able to be found in practice. For example, for the Intel Xscale processor model [19], if the leakage power is four times higher than the dynamic power, the normalized theoretical critical speed will be around 1.09, which is greater than the maximal speed 1 and will not be achievable in practice. To incorporate such kind of processor models, instead of adopting s_{crit} directly, we should use the smaller one between the theoretical critical speed and the maximal available speed, *i. e.,* $\min\{s_{crit}, s_{max}\}$, as the practical speed for minimizing $E_{act}(s)$. Without loss of generality, we call this speed the *practical critical speed*, and represent it with s_{prcs}.

18.2.5. Tradeoff between the Practical Critical Speed and Pure DVS Speed

To develop a real-time schedule that can minimize the overall energy consumption, both the dynamic energy and static energy need to be considered with the deadlines of the tasks guaranteed. Based on the concept of practical critical speed s_{prcs} and break even time t_{be}, a widely adopted strategy [48, 21, 36] is to scale the job speed as low as s_{prcs} and then shut down the processor during the idle intervals whenever the predicted idle interval length is greater than t_{be}. However, this strategy might not always be most energy-efficient.

Consider a job $J_1 = (0, 5.7\ ms, 30\ ms)$. Assume the job will be executed on the Intel XScale processor model [19]. According to [12], the power consumption function for Intel XScale [19] can be modeled approximately as $P_{act}(s)=1.52s^3+0.08$ Watt by treating 1 GHz as the reference speed 1. And the normalized practical critical speed s_{prcs} in such a model is about 0.3 (at 297 MHz) with power consumption 0.12 W. We assume the shutdown overhead to be $E_o = 0.8$ mJoule as did in [12]. Assuming the minimal speed is 0, the idle power consumption of the processor is 0.08 Watt and the corresponding break even time is $t_{be} = 10\ ms$.

The job execution time after it is scaled to s_{prcs} is 19 *ms* and the corresponding schedule is shown in Fig. 18.1(a). Since J_1 finishes at $f_1 = 19\ ms$, the idle period between its completion time and deadline is 11 *ms* ($> t_{be}$). So the processor can be shut down between f_1 and its deadline. The energy consumption under this schedule is $(1.52 \times 0.3^3 + 0.08) \times 19 + 0.8 =$ = 3.1 mJoule. However, if we apply the traditional DVS strategy between its arrival time and deadline, as shown by the schedule in Fig. 18.1(b), the scaled speed for job J_1 will be 0.19. The energy consumption under the new schedule in Fig. 18.1(b) will be $(1.52 \times 0.19^3 + 0.08) \times 30 = 2.71$ mJoule, which is 12.4 % lower than the previous case. On the other hand, if we consider a similar job $J_2 = (0.3\ ms, 30\ ms)$ with same arrival and deadline as J_1 but with different execution time of 3 *ms*, as shown in Fig. 18.1(c), the schedule based on the critical speed strategy will have total energy consumption of $(1.52 \times 0.3^3 + 0.08) \times 10 + 0.8 = 2.01$ mJoule while the schedule in Fig. 18.1(d) based on the pure DVS strategy will generate total energy consumption of $(1.52 \times 0.1^3 + 0.08) \times 30 =$ = 2.45 mJoule. In this case, the pure DVS strategy consume 22 % more energy than the critical speed strategy.

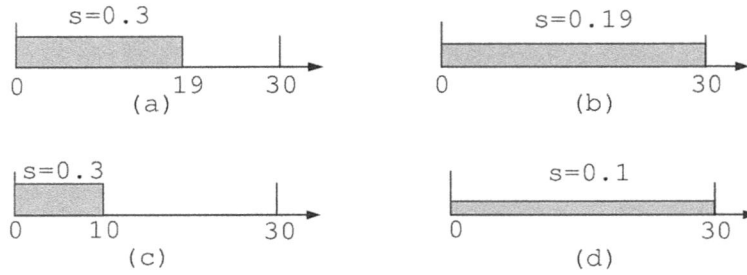

Fig. 19.1. (a) J1 = (0, 5.7 ms, 30 ms) scheduled with practical critical speed;
(b) J1 scheduled with DVS; (c) J2 = (0, 3 ms, 30 ms) scheduled with practical critical speed;
(d) J2 scheduled with DVS.

The example in Fig. 18.1 shows that neither of the two widely used strategies, *i. e.,* DVS strategy and the critical speed strategy, will always outperform the other and there is a tradeoff between them depending on the job workload to be finished within the interval. Then the problem is how to find the most energy-efficient speed for a given job *J*?

18.2.6. Computing the Overall Energy-Efficient Speed

Given the information above, in this part, we will try to determine the most energy-efficient speed for job *J* analytically, based on the concept of *feasible interval* defined as followed:

Definition 1 Given a job *J* = (*r, w, d*), the **feasible interval** of job *J* is a close interval [t_s, t_e] during which *J* can be executed without being interrupted. And without loss of generality, we assume $t_s \geq r$ and $t_e \leq d$.

With the definition of *feasible interval*, we have the following lemmas (the proofs can be found in [27] and are thus omitted):

Lemma 1 *Given a job J =(r, w, d) to be scheduled continuously in a feasible interval* [t_s, t_e], *if the processor is always shut down during the idle interval, the most energy-efficient speed for executing J within the feasible interval is* s_{prcs}.

Lemma 2 *Given a job J =(r, w, d) to be scheduled continuously in a feasible interval* [t_s, t_e], *if the processor is never shut down, the most energy-efficient speed for executing J is the lowest possible speed (DVS speed), i. e.,* $\frac{w}{(t_e - t_s)}$.

From Lemma 1 and Lemma 2, the most energy-efficient speed for job *J* within the feasible interval should be chosen between the practical critical speed and the pure DVS speed. Since when the workload *w* of job *J* is larger than or equal to $s_{prcs} \times (t_e - t_s)$, the feasible interval will be fully utilized by both strategies and there is no difference between them (in this scenario the unique optimal speed will be $\frac{w}{(t_e - t_s)}$, we only need to make choice in the case when $w < s_{prcs} \times (t_e - t_s)$. In the latter case, if we execute the job with s_{prcs} and always shut down the processor during the idle interval, the total energy consumption within the feasible interval will be

$$E_1 = \left(\alpha s_{prcs}^3 + P_{ind}\right) \times \frac{w}{s_{prcs}} + E_0. \qquad (18.2)$$

On the other hand, if we always execute the job with traditional DVS speed and let $l = (t_e - t_s)$, the total energy consumption within the feasible interval will be

$$E_2 = \left(\alpha \left(\frac{w}{l}\right)^3 + P_{ind}\right) \times l \qquad (18.3)$$

As seen in the motivation example, whether E_1 is larger than E_2 will depend on the value of the variable w, *i. e.*, the work-demand of job J within the interval. In the following, we will formulate this observation into a theorem (The detailed proof is provided in [27]):

Theorem 1 *Let job $J = (r, w, d)$ $(0 < w < l \times s_{prcs})$ be executed continuously in a given feasible interval with length $l = [t_s, t_e]$. Let E_1 and E_2 be defined by Eq. (18.2) and (18.3) correspondingly. There exists a threshold value w_{th} within the range of $[0, l \times s_{prcs})$ such that if $w \geq w_{th}$, $E_2 \leq E_1$; Otherwise $E_2 \geq E_1$.*

As shown in [27], given a feasible interval with length l, the threshold value w_{th} for it can be computed in constant time [40]. Moreover, according to the proof for Theorem 1 in [27], so long as the workload w of job J is not below w_{th}, the pure DVS strategy will always consume no more energy than the critical speed strategy in this interval. To be convenient, we call w_{th} the *threshold work-demand*.

Note that the conclusion above is still true even when $w \geq l \times s_{prcs}$ because in this case the feasible interval will be fully utilized by either strategy and both will consume the same energy. Based on it, given a job J and its feasible interval $[t_s, t_e]$, whether we should adopt the DVS speed or the practical critical speed for energy saving purpose will depend on the relationship between its workload w and the w_{th} for $[t_s, t_e]$. If $w \geq w_{th}$, the DVS speed should be adopted. Otherwise, the practical critical speed should be adopted.

18.2.7. Meeting the (*m, k*)-constraint

A key problem for meeting the (*m, k*)-constraint is to judiciously partition the jobs into *mandatory* jobs and *optional* jobs [34]. The partitioning can be done statically or dynamically. Two well-known partitioning strategies proposed are the *evenly distributed pattern* (or E-pattern) [28] and the *deeply-red pattern* (or R-pattern). The mandatory/optional job partitioning according to E-pattern has the property that it helps to spread out the mandatory jobs in each task evenly along the time. And it has been proved [28] that the mandatory job set assigned according to E-pattern has better schedulability than that by R-pattern. Interesting readers can refer to [28] for more technical details about the E-pattern and R-pattern. Moreover, E-pattern has the following important property:

Lemma 3 *[38] Let the mandatory job set of task set T be determined with E-pattern and scheduled under FP scheme, the first job of each task τ_i is the critical instance, i. e., the mandatory job instance with the longest response time.*

With lemma 3, if the first mandatory job of each task can meet its deadline, the whole task set is schedulable, which is very useful for checking the schedulablity of the task set under (m, k)-constraint.

Note that the job patterns defined with E-pattern also have the property that they define a *minimal set* of mandatory jobs that "just" satisfies the given (m, k)-constraint in each sliding window. In other words, if any mandatory job failed, the (m, k)-constraint will be violated. Due to this property, before we consider scaling the speed for any task, we need to reserve a recovery job for each mandatory job of the task in order to ensure the reliability of the same task. Moreover, the recovery job needs to be reserved within the same time frame of their corresponding "primary job" [48]. In addition, if the primary job failed, to preserve the system reliability, the recovery job should always be executed with the maximal speed s_{max} [48].

One of the most significant advantages of applying static partitioning strategies is that they enable the application of theoretic real-time techniques to analyze system feasibility. The problem, however, is its poor adaptivity in dealing with the run-time variations in execution times, which is inherent in many real-time applications. To exploit such kind of dynamic nature in real-time systems, dynamic partitioning strategy was proposed in [28], which is based on executing the optional jobs and vary the patterns dynamically. However, since the approach in [28] only considers dynamic power and the optional jobs are scheduled in a greedy manner, when static power is included, its overall energy efficiency could be limited. On the other hand, no system reliability is considered in [28]. With leakage power and system reliability in mind, in this chapter, we will explore how to reduce the overall energy consumption (including both static and dynamic power) while ensuring the (m, k)-constraints.

Based on the above discussions, in the following sections, we introduce our dynamic scheduling algorithms for reducing the overall energy consumption while ensuring the (m, k)-constraints and preserving the system reliability.

18.3. The General Algorithm

Our algorithm consists of two phases: an off-line phase followed by an online phase. During the off-line phase, we first partition the jobs based on E-pattern and determine the subset Ψ of the tasks which could be managed with recovery jobs. This could be done by using branch-and-bound method based on the schedulability condition of E-pattern provided in [38]. Note that the static speed of each task τ_i in Ψ will also be determined during the procedure of branch-and-bound. Next, we further reduce the energy consumption during the on-line phase.

During the online phase, we selectively execute the jobs and vary the job/recovery patterns dynamically. For presentation convenience, we have the following definitions.

Definition 2 [36] **(latest starting time)** Given at time t, the *latest starting time (LST) of a mandatory/recovery job set $\mathcal{J}$* (denoted as $LST(t)$) is the latest time such that, if the execution of any job in $\mathcal{J}$ starts no later than $LST(t)$, all jobs will meet their deadlines.

In Section 18.3.1, we will introduce in detail how to compute LST for a given mandatory/recovery job set.

Definition 3 (pattern rotation point) Given at time t, the *pattern rotation point* of task τ_i (denoted as $PRP_i(t)$) is defined as the earliest time point t^0 ($t^0 > t$) such that at t^0 the number of consecutive optional jobs for task τ_i is maximized.

According to [26], under E-pattern, there can be at most $[\frac{k_i - m_i}{m_i}]$ consecutive optional jobs for task τ_i. So $PRP_i(t)$ is the starting time point of the first subsequence of $[\frac{k_i - m_i}{m_i}]$ consecutive optional jobs for task τ_i after time t. Note that during run time we could use a pointer to keep track of the *Pattern Rotation Point $PRP_i(t)$* dynamically.

Moreover, the following concept will be used to capture the characteristics for each job.

Definition 4 The *tolerability* of task τ_i at time t (denoted as $Tr_i(t)$) is defined as the maximal number of future deadline misses that can be tolerated by task τ_i without causing any violation on its (m, k)-constraint.

Based on the definitions above, the salient part of our online algorithm is presented in Algorithm 1.

As shown in Algorithm 1, during the online phase: two job containers, *i. e.,* the mandatory job container (MC) and the optional job container (OC), will be maintained. Upon arrival, a job is designated as mandatory job or optional job based on its current (m, k)-pattern and put to the MC or OC correspondingly. Note that a recovery job will be inserted into MC only if its "primary" mandatory job has failed. Otherwise the recovery job will be simply dropped and become slack time. The jobs in MC always have higher priority than those in OC. If the MC is not empty, the jobs in MC will be executed following the fixed-priority scheme with their speeds predetermined at the offline phase. Otherwise the jobs in OC will be executed.

Assuming at current time t, the MC is empty and the jobs in OC get chance to be executed. It is not difficult to see that there may be more than one jobs in OC and selecting which one to execute could affect the overall energy efficiency of the system directly. To address this issue, we propose a heuristic based on our concept of feasible interval in Section 18.2.6. Specifically, each optional job J_i could be regarded as having a dedicated feasible interval $[t, \min\{d_i, LST(t)\}]$, where d_i is the absolute deadline of J_i and $LST(t)$ is the latest starting time for the upcoming mandatory jobs. Then the energy-efficient speed for running J_i within its feasible interval could be determined based on Theorem 1 and the corresponding energy consumption $E(J_i)$ during its feasible interval could also be computed.

Algorithm 1 The online phase of scheduling algorithm.

1: **Upon job dispatching at time t:**
2: **if** MC is not empty **then**
3:    **if** J_i is a recovery job or $\tau_i \notin \Psi$ **then**
4:       Run J_i with maximal speed 1;
5:    **else**
6:       Run J_i with predetermined speed s_i;
7:    **end if**
8: **else**
9:    Select the job J_i in OC that have the maximum energy-gain/tolerability ratio $\Delta E(J_i)/Tr(J_i)$;
10:    Let $t_e = \min\{LST(t), d_i\}$;
11:    let w_{th} be the threshold work-demand for J_i's feasible interval $[t, t_e]$ according to Theorem 1;
12:    **if** $aw_i^{rem} \geq w_{th}$ **then**
13:       $s_i' = \min\{\frac{aw_i^{rem}}{t_e - t}, s_i\}$; //adopt DVS strategy
14:    **else**
15:       $s_i' = s_{prcs}$; //adopt *critical speed* strategy
16:       **if** the processor is busy at t^- **then**
17:          Execute J_i non-preemptively with s_i' at t;
18:          When J_i finished at f_i, shut down the processor/devices and set up the wake-up timer to be $(LST(t) - f_i)$;
19:       **else**
20:          Shut down the processor at t and set up the wake-up timer to be $(t_e - t - \frac{aw_i^{rem}}{s_i'})$;
21:       **end if**
22:    **end if**
23: **end if**
24:
25: **Upon job completion at time t:**
26: **if** Job J_i failed with transient faults **then**
27:    **if** J_i is a mandatory job **then**
28:       Invoke its recovery job with maximal speed 1;
29:    **end if**
30: **else**
31:    **if** J_i is an optional job **then**
32:       Modify the E-pattern for τ_i following the Pattern Adjustment Rule;
33:    **end if**
34: **end if**
35: **if** Both MC and OC are empty at time t **then**
36:    t_a = the arrival time for the next upcoming job;
37:    **if** $(t_a - t) > t_{be}$ **then**
38:       shut down the processor and set up the wake-up timer to be $(t_a - t)$;
39:    **end if**
40: **end if**

Thereafter, the energy-gain for J_i, represented as $\Delta E(J_i)$, will be calculated, where $\Delta E(J_i)$ is the difference between $E(J_i)$ and the energy consumption for executing J_i under the predetermined speed for task τ_i. Only those optional jobs with positive energy-gain, *i. e.*, $\Delta E(J_i) > 0$, will be chosen as candidate jobs. After that the candidate jobs are sorted according to their ***energygain/tolerability ratio*** $\Delta E(J_i)/Tr_i(t)$, where $Tr_i(t)$ is the tolerability of task τ_i at time t. The candidate job with the maximal value of $\Delta E(J_i)/Tr(J_i)$ will be chosen to be executed. And to guarantee the selected optional job could be completed in time, once started, it should be executed non-preemptively until finished.

One thing to be noted is that if the selected optional job J_i is determined to adopt the (practical critical speed + processor shut-down) strategy during its feasible interval, whether the processor should be shut down first or not will depend on the status of the processor at time t^-. If the processor is busy at t^-, then we should execute J_i with s_{prcs} at time t first and then shut down the processor later. Otherwise we should let the processor be shut down at time t first to reduce the overhead of switching processor status (line 16-20).

Once the selected optional job J_i is completed successfully, the (m, k)-pattern of task τ_i could be adjusted by rotating the future patterns of τ_i. Based on our definition of Pattern Rotation Point, the ***pattern adjustment rule*** could be stated as followed:

Pattern Adjustment Rule: at time t, rotate the future patterns of task τ_i such that its Pattern Rotation Point $PRP_i(t)$ overlap with the arrival time of next upcoming job, *i. e.*, $[\frac{t}{T_i}]T_i$.

The main purpose of the pattern adjustment rule is to insert as many optional jobs as possible after the current optional job met its deadline such that there could be more space for executing more optional jobs. The advantage is obvious: the optional jobs could be executed with arbitrary low speed(s) without requiring recovery jobs reserved for them. It is worthy to note that such kind of pattern adjustment rule will not cause any violation on (m, k) constraints or any deadline missing on future mandatory jobs, which will be formulated into the following theorem (the proof is provided in the Appendix part B).

Theorem 2 *Algorithm 1 can ensure the (m, k)-constraints for $\mathcal{T}$ if $\mathcal{T}$ is schedulable under E-pattern.*

18.3.1. Computing the LST for the Upcoming Mandatory/Recovery Jobs

In this section, we present how to compute the LST for the mandatory/recovery jobs under the procrastination scenario. Please note that since only mandatory jobs (and their recovery jobs, if any) are considered for procrastination, in this part, the terms "job" and "mandatory job (together with its recovery job, if any)" have the same meaning and can be used interchangeably for convenience of presentation.

Quan *et al.* [36] introduced an off-line approach that can be applied to find the LST for general job sets scheduled by FP scheme. Although the approach in [36] could also be applied to periodic task sets, the computation and storage costs for it could be expensive. In the following, we will introduce our new approaches which could be applied during online phase to find the LST for periodic task sets scheduled by FP scheme. Firstly, we introduce our approach based on exact timing analysis as followed.

18.3.1.1. Computing LST with Exact Timing Analysis

Definition 5 [35] (J_i-**scheduling point**) Time t is called a J_i - *scheduling point* if $t = d_i$ or $t = r_q$, $q < i$ and $r_i < r_q < d_i$.

Definition 6 (Level-i latest starting time) Given at time t, the *level-i latest starting time* (denoted as $LST_i(t)$) is defined to be the maximal time instance after t such that, if the execution of J_i and any other jobs with priorities higher than J_i and arrival times later than t starts no later than $(LST_i(t))$, J_i can still meet its deadline.

Note that for periodic task sets, it is sufficient to just consider J_i as either the current job of task τ_i (if τ_i is not finished yet) or the next upcoming instance of τ_i (if τ_i is finished) for each task priority level i.

Based on the J_i-scheduling point, the level-i latest starting time $LST_i(t)$ can be computed by the following lemma:

Lemma 4 *Given at time t, let J_i be either the current job of task τ_i (if τ_i is not finished yet) or the next upcoming instance of τ_i (if τ_i is finished) and $SP(J_i)$ be all the J_i-scheduling points. Then,*

$$LST_i(t) = \max\{\hat{t} - \sum_{y_k \in hp(j_i)}^{t < r_k < \hat{t}} \frac{w_k}{s_k} - \frac{w_i^{rem}}{s_i}, \hat{t} \in \mathcal{SP}(j_i)\}, \quad (18.4)$$

where w_i^{rem} is the remaining workload of job J_i at time t and hp(J_i) are the jobs with priorities higher than J_i.

The basic rationale behind Lemma 4 is that if there exists a scheduling point $\hat{t}$ of J_i in $SP(J_i)$ such that the remaining workload from J_i and all the upcoming higher priority jobs with arrival times earlier than $\hat{t}$ can be finished by it, then J_i is guaranteed to be schedulable. And the goal of Eq. (18.4) is to compute the maximal time t to which J_i and all the upcoming higher priority jobs can be procrastinated without violating this requirement.

However, the level-i latest starting time $LST_i(t)$ computed by Eq. (18.4) can only guarantee the feasibility of job J_i but not necessarily any other lower priority jobs. The reason is that, if J_i and the higher priority jobs are procrastinated to $LST_i(t)$, they will potentially block the executions of other lower priority jobs and cause them to miss their deadlines. To guarantee the feasibility for all job deadlines, we need to compute the level-i latest starting time $LST_i(t)$ in a similar way for all priority levels and choose the smallest one as the LST for the whole job/task set, which is stated in the following theorem (The formal proof can be done in the same way as Theorem 1 in [36] and is thus omitted):

Theorem 3 *Given a mandatory job set $\mathcal{M}$ for task set $\mathcal{T}$ according to E-pattern, the latest starting time for $\mathcal{M}$ at time t can be computed as*

$$LST(t) = \min_i\{LST_i(t)\}, \quad (18.5)$$

where $LST_i(t)$ is computed according to Eq. (18.4).

Note that the computation complexity of LST in Theorem 3 is $O(N^2)$, which is more efficient than that ($O(N^3)$) in [36] as it takes one less loop. Moreover, Theorem 3 is different from the condition in [36] in that it can be used to compute the LST for the upcoming job set not only when the processor is idle, but also when the processor is busy at time t, which will be very useful for our dynamic slack reclamation algorithms during the online phase (more details are provided in Section 18.4).

18.3.1.2. More Efficient Conditions in Computing LST

The time complexity of the above approach in computing LST mainly comes from computing $LST_i(t)$ for each priority level, which can be $O(N^2)$ according to Theorem 3. In order to reduce the time complexity during the online-phase, in this section, we also develop two sufficient conditions for computing LST:

Condition 1: The first sufficient condition is based on the observation that, under Epattern, the first job of each task τ_i is the *critical instance* (Lemma 3). So if we let J_i be the critical instance at time $t = 0$ for priority level i, it will suffer the maximal interference from higher priority jobs. And the maximal time that J_i can be delayed safely at priority level i, represented as δ_i, can be computed according to Eq. (18.4). Based on it, the first sufficient condition can be formulated as follows:

Theorem 4 *Let M be the mandatory job set for task set T according to E-pattern. Assuming at current time t, let J' be the upcoming mandatory job set (i. e. with arrival time r_i no earlier than t). Then the latest starting time for J' at time t can be computed as:*

$$LST(t) = \min_{J_i \in J'}(r_i + \delta_i). \qquad (18.6)$$

Since δ_i can be computed off-line, the online complexity for Theorem 4 is $O(N)$.

Theorem 4 allows us to determine the maximal delay for mandatory jobs based on worst case interference assumption for each priority level, which is available off-line. The advantage is its small run-time overhead. Unfortunately, similar to any other off-line approach, it suffers the pessimistic estimation due to its assumption of the worst case interference. Regarding that, we also developed the second sufficient condition in computing $LST(t)$.

Condition 2: Our second sufficient condition is based on limiting the number of jobs to be checked in computing $LST_i(t)$ in Eq. (18.4). Recall that the processor must be activated before the deadline of any incoming job. Therefore, we can immediately set up an upper bound for procrastinating the mandatory jobs. We call this bound the *procrastination bound* and denote it as T_B. Specifically, we have

$$T_b = \min_{r_n \geq r_i}\{d_n\}. \qquad (18.7)$$

With T_B, one intuitive method is to use the minimum of the latest starting times of the jobs arriving before T_B as the $LST(t)$ for the incoming jobs. However, the latest starting time computed by employing Eq. (18.4) only for jobs arriving before T_B may not be valid since the deadlines of lower priority jobs arriving after T_B cannot be guaranteed. In order to guarantee the feasibility of all jobs, the lower priority jobs arriving after T_B should also be considered in determining the $LST(t)$.

Note that Eq. (18.4) guarantees the schedulability of job J_i by requiring J_i be finished just *at* its deadline. However, even though J_i can meet its deadline, other lower priority jobs may miss their deadlines if J_i completes exactly at d_i. This indicates that to guarantee the deadlines for the lower priority jobs, a higher priority job may need to finish earlier than its deadline. In addition, since the deadlines of the jobs arriving earlier than T_B can be guaranteed by Eq. (18.4), only the jobs arriving after T_B may possibly miss their deadlines. In what follows, we want to identify the time point when a higher priority job has to be finished, *i. e., the effective deadline*, such that other lower priority jobs arriving after T_B can also meet their deadlines. Specifically, we have the following definition.

Definition 7 Let $J_i \in \mathcal{J}$ be any job that arrives before T_B. The *effective deadline* for J_i is the time $t = d_i^{**}$ such that if J_i finishes by d_{i^*}, J_i and all the jobs that arrive later than T_B can meet their deadlines.

With the definition above, given a job J_i arriving before T_B, its effective deadline d_{i^*} can be computed as follows:

$$d_i^* = \min_p\{d_i, r_p + \delta_p\}, \forall J_p \text{ with } r_p > T_B \text{ and } p > i. \qquad (18.8)$$

With these effective deadlines, we can therefore compute the corresponding latest starting times for the jobs arriving earlier than T_B, and take the minimum as the latest starting time for the job sets. Specifically, we have the following theorem (The proof is provided in Appendix A).

Theorem 5 Given a job set $\mathcal{J}$, let the speed for each job $J_k \in \mathcal{J}$ be s_k. Assuming at time t, let $\mathcal{J}$ be the job set containing all jobs yet to finish at t or with arrival times later than t. And let the procrastination bound T_B be defined according to Eq. (18.7) for $\mathcal{J}$. Then the execution of $\mathcal{J}$ can be delayed to LST(t) with no job missing its deadline, where

$$LST(t) = \min_{J_i \in \mathcal{J}_s}\{\max\{\hat{t} - \textstyle\sum_{J_k \in hp(J_i)}^{t < r_k < \hat{t}} \frac{w_k}{s_k} - \frac{w_i^{rem}}{s_i}, \hat{t} \in \mathcal{SP}^*(J_i)\}\}, \qquad (18.9)$$

where J_s consists of the jobs from $\mathcal{J}$ with arrival times earlier than T_B, $hp(J_i)$ are the jobs with priorities higher than J_i and $SP^*(J_i)$ is the set of effective scheduling points of job J_i based on its effective deadline d_{i^*} defined in Eq. (18.8).

Note that in the theorem above, the set of effective scheduling points $SP^*(J_i)$ of job J_i can be got by replacing d_n in Definition 5 with d_{n^*} defined in Eq. (18.8).

In Theorem 5, we only need to check at each of the scheduling points for all jobs in J_s, which has a complexity of $O(N^0 \times M^0)$, where N^0 is the total number of jobs in J_s and M^0 is the number of scheduling points for each job in J_s, which is no larger than the number of jobs arriving within the interval between the earliest arrival time and the maximal deadline of the jobs in J_s. Since N^0 and M^0 are usually very small for periodic task sets, the computation of $LST(t)$ in Theorem 5 will have a very low computation complexity.

Finally, it is worthy to mention that both Theorem 1 and Theorem 2 are sufficient conditions. Therefore, the larger one from Eq. (18.6) and (18.9) should be used as LST for the job set considered.

18.4. Online Slack Reclaiming

During the online phase, some mandatory jobs may present actual execution times shorter than their worst case and slack times can be available. Moreover, under fixed-priority scheduling, there could also be some "interval slack" due to processor time under utilized even when all jobs present their worst case execution times. How to utilize the slack time most efficiently is not a trivial a problem. For example, the jobs with recovery jobs reserved might want to use the slack time to help reduce their speeds, while the jobs without recovery jobs might want to use the slack time to reserve recovery jobs for them first before considering speed scaling. Moreover, when the processor is idle, it might be more energy beneficial to shut down the processor during the slack time.

In [21], dynamic reclaiming approach was proposed to reduce the overall energy consumption including leakage. However, as shown in [27], even without considering the system reliability, this approach is not necessary most energy-efficient partially because it never scales the job speeds below the practical critical speed s_{prcs}. With system reliability in mind, more advanced dynamic reclaiming techniques were proposed in [48]. The main idea[1] is: assuming at time t, job J_i is released with an amount of reclaimable slack time $S_i(t)$ from jobs with priorities higher than or equal to it, the slack time $S_i(t)$, if long enough, could be used to reserve a recovery job for J_i first if J_i does not have a recovery job reserved for it yet. The residue part can be used to scale the speed of J_i to be not below the practical critical speed s_{prcs}. The limitation of this approach is that it can only reclaim slack time due to early completion of jobs with priorities higher than or equal to the current job. So the slack estimation in it could be rather conservative. Moreover, it cannot reclaim such kind of "interval slack" under FP scheme when all jobs present their worst case execution times. The situation can become worse when additional QoS requirements such as (m, k)-constraints are imposed on the system. On the other hand, since it does not consider job procrastination, it might not be able to merge the idles intervals to facilitate processor shut-down to save more energy. Similar to the approach in [21], it also adopted practical

---

[1] Note that although the approach in [48] is for real-time systems based on EDF scheme, its principle can also be applied to systems with fixed-priority scheme

critical speed as the lower bound for speed scaling, which could limit its energy efficiency as well.

Consider an FP task set consisting of three tasks $\{\tau_1 = (4\ ms,\ 20\ ms,\ 20\ ms,\ 1,\ 1);$ $\tau_2 = (56\ ms,\ 80\ ms,\ 80\ ms,\ 1,\ 1);\ \tau_3 = (8\ ms,\ 80\ ms,\ 80\ ms,\ 1,\ 1)\}$ which will be executed on the same Intel XScale processor model as in Section 18.2.5. Here we assume all mandatory jobs of τ_1 present their worst case execution times while the mandatory jobs of τ_2 and τ_3 present actual execution times less than their worst case, *i. e.*, $aw_{21} = 4\ ms$ and $aw_{31} = 2.4\ ms$. The task worst case execution schedule, actual execution times, and slack times are shown in Fig. 18.2(a). It is easy to see that when all mandatory jobs present their worst case execution times, there is no slack time available in the system for reserving recovery jobs. So the static speeds for all mandatory jobs of each task should be determined as maximum speed 1.

As shown in Fig. 18.2 (b), according to the approach in [48], at time $t = 8$, job J_{31} began to execute with slack time $S_3(8) = 52$ time units from higher priority job(s), which is larger than w_{31}. So 8 time units from the slack could be used to reserve a recovery job R_{31} for J_{31} first. Since there is still unused slack time from $S_3(8)$ left, the remaining part could be used to scale s_{31} to be as low as the practical critical speed $s_{prcs} = 0.3$. However, as shown in Fig. 18.2(b), assuming J_{31} is executed with s_{prcs} and completed successfully at $t = 15$, the slack time between [15, 20], due to its lower priority than any other jobs, cannot be utilized by any of them according to [48]. On the other hand, it is also smaller than $t_{be} = 10$ and cannot be used to shut down the processor, either. So this part of slack time will simply be wasted. Moreover, when J_{12} arrived at $t = 20$, although part of the slack time from J_{31} is not consumed up yet, it cannot be utilized by J_{12} either for the same reason as stated above. The situations at $t = 40$ and $t = 60$ are also the same. As shown in Fig. 18.2(b), the complete schedule within LCM (= 80) generated at least 4 idle intervals and part of the slack time is wasted. Moreover the processor needs to be shut down more than once. As a result, the total energy consumption within the interval [0, LCM] is 36.09 *mJ*.

However, if we apply our procrastination approach at $t = 8$, J_{31} and all the upcoming mandatory jobs can be procrastinated to as late as $LST(t) = 36$. So the slack time between $[t,\ LST(t)]$, *i. e*, [8, 36], could be fully utilized by J_{31}. As shown in Fig. 18.2 (c), 8 time units from it could be used to reserve a recovery job R_{31} for job J_{31}. And there is still 20 time units left. How to utilize this residue part of slack time most efficiently could be a problem as it could be used either to scale the speed of J_{31} or to shut down the processor. Based on our concept of feasible interval in Section 18.2, we could determine the speed of J_{31} by constructing a "virtual" feasible interval with length $(LST(t)-t-C_{31}+w_{31})^1$. Note that here we name the feasible interval as "virtual" because J_{31} might potentially be preempted by other higher priority jobs (such as J_{12}) and therefor this feasible interval might not be a continuous one. However, in constructing the virtual feasible interval, we can imagine those higher priority jobs be shrunk into a single point, which will not affect the correctness of our approach. With this virtual feasible interval, the speed of J_{31} could

---

[1] Here C_{31} time units has been reserved for recovery job R_{31} and thus should be excluded when constructing the "virtual" feasible interval

be determined based on our threshold work-demand analysis in Section 18.2. In this particular example, the output is that the *critical speed* strategy should be adopted on J_{31} during the virtual feasible interval. So s_{31} will be set as 0.3. Note that in this case, in order to make the processor shut down interval overlapped with the remaining slack interval, we should shut-down the processor first during the time interval [8, 28] and wake it up at the end of the shut-down interval. The virtual schedule within the virtual feasible interval of J_{31} is shown in Fig. 18.2 (c).

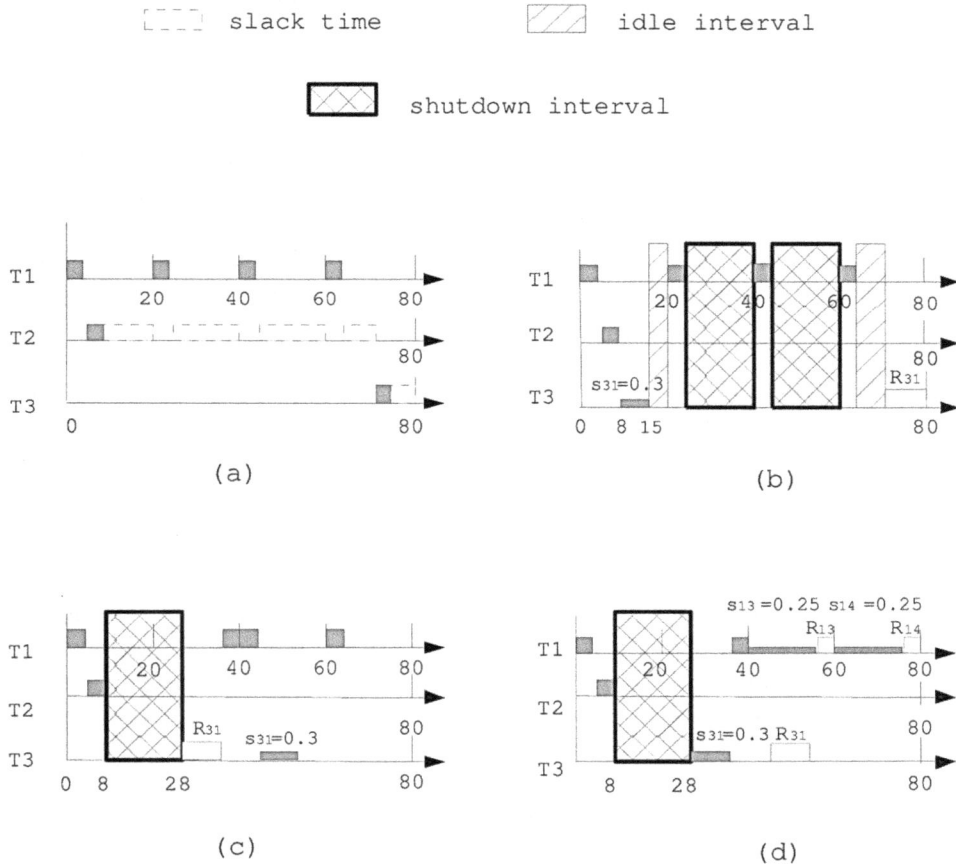

Fig. 18.2. (a) The task set $\{\tau_1 = (4\ ms,\ 20\ ms,\ 20\ ms,\ 1,\ 1);\ \tau_2 = (56\ ms,\ 80\ ms,\ 80\ ms,\ 1,\ 1);\ \tau_3 = (8\ ms,\ 80\ ms,\ 80\ ms,\ 1,\ 1)\}$ schedule with actual execution times ($aw_{21} = 4\ ms$ and $aw_{31} = 2.4\ ms$) and slack times; (b) The slack reclaiming for the same task set based on the algorithm in [48]; (c) The expected (virtual) schedule with slack reclaiming with our approach in Algorithm 2 at time $t = 8$; (d) The actual schedule within [0, LCM], with the time budget of J_{31} and its reserved recovery job R_{31} switched.

One thing to be noted is that, in the virtual schedule in Fig. 18.2(c), it seems as if the recovery job R_{31} were reserved before J_{31}. However, we can always "switching" execution time budget between R_{31} and J_{31} to let J_{31} be executed first. It is not hard to see that such kind of "time switching" is safe so long as the amount of switched time does not exceed their individual time budget. The actual schedule is shown in Fig. 18.2 (d) (note that in the actual schedule, since J_{31} completes successfully, the time budget for R_{31} is freed and can be reclaimed by other jobs). The situation for J_{13} at $t = 40$ (and also J_{14} at $t = 60$) is similar but in this case the energy-efficient speed of J_{13} is determined to be 0.25, which is below the practical critical speed. From the complete schedule in Fig. 18.2(d), the total energy consumption within the interval [0, LCM] is 24.9 *mJ*, which is 31 % lower than that from Fig. 18.2 (b).

From the example we can see that slack reclaiming for mandatory jobs under reliability requirement can affect the overall energy consumption significantly. Based on the analysis above, we propose a reliability-aware dynamic reclaiming approach as sketched in Algorithm 2.

As shown in Algorithm 2, at time t, if job J_i becomes current and is not a recovery job, in order to estimate the slack time reclaimable by job J_i, we try to delay J_i and all incoming mandatory jobs to $LST(t)$. Then the time interval $(LST(t)-t)$ could be regarded as reclaimable slack time for J_i. This part of slack time could have a triple-purpose use: reserving recovery job for J_i (if J_i does not have one yet), reducing speed for J_i, and/or shut down the processor if necessary to save energy. In order to do so, depending on whether J_i already has recovery job or not, we can construct the virtual feasible interval for J_i correspondingly (line 7 and line 22). Based on the virtual feasible interval, the speed of J_i can be determined with Theorem 1 correspondingly (line 8-16 and line 23-31). Also note that in Algorithm 2, the time budget for a recovery job and its "primary job" could be switched when necessary following the principle that the "primary job" should always be executed before its recovery job.

It is not hard to see that the efficiency of Algorithm 2 largely depends on the estimation of the slack time $(LST(t)-t)$, which is based on the computation of $LST(t)$. In Section 18.3.1, we have introduced two sufficient conditions for computing $LST(t)$ by assuming worst case execution time of the current job J_i. Since in our systems model, we assume the actual execution time aw_i of the current job J_i could be known after its arrival, both conditions in Section 18.3.1 could be applied here by replacing w_i in them with aw_i. Moreover, with the information of job slack time at each priority level, we also propose the third condition of computing $LST(t)$ as followed.

Condition 3: Assuming at time t, task τ_j has an amount of slack time $S_j(t)$ from early completed jobs with priorities higher than or equal to it, then obviously this part of slack time $S_j(t)$ can be used to procrastinate τ_j safely. Moreover, based on our concept of level-i latest starting time, if the current job of task τ_j has already been finished, the next upcoming mandatory job J_α of τ_j can be delayed to $\widetilde{LST}_j(t) = r_\alpha + \delta_j$. So the maximal between $\widetilde{LST}_j(t)$ and $(t + S_j(t))$ can be used as the latest starting time for the upcoming mandatory job of task τ_j.

Algorithm 2 Algorithm for dynamic reclamation

1: **Input:** the current time t, job J_i with the highest priority in MC
2: **if** J_i is a recovery job **then**
3:      Execute J_i with maximal speed 1;
4: **else**
5:      compute $LST(t)$ for the whole task set at time t;
6:      **if** J_i has a recovery job **then**
7:          let w_{th} be the threshold work-demand based on J_i's virtual feasible interval length $L = (LST(t) - t) + \frac{aw_i^{rem}}{s_i}$;
8:          **if** $aw_i^{rem} < w_{th}$ **then**
9:              // adopt *shut down + critical speed* strategy
10:             $s_i' = s_{prcs}$;
11:             recompute $LST(t)$ with J_i's updated speed;
12:             $t_{wp} = \min\{t + (L - \frac{aw_i^{rem}}{s_i}), LST(t)\}$; //set safe wake-up time point
13:             Shut down the processor at t and set the wake-up timer to be $(t_{wp} - t)$;
14:         **else**
15:             $s_i' = \frac{aw_i^{rem}}{L}$; // use slack time to scale the speed of J_i
16:         **end if**
17:     **else if** $(LST(t) - t) < C_i$ **then**
18:         //slack time not enough for reserving recovery job
19:         Execute J_i with maximal speed 1;
20:     **else**
21:         Reserve a recovery job R_i for J_i;
22:         let w_{th} be the threshold work-demand based on J_i's virtual feasible interval length $L = (LST(t) - t) - C_i + aw_i^{rem}$;
23:         **if** $aw_i^{rem} < w_{th}$ **then**
24:             // adopt *shut down + critical speed* strategy
25:             $s_i' = s_{prcs}$;
26:             recompute $LST(t)$ for the job at all priority levels including R_i;
27:             $t_{wp} = \min\{t + (L - \frac{aw_i^{rem}}{s_i}), LST(t)\}$; //set safe wake-up time
28:             Shut down the processor at t and set up the wake-up timer to be $(t_{wp} - t)$;
29:         **else**
30:             $s_i' = \frac{aw_i^{rem}}{L}$; // use slack time to scale the speed of J_i
31:         **end if**
32:     **end if**
33: **end if**

It is not hard to see that whenever a mandatory job J_i becomes current at time t, all the higher priority jobs that arrived before t must already have been finished. Moreover, all the other mandatory jobs in MC must have lower priorities than J_i and therefore can also be delayed by $S_i(t)$ time units safely. So the latest starting time $LST(t)$ for the whole task set at t can be computed as

$$LST(t) = \min\{\min_{j<i}\{\max\{\widetilde{LST}_j(t), t + S_j(t)\}\}, t + S_i(t)\} \qquad (18.10)$$

Note that since all the three conditions in Eq. (18.6), (18.9), and (18.10) are sufficient conditions in computing $LST(t)$, the maximal value computed by them should be used as $LST(t)$ in Algorithm 2.

Compared with the reliability-aware dynamic reclaiming algorithm in [48], the major difference in Algorithm 2 is that it can reclaim slack not only from higher priority jobs, but also from lower priority jobs, as shown in Fig. 18.2 (c). Moreover, by integrating task/job procrastination in it, the "interval" slack under FP scheme could also be reclaimed and the idle intervals could be merged to facilitate processor shut-down. And the job speed could be scaled below the practical critical speed when necessary.

In next section, we use experimental results to demonstrate the effectiveness of our newly proposed approach.

18.5. Experimental Results

In this section, we evaluate the proposed techniques using simulations. In general, we compare the performance of four different approaches.

– *NoEM* The task sets are partitioned with E-pattern, and the mandatory jobs of all tasks are executed following the FP scheme but without energy management, *i. e.*, always with the highest speed. We use its results as the reference results.

– *FPRPM$_E$* The task sets are partitioned with E-pattern to satisfy the given (m, k)-constraints. Then the mandatory jobs are scheduled with the approach from [48] but customized to FP scheme.

– *FPMK$_{HYB}$* The task sets are partitioned with E^R-pattern [28] first, then the approach in [28] (also customized to FP scheme) was applied to schedule the jobs but with practical critical speed s_{prcs} as the lower bound for speed scaling (on optional jobs). Note that although the original approach in [28] did not consider system reliability, it can still preserve the system reliability if all mandatory jobs except the optional jobs are executed with the maximum speed.

– *FPRMK-EM* This approach is based on our newly proposed approach in Section 18.3.

We conducted two groups of experiments to evaluate the performance of the different approaches. The first group was based on synthesized task sets. And the second group was based on test cases drawn from real-world applications. The experiments and results are discussed as follows.

18.5.1. Experimental Results from Synthesized Task Sets

The periods of the synthesized tasks were randomly chosen in the range of [50 *ms*, 200 *ms*] and the deadlines were assumed to be less than or equal to their periods. The worst case execution time (WCET) was set to be uniformly distributed from 1 to its deadline. The m_i and k_i for the (m, k)-constraints were also randomly generated such that k_i is uniformly distributed between 2 to 10, and $m_i \leq k_i$. To investigate the energy performance of the different approaches under different workload, we divided the total (m, k)-utilization, *i. e.*, $\sum_i \frac{m_i C_i}{k_i T_i}$ into intervals of length 0.1. To reduce the statistical errors, we required that each interval contain at least 20 task sets schedulable with E-pattern, or until at least 5000 task sets within each interval had been generated. For the processor model we adopted the same processor model as in [12], *i. e.,* the Intel XScale processor [19]. The parameters are the same as used in Section 18.2.5.

For the fault and recovery model we adopted the same model as in [50, 49], *i. e.,* the transient faults are assumed to follow the Poisson distribution with an average fault rate of $\lambda_0 = 10^{-6}$ at the maximum speed s_{max} (and corresponding supply voltage). For the fault rates at lower speeds/voltages, we adopt the exponential fault rate model $g(s) = \lambda_0 \, 10^{\frac{d(1-s)}{1-s_{min}}}$ and assume that $d = 3$ [49]. That is, the average fault rate is 1000 times higher at the lowest speed s_{min} (and corresponding supply voltage).

Note that, under RAPM schemes [48], it is expected that the reliability of any executed job assuming recovery job could be improved [47]. However, in our newly proposed approaches, the selectively executed optional jobs do not have recovery jobs. They are only executed to save energy without incurring (m, k)-violations. Therefore we are interested in exploring the overall reliability of the system in terms of the capability of satisfying the (m, k)-constraints by the different approaches. Following the idea in [50], we define the *probability of job failure* (denoted as PrJF) as the ratio of the number of failed jobs over the total number of jobs executed. Moreover, with the (m, k)-constraints in mind, we also define the *probability of (m, k)-violation* (denoted as PrMKV), as the ratio of the number of (m, k) violations (due to job failure) over the total number of dynamic sliding windows inspected (for checking the (m, k)-constraints). The results in different (m, k)-utilization intervals are shown in Fig. 18.3(a) and Fig. 18.3(b).

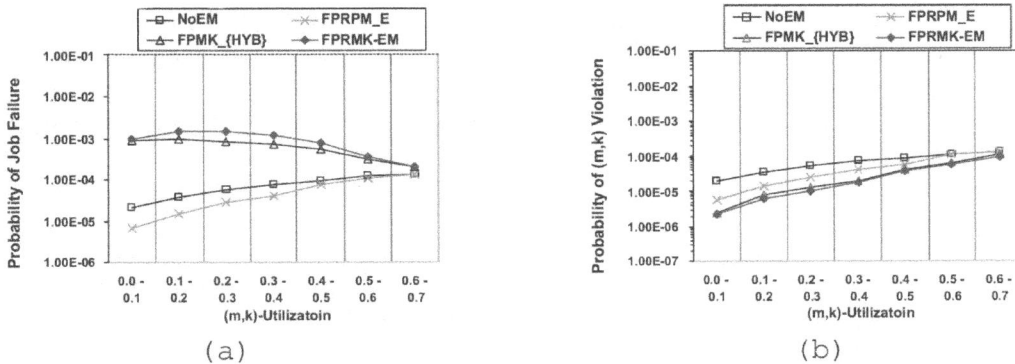

(a)   (b)

Fig. 18.3. (a) Comparison of the probability of job failure by the different approaches; (b) Comparison of the probability of (m, k)-violation by the different approaches.

As can be seen from Fig. 18.3(a), in all utilization intervals the PrJF of *FPRPM$_E$* is slightly lower than that by *NoEM*, which conforms to the conclusion in [47]. Moreover, as the (m, k)-utilization decreases, the PrJFs of *NoEM* and *FPRPM$_E$* also decrease correspondingly because less workload from mandatory jobs needs to be executed. The tendencies for *FPMK$_{HYB}$* and *FPRMK-EM* are different as their PrJFs will increase first with the decrease in (m, k)-utilization. This is mainly because they executed a number of optional jobs (at relative lower speeds) for which no recovery jobs are assumed to be reserved. If any optional job failed, it is simply discarded. For statistical purpose we also count it as a job failure (although it will not cause any (m, k)-violation). It is not surprising that, as the (m, k)-utilization decreases, more optional jobs will be available for execution in $FPMK_{HYB}$ and $FPRMKEM$, causing higher PrJFs in them. That is also why the average PrJFs of $FPMK_{HYB}$ and $FPRMK$-EM are obviously higher than those of $FPRPM_E$ and $NoEM$ in low utilization intervals. Meanwhile, when the (m, k)-utilization is relatively low, the PrJF of $FPRMK$-EM is a little higher than that of $FPMK_{HYB}$. This is because, unlike $FPMK_{HYB}$ which adopts the practical critical speed as the lower bound for speed scaling, $FPRMK$-EM could scale the speed for some optional jobs below the practical critical speed when necessary, which could result in higher PrJF on optional jobs due to the negative effect of speed reduction on the transient fault rate. Moreover, when the (m, k)-utilization is around 0.1-0.2, the PrJF of $FPRMK$-EM reaches its maximal value. This is because around this interval, according to its speed determination strategy, $FPRMK$-EM could execute the largest number of optional jobs with speed below the critical speed. Correspondingly, the PrJF of $FPRMK$-EM is also the maximum around this interval. When the (m, k)-utilization becomes still smaller, the speed determination strategy of $FPRMK$-EM will choose to execute most of the optional jobs with critical speed (together with processor shut-down) to save energy. As a result its PrJF will not be higher than that during the interval 0.1-0.2. On the other hand, the higher PrJF in $FPRMK$-EM (mainly from optional jobs) does not hurt its capability of satisfying the (m, k)-constraints for the tasks. As shown in Fig. 18.3(b), in all cases, the average PrMKVs of $FPRMK$-EM and $FPMK_{HYB}$ are not higher than the other approaches. This is because, once completed successfully, the optional jobs also become effective jobs, which could help improve the chance for the tasks to satisfy the (m, k)-constraints. Moreover, it is interesting to note that when the (m, k)-utilizations of the task sets are small, the average PrMKV of $FPRMK$-EM could be slightly lower than $FPMK_{HYB}$. This is because in this case the system has more space to reserve recovery jobs for more mandatory jobs in $FPRMKEM$ during the off-line phase. As shown in [47], reserving recovery job for executed job could help improve its reliability and reduce the failure rate at the job level (because failed job could be compensated by its recovery job). Correspondingly, it could help reduce the chance for the task(s) to violate the (m, k)-constraint(s).

Next, we inspect the energy consumption of the different approaches. Note that under this scenario, we evaluate the performances of the different approaches over a large spectrum of (m, k)-utilizations as well as variability in actual workloads. In particular, we performed two sets of simulations.

In the first set, we randomly picked the actual execution times of the jobs from [0.2 WCET, WCET] and checked the energy consumption of the different approaches within each (m, k) utilization interval. The results are normalized to that by $NoEM$ and shown in Fig. 18.4(a).

From Fig. 18.4(a), our newly proposed approach, *i. e., FPRMK-EM* can have much better energy performance than the other approaches in most utilization intervals. In particular, compared with $FPRPM_E$, the average energy reduction is more than 20 %. The average energy improvement over $FPMK_{HYB}$ could be even higher for relatively lower (m, k) utilization intervals. This is because, by adopting dynamic job/recovery pattern and determining the job speed based on the threshold work-demand analysis, *FPRMK-EM* can schedule the jobs more adaptively and reduce the job speed below the practical critical speed when necessary. On the other hand, with our newly proposed mandatory job procrastination methods, *FPRMK-EM* can help extend/merge the processor sleeping time more effectively, which could not only help reduce the idle energy and/or processor shut-down overhead but also limit the number of optional jobs to be scheduled effectively. Note that although $FPMK_{HYB}$ also adopted dynamic pattern in scheduling mandatory/optional jobs, since it executed optional jobs in a greedy manner, it is only efficient in reducing the dynamic energy [28]. When static energy (such as leakage) is taken into consideration, the performance of $FPMK_{HYB}$ degraded significantly because it could execute an excessive number optional jobs when using the practical critical speed as the low bound for speed scaling. As a consequence its effect in reducing dynamic energy could be counteracted significantly and its overall energy consumption could be even higher than the approach based on static pattern, *i. e.,* $FPRPM_E$. For example, when the (m, k)-utilization of the system is relatively low, the energy consumption of $FPMK_{HYB}$ is obviously higher than that of $FPRPM_E$.

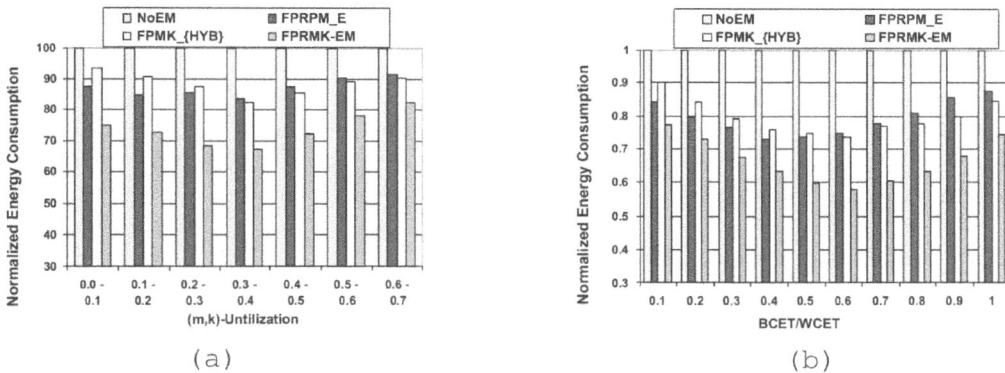

Fig. 18.4. (a) The total energy comparison by different approaches;
(b) The energy comparison with variation of $\frac{BCET}{WCET}$ ratio.

In the second set of experiments, we performed simulations on some representative utilization intervals, *i. e.,* between 0.2-0.4 (the results in other utilization intervals are fairly similar), but varied the value of $\frac{BCET}{WCET}$ (BCET stands for the best case execution time) ratio from 0.1 to 1.0 to compare the energy consumption for the different approaches. The normalized results are shown in Fig. 18.4 (b).

As shown in Fig. 18.4 (b), with efficient reliability-aware dynamic slack reclaiming, the energy performance of our newly proposed approach, *i. e.,* *FPRMK-EM*, still outperformed the other approaches in all cases. For example, when the $\frac{BCET}{WCET}$-ratio is relatively high, the energy reduction of *FPRMK-EM* over *FPRPM$_E$* could be nearly 25 %, while when the $\frac{BCET}{WCET}$-ratio is relatively low, the energy reduction of *FPRMK-EM* over *FPMK$_{HYB}$* could be nearly 22 %. This is mainly due to the fact that our approach could combine the pure DVS strategy and the critical speed strategy in reclaiming the slack time and determining the job speed more adaptively. At the same time, different from the approach in *FPRPM$_E$* which could only reclaim slack time from higher priority jobs, our newly proposed slack reclaiming approach could also reclaim slack time from lower priority jobs as well as some "internal slack time" under FP scheme (due to under-utilized processor time). Moreover, with efficient mandatory job procrastination, the virtual feasible interval of the current job could be extended effectively, which provided more chance in choosing the most energy-efficient speed for executing the job and facilitate the processor shutdown when necessary (to save more energy).

18.5.2. Experimental Results from Practical Applications

Next, we evaluate the different approaches in a more practical environment. The test cases contained two real world applications: INS (Inertial Navigation System) [1] and CNC (Computerized Numerical Control) machine controller [32]. The timing parameters such as the deadlines, periods, and execution times were adopted from these practical applications directly.

The processor model adopted in this group is the same as that used in Section 18.5.1, *i. e.* the Intel XScale processor [19], but with only five discrete speeds [12], *i. e.* : (0.15; 0.4; 0.6; 0.8; 1) GHz with corresponding power consumption (80, 170, 400, 900, 1600) mW. The other parameters are the same as used in Section 18.5.1.

Similar to the tests on synthesized task sets, we also collected the PrJFs and PrMKVs by the different approaches on INS and CNC. The results in different (*m, k*)-utilization intervals are shown in Fig. 18.5 and Fig. 18.6.

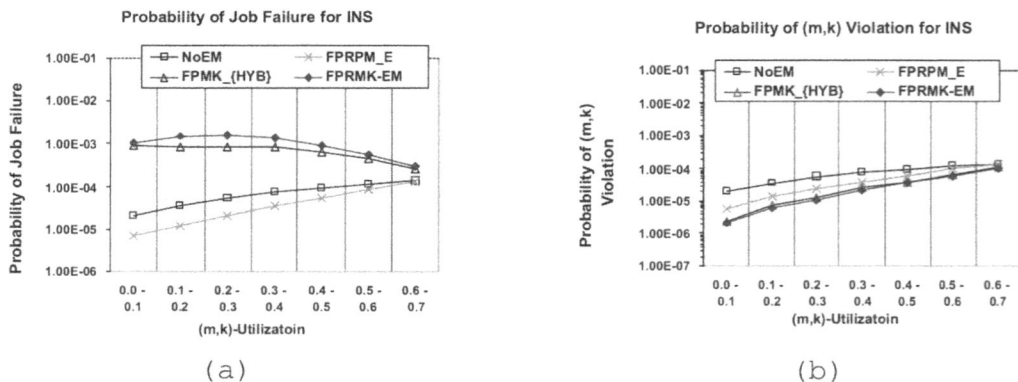

Fig. 18.5. (a) Comparison of the probability of job failure by the different approaches for INS; (b) Comparison of the probability of (*m, k*)-violation by the different approaches for INS.

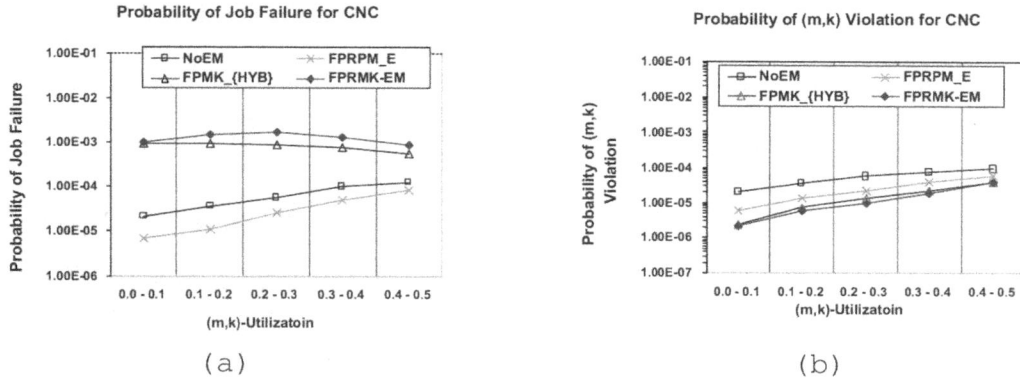

Fig. 18.6. (a) Comparison of the probability of job failure by the different approaches for CNC; (b) Comparison of the probability of (m, k)-violation by the different approaches for CNC.

As shown in Fig. 18.5 (a) and Fig. 18.6 (a), the tendencies for PrJFs by the different approaches on both INS and CNC are quite similar to the case on the synthesized task sets in Section 18.5.1. The main difference is that for these two applications, the PrJF of *FPRMK-EM* reaches its maximal value at around the (m, k)-utilization interval 0.2-0.3. This is because the periods for the tasks in both INS and CNC are close to harmonic (*i. e.*, longer periods are the multiples of shorter periods), which could result in longer feasible intervals for *FPRMK-EM* to schedule the jobs (with relative lower speeds). The PrJF of $FPMK_{HYB}$ keeps increasing with the decrease in (m, k)-utilization because more optional jobs will be executed. But its curve will be more flat after the interval 0.3-0.4 because $FPMK_{HYB}$ never reduces the job speed below the critical speed.

The tendencies for PrMKVs by the different approaches are also similar to the results on the synthesized task sets. As shown in Fig. 18.5 (b) and Fig. 18.6 (b), for both INS and CNC, the PrMKV by *FPRMK-EM* still stays at the lowest level compared with the other approaches, which means the reliability can be preserved by *FPRMK-EM* under the context of (m, k)-constraint.

With energy conservation as our major goal for the propose work, we also collected the average energy consumption for each approach and normalized it to that by *NoEM*. The results are shown in Fig. 18.7.

From Fig. 18.7, the above analysis also conforms to the results on energy consumption. In nearly all (m, k)-utilization intervals, *FPRMK-EM* still have much better energy performance than the other approaches for both real world applications. As shown in Fig. 18.7, the maximal energy reduction by *FPRMK-EM* over $FPRPM_E$ can be nearly 26 % (20 %) for INS (CNC), respectively. And the maximal energy reduction by *FPRMK-EM* over $FPMK_{HYB}$ could be 28 % (25 %) for INS (CNC), respectively.

In summary, the experimental results based on both synthesized task sets as well as practical real-time applications have clearly demonstrated the effectiveness of our

501

approach in reducing energy while preserving reliability for real-time embedded systems with weakly hard QoS-constraint.

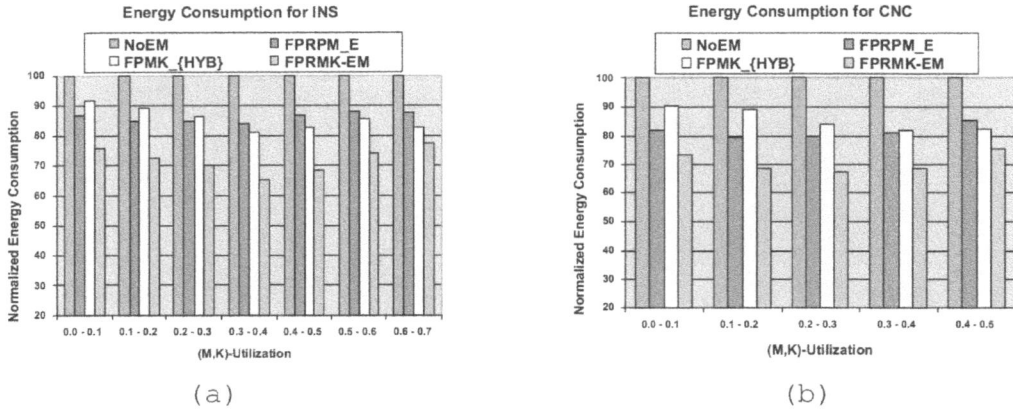

Fig. 18.7. (a) The energy comparison for INS; (b) The energy comparison for CNC.

18.6. Related Work

In recent years, there has been increasing interest to incorporate QoS requirement into power/energy conservation for real-time embedded systems. These approaches can be classified into *statistic* QoS-constrained model (e. g., [42]) and *deterministic* QoS-constrained model (e. g., [3, 28]). The statistic model tries to ensure the overall QoS statistically using certain overall deadline missing rate. However, even a very low overall deadline missing rate tolerance cannot prevent deadline misses bursting within a short period of time such that critical information during it could be lost. In contrast to that, the deterministic QoS-constrained model could ensure the QoS requirement not only statistically, but also deterministically at each of the user specified time interval to ensure the data integrity.

A widely adopted deterministic QoS-constrained model is the weakly hard model [7]. In the weakly hard QoS-constrained model, a repetitive task must have at least m out of any k consecutive job instances of the task meet their deadlines in order to satisfy the user specified QoS-constraint. Obviously the (m, k)-model proposed by Hamdaoui *et al.* [16] can serve well for this purpose. Based on the (m, k)-model, Ramanathan *et al.* [38] proposed to use partitioning strategy to guarantee the given (m, k)-constraints. West *et al.* [41] introduced another weakly hard QoS-constrained model called the *window-constrained* model, which requires that within any *non-overlapped* and *consecutive* windows containing k jobs, at least m of them meet their deadlines. From their definitions, it is not hard to see that *window-constrained* model is a relaxed version of (m, k)-model and the latter one could be transferred into the former one directly.

With the weakly hard QoS-constraint in mind, some previous researches have been conducted on providing QoS guarantee for soft real-time systems (e. g., [34, 13]). As

502

energy consumption has been a major concern for real-time embedded systems, it is becoming increasingly important to combine QoS-constraints and power/energy saving technologies to achieve better energy efficiency for embedded systems while satisfying their QoS requirements. In [28], Niu *et al.* introduced a hybrid approach which can reduce the dynamic energy consumption for real-time systems with (m, k)-guarantee. In [3], Alenawy *et al.* proposed an approach to minimize the number of dynamic failures for (m, k)-firm real-time systems with fixed energy budget constraint. In [30], Niu *et al.* proposed approaches to reduce the system wide energy consumption for weakly hard real-time systems with peripheral devices. In [29, 26], leakage-aware DVS approaches were proposed to reduce the energy consumption for real-time systems with (m, k) or *window-constraints*

Recently, reliability and fault tolerance have also been studied for the design of dependable computer systems. In [49], Zhu et al. formulated the reliability of a real time system into the probability of successful completion of all real-time tasks in it. They also showed that voltage reduction could affect the system reliability adversely. To address this issue, they proposed to use recovery job to compensate the loss of reliability in real-time jobs when applying DVS on them. The recovery job should have the same worst case execution time as the original job and will be reserved within the same time frame as the original one. Moreover, the recovery job will account for part of the total processor utilization as well. To utilize the processor resource more efficiently, Zhao et al. [44] proposed to reserve recovery jobs in a way that they could be shared among different tasks. However, this technique can only be applied to "frame-based" real-time systems in which all tasks have the same deadlines and periods. In [23], Li et. al improved the shared recovery technique by reducing unnecessary resource reservations in [44] such that more slack time could be utilized on voltage scaling to reduce energy further. In [24], they further proposed an adaptive checkpointing scheme to reduce energy consumption for "frame-based" real-time systems while preserving the reliability. For more general real-time system models, Zhao et al. [45] proposed a reliability-aware scheduling scheme to deal with energy reduction for periodical task sets in which different tasks may have different periods (but with deadlines equal to the periods). Their approach can deal with energy minimization for task sets with individual reliability requirement at the task level. With the negative impact of voltage scaling on system reliability in mind, their approach can have the reliability for each individual task quantified. In [18], a different job level scheduling framework was introduced to minimize the energy consumption for real-time systems with arbitrary arrival times and deadlines. The reliability model in it is a little different from the previous ones in that the system can tolerate at most K transient faults.

All of the reliability-aware approaches above only target hard real-time systems. And most of them are based on EDF scheme, whereas existing works based on FP schemes are much fewer. In [31], scheduling framework was developed for reducing energy for realtime systems under window-constraints, which are weaker than the (m, k)-constraints used in this chapter. In [50], reliability-aware scheduling algorithms were proposed to reduce the energy for weakly hard real-time systems based on "deeply-red" pattern. In this chapter, we target minimizing the energy consumption and preserving the system reliability for weakly hard real-time systems based on FP scheme. And our approach will adopt "evenly distributed" pattern [38] as generally it has better schedulability than the "deeply-red" pattern [28].

18.7. Conclusions

Energy consumption and Quality of Service are two primary concerns for the design of real-time embedded systems. At the same time, reliability requirement is also becoming increasingly important with the scaling of IC technology. In this chapter, we present a reliability-aware energy management (RAEM) scheme for reducing energy consumption for fixed-priority real-time embedded systems with weakly hard QoS requirement in terms of (m, k)-constraint. We first show that the widely adopted strategy which uses the critical speed as the lower bound for speed/voltage scaling might not always be more energy efficient than the traditional DVS strategy and there is a tradeoff between these two strategies depending on the job workload to be finished within certain interval. In order to ensure the (m, k)-constraints while preserving the system reliability, we propose to dynamically partition real-time jobs into mandatory ones and optional ones as well as to reserve recovery space for mandatory jobs in an adaptive way. Based on it, efficient energy management algorithms were proposed to reduce the energy consumption for real-time embedded systems under the requirement of (m, k)-constraints. Considering the dynamic nature of real-time embedded systems, we also presented efficient online slack reclaiming techniques to incorporate the variations in job execution times at runtime. Through extensive simulations, our experiment results demonstrate that the proposed techniques significantly outperformed the previous research in reducing energy consumption while preserving the system reliability and satisfying the (m, k)-constraints.

References

[1]. A. Burns, K. Tindell, A. Wellings, Effective analysis for engineering real-time fixed priority schedulers, *IEEE Transactions on Software Engineering*, May 1995, Vol. 21, pp. 920-934.
[2]. S. Albers, A. Antoniadis, Race to idle: new algorithms for speed scaling with a sleep state, *ACM Trans., Algorithms*, Vol. 10, Issue 2, Feb. 2014, pp. 1-9.
[3]. T. A. Al Enawy, H. Aydin, Energy-constrained scheduling for weakly-hard real-time systems, in *Proceedings of the IEEE Real-Time Systems Symposium (RTSS'05)*, 2005.
[4]. A. Antoniadis, C.-C. Huang, S. Ott, A fully polynomial-time approximation scheme for speed scaling with sleep state, In *Proceedings of the 26th Annual ACM-SIAM Symposium on Discrete Algorithms (SODA'15)*, 2015, pp. 1102-1113.
[5]. N. C. Audsley, Deadline monotonic scheduling, PhD Thesis, YCS 146, Dept. of Comp. Sci., *Univ. of York*, 1990.
[6]. M. A. Awan, S. M. Petters, Race-to-halt energy saving strategies, *Journal of Systems Architecture*, Vol. 60, Issue 10, 2014, pp. 796-815.
[7]. G. Bernat, A. Burns, Weakly hard real-time systems, *IEEE Trans. on Comp.*, Vol. 50, Issue 4, April 2001, pp. 308-321.
[8]. G. Buttazzo, Rate monotonic vs. EDF: judgement day, *Real-Time Systems*, Vol. 29, Issue 1, 2005, pp. 5-26.
[9]. X. Castillo, S. R. McConnel, D. P. Siewiorek, Derivation and calibration of a transient error reliability model, *IEEE Trans. Comput.*, Vol. 31, July 1982, pp. 658-671.

[10]. A. P. Chandrakasan, S. Sheng, R. W. Brodersen, Low-power CMOS digital design, *IEEE Journal of Solid-State Circuits*, Vol. 27, Issue 4, April 1992, pp. 473-484.

[11]. G. Chen, K. Huang, A. Knoll, Energy optimization for real-time multiprocessor system-on-chip with optimal DVFS and DPM combination, *ACM Trans. Embed. Comput. Syst.*, Vol. 13, Issue 3s, Mar 2014, pp. 1-111.

[12]. J.-J. Chen, T.-W. Kuo, Procrastination determination for periodic real-time tasks in leakage-aware dynamic voltage scaling systems, in *Proceedings of the International Conference on Computer Aided Design (ICCAD'07)*, 2007, pp. 289-294.

[13]. H. Cho, Y. Gong, A guaranteed real-time scheduling algorithm for (m, k)-firm deadline-constrained tasks on multiprocessors, *IETE Journal of Research*, Vol. 59, Issue 5, 2013, pp. 597-603.

[14]. V. Degalahal, L. Li, V. Narayanan, M. Kandemir, M. J. Irwin, Soft errors issues in low-power caches, *IEEE Trans. Very Large Scale Integr. Syst.*, Vol. 13, October 2005, pp. 1157-1166.

[15]. D. Ernst, S. Das, S. Lee, D. Blaauw, T. Austin, T. Mudge, N. S. Kim, K. Flautner, Razor: circuit-level correction of timing errors for low-power operation, *IEEE Micro*, Vol. 24, Issue 6, 2004, pp. 10-20.

[16]. M. Hamdaoui, P. Ramanathan, A dynamic priority assignment technique for streams with (m, k)-firm deadlines, *IEEE Transactions on Computers*, Vol. 44, December 1995, pp. 1443-1451.

[17]. J.-J. Han, M. Lin, D. Zhu, L. Yang, Contention-aware energy management scheme for noc-based multicore real-time systems, *IEEE Transactions on Parallel and Distributed Systems*, Vol. 26, Issue 3, March 2015, pp. 691-701.

[18]. Q. Han, L. Niu, G. Quan, S. Ren, S. Ren, Energy efficient fault-tolerant earliest deadline first scheduling for hard real-time systems, *Real-Time Syst.*, Vol. 50, Issue 5-6, November 2014, pp. 592-619.

[19]. Intel-Xscale, http://developer.intel.com/design/xscale/

[20]. J. Srinivasan, S. V. Adve, P. Bose, J. Rivers, C.-K. Hu, Ramp: A model for reliability aware microprocessor design, *IBM Research Report*, RC23048, 2003.

[21]. R. Jejurikar, C. Pereira, R. Gupta, Dynamic slack reclamation with procrastination scheduling in real-time embedded systems, in *Proceedings of the 42nd ACM/IEEE Design Automation Conference (DAC'05)*, 2005, pp. 111-116.

[22]. J. Li, L. Shu, J.-J. Chen, G. Li, Energy-efficient scheduling in nonpreemptive systems with real-time constraints, *IEEE Transactions on Systems, Man, and Cybernetics - Part A: Systems and Humans*, Vol. 43, Issue 2, 2013, pp. 332-344.

[23]. Z. Li, L. Wang, S. Li, S. Ren, G. Quan, Reliability guaranteed energy-aware frame-based task set execution strategy for hard real-time systems, *Journal of Systems and Software*, Vol. 86, Issue 12, December 2013, pp. 3060-3070.

[24]. Z. Li, L. Wang, S. Ren, G. Quan, Energy minimization for checkpointing-based approach to guaranteeing real-time systems reliability, in *Proceedings of the IEEE 16th International Symposium on Object/Component/Service-Oriented Real-Time Distributed Computing (ISORC'13)*, Paderborn, Germany, 2013.

[25]. J. Liu, Real-Time Systems, *Prentice Hall*, NJ, 2000.

[26]. L. Niu, Energy efficient scheduling for real-time systems with QoS guarantee, *Journal of Real-Time Systems*, Vol. 47, Issue 2, 2011, pp. 75-108.

[27]. L. Niu, W. Li, Energy-efficient fixed-priority scheduling for real-time systems based on threshold work-demand analysis, in *Proceedings of the International Conference on Hardware/Software Codesign and System Synthesis (CODES+ISSS'11)*, 2011.

[28]. L. Niu, G. Quan, Energy minimization for real-time systems with (m, k)-guarantee, *IEEE Trans. on VLSI, Special Section on Hardware/Software Codesign and System Synthesis*, July 2006, pp.717-729.

[29]. L. Niu, G. Quan, Leakage-aware scheduling for embedded real-time systems with (m, k)-constraints, *International Journal of Embedded Systems*, Vol. 5, Issue 4, 2013, pp. 189-207.

[30]. L. Niu, G. Quan, Peripheral-conscious energy-efficient scheduling for weakly hard real-time systems, *International Journal of Embedded Systems*, Vol. 7, Issue 1, 2015, pp. 11-25.

[31]. L. Niu, J. Xu, Improving schedulability and energy efficiency for window-constrained real-time systems with reliability requirement, *Journal of Systems Architecture*, Vol. 61, Issue 5-6, May-June 2015, pp. 210-226.

[32]. N. Kim, M. Ryu, S. Hong, M. Saksena, C. Choi, H. Shin, Visual assessment of a real-time system design: a case study on a CNC controller, in *Proceedings of the IEEE Real-Time Systems Symposium (RTSS'96)*, Dec 1996, pp. 300-310.

[33]. K. Pradhan, Fault-tolerant Computing: Theory and Techniques; Vol. 2, *Prentice-Hall Inc.,* Upper Saddle River, NJ, USA, 1986.

[34]. G. Quan, X. Hu, Enhanced fixed-priority scheduling with (m, k)-firm guarantee, in *Proceedings of the IEEE Real-Time Systems Symposium (RTSS'2000)*, 2000, pp. 79-88.

[35]. G. Quan, X. Hu, Energy efficient DVS schedule for fixed-priority real-time systems, *ACM Transactions on Embedded Computing Systems*, Vol. 6, Issue 4, Sept. 2007, pp. 1-31.

[36]. G. Quan, L. Niu, X. S. Hu, B. Mochocki, Real time scheduling for reducing overall energy on variable voltage processors, *International Journal of Embedded System: Special Issue on Low Power Embedded Computing*, Vol. 4, Issue 2, 2009, pp. 127-140.

[37]. K. Ramamritham, J. A. Stankovic, Scheduling algorithms and operating system support for real-time systems, *Proceedings of the IEEE*, Vol. 82, Issue 1, January 1994, pp. 55-67.

[38]. P. Ramanathan, Overload management in real-time control applications using (m, k)-firm guarantee, *IEEE Trans. on Paral. and Dist. Sys.,* Vol. 10, Issue 6, Jun 1999, pp. 549-559.

[39]. N. Seifert, D. Moyer, N. Leland, R. Hokinson, Historical trend in alpha-particle induced soft error rates of the alpha microprocessor, in *Proceedings of the 39th Annual IEEE International Reliability Physics Symposium*, 2001, pp. 259-265.

[40]. H. W. Turnbull, Theory of Equations, *Oliver and Boyd,* London, 1947.

[41]. R. West, Y. Zhang, K. Schwan, C. Poellabauer, Dynamic window-constrained scheduling of realtime streams in media servers, *IEEE Trans. on Computers*, Vol. 53, Issue 6, June 2004, pp. 744-759.

[42]. W. Yuan, K. Nahrstedt, Energy-efficient soft real-time CPU scheduling for mobile multimedia systems, in *Proceedings of the ACM Symposium on Operating Systems Principles (SOSP'03)*, 2003, pp. 149-163.

[43]. Y. Zhang, K. Chakrabarty, V. Swaminathan, Energy-aware fault tolerance in fixed-priority real-time embedded systems in *Proceedings of the International Conference on Computer Aided Design, (ICCAD'03)*, 2003, pp. 209-213.

[44]. B. Zhao, H. Aydin, D. Zhu, Generalized reliability-oriented energy management for real-time embedded applications, In *Proceedings of the 48th Design Automation Conference (DAC'11)*, 2011, pp. 381-386.

[45]. B. Zhao, H. Aydin, D. Zhu, Energy management under general task-level reliability constraints, in *Proceedings of the IEEE 18th Real Time and Embedded Technology and Applications Symposium (RTAS'12)*, Washington, DC, USA, 2012, pp. 285-294.

[46]. D. Zhu, Reliability-aware dynamic energy management in dependable embedded real-time systems, in *Proceedings of the Real-Time and Embedded Technology and Applications Symposium (RTAS'06)*, 2006.

[47]. D. Zhu, H. Aydin, Energy management for real-time embedded systems with reliability requirements, in *Proceedings of the IEEE/ACM International Conference on Computer-aided Design (ICCAD'06),* New York, NY, USA, 2006, pp. 528-534.

[48]. D. Zhu, H. Aydin, Reliability-aware energy management for periodic real-time tasks, *IEEE Transactions on Computers,* Vol. 58, Issue 10, 2009, pp. 1382-1397.

[49]. D. Zhu, R. Melhem, D. Mosse. The effects of energy management on reliability in real-time embedded systems, in *Proceedings of the IEEE/ACM International Conference on Computer-aided Design (ICCAD'04)*, 2004, pp. 35-40.

[50]. D. Zhu, X. Qi, H. Aydin, Energy management for periodic real-time tasks with variable assurance requirements, In *Proceedings of the 14th IEEE International Conference on Embedded and Real-Time Computing Systems and Applications (RTCSA''08)*, 2008, pp. 259-268.

[51]. Y. Zhu, F. Mueller, DVSleak: combining leakage reduction and voltage scaling in feedback EDF scheduling, in *Proceedings of the ACM SIGPLAN/SIGBED Conference on Languages, Compilers, and Tools for Embedded Systems (LCTES'07)*, 2007, pp. 31-40.

Appendix A. Proof for Theorem 5

Proof Assume the processor resumes its execution at $LST(t)$, as defined in Eq. (18.9). Let J_p be the first job that is executed. From Lemma 4, it is easy to see that J_p and all higher priority jobs are schedulable. For any job J_i with priority lower than that of J_p, we consider two cases: (a) $r_i < T_B$, (b) $r_i \geq T_B$. When $r_i < T_B$, similarly as J_p, its schedulability is guaranteed. When $r_i \geq T_B$, note that the jobs are delayed to no later than $min_i(r_i + \delta_i)$ and are therefore schedulable according to Theorem 4. Thus, all the mandatory jobs can meet their deadlines.

Appendix B. Proof for Theorem 2

Proof When there is no optional job to be executed or meet the deadline, the (m, k)-constraints are satisfied based on the schedulability testing condition of E-pattern provided in [38]. Hence, we only need to consider the case when an optional job is executed and meets its deadline, which results in the (m, k)-pattern adjustment. Here we need to consider two aspects:

a) The pattern adjustment will not cause any (m, k) violation on task τ_i;

b) The schedulability of the entire job set can still be guaranteed.

Part a) can be proved in a similar way to that of Lemma 3 in [28].

To prove part b), it is easy to see that after pattern rotation, the worst case interference that each mandatory job can suffer is still no more than that for the critical instance under E-pattern, whose schedulability is guaranteed by the testing condition in [38].

Chapter 19

Faking Countermeasure Against Side-Channel Attacks

Rubén Lumbiarres-López, Mariano López-García and Enrique Cantó-Navarro

19.1. Introduction

Any cryptographic device is susceptible to being fraudulently attacked for the purpose of obtaining the cryptographic key. Such a key is used to cipher or decipher confidential data that must be protected. Side-channel attacks (SCAs) were proposed by Paul Kocher et al. [1] at the end of the 1990s (they were known as differential analysis of power consumption). These attacks have become a powerful tool that is suitable for obtaining a cryptographic key in a short period of time using low-cost equipment. This equipment basically consists of a data-acquisition set, needed for capturing the current traces that represent the consumed power, and a desk computer that is used for processing the information contained in such traces. These attacks exploit the existing relationship between the energy consumed by an electronic device and the data that are being processed at a specific instant of time. As data are related with the cryptographic key, this analysis could potentially reveal the secret key.

So far, techniques used by engineers to conceal the cryptographic key, usually known as countermeasures, are focused in two different ways: hiding and masking. The aim of hiding is to design systems whose power consumption is constant and independent of data. Yet, masking is intended for devices in which the consumed power depends not only on data, but also on a random mask that is unknown to the attacker. In both cases, the consumed power is not directly linked with the processed data, hence the secret key is safe. Although hiding and masking make a successful attack more difficult, the use of sophisticated algorithms, or the fact of including a high number of traces in the analysis, allows the attacker to breach the security of the system.

Rubén Lumbiarres-López
Electronic Engineering, Universidad Politècnica de Cataluña,
Avda. Victor Balaguer, 08800, Vilanova i la Geltrú, Spain

The countermeasure based on faking represents a paradigm shift towards the protection of cryptographic devices: the aim is not to design a non-vulnerable system, but to profit from the vulnerability of cryptographic devices in revealing a fake key when SCA analysis is performed.

This chapter shows the theoretical basis for applying the faking countermeasure to the AES 128-bit cryptographic algorithm [2]. Initially, such an algorithm is analysed and its weak points for performing a successful SCA analysis are noted. Afterwards, several modifications are proposed and applied for implementing the proposed faking countermeasure. Experimental results were obtained for a software implementation using a Sasebo-GII [3] development board and the set of measures proposed in [4] by E. Oswald.

19.2. Fundamentals

The correlation and the differences-in-means are two of the main techniques used to reveal cryptographic keys by means of SCA analysis [4]. These kinds of attacks do not need any previous knowledge on the internal structure of the system; only the power consumption and the plain-text that is being encrypted need be known. Therefore, it is assumed that a number of T current traces, and their respective T plain-texts that are chosen by the attacker, are available to perform the analysis. A simple expression to know the minimum number of necessary traces can be found in [4].

The consumed power is measured when the device is executing a set of instructions related to a specific part of the cryptographic algorithm. Data related with such a selected set of instructions should have the property that only depends on a few bits of the cryptographic key. Hence, only a subset of the bits that form the key need to be analysed, which makes the processing easier. The attack is based on comparing the actual power consumption of the electronic device with a theoretical model. Such a model is used to predict the expected power that is consumed by the device. As this consumption depends on the key, the model is evaluated for every possible key whose bits may affect the power consumption. This is the reason why subsets of 8 bits are chosen, which reduces the number of guessed hypotheses to 256. The hypothesis whose theoretical model has the higher similarity with the actual power consumed will be the real key [5, 6]. The Hamming weight (HW) and the Hamming distance (HD) are the most used power models [4]. The Hamming weight corresponds to the number of bits set to 1 (or zero) in the data that are being analysed. The Hamming distance is the difference between the number of bits set to 1 (or zero) at two consecutive instants of time t_1 and t_2, respectively. For instance $HW(11010001) = 4$, whereas $HD(11010001, 01011001) = 2$.

These attacks are known as first order attacks, because they are focused on a particular point related to a specific instant of time. Second order attacks are more sophisticated, because they are based on processing the power consumption of two points that may take place in two different instants of time. Additionally, second order attacks are able to eliminate the protection provided by masking using a simple logical operation. Indeed, let u and v be two sets of bits that should be masked by using mask m. Then, this operation could be performed as follows:

$$u_m = u \oplus m \text{ and } v_m = v \oplus m, \qquad (19.1)$$

where u_m and v_m are the masked values associated with the intermediate points u and v, and $\oplus$ represents the bit-wise logical exclusive-OR operator. Assuming that the Hamming weight model is accurate for predicting the real power consumption, from (19.1) the following expression could be obtained as:

$$HW(u_m \oplus v_m) = HW(u \oplus m \oplus v \oplus m) = HW(u \oplus v). \qquad (19.2)$$

As can be seen, the effect of the mask is removed, since the HW for both masking and non-masking data are identical. Now, the system is potentially vulnerable, because the attacker knows the values of u and v (such values depend on the plain-text) and therefore the attacker is capable of predicting the theoretical power consumed by the device.

The following step is to find a suitable function for processing the power consumed by the pair values (u_m, v_m) produced at the instants of time (C_{ti}, C_{tk}), respectively. Such a function should have a significant degree of correlation with the model of power consumption described in (19.1). In [7], a function based on the absolute value of both points is proposed, represented by Eq. (19.3), but other authors propose the use of products, the square of the sum or even the square of the absolute value [8, 4].

$$F_{pre-processing} = abs(C_{ti} - C_{tk}). \qquad (19.3)$$

The success of this process depends on the detailed knowledge held by the attacker of the instants of time in which points (C_{ti}, C_{tk}) are produced. When such points are unknown, the process should be applied to all the pairs of values that arise from the combination of all possible points that form the current trace. Therefore, each processed trace consists in $n!$ values, where n represents the number of points included in the original captured traces.

19.2.1. AES Algorithm

The AES 128-bit algorithm [2] consists of 4 basic functions (*AddRoundKey*, *ShiftRows*, *SubBytes* and *MixColumns*) that are executed sequentially during several rounds following the diagram shown in Fig. 19.1. The number of rounds depends on the key-length, being 11 rounds (0 to 10) for the case of AES128-bit. The input and output of each function is a matrix of 4×4 bytes known as state. A basic description of these functions is as follows:

AddRoundKey: It performs the exclusive-OR bit-wise operation between each byte of the state and each byte of the key (its output is St_A).

ShiftRows: This function makes a shift between the bytes that form the rows of the state (its output is St_R).

SubBytes: It makes a SBOX substitution, which is defined in the AES 128-bit algorithm. The most usual way for performing these operations is by using a look-up table that is stored in memory (its output is St_B).

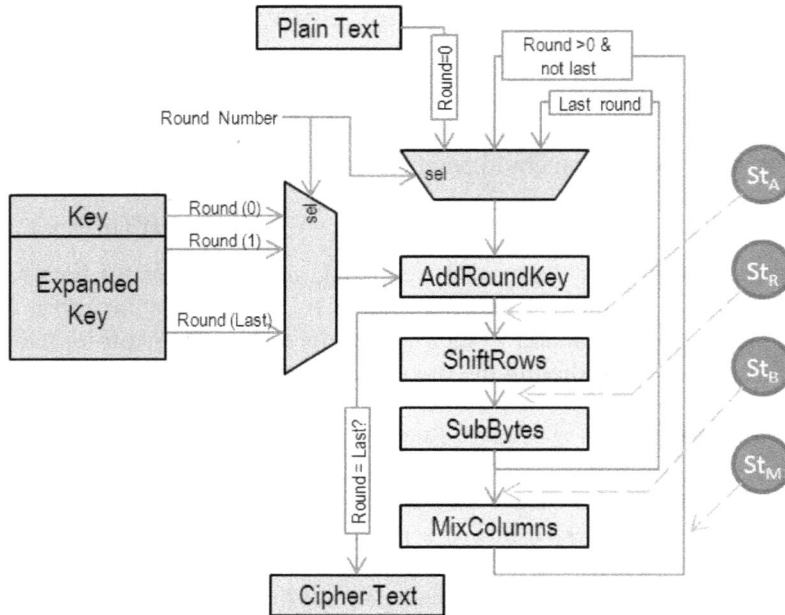

Fig. 19.1. AES algorithm structure.

MixColumns: Each column of state is treated as a polynomial in a Galois finite field, $GF(2^8)$. Each column is multiplied (modulus x^4+1) with the fixed polynomial:

$$C(x) = 3x^3 + x^2 + x + 2, \tag{19.4}$$

or the polynomial:

$$C^{-1}(x) = 11x^3 + 13x^2 + 9x + 14, \tag{19.5}$$

when the cipher text is decrypted (the output of *MixColumns* is St_M).

KeyExpand: is another important function in the algorithm that only operates on the key. The key is subjected to an expansion process by which, starting with the original key, several keys of identical length are obtained. Each of these expanded keys are used in each of the 11 rounds performed during the execution of the AES128-bit algorithm.

19.2.2. AES Vulnerabilities

The fortress of the algorithm increases when the key is scattered through the matrix state. The *Mixcolumns* function performs efficiently the diffusion of the key, since after the first round each byte of the state at the output of *MixColumns* depends on both the 4 bytes of the key and the 4 bytes of the plain-text. This key dependence increases for each round.

This effect can be appreciated in Table 19.1, which represents the variation of the values of the matrix state (bytes B0 to B15) during two rounds (R0 to R2) and in two different cases. In the first case, a random plain-text was chosen. In the second case, the same text was used, but changing only the less significant bit of byte 15 in the key. The shadowed values represent changes in the state matrix. As can be seen, all bytes of state at the output of function *MixColumns* in round 2 have changed, so that from this point any value of state depends on the complete value of the key.

Table 19.1. AES128-bit algorithm key dispersion.

B0	B1	B2	B3	B4	B5	B6	B7	B7	B8	B10	B11	B12	B13	B14	**B15**		
														State bytes			
15	4	83	250	3	140	142	46	132	46	7	109	99	162	137	**103**	AdKey	R0
118	242	237	45	123	100	25	49	95	49	197	60	251	58	167	**133**	SBytes	
118	100	197	**133**	123	49	167	45	95	58	237	49	251	242	25	60	SRow	R1
0	111	23	42	47	198	104	65	44	54	247	84	197	19	127	133	MCol	
220	255	32	154	180	143	**183**	168	187	200	**133**	107	253	146	**106**	34	AdKey	
134	22	183	184	141	115	**169**	194	234	232	**151**	127	84	79	**2**	147	SBytes	
134	115	**151**	**147**	141	232	**2**	184	234	79	**183**	194	84	**22**	**169**	127	SRow	R2
134	81	110	72	152	248	178	13	107	116	141	66	68	231	138	189	MCol	
84	152	5	255	209	120	6	83	181	10	75	35	162	24	89	123	AdKey	

The conclusion is that the main vulnerability of the algorithm is found in rounds 0 and 1 (it can be demonstrated that the same vulnerability arises in rounds 9 and 10) [5]. In these rounds the key is not scattered through the state and then it is more susceptible to attack.

In the first round, the state at the output of *AddRoundkey* corresponds to the combination of the plain-text with the non-expanded original key. The operation performed is:

$$St_{A(i,j)} = T_{(i,j)} \oplus K_{(i,j)}, \qquad (19.6)$$

where $St_{A(i,j)}$ ($i = 1..4, j = 1..4$) represents one of the 16 bytes that form the state, and $T_{(i,j)}$ and $K_{(i,j)}$ refer to the plain-text and the key, respectively. As can be observed in (19.6), the state only depends on one byte of the plain-text and one byte of the key.

Although an attack on such a function can be addressed, it is difficult to discriminate between the true hypothesis and the false hypothesis, since the bit-wise exclusive-OR operator is linear. Consequently, it is expected that the correlations obtained for those hypotheses that have a similar Hamming weight will also be quite similar. Table 19.2 shows the result for the correlation when an attack over byte 1 of the key is performed. The maximum correlation is obtained for the real key (211) and corresponds to $\rho = 0.6692$. However, it is also observed that such a value is very close to other correlations obtained for key hypotheses that have identical or similar HWs.

The function *ShiftRows* only moves the bytes of state, but without performing any operation between them. Thus, any attack on this function is conceptually identical to an attack on *AddRoundKey*.

Table 19.2. Maximum correlations obtained when attacking function *AddRoundKey*.

Key hypothesis and Hamming weight (HW)		Max correlation
209	HW(11010001) = 4	0.6282
210	HW(11010010) = 4	0.6002
211	HW(11010011) = 5	0.6692
196	HW(11000011) = 4	0.6545
148	HW(10010011) = 4	0.6236

The output of function *SubBytes* is widely used as a target in SCAs analysis, because the non-linearity of such a function makes its output very different for key hypotheses having similar HWs. Its definition can be found in [2], and it will be represented by *SBOX*, as follows:

$$St_{B(i,j)} = SBOX(St_{R(i,j)}), \qquad (19.7)$$

where $St_{R(i,j)}$ and $St_{B(i,j)}$ refer to the value of matrix state at the output of *ShiftRows* and *SubBytes*, respectively. It can be demonstrated that the elements of matrix state $St_{B(i,j)}$ only depend on one byte of the key, and therefore only a reduced number of hypotheses must be guessed. Table 19.3 represents the value of the correlation for several keys. As can be seen, the maximum correlation is given for the true key ($\rho = 0.8143$) and is clearly identifiable (one order of magnitude) even against those keys whose HWs are quite similar.

Table 19.3. Maximum correlations obtained when attacking the output of function *SubBytes*.

Key hypothesis		Max correlation
209	HW(11010001) = 4	0.0632
210	HW(11010010) = 4	0.0747
211	HW(11010011) = 5	0.8143
196	HW(11000011) = 4	0.0591
148	HW(10010011) = 4	0.0729

After the calculation of function *MixColumns* in the first round, each value of the matrix state depends on 4 bytes of the key. This means that the number of possible keys to be guessed for predicting the theoretical model is 2^{32}. The *Mixcolumns* operation is applied on the four columns of the matrix state and it is defined as:

$$\begin{pmatrix} St_{M(i,1)} \\ St_{M(i,2)} \\ St_{M(i,3)} \\ St_{M(i,4)} \end{pmatrix} = \begin{pmatrix} 2 & 3 & 1 & 1 \\ 1 & 2 & 3 & 1 \\ 1 & 1 & 2 & 3 \\ 3 & 1 & 1 & 2 \end{pmatrix} \cdot \begin{pmatrix} St_{B(i,1)} \\ St_{B(i,2)} \\ St_{B(i,3)} \\ St_{B(i,4)} \end{pmatrix}, \qquad (19.8)$$

where $St_{B(i,j)}$ and $St_{M(i,j)}$ represent the bytes of state at the output of *SubBytes* and *MixColumns*, respectively. From (19.8), it can be concluded that the first byte of state can be evaluated as:

$$St_{M(1,1)} = 2 \cdot St_{B(1,1)} \oplus 3 \cdot St_{B(1,2)} \oplus St_{B(1,3)} \oplus St_{B(1,4)}. \qquad (19.9)$$

Taking into account that the attacker has full control over the plain-text L., *Pan et al.* in [5] proposed the use of a plain-text in which all of its bytes are identical, except the byte that is the target of the attack. Assuming that in Eq.(19.9) such a byte is $St_{B(1,4)}$, then it is satisfied that:

$$St_{M(1,1)} = Constant \oplus St_{B(1,4)}. \qquad (19.10)$$

Note that in (19.10), the expression of $St_{M(1,1)}$ is equivalent to masking the input byte $St_{M(1,1)}$ using a non-variable constant mask. Masking is only an effective technique if the mask changes its value randomly. Thus, although the value of the correlation would be affected, the output of function *MixColumns* will be vulnerable in an identical way as it is function *SubBytes*.

In round 2, the key is completely scattered in such a way that any byte of the matrix state depends on the 128 bits of the key. Thus, any attack in this round or the following rounds is equivalent to performing a brute force attack.

19.3. The FAKING Countermeasure

The basic idea of the proposed faking countermeasure is to carry out the encryption process using a false key. Such a false key is obtained by operating with an exclusive-OR operator, the real key and a mask-key that is randomly chosen:

$$Key_{FAKE(i,j)} = Key_{REAL(i,j)} \oplus Key_{MASK(i,j)}. \qquad (19.11)$$

In (19.11), Key_{FAKE} refers to the false key, Key_{REAL} is the true key and Key_{MASK} is the key-mask. All of the keys are organized as a matrix of 4×4 bytes in such a way that they can be properly operated with the state. As the operations of the AES 128-bit algorithm are performed using the false key, and no additional countermeasures are taken, then any SCA analysis will lead to revealing the false key [9]. Fig. 19.2 shows the block diagram for implementing the faking countermeasure.

Note that, the real and the expanded keys are both masked with Key_{MASK}, so that in any round the false key is used for processing function *AddRoundKey*. Afterwards, the encryption follows the usual steps for processing the rest of the functions included in the standard AES128-bit algorithm. Obviously, although the system is effectively protected, if no additional actions are taken then the plain-text will be incorrectly encrypted with the false key. Then, to obtain the expected results the process should be reverted at some point. Functions described in Fig. 19.2 as *SboxTrans* and *MixColumns* are included to remove the faking countermeasure before a new round is started.

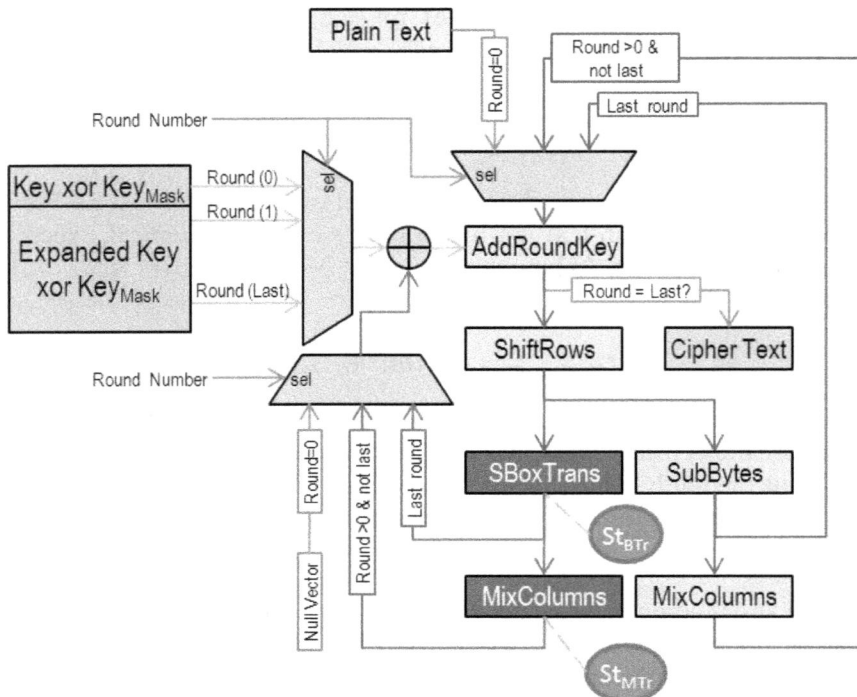

Fig. 19.2. Implementation of the faking countermeasure applied on the AES128-bit encryption algorithm.

Let *State(i,j)* be the output of function *ShiftRows*, then the effect of the non-linear function *SubBytes* over the matrix state will be:

$$St_{b(i,j)} = SBOX\left(St_{r(i,j)}\right) = SBOX(Key_{FAKE(i,j)} \oplus State_{(i,j)}). \quad (19.12)$$

By substituting (19.11) in (19.12), and developing the result in terms of the output that would be obtained if the real key was used, (19.12) can be expressed as:

$$St_{b(i,j)} = SBOX(Key_{REAL(i,j)} \oplus Key_{MASK(i,j)} \oplus State_{(i,j)})$$
$$St_{b(i,j)} = SBOX(Key_{REAL(i,j)} \oplus State_{(i,j)}) \oplus St_{b\,TRANS(i,j)}. \quad (19.13)$$
$$St_{b(i,j)} = St_{b\,REAL(i,j)} \oplus St_{b\,TRANS(i,j)}$$

Note that, the output of *SubBytes* can be obtained by operating with an exclusive-OR, the matrix state encrypted with $Key_{REAL(i,j)}$ and a new matrix denoted as $St_{b\,TRANS(i,j)}$, which represents the output of *SboxTrans* (see Fig. 19.2).

Finally, the result of (19.13) is processed by using function *MixColumns*. This function operates in the Galois finite field $GF(2^8)$ and it only includes additions as products. An interesting property of such an algebraic structure is that it satisfies the distributive property of multiplication with respect to the addition, and thus:

$$A \cdot (B + C) = A \cdot B + A \cdot C. \tag{19.14}$$

Using (19.13) and (19.14) it can be found that:

$$\begin{pmatrix} St_{M(i,1)} \\ St_{M(i,2)} \\ St_{M(i,3)} \\ St_{M(i,4)} \end{pmatrix} = \begin{pmatrix} 2 & 3 & 1 & 1 \\ 1 & 2 & 3 & 1 \\ 1 & 1 & 2 & 3 \\ 3 & 1 & 1 & 2 \end{pmatrix} \cdot \begin{pmatrix} St_{b\,REAL\,(i,j)} + St_{b\,TRANS(i,j)} \\ St_{b\,REAL\,(i,j)} + St_{b\,TRANS(i,j)} \\ St_{b\,REAL\,(i,j)} + St_{b\,TRANS(i,j)} \\ St_{b\,REAL\,(i,j)} + St_{b\,TRANS(i,j)} \end{pmatrix}. \tag{19.15}$$

Afterward, the distributive property is developed:

$$\begin{pmatrix} St_{M(i,1)} \\ St_{M(i,2)} \\ St_{M(i,3)} \\ St_{M(i,4)} \end{pmatrix} = \begin{pmatrix} 2 & 3 & 1 & 1 \\ 1 & 2 & 3 & 1 \\ 1 & 1 & 2 & 3 \\ 3 & 1 & 1 & 2 \end{pmatrix} \cdot \begin{pmatrix} St_{b\,REAL(i,j)} \\ St_{b\,REAL(i,j)} \\ St_{b\,REAL(i,j)} \\ St_{b\,REAL(i,j)} \end{pmatrix} + \begin{pmatrix} 2 & 3 & 1 & 1 \\ 1 & 2 & 3 & 1 \\ 1 & 1 & 2 & 3 \\ 3 & 1 & 1 & 2 \end{pmatrix} \cdot$$

$$\begin{pmatrix} St_{b\,TRANS(i,j)} \\ St_{b\,TRANS(i,j)} \\ St_{b\,TRANS(i,j)} \\ St_{b\,TRANS(i,j)} \end{pmatrix}. \tag{19.16}$$

The conclusion is that:

$$St_M = MixColumns\,(St_{b\,REAL}) \oplus MixColumns\,(St_{b\,TRANS}),$$
$$St_M = St_{M\,REAL} \oplus St_{M\,TRANS} \tag{19.17}$$

Where St_M is the output of function *MixColumns*, $St_{M\,REAL}$ is the output of function *MixColumns* if it was processed with the true key, and $St_{M\,TRANS}$ is a masking matrix that will be used at the end of each round.

Additionally, as the AES algorithm is executed in several rounds, the effects of the faking countermeasure should be reverted for each round. Thus, matrix $ST_{M\,TRANS}$, defined in (19.14), should be calculated in order to neutralize such an effect, since

$$St_M \oplus St_{M\,TRANS} = St_{M\,REAL} \oplus St_{M\,TRANS} \oplus St_{M\,TRANS} = St_{M\,REAL}. \tag{19.18}$$

On the other hand, matrix $ST_{M\,TRANS}$ could be obtained by applying function *MixColumns* over matrix $St_{b\,TRANS}$, which can by deduced by using (19.13) as:

$$St_{b\,TRANS(i,j)} = St_{b\,REAL(i,j)} \oplus St_{b\,(i,j)}$$
$$St_{b\,TRANS(i,j)} = SBOX\big(Key_{REAL(i,j)} \oplus State_{(i,j)}\big) \oplus Sb_{(i,j)}. \tag{19.19}$$

by substituting (19.12) in (19.19):

$$St_{b\,TRANS(i,j)} = SBOX\big(Key_{REAL(i,j)} \oplus State_{(i,j)}\big) \oplus$$

$$\oplus\ SBOX\big(Key_{FAKE(i,j)} \oplus State_{(i,j)}\big). \tag{19.20}$$

Since elements of state are bytes, the 256 possible values for Sb_{TRANS} can be pre-calculated forming a table of 256 elements named $SBOX_{TRANS}$. For calculating such a table, the following substitution is proposed:

$$val\ =\ Key_{FAKE} \oplus State. \tag{19.21}$$

By using (19.11), Eq. (19.20) could be generalized for any value of the elements of state, as follows:

$$SBOX_{TRANS}(val)\ =\ SBOX(val \oplus Key_{MASK}) \oplus SBOX(val). \tag{19.22}$$

The use of this expression allows to calculate all the elements of $SBOX_{TRANS}$ by using the pseudocode shown in (19.23). Finally, a suitable value for a specific input could be calculated in a similar way to that which is performed when searching in $SBOX$ during the execution of function $SubBytes$ included in AES.

$$\begin{aligned} &for\ j\ =\ 0\ to\ 255 \\ &\quad SBOX_{TRANS}(j)\ =\ SBOX(j \oplus Key_{MASK}) \oplus SBOX(j) \\ &next\ j \end{aligned} \tag{19.23}$$

19.3.1. Added Vulnerabilities

It is necessary to analyze if the real implementation of the faking countermeasure adds itself new vulnerabilities that can be used to reveal the true key. Note that, the application of SCA analysis allows to find the false key Key_{FAKE}. So that, if the attacker does focus its target on revealing the key-mask Key_{MASK}, then it would be easy to find Key_{REAL} just by using (19.11). There are two weak points in the implementation presented in Fig. 19.2, which can be used to reveal Key_{REAL}.

19.3.1.1. Second Order Attack

A second order attack between the output of functions $SboxTrans$ and $SubBytes$ can be used to find Key_{REAL}. The model to be used is based on the combination, by means of an exclusive-OR operator, between the output of $SubBytes$ (19.13) and the output of $SboxTrans$ (19.19):

$$St_{b(i,j)} \oplus SBOX_{TRANS(i,j)}\ =\ SBOX\big(Key_{REAL(i,j)} \oplus State_{(i,j)}\big). \tag{19.24}$$

In the first round, the state corresponds directly to the plain-text. Additionally, SBOX is defined by the algorithm, so that both data are known by the attacker and can be used to perform SCA analysis for revealing Key_{REAL}.

The power consumption can be pre-processed by using the function defined in (19.3), being the pair $(C_{ti},\ C_{tk})$, the power consumed at the output of functions $SubBytes$ and $SboxTrans$, respectively.

Fig. 19.3 shows the result of a theoretical second order attack following this procedure and which is performed over the first byte of the key. It can be observed as the obtained correlations are quite reduced, but the maximum is perfectly distinguishable among the rest of the values and it is in accordance with the real key.

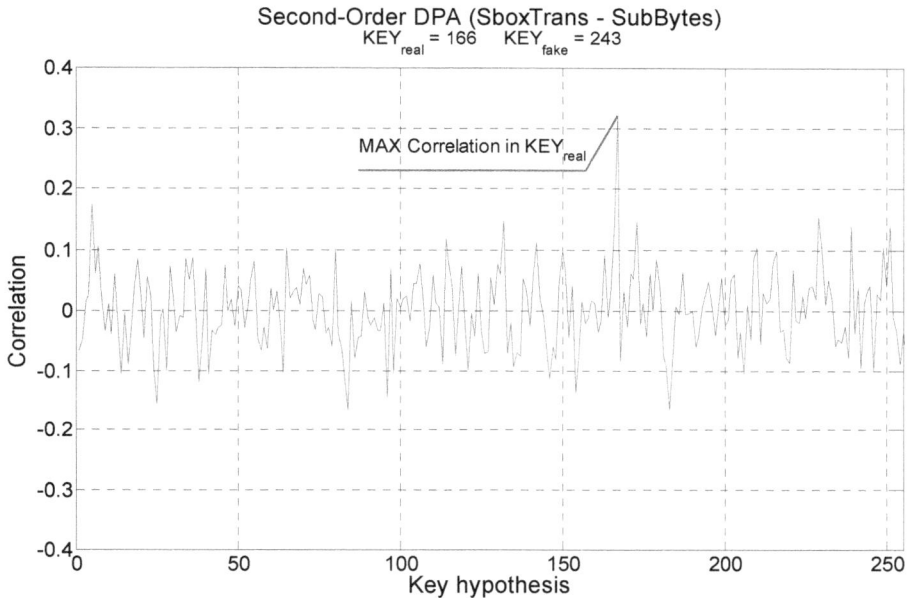

Fig. 19.3. Simulation of a second order attack based on the correlation.

19.3.1.2. Attack on the Output of *SboxTrans*

Eq. (19.20) depends only on the state, the false key and the real key. Note that, if both function *SBOX* and the KEY_{FAKE} are known by the attacker and then it is possible to perform SCA on *SboxTrans* to reveal the real key.

19.3.2. Protecting the StM $_{TRANS}$ Array

The solution for avoiding the attacks presented above is to protect the matrix St_{MTRANS} using a random mask. Such a matrix is combined, by using an exclusive-OR operator, with a mask that is changing its bytes at every encryption cycle.

Masking St_{MTRANS} has a significant effect on function *MixColumns*, since this function operates over different bytes that are located in several rows. It is not recommended to use the same mask for masking the 16 bytes of state, since the protection can be removed momentarily when one of the exclusive-OR operations involved in *Mixcolumns* is performed. In [4], it is proposed to use different masks for each row of state, in order to protect the integrity of data when *Mixcolumns* is executed.

Then, the values of bytes included in the mask are randomly generated for every encryption (decryption) cycle. Such values form a matrix named MT_J and defined as:

$$MT_J = \begin{pmatrix} m_0 & m_0 & m_0 & m_0 \\ m_1 & m_1 & m_1 & m_1 \\ m_2 & m_2 & m_2 & m_2 \\ m_3 & m_3 & m_3 & m_3 \end{pmatrix}, \tag{19.25}$$

where $[m_0, m_1, m_2, m_3]$ represent the mask values for every row of matrix $St_{M\ TRANS}$.

If matrix MT_J is included, then the calculation of $St_{M\ TRANS}$ would be:

$$St_{b\ TRANS(i,j)} = St_{b\ REAL(i,j)} \oplus St_{b\ (i,j)} \oplus MT_J. \tag{19.26}$$

Note that, the system is protected since now a random term that is unknown by the attacker is included. Besides, it should be pointed out that this random term cannot be cancelled by means of a second order attack, since it is only applied on function *SboxTrans*.

If in this new scenario a second order attack is performed, the results shown in Fig. 19.4 demonstrate the effectiveness of the protection that is added by the mask. It is impossible to distinguish between the real hypothesis and the false hypothesis that correspond to the true key.

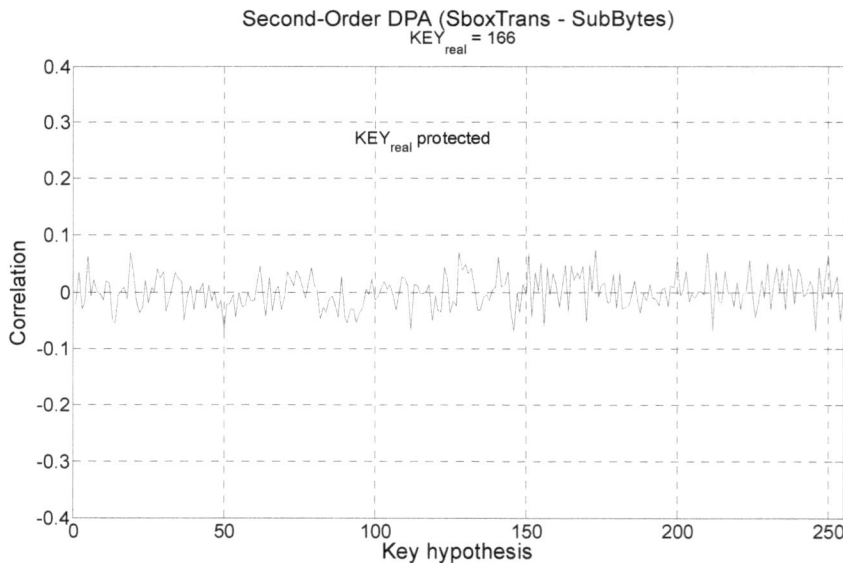

Fig. 19.4. Second order attack including masking.

Obviously, before starting a new round, the effect of masking should be removed in order to perform the encryption process correctly. The cancelation matrix is named MT_K.

After the application of function *SboxTrans*, Eq.(19.27) is obtained

$$St_{b\,TRANS(i,j)} \oplus MT_J. \tag{19.27}$$

Afterwards, function *MixColumns* is applied to (19.27), obtaining:

$$MixColumns\left(St_{b\,TRANS(i,j)} \oplus MT_J\right) =$$

$$= MixColumns\left(St_{b\,TRANS(i,j)}\right) \oplus MixColumns\left(MT_J\right). \tag{19.28}$$

From (19.28), it can be deduced that the elements of matrix MT_K can be calculated by using expression (19.29):

$$MT_K = \begin{pmatrix} 2 & 3 & 1 & 1 \\ 1 & 2 & 3 & 1 \\ 1 & 1 & 2 & 3 \\ 3 & 1 & 1 & 2 \end{pmatrix} \cdot \begin{pmatrix} m_0 & m_0 & m_0 & m_0 \\ m_1 & m_1 & m_1 & m_1 \\ m_2 & m_2 & m_2 & m_2 \\ m_3 & m_3 & m_3 & m_3 \end{pmatrix} =$$

$$= \begin{pmatrix} m_a & m_a & m_a & m_a \\ m_b & m_b & m_b & m_b \\ m_c & m_c & m_c & m_c \\ m_d & m_d & m_d & m_d \end{pmatrix} \tag{19.29}$$

Finally, Fig. 19.5 shows the complete structure of the system, including the random number generator used to calculate matrix MT_J, and the operations needed to perform the cancelation of masks before starting a new round.

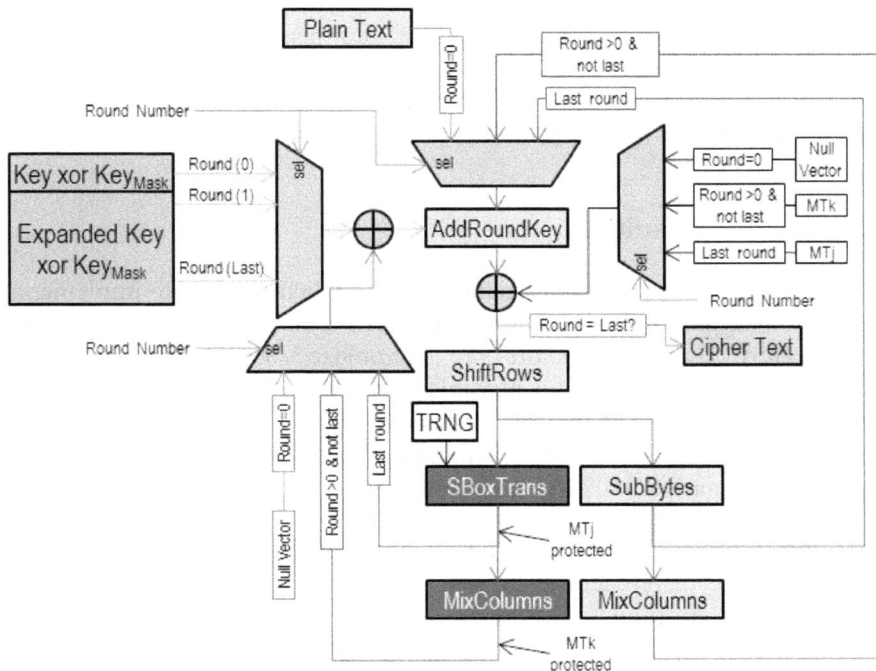

Fig. 19.5. AES algorithm structure with faking countermeasure and masking protection.

19.3.3. Computational Cost

Additional operations included to perform the faking countermeasure increase the processing time. These operations can be divided into three groups:

- Those operations that are carried out at the beginning of the encryption process for changing KEY_{REAL} or KEY_{MASK};

- Operations that are performed at every encryption cycle;

- Operations made at every round of the algorithm.

In the first group, only the masking of the expanded key is involved. The impact of this operation is not very important, since the execution time is distributed among the plaintexts that are processed using the same key.

The operations included in the second group are intended for calculating matrix MT_J and matrix MT_K. Eq. (19.27) is applied on the 16 bytes that form the state, so that it is necessary to calculate 16 $SBOX_{TRANS}$ tables for calculating $SboxTrans$. In order to minimize the impact of these operations, it is possible to perform their calculation concurrently with the communication of the device when reading or writing the original and cipher texts, respectively.

The third group of operations is necessary for calculating matrix $St_{M\ TRANS}$. Since these operations are made in each round, they have an important impact on the execution time. As the key is completely scattered after round 2, and then performing a successful SCA is almost impossible, the time could be reduced if after round 2 the countermeasure is disabled (the countermeasure should be activated again after round 9, because SCA analysis could be focus on the last rounds [5]).

19.4. Experimental Results

This section shows the experimental results obtained when a software implementation of the proposed countermeasure is included in the AES 128-bit encryption algorithm. The execution is performed on a 32-bit soft-core Microblaze V8.10 microprocessor provided by Xilinx. In [10] it is shown that it is possible to undertake a successful SCA analysis on this microprocessor. The system was implemented on a Virtex-5 FPGA clocked at 24 MHz [3]. The current traces were captured using a Tektronic CT-1 current probe connected to an Agilent DSO1024A oscilloscope.

The presented results are based on the analysis of the correlation and the attack proposed in [11], based on the differences-in-means. In order to observe the efficiency of the proposal, both results are shown when the countermeasure is activated and disabled. The target of the attack was focussed on function $SubBytes$, when byte 0 of the key is processed in the first round. The value chosen for KEY_{REAL} is 134, KEY_{MASK} is 85, and therefore we take into account that in (19.27) KEY_{FAKE} is 211.

The upper traces in Fig. 19.6 shows SCA analysis based on the value of the correlation when the countermeasure is disabled. Clearly, the maximum value for such correlation is about 0.8 and it corresponds with the true key. When the countermeasure is activated (lower traces of Fig. 19.6), the maximum correlation is given for the false key. The true key is completely protected and has a very low correlation that is non-distinguishable between the rests of the keys.

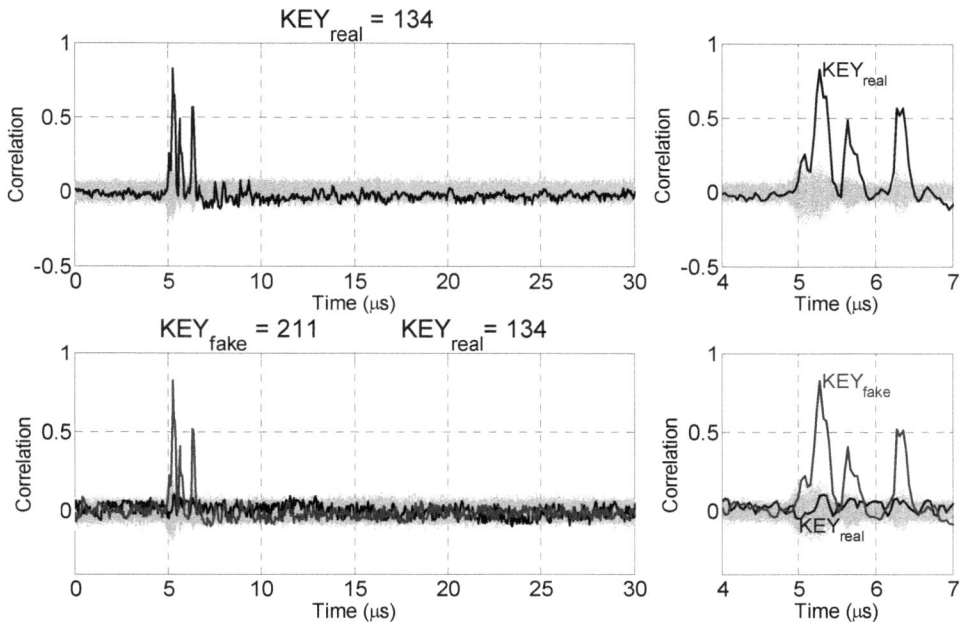

Fig. 19.6. Correlation attack. Upper traces correspond to an unprotected system; lower traces are obtained by activating the faking countermeasure.

Fig. 19.7 shows the values of the correlation against an increasing number of captured traces. The SCA is able to reveal the key with only 50 traces, independently, if the system is protected or if the countermeasure is disabled.

The attack based on differences-in-means was performed following the proposal of R. Lumbiarres in [11], which is based partially on the original publication of P. Kocher in [1]. Results shown in Fig. 19.8 again demonstrate that the system is completely protected when the faking countermeasure is activated by revealing a false key.

Generally, any countermeasure leads to increasing the number of hardware resources and/or the total execution time. Table 19.4 shows the penalty generated due to the inclusion of the proposed countermeasure. The execution time is increased by about 39 %, whereas, in accordance with this value, the throughput is reduced by 28 %. On the other hand, the memory needed to execute the process is also increased due to the extra memory required to store the 16 $SBOX_{TRANS}$ tables. Compared with other implementations, these penalties are quite acceptable. For instance, a masked implementation of an AES

128-bit algorithm that is executed on a microprocessor is also presented in [4]. The difference in the execution time between the non-protected and masked implementations was doubled.

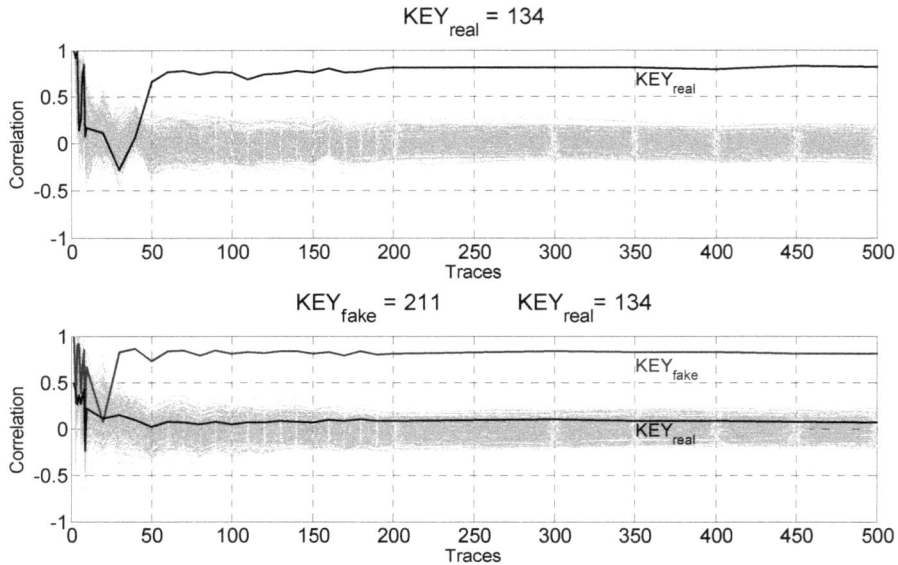

Fig. 19.7. Correlation attack over an increasing number of traces. Upper traces correspond to a non-protected system; lower traces are obtained by activating the faking countermeasure.

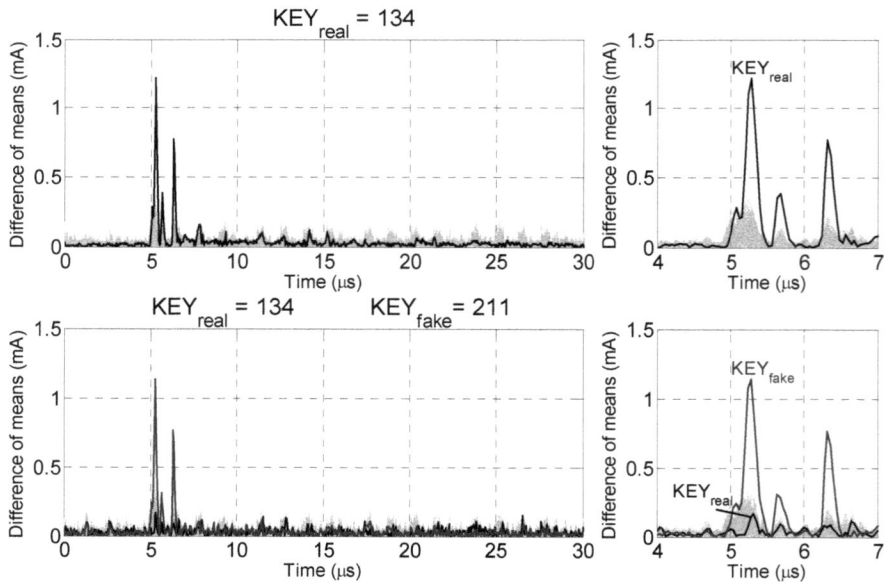

Fig. 19.8. Differences-in-means. Upper traces correspond to a non-protected system; lower traces are obtained by activating the faking countermeasure.

Table 19.4. Execution time and memory used.

Parameter	Non-protected	Fake protected	Improvement
Execution time	1.14 ms	1.59 ms	39 %
Time expended in pre-calculating the tables $SBOX_{TRANS}$	----	1.90 ms	---
Throughput	112 kb/s	80kb/s	-28 %
Memory needed for executing the algorithm	11770 bytes	19094 bytes	62 %

19.5. Conclusions

The proposed countermeasure conceals the true key by presenting a strong correlation related to a false key. Experimental results showed that the method is effective when either the correlation or differences-in-means attacks are performed.

When compared with a non-protected system, the execution time and the memory needed to execute the encryption algorithm are increased. However, such an increase is lower than the penalty related to previous proposals made by different authors.

Another interesting feature of the faking countermeasure is that its particular structure allows to include additional actions that increase the difficulty to find the true key. Thus, it is possible to modify the real key without introducing any modification in the false key. Then, the maximum correlation is the same although the encrypting key was changed. Also, it is possible to introduce modifications into the false key, while keeping the value of the real key. In this case, the attacker may conclude that the encryption key has changed.

Acknowledgements

This work was supported by the Ministerio de Economía y Competitividad in the framework of the Programa Nacional de Proyectosde Investigación Fundamental, project TEC2015-68784-R.

References

[1]. J. Jaffe, B. Jun, P, Kocher, Differential power analysis, in *Proceedings of the 19th Annual International Cryptology Conference (CRYPTO'99)*, Santa Barbara, California, 1999, pp. 388-397.
[2]. Advanced Encryption Standard (AES), FIPS PUB 197, *National Institute of Standards and Technology*, 2001.
[3]. Side-channel Attack Standard Evaluation Board SASEBO-GII Specification, Research Center for Information Security, *National Institute of Advanced Industrial Science and Technology*, 2009, http://satoh.cs.uec.ac.jp/SAKURA/index.html

[4]. E. Oswald, T. Popp, S. Mangard, Power Analysis Attacks - Revealing the Secrets of Smart Cards, *Springer Science+Busines Media*, Graz, Austria, 2007.

[5]. L. Pan, J. den Hartog, J. Lu, Security of AES against first and second-order differential power analysis, in *Proceedings of the 4th Benelux Workshop on Information and System Security (WISSec'09)*, Louvain-la-Neuve, Belgium, 2009.

[6]. K. Wu, B. Peng, Y. Zhang, X. Zheng, F. Yu, H. Li, Enhanced correlation power analysis attack on Smart Card, in *Proceedings of the 9th International Conference for Young Computer Scientists (ICYCS'08)*, 2008, pp. 2143-2148.

[7]. T. S. Messerges, Using second-order power analysis to attack DPA resistant software, in *Proceedings of the Cryptographic Hardware and Embedded Systems (CHES'2000)*, Vol. 1, Worcester, 2000, pp. 238-251.

[8]. M. Rivain, R. Bévan, E. Prouff, Statistical analysis of second order differential power analysis, *IEEE Transactions on Computers*, Vol. 58, No. 6, June 2009, pp. 799-811.

[9]. M. López-García, E. F. Cantó-Navarro, R. Lumbiarres-López, Implementation on MicroBlaze of AES algorithm to reveal fake keys against side-channel attacks, in *Proceedings of the IEEE 23rd International Symposium on Industrial Electronics (ISIE'14)*, Istanbul, 2014, pp. 1882-1887.

[10]. M. López-García, E. F. Cantó-Navarro, R. Lumbiarres-López, Ataques por canal lateral sobre el algoritmo de encriptación AES implementado en MicroBlaze, in *Proceedings of the XIII Jornadas de Computación Reconfigurable y Aplicaciones (JCRA'13)*, Madrid, 2013, pp. 105-112.

[11]. M. Lopez-Garcia, E. Canto-Navarro, R. Lumbiarres-Lopez, Hardware architecture implemented on FPGA for protecting cryptographic keys against side-channel attacks, *IEEE Transactions on Dependable and Secure Computing*, Vol. 99, September 2016, pp. 1-10.

Index

operation, 467

Q

QoS guarantee, 502

R

Random Dopant Fluctuation Effect, 391
ransconductance, 283
raw data, 438, 445
RDL. *See* Redistribution Layer
real-time
    embedded systems. *See* real-time systems
    scheduling, 477, 505
    systems, 478, 484, 491, 502-506
recovery job, 479, 484, 485, 487, 491-494,
    497, 498, 503
recurrent bending cycles, 321
Redistribution layer (RDL), 229
    Chip first, 230
    Chip last, 230
redistribution layers (RDL), 324
regulated cascode, 108
reliability, 191, 192, 220, 221, 427, 477,
    478, 484, 491, 494-499, 501-507
reliability-aware energy management, 504
remote
    programming, 446, 447
    sensor, 443, 444
Residue, 244, 245
resistive feedback, 108, 110, 111, 115, 118
resistivity, 191, 193-195, 197, 199, 202-
    204, 208, 212, 213, 218, 219, 222, 223
resonant tunneling, 85
Results and Comparison, 61
ripple carry adder, 39
Roll-to-Roll (R2R), 324
R-pattern. *See* deeply-red pattern

S

Safe state, 472
Safety process, 459
Saleh model, 169
sampling function, 24
SBOX, 511, 514, 518, 519
Scan-based testing, 124
scanning
    electron microscope, 102

SEM, 93, 102, 103, 320
    ion microscope, 95
    SIM, 95-97, 99-102
scheduling point, 488
Second order attacks, 510
self-formed, 193, 199, 201, 209, 220, 221
Self-heating, 261
sensor
    hub, 448
    Network, 79
Sharing Multiplication Technique, 35
sheet resistance, 195, 206, 207
Shift
    mode of operation, 132
    Power Consumption, 144
ShiftRows, 511, 513, 514, 516
Short-channel effects, 83
Si substrates, 191, 193, 194
Silicon IC
    thinned, ultra-thin, 313
silicon-on-insulator (SOI), 79
silicon-on-Sapphire (SOS), 79
silicon-on-SiO_2 (SOI), 80
size effect, 94, 98, 105
slack, 478, 485, 489, 491-495, 499, 503-505
small-signal
    impedance, 108
    model, 108
SOI devices, 82
SOI MOSFET, 80, 364
sol-gel
    oxides
      ZnO, MgO, 293, 294
    spin-coating method, 294, 296
source potential floating effect, 83
stable, 193, 197, 206, 221
static
    noise margin, 384, 393
    power, 480, 484
step coverage, 192, 193
stress measurements, 362
stress-strain, 97, 98, 104
SubBytes, 511, 514-516, 518, 522
sulfite-based
    electrolyte, 95, 97, 99, 101
    gold
      sulfite-gold, 95-98, 104
      sulfite-gold pillar, 97
supercritical CO_2, 99, 106
surface mount devices (SMD), 313
System, types of, 458

www.ingramcontent.com/pod-product-compliance
Lightning Source LLC
Chambersburg PA
CBHW060954210326
41598CB00031B/4826